New Perspectives in Theoretical and Applied Statistics

Probability and Mathematical Statistics (Continued)

PURI, VILAPLANA, and WERTZ • New Perspectives in Theoretical and Applied Statistics
RANDLES and WOLFE • Introduction to the Theory of Nonparametric Statistics
RAO • Linear Statistical Inference and Its Applications, *Second Edition*
RAO • Real and Stochastic Analysis
RAO and SEDRANSK • W.G. Cochran's Impact on Statistics
RAO • Asymptotic Theory of Statistical Inference
ROHATGI • An Introduction to Probability Theory and Mathematical Statistics
ROHATGI • Statistical Inference
ROSS • Stochastic Processes
RUBINSTEIN • Simulation and The Monte Carlo Method
SCHEFFE • The Analysis of Variance
SEBER • Linear Regression Analysis
SEBER • Multivariate Observations
SEN • Sequential Nonparametrics: Invariance Principles and Statistical Inference
SERFLING • Approximation Theorems of Mathematical Statistics
SHORACK and WELLNER • Empirical Processes with Applications to Statistics
TJUR • Probability Based on Radon Measures

Applied Probability and Statistics

ABRAHAM and LEDOLTER • Statistical Methods for Forecasting
AGRESTI • Analysis of Ordinal Categorical Data
AICKIN • Linear Statistical Analysis of Discrete Data
ANDERSON, AUQUIER, HAUCK, OAKES, VANDAELE, and WEISBERG • Statistical Methods for Comparative Studies
ARTHANARI and DODGE • Mathematical Programming in Statistics
ASMUSSEN • Applied Probability and Queues
BAILEY • The Elements of Stochastic Processes with Applications to the Natural Sciences
BAILEY • Mathematics, Statistics and Systems for Health
BARNETT • Interpreting Multivariate Data
BARNETT and LEWIS • Outliers in Statistical Data, *Second Edition*
BARTHOLOMEW • Stochastic Models for Social Processes, *Third Edition*
BARTHOLOMEW and FORBES • Statistical Techniques for Manpower Planning
BECK and ARNOLD • Parameter Estimation in Engineering and Science
BELSLEY, KUH, and WELSCH • Regression Diagnostics: Identifying Influential Data and Sources of Collinearity
BHAT • Elements of Applied Stochastic Processes, *Second Edition*
BLOOMFIELD • Fourier Analysis of Time Series: An Introduction
BOX • R. A. Fisher, The Life of a Scientist
BOX and DRAPER • Empirical Model-Building and Response Surfaces
BOX and DRAPER • Evolutionary Operation: A Statistical Method for Process Improvement
BOX, HUNTER, and HUNTER • Statistics for Experimenters: An Introduction to Design, Data Analysis, and Model Building
BROWN and HOLLANDER • Statistics: A Biomedical Introduction
BUNKE and BUNKE • Statistical Inference in Linear Models, Volume I
CHAMBERS • Computational Methods for Data Analysis
CHATTERJEE and PRICE • Regression Analysis by Example
CHOW • Econometric Analysis
CLARKE and DISNEY • Prob
Course with Applications, *Seco*
COCHRAN • Sampling Techniq
COCHRAN and COX • Experim
CONOVER • Practical Nonpara
CONOVER and IMAN • Introd
CORNELL • Experiments with
of Mixture Data

Applied Probability and Statistics (Continued)

COX • Planning of Experiments
COX • A Handbook of Introductory Statistical Methods
DANIEL • Biostatistics: A Foundation for Analysis in the Health Sciences, *Fourth Edition*
DANIEL • Applications of Statistics to Industrial Experimentation
DANIEL and WOOD • Fitting Equations to Data: Computer Analysis of Multifactor Data, *Second Edition*
DAVID • Order Statistics, *Second Edition*
DAVISON • Multidimensional Scaling
DEGROOT, FIENBERG and KADANE • Statistics and the Law
DEMING • Sample Design in Business Research
DILLON and GOLDSTEIN • Multivariate Analysis: Methods and Applications
DODGE • Analysis of Experiments with Missing Data
DODGE and ROMIG • Sampling Inspection Tables, *Second Edition*
DOWDY and WEARDEN • Statistics for Research
DRAPER and SMITH • Applied Regression Analysis, *Second Edition*
DUNN • Basic Statistics: A Primer for the Biomedical Sciences, *Second Edition*
DUNN and CLARK • Applied Statistics: Analysis of Variance and Regression
ELANDT-JOHNSON and JOHNSON • Survival Models and Data Analysis
FLEISS • Statistical Methods for Rates and Proportions, *Second Edition*
FLEISS • The Design and Analysis of Clinical Experiments
FOX • Linear Statistical Models and Related Methods
FRANKEN, KÖNIG, ARNDT, and SCHMIDT • Queues and Point Processes
GALLANT • Nonlinear Statistical Models
GIBBONS, OLKIN, and SOBEL • Selecting and Ordering Populations: A New Statistical Methodology
GNANADESIKAN • Methods for Statistical Data Analysis of Multivariate Observations
GREENBERG and WEBSTER • Advanced Econometrics: A Bridge to the Literature
GROSS and HARRIS • Fundamentals of Queueing Theory, *Second Edition*
GUPTA and PANCHAPAKESAN • Multiple Decision Procedures: Theory and Methodology of Selecting and Ranking Populations
GUTTMAN, WILKS, and HUNTER • Introductory Engineering Statistics, *Third Edition*
HAHN and SHAPIRO • Statistical Models in Engineering
HALD • Statistical Tables and Formulas
HALD • Statistical Theory with Engineering Applications
HAND • Discrimination and Classification
HOAGLIN, MOSTELLER and TUKEY • Exploring Data Tables, Trends and Shapes
HOAGLIN, MOSTELLER, and TUKEY • Understanding Robust and Exploratory Data Analysis
HOEL • Elementary Statistics, *Fourth Edition*
HOEL and JESSEN • Basic Statistics for Business and Economics, *Third Edition*
HOGG and KLUGMAN • Loss Distributions
HOLLANDER and WOLFE • Nonparametric Statistical Methods
IMAN and CONOVER • Modern Business Statistics
JAGERS • Branching Processes with Biological Applications
JESSEN • Statistical Survey Techniques
JOHNSON • Multivariate Statistical Simulation
JOHNSON and KOTZ • Distributions in Statistics
Discrete Distributions
Continuous Univariate Distributions—1
Continuous Univariate Distributions—2
Continuous Multivariate Distributions

(*continued on back*)

New Perspectives in Theoretical and Applied Statistics

Edited by

MADAN LAL PURI
Indiana University, Bloomington, Indiana

JOSÉ PÉREZ VILAPLANA
University of the Basque Country, Bilbao, Spain

WOLFGANG WERTZ
Technical University, Vienna, Austria

JOHN WILEY & SONS

New York • Chichester • Brisbane • Toronto • Singapore

***Library of Congress Cataloging in Publication Data*:**

New perspectives in theoretical and applied statistics.

(Wiley series in probability and mathematical statistics. Probability and mathematical statistics)

"Selection of the papers presented at the Third International Meeting of Statistics in the Basque Country, held in Bilbao, Spain, from August 7 to 9, 1986"—Pref.

Includes bibliographies and indexes.

1. Mathematical statistics—Congresses. 2. Probabilities—Congresses. I. Puri, Madan Lal. II. Pérez Vilaplana, José. III. Wertz, Wolfgang. IV. International Meeting of Statistics in the Basque Country (3rd: 1986: Bilbao, Spain) V. Series.

QA276.A1N49 1986 519.5 86-22401
ISBN 0-471-84800-X

Printed in the United States of America

10 9 8 7 6 5 4 3 2 1

Contributors

EMAD-ELDIN A. A. ALY, Department of Statistics and Applied Probability, The University of Alberta, Edmonton, Alberta, Canada

N. R. ARGHAMI, Department of Mathematics, Statistics and Computer Science, Faculty of Science, Mashhad University, Mashhad, Iran

ADELCHI AZZALINI, Dipartimento di Scienze Statistiche, Universítà di Padova, Padua, Italy

R. N. BHATTACHARYA, Department of Mathematics, Indiana University, Bloomington, Indiana

L. BILLARD, Department of Statistics, University of Georgia, Athens, Georgia

RICHARD C. BRADLEY, Department of Mathematics, Indiana University, Bloomington, Indiana

WŁODZIMIERZ BRYC, Department of Mathematics, University of California at San Diego, La Jolla, California

M. SALOMÉ E. CABRAL NETO, Departamento de Estatística, Centro de Estatística e Aplicações, Faculdade de Ciências, Universidade de Lisboa, Lisbon, Portugal

GIORGIO CANARELLA, Department of Economics and Statistics, California State University, Los Angeles, Los Angeles, California

DAVID R. COX, Department of Mathematics, Imperial College, London, England

MIKLÓS CSÖRGŐ, Department of Mathematics and Statistics, Carleton University, Ottawa, Ontario, Canada

ANTONIO CUEVAS, Departamento de Estadistica E Investigacion Operativa, Facultad de Ciencias Matematicas, Universidad Complutense de Madrid, Madrid, Spain

PAUL DEHEUVELS, Institut de Statistique de Paris, Université de Paris VI, Paris, France

MANFRED DENKER, Institut für Mathematische Stochastik, Universität Göttingen, Göttingen, Federal Republic of Germany

EDWARD J. DUDEWICZ, Department of Mathematics, Syracuse University, Syracuse, New York

P. FARALDO ROCA, Departamento de Estadistica Matematica, Universidad de Santiago de Compostela, Santiago de Compostela, Spain

M. IVETTE GOMES, Departamento de Estatística, Centro de Estatística e Applicações, Faculdade de Ciências, Universidade de Lisboa, Lisbon, Portugal

W. GONZÁLEZ MANTEIGA, Departamento de Estadistica Matematica, Universidad de Santiago de Compostela, Santiago de Compostela, Spain

M. E. GRAÇA MARTINS, Departamento de Estatística, Centro de Estatística e Aplicações, Faculdade de Ciências, Universidade de Lisboa, Lisbon, Portugal

JAMES E. GRIZZLE, Department of Biostatistics, School of Public Health, University of North Carolina at Chapel Hill, Chapel Hill, North Carolina

GEORGE HAIMAN, Institut de Recherche des Transports, Arcueil, France

MARC HALLIN, Institut de Statistique, Université Libre de Bruxelles, Brussels, Belgium

LAJOS HORVÁTH, Bolyai Institute, Szeged University, Szeged, Hungary

S. RAO JAMMALAMADAKA, Department of Mathematics, University of California at Santa Barbara, Santa Barbara, California

SVANTE JANSON, Department of Mathematics, Uppsala University, Uppsala, Sweden

K. G. JINADASA, Department of Mathematics, University of Windsor, Windsor, Ontario, Canada

PETER W. M. JOHN, Department of Mathematics, The University of Texas, Austin, Texas

ROBERT KEENER, Department of Statistics, The University of Michigan, Ann Arbor, Michigan

GERHARD KELLER, Institut für Angewandte Mathematik, Universität Heidelberg, Heidelberg, Federal Republic of Germany

LUDOVIC LEBART, Centre National de la Recherche Scientifique, Paris, France

JEAN-PIERRE LECOUTRE, Laboratoire de Statistique Théoretique et Appliquée, Université de Paris VI, Paris, France

LUIS J. LÓPEZ MARTIN, Departamento de Estadistica y Econometria, Facultad Ciencias Económicas, Universidad de la Laguna, Canary Islands, Spain

HARUO MIYAZAKI, Department of Computer Science, Gunma University, Gunma, Japan

DANIEL PEÑA, Departamento de Estadistica, E.T.S. Ingenieros Industriales, Universidad Politécnica de Madrid, Madrid, Spain

DINIS D. PESTANA, Departamento de Estatística, Centro de Estatística e Applicações, Faculdade de Ciências, Universidade de Lisboa, Lisbon, Portugal

D. PFEIFER, Institut für Statistik, RWTH, Aachen, Federal Republic of Germany

STEPHEN K. POLLARD, Department of Economics and Statistics, California State University, Los Angeles, Los Angeles, California

M. B. PRIESTLEY, Department of Mathematics, The University of Manchester, Institute of Science and Technology, Manchester, England

MADAN L. PURI, Department of Mathematics, Indiana University, Bloomington, Indiana

PREM S. PURI, Department of Statistics, Purdue University, West Lafayette, Indiana

J. N. K. RAO, Department of Mathematics and Statistics, Carleton University, Ottawa, Ontario, Canada

PAL RÉVÉSZ, Mathematical Institute, Hungarian Academy of Sciences, Budapest, Hungary*

J. ROBINSON, Department of Mathematical Statistics, University of Sydney, Sydney, New South Wales, Australia

VIJAY K. ROHATGI, Department of Mathematics and Statistics, Bowling Green State University, Bowling Green, Ohio

PEDRO SAAVEDRA SANTANA, Departamento de Estadistica E Invastigacion Operativa, Facultad de Matematicas, Universidad de la Laguna, Canary Islands, Spain

*Present address: Institut für Statistik und Wahrscheinlichkeitstheorie, Technische Universität, Vienna, Austria

MIGUEL SÁNCHEZ GARCÍA, Departamento de Estadistica E Invastigacion Operativa, Facultad de Matematicas, Universidad de la Laguna, Canary Islands, Spain

ARMIN SCHILLING, Mathematisches Institut, Universität Giessen, Giessen, Federal Republic of Germany

WINFRIED STUTE, Mathematisches Institut, Universität Giessen, Giessen, Federal Republic of Germany

RAM C. TIWARI, Department of Mathematics, Indian Institute of Technology, Bombay, India

D. S. TRACY, Department of Mathematics, University of Windsor, Windsor, Ontario, Canada

EDWARD C. VAN DER MEULEN, Department of Mathematics, Katholieke Universiteit Leuven, Leuven, Belgium

JOSÉ PÉREZ VILAPLANA, Seminar of Statistics, Department of Applied Mathematics, Faculty of Science, University of the Basque Country, Bilbao, Vizcaya, Spain

HANS JOACHIM WERNER, Institute for Econometrics and Operations Research, Econometrics Unit, University of Bonn, Bonn, Federal Republic of Germany

WOLFGANG WERTZ, Institut für Statistik und Wahrscheinlichkeitstheorie, Technische Universität, Vienna, Austria

ANDERS H. WESTLUND, Department of Economic Statistics, University of Stockholm School of Economics, Stockholm, Sweden

MICHAEL B. WOODROOFE, Department of Statistics, The University of Michigan, Ann Arbor, Michigan

Preface

There have been numerous important developments in statistical theory in recent years. New methodologies have emerged, accompanied by new and improved theoretical tools. As a result, past restrictions have been removed and the domains of potential applications have been vastly broadened. In this book we present many of these developments, some of which have already had significant impacts on modeling, design, and analysis of statistical experiments.

The chapters included in this book cover a wide range of topics of current interest in applied as well as theoretical statistics and probability. They include some aspects of the design of experiments in which there are current developments, data analysis, sampling theory, growth curves analysis, theoretical as well as applied aspects of time-series analysis, nonparametric theory, the Edgeworth expansions in statistics and probability, the bootstrap and randomization methods, density estimation, hypotheses testing, pseudorandom numbers, and sequential estimation. Also included are chapters on some aspects of probability theory that, apart from their intrinsic mathematical interest, have significant applications in statistics. The book provides a wealth of ideas for advanced graduate students and research workers in statistics and probability; it should also be useful to statisticians in industry, agriculture, engineering, medical sciences, and other fields.

The book contains a selection of the papers presented at the Third International Meeting of Statistics in the Basque Country, held in Bilbao, Spain, from August 7 to 9, 1986. Also included are a few additional invited papers not presented in person at the meeting. The aim of this meeting was to provide an opportunity for personal contacts among scholars from different intellectual centers of the world who could represent a comprehensive cross section of contemporary thinking on some of the areas of theoretical as well as applied statistics and probability. In selecting the papers for the volume, we were guided by the advice of the referees whose names appear in the Acknowledgments following this Preface. It is a pleasure to express our appreciation to them as well as to the authors for their help and cooperation.

The Scientific Committee of the conference consisted of E. Chacón, S.J. (Spain, Chairman), Sir David R. Cox (U.K.), P. Deheuvels (France), E. J. Dudewicz (U.S.A.), J. E. Gentle (U.S.A.), M. Hofri (Israel), L. Lebart (France), M. B. Priestley (U.K.), Yu. V. Prohorov (USSR), M. L. Puri (U.S.A.), J. L. Sanchez-Crespo (Spain), H. Vasquez (Peru), and W. Wertz (Austria). We take this opportunity to express our sincere thanks to the members of this committee for their overall help in making the conference a great success. These efforts resulted in the rich and active participation of about 200 scholars from 29 countries. Special thanks are due to members of the organizing committee (its chairman was J. P. Vilaplana) for providing an excellent atmosphere of warmth, friendliness, and generous hospitality. This event was organized by the Seminar of Statistics, Department of Applied Mathematics, University of the Basque Country at Lejona to pay homage to the International Statistical Institute on its hundreth anniversary.

We would also like to take this opportunity to express our thanks to the Basque Government and IBM-ESPANA for their financial support and to the Catholic University of Deusto, the mayors of Bilbao and Guernica, the British Council, and the Service Scientifique de France in Spain for their collaboration.

MADAN L. PURI
JOSÉ PÉREZ VILAPLANA
WOLFGANG WERTZ

Bloomington, Indiana
Bilbao, Spain
Vienna, Austria
January 1987

Acknowledgments

More than 150 papers were submitted for publication in this volume. We regret that we could not publish all of them. In selecting papers for publication in this book, we were guided by the advice of the following referees, who were an indispensable part of our editorial process. We express our deepest gratitude to them for their help.

I. A. Ahmad (U.S.A.), T. Aluja Banet (Spain), A. J. B. Anderson (U.K.), K. B. Athreya (U.S.A.), A. Azzalini (Italy), V. Barnett (U.K.), I. Basawa (Australia), A. P. Basu (U.S.A.), P. Bauer (Austria), J. Behoboodian (Iran), J. O. Berger (U.S.A.), J. M. Bernardo (Spain), R. N. Bhattacharya (U.S.A.), R. C. Bradley (U.S.A.), K. R. W. Brewer (Australia), D. R. Cox (U.K.), P. Deheuvels (France), M. Denker (Germany), L. Devroye (Canada), N. R. Draper (U.S.A.), E. J. Dudewicz (U.S.A.), R. Dutter (Austria), B. Efron (U.S.A.), J. D. Esary (U.S.A.), L. F. Escudero (Spain), R. Fay (U.S.A.), X. L. Fernique (France), N. R. J. Fieller (U.K.), S. E. Fienberg (U.S.A.), D. J. Finney (U.K.), B. Flury (Switzerland), J. Gani (U.S.A.), J. E. Gentle (U.S.A.), A. Giovagnoli (Italy), St. Gnot (Poland), V. Goodman (U.S.A.), J. Gordesch (Germany), J. E. Grizzle (U.S.A.), W. Grossman (Austria), F. M. Guess (U.S.A.), M. Hallin (Belgium), D. K. Hildebrand (U.S.A.), M. Hollander (U.S.A.), J. Hurt (Czechoslovakia), M. Hušková (Czechoslovakia), H. Joe (Canada), N. L. Johnson (U.S.A.), G. Koch (U.S.A.), F. Konecny (Austria), W. Kruskal (U.S.A.), J. Kuelbs (U.S.A.), N. Kusolitsch (Austria), P. A. Lachenbruch (U.S.A.), L. Lebart, (France), J. Llena Sitges (Spain), D. O. Loftsgaarden (U.S.A.), A. Martinsek (U.S.A.), A. M. Mathai (Canada), P. McCullagh (U.S.A.), G. McLachlan (Australia), F. J. Montero (Spain), P. Naeve (Germany), H. Neudecker (Netherlands), G. Neuhaus (Germany), E. Neuwirth (Austria), H. Niederreiter (Austria), J. K. Ord (U.S.A.), S. Panchapakesan (U.S.A.), G. Patil (U.S.A.), R. W. Payne (U.K.), G. Pflug (Germany), D. Plachky (Germany), B. Pötscher (Austria), B. L. S. Prakasa Rao (India), Z. Prašková (Czechoslovakia), M. B. Priestley (U.K.),

F. Proschan (U.S.A.), D. Raghavarao (U.S.A.), C. R. Rao (U.S.A.), J. N. K. Rao (Canada), P. Révész (Austria, Hungary), H. Rieder (Germany), H. Riedwyl (Switzerland), J. M. Rincon (Spain), S. Rios Garcia (Spain), V. K. Rohatgi (U.S.A.), J. Rosenblatt (U.S.A.), F. H. Ruymgaart (Netherlands), P. K. Sen (U.S.A.), R. J. Serfling (U.S.A.), P. W. Stewart (U.S.A.), M. Thompson (Canada), L. T. Tran (U.S.A.), A. S. Üstünel (France), R. Viertl (Austria), H. Witting (Germany), D. A. Wolfe (U.S.A.), M. B. Woodroofe (U.S.A.), N. Zagorniko (U.S.S.R.), R. Zmyślony (Poland).

We apologize for any omissions.

M. L. P.
J. P. V.
W. W.

Contents

New Perspectives in Theoretical and Applied Statistics

PART 1

Design of Experiments and Sampling Theory

Some Aspects of the Design of Experiments

David R. Cox
Department of Mathematics
Imperial College
London, England

SUMMARY

A general nontechnical review is given of some aspects of the design of experiments in which there are current developments. Topics considered include the relation between experiments and observational studies, randomization theory, optimal design, serial designs, sequential designs, and combinatorial matters.

1. INTRODUCTION

Statistical study of experimental design has a long history and various principles are well established, but nevertheless there are many points where further study is needed and also areas of application where greater involvement of statistical notions is needed. This chapter is a selective review of and commentary on some aspects in which there is current research work of fairly direct concern in applications.

The key ideas in the design of experiments are those of

Experimental unit
Treatment
Response

one treatment being applied to each experimental unit, the allocation of treatments being under the investigator's total control.

An experimental unit is defined as the smallest subdivision of the experimental material such that two different units may receive different treatments. Estimation of error is directly or indirectly by the comparison of distinct units receiving different treatments, rather than from "sampling variation" within a unit. Of course this can raise major difficulties in, for instance, investigations in which a "treatment" refers to the regime in an educational or other institution. Then, even though the number of individual people taking part may be large, the number of experimental units, that is, institutions, may be very small.

Usually the objective of the investigation is best formulated as the estimation of treatment contrasts regarded as measures of causal effect, although sometimes there may be more specific objectives, such as the determination of conditions of maximum response.

General reviews of recent work on experimental design have been given by Atkinson (1982b) and, with some emphasis on industrial applications, by Steinberg and Hunter (1984). Federer (1981) and, earlier, Herzberg and Cox (1969) have provided bibliographies.

2. DISTINCTION BETWEEN EXPERIMENTS AND OBSERVATIONAL STUDIES

In principle there is a clearcut distinction between experiments and observational studies. In particular, with randomized treatment allocation a conclusion about causal differences can be drawn with statistically assessed uncertainty, whereas in a comparable observational study there are always additional uncertainties of interpretation.

While this distinction is of fundamental importance, in practice there are often intermediate situations, the study of which is very important. For example, in a randomized clinical trial in which there are historical controls as well as a concurrent control group, what use, if any, should be made of the former? More broadly, the development of techniques for use when the investigator's control over treatment allocation is less than total is of considerable interest, especially for the social sciences. Furthermore, when detailed interpretation of the reasons for treatment differences is attempted, for example, by the effect on intermediate response variables, uncertainties of interpretation arise similar to those in observational studies. For example, if in a clinical trial on analysis is made in terms of the (probably complicated) treatment regime actually employed for each patient, rather than in terms simply of the arms of the randomization, difficult problems of interpretation will arise.

For a penetrating discussion of some of the difficulties of interpreting observational studies, see Pratt and Schlaifer (1984).

3. STEPS IN DESIGN

The main steps in design, all of which raise both subject-matter and technical statistical issues, are the following:

The choice of experimental material (protocol)

The size of experimental units (plots, etc.)

The selection of units, the definitions of intrinsic factors associated with the units, the scale of effort appropriate, the use of subsampling within units

The choice of treatments, their precise definition and possible factorial structure

The measurement of preliminary variables on each unit

Preliminary processing of units

Measurement of further variables; establishment of "external" conditions for each unit

Assignment of treatments to units

Application of treatments

Measurement of treatment monitoring variables

Further processing

Measurement of intermediate response variables

Measurement of further treatment monitoring variables (compliance)

Measurement of final response

One purpose in listing so much detail is to stress the possible complexity of "error." In practice we must, of course, pay most attention to the major sources of variation for the particular investigation under study, and rough preliminary estimation of components of variance will often aid in this.

4. FOCUS OF INTEREST

There is great variation between different fields of application in the relative emphasis to be put on various aspects of this process. Possibilities are to have

Fairly simple treatment structure (possibly factorial) with careful attention to error structure essential, as in traditional agricultural field trials

Relatively complex treatment structure with fairly simple error structure as in response surface designs

Very simple treatment structure, with very complex preliminary and intermediate information available, as in some randomized clinical trials

5. A CLASSICAL TREATMENT OF ERROR

One of the key themes of "classical" work on experimental design is that "error," that is, haphazard variation, may have complex structure and may need careful specification. Analysis of variance is of central importance in this. It is important to distinguish several roles for analysis of variance, in particular, the following:

1. The form of an analysis of variance table can give a concise summary of the logical structure of a design by the effects and interactions that can be estimated and the appropriate source of error for each.
2. The decomposition of sum of squares is a descriptive partition of total variability.
3. In the context of the normal theory linear model, analysis of variance provides a convenient algorithm for the estimation of residual variance and for the construction of confidence limits, F tests, and so on.
4. In hierarchical error systems (e.g., split plot designs), estimation of the various error variances is clarified.
5. The estimation of components of variance and the resulting possibilities of synthesis of variance illustrate what is in one sense the most literal meaning of the term *analysis of variance*.
6. Many of the calculations of normal theory multivariate analysis are most conveniently set out in an analysis of variance table.
7. Careful inspection of mean squares in a complex analysis of variance can provide a powerful diagnostic tool.

Of these, 1 and 4 are probably the most important in the present context. For the standard designs, the analysis of variance table and its second-order properties are a consequence of the randomization and an assumption of unit-treatment additivity: the response that would be observed if treatment s were applied to experimental unit i is $\tau_s + \eta_i$ regardless of the allocation of treatments to other experimental units, where $\{\eta_i\}$ are constants characterizing the experimental units and $\{\tau_s\}$ are constants characterizing the treatments.

The conceptual importance of this is that the analysis of variance table, and hence implicitly a linear model, can be regarded as a consequence of a single assumption, the same for all designs. Deduction from an "assumed" linear model gives a spurious impression that a special physical assumption about the form of the random variability is needed, specific to each design, and thus that, for example, a Latin square or a split-plot design requires different and stronger assumptions than a completely randomized design.

Thus the randomization analysis provides a "default" analysis for each design to be used in the absence of specific considerations to the contrary.

Randomization is effective to the extent that there is no important scientifically relevant feature in which the design is different from the reference set of designs on which the randomization distribution is based. This is the usual requirement for hypothetical long-run frequency calculations to be appropriate for the interpretation of a particular investigation, namely, that suitable conditioning is essential.

If some relevant feature is detected at the design stage, then either the basis of the design needs reconsideration or the randomization should be conditioned on holding the feature at or near its optimal value. If the feature is not detected until the analysis stage, then a special model will be desirable to adjust for the discrepancy in question. Sometimes this will be approximately equivalent to taking a conditional randomization distribution. Of course such modifications to the "standard" analysis are worthwhile only if they make a noticeable difference to the resulting confidence limits for the contrasts of interest.

In clinical trials, these points are relevant in that there are often a large number of potentially important features with respect to which approximate balance is desirable. In agricultural field trials, aspects of spatial arrangement, in particular the configuration of nearest neighbors, may be relevant in view of presumed patterns of spatial correlation and are the object of much current controversy, partly about the precise technique of analysis to be used and partly about the real gain in precision achievable.

The general view of randomization given in this section is a traditional one going back to the pioneer work of R. A. Fisher and F. Yates. Its mathematical formalization is the subject of current work, especially by R. A. Bailey and T. P. Speed. For discussion of balancing in clinical trials, see J. Grizzle's chapter in this book and for work on nearest-neighbor designs, see Wilkinson et al. (1983), Williams (1986), and Martin (1986). For the theory of conditional randomization, see Cox (1982a) and Bellhouse (1986).

6. OPTIMAL EXPERIMENTAL DESIGN

A major aspect in recent years has been the development of a very elegant theory of optimal design serving both to unify previous fragmentary ideas and to provide a systematic basis for tackling nonstandard problems (e.g., mixture designs with nonstandard design spaces). Practical methods for nonlinear problems have to be linked with sequential experimentation.

Like other supposedly "optimal" solutions of practical problems, the results have to be applied with caution because the theory requires the precise prior formulation of matters that in applications are usually only rather imprecisely known, namely,

The model to be used in analysis
The space of allowable treatments (design region)
The criterion to be optimized

For example, it is frequently the case that an optimal design for a given model does not allow the testing of the adequacy of that model nor, in general, the fitting of models richer than that initially contemplated. Also the design points will tend to be maximally dispersed over the allowable design region, and this will not be appropriate when the region of primary interest is inside the allowable region. For an interesting analysis of these and related points from a Bayesian viewpoint, see Steinberg (1985).

For a systematic account of optimal design theory, see Silvey (1980) and for more recent developments Atkinson (1982b). For an algorithm for the construction of optimal designs in nonstandard situations, see Welch (1982).

7. SERIAL DESIGNS

A serial design is one in which more than one treatment is applied to the same physical material (animal, plot, etc.), so that the same material is used to form more than one experimental unit. There is thus the possibility of carryover, that is, that the response on one unit may depend on the treatment applied to another unit, although usually strenuous efforts are made to avoid such effects. There is considerable current discussion of the role of such designs in clinical trials, and in the past there have been rather similar discussions in animal nutrition and experimental psychology.

Incidentally, serial designs in this sense should be distinguished from repeated-measures designs where a time series of measurements is made on each unit, even though the treatment is held constant on each unit. Such arrangements may raise special problems of analysis but not especially of design.

To an appreciable extent ambiguities of interpretation in serial designs can be overcome by the use of more periods per subject than there are treatments; see, for example, Ebbutt (1984) and Cox (1984). For example, treatment × period interaction and occurrence of residual effects can be so disentangled. The central difficulty remains, however, that unless there is intrinsic interest in, say, the nature of residual effects, their presence makes it hard to draw any clear conclusion about the treatment effects that would be present in a monotreatment regime: it is this last that would be often of main interest.

8. SEQUENTIAL DESIGN

It is reasonable to draw a distinction between serial designs, where the same material is used as an experimental unit several times, and sequential design, in which key aspects of the investigation may be changed as the experiment progresses.

Aspects of sequential design include the following.

1. There may be the possibility of sequential stopping, which may be appropriate from considerations of economy or, in clinical trials, from ethical considerations. This last raises some difficult issues, in particular because there may be several different responses under study and because of the need to avoid premature stopping because of an apparent treatment effect in a subset of patients.

2. Sequential changes in the set of treatments under study may be valuable. This has been investigated to some extent in connection with agricultural varietal trials. In the context of response surface designs, changes in design region can be accommodated within the computer algorithms for optimal design mentioned in Section 6.

3. Sequential rules for treatment allocation may be needed for use when such allocation is made in sequence in time. There are two sorts of such rule, those depending on the treatment allocations before the current one and those depending also on the responses before the current one. For a discussion of the former, see, for example, Efron (1971), Cox (1982b), Atkinson (1982a) and Smith (1984). There is an extensive theoretical literature on the latter, much stemming from the "play the winner" rule: see Bather (1985) and Armitage (1985) for a discussion of the role in applications, especially clinical trials. Difficulties inhibiting such application

include the multidimensional character of response in many such investigations and the fact that, while such trials have often have a strong decision-influencing role, any one trial is unlikely to be decisive, at least in more than a very temporary sense. A formulation in terms of estimating the magnitude of effects seems on the whole more fruitful. Adaptive schemes may, however, have a very fruitful role in contexts more akin to control theory.

4. Sequential changes in response definition and measurement, including, for example, changes in any subsampling schemes used, may be called for. There does not seem to have been systematic discussion of such issues.

5. Sequential changes in the criteria for the eligibility of experimental units (Cox, 1982c) may be desirable, for example, if experience suggests that the criteria are too restrictive.

6. Preliminary analysis may suggest a change in the detailed method of analysis to be used (i.e., a change in the detailed model), or in extreme cases a change in the whole focus of the investigation, either of which may well have consequences for design.

7. Special sequential considerations enter into design when the fitting of nonlinear models is involved. In particular the optimal design will depend on the unknown parameters, at least preliminary estimates of which are therefore required.

9. SOME TRADITIONAL TOPICS

A substantial part of the literature on the design of experiments is concerned with combinatorial problems, that is, with the construction of design patterns that will fulfill various desirable criteria. This aspect remains active and of considerable interest, the main emphasis being on the provision of flexible families of designs that will enable the provision of designs of full or high efficiency for a wide range of situations.

Examples are cyclic and generalized cyclic block designs regarded as generalizations of balanced incomplete block designs. Lists of reasonable designs are given, in particular, by John et al. (1972), Patterson et al. (1978), and Hall and Jarrett (1981). For factorial experiments, a primary question is the systematic construction of schemes of confounding and fractional replication for mixed systems, that is, those with factors at unequal numbers of levels. There are two main approaches, the DSIGN method of Patterson (1976) and methods based on generalized cyclic designs (John and Dean, 1975; Dean and John, 1975), these being quite different from generalized cyclic incomplete block designs. There is scope for further work on problems in which some factors have levels specified qualitatively and others are quantitative.

One very active field of application of highly fractional experiments is in industrial experiments with emphasis on quality and reliability, this current interest stemming in large part from the very influential work in Japan by G. Taguchi. Special features of his work are the systematic use of orthogonal arrays and the notion that minimization of the variability of response under practical perturbations of the system is of great practical importance. By contrast, in most applications the interest is primarily in the mean response with variability regarded as "noise" polluting information about the aspect of main concern. Of course to some extent the difference of emphasis is taken account of by regarding, say, log variance as a derived response variable.

REFERENCES

Armitage, P. (1985). The search for optimality in clinical trials. *Int. Statist. Rev.* **53**, 15–24.

Atkinson, A. C. (1982a). Optimum biased coin designs for sequential clinical trials with prognostic factors. *Biometrika* **69**, 61–67.

Atkinson, A. C. (1982b). Developments in the design of experiments. *Int. Statist. Rev.* **50**, 161–177.

Bather, J. A. (1985). On the allocation of treatments in sequential medical trials. *Int. Statist. Rev.* **53**, 1–13.

Bellhouse, D. R. (1986). Randomization in the analysis of covariance. *Biometrika* **73**, 207–211.

Cox, D. R. (1982a). Randomization and concomitant variables in the design of experiments. In *Statistics and Probability: Essays in Honor of C. R. Rao*, G. Kallianpur, P. R. Krishnaiah, and J. K. Ghosh, Eds.), Amsterdam: North-Holland, pp. 197–202.

Cox, D. R. (1982b). A remark on randomization in clinical trials. *Utilitas Math.* **21A**, 245–252.

Cox, D. R. (1982c). On the role of data of possibly lowered sensitivity. *Biometrika* **69**, 215–219.

Cox, D. R. (1984). Interaction. *Int. Statist. Rev.* **52**, 1–31.

Dean, A. M. and John, J. A. (1975). Single replicate factorial experiments in generalized cyclic designs. II. Asymmetrical arrangements. *J. R. Statist. Soc. B* **37**, 72–76.

Ebbutt, A. W. F. (1984). Three-period cross-over designs for two treatments. *Biometrics* **40**, 219–224.

Efron, B. (1971). Forcing a sequential experiment to be balanced. *Biometrika* **58**, 403–417.

Federer, W. T. (1981). Some recent results in experimental design with bibliography. II Bibliography A–K, III Bibliography L–Z. *Int. Statist. Rev.* **49**, 95–109, 185–197.

Hall, W. B. and Jarrett, R. J. (1981). Nonresolvable incomplete block designs with few replicates. *Biometrika* **68**, 617–627.

Herzberg, A. M. and Cox, D. R. (1969). Recent work on the design of experiments: a bibliography and a review. *J. R. statist. Soc. A* **132**, 29–67.

John, J. A. and Dean, A. M. (1975). Single replicate factorial experiments in generalized cyclic designs: I. Symmetrical arrangements. *J. R. Statist. Soc. B* **37**, 63–71.

John, J. A., Wolcock, F. W. and David, H. A. (1972). Cyclic designs. *Appl. Math. Ser.* **62**. Washington, D. C.: National Bureau of Standards.

Martin, R. M. (1986). On the design of experiments under spatial correlation. *Biometrika* **73**, 247–277.

Patterson, H. D. (1976). Generation of factorial designs. *J. R. Statist. Soc. B* **38**, 175–179.

Patterson, H. D., Williams, E. R. and Hunter, E. A. (1978). Block designs for variety trials. *J. Agric. Sci.* **90**, 395–400.

Pratt, J. W. and Schlaifer, R. (1984). On the nature and discovery of structure. *J. Amer. Statist. Soc.* **79**, 9–21.

Silvey, S. D. (1980). *Optimal Design*. London: Chapman and Hall.

Smith, R. L. (1984). Properties of biased coin designs in sequential clinical trials. *Ann. Statist.* **12**, 1018–1034.

Steinberg, D. M. and Hunter, W. G. (1984). Experiment design: review and comment (with discussion). *Technometrics* **26**, 71–130.

Steinberg, D. M. (1985). Model robust response surface designs: scaling two-level factorials. *Biometrika* **72**, 513–526.

Welch, W. J. (1982). Branch-and-bound search algorithms for experimental designs based on *D*-optimality and other criteria. *Technometrics* **24**, 41–48.

Wilkinson, G. N., Eckert, S. R., Hancock, T. W. and Mayo, O. (1983). Nearest neighbour (NN) analysis of field experiments. *J. R. Statist. Soc. B* **45**, 151–211.

Williams, E. R. (1986). A neighbour model for field experiments. *Biometrika* **73**, 279–287.

Completely Random Assignment versus Stratification in Clinical Trials

James E. Grizzle
Department of Biostatistics
School of Public Health
University of North Carolina at Chapel Hill
Chapel Hill, North Carolina

SUMMARY

The efficiency of stratifying on a risk factor before randomization as opposed to complete randomization and adjustment for the risk factor by postrandomization analysis has been examined by several investigators. The analyses show that in small samples, relying on postrandomization stratification can lead to significant loss in efficiency when compared to prerandomization stratification. In moderate and large samples the two design strategies should result in very similar efficiency.

1. INTRODUCTION

Recent trends in the design of clinical trials have been toward more complex designs, at least with respect to theoretical investigation. The increasing complexity has been evolving in several different directions. The

---

Sponsored by National Heart, Lung, and Blood Institute, Contract Number 1-HV12243-L.

simplest has been to increase the number of strata by taking into consideration in the design numerous patient characteristics that may affect their response. This may even take into account the time of entry into the study through the use of permuted block randomization. More complex design may require calculating the projected variance of the estimated treatment difference before each patient is randomized and then assigning the patient to the treatment that minimizes the projected variance of the treatment difference. Still other plans attempt to minimize the number of patients assigned to the apparently inferior treatment.

The design and administration of clinical trials are important interrelated aspects that require many compromises between the art of the possible and the theoretical ideal. An important part of the design is the decision about which patient characteristics that are prognostic of study outcomes should be used in randomization and which ones should be left for the act of randomization to achieve balance.

Some statisticians believe that randomization can be trusted to achieve balance of prognostic factors among treatments, and they then attempt to account for the effects of the prognostic factors by poststratification or by the analysis of covariance. Others believe in being more purposeful in controlling the experimental situation. They ensure balance by stratifying on prognostic factors and then include additive strata effects in the analytical model. The main conflict that arises between the two points of view is whether the added complexity of many strata is preferred over the possibility that the efficiency of a study will be reduced significantly by chance imbalance.

Peto et al. (1976) advocated poststratification to adjust for the effect of potential stratification characteristics and to control variance. This represents a move in the direction of simplicity in design, but, perhaps, with some increase in the complexity in the analysis. Several investigators have been prompted to examine the issue theoretically and by simulation in simple and relatively complex situations for a variety of possible treatment responses. These investigations do not employ parallel techniques in their approaches to the problem, but a remarkable similarity of results has been found.

2. RESULTS

The simplest case was examined algebraically by Grizzle (1982). He assumed that a single possible dichotomous stratification variable was ignored in the design, but that this variable was included in the analytical model. He assumed that the potential stratification variable had a prevalence g in one treatment and h in the other. Simple calculations show that the efficacy of poststratification, relative to prestratification, conditional on g and h, is

$$RE = \left\{ \frac{n_1 g + n_2 h}{n_1 g(1-g) + n_2 h(1-h)} \left[1 - \frac{n_1 g + n_2 h}{n_1 + n_2} \right] \right\}^{-1} \leq 1$$

with $RE = 1$ when $g = h$.

A special case is $n_1 = n_2$, in which case

$$RE = \left\{ \frac{g + h}{g(1-g) + h(1-h)} \left[1 - \frac{g + h}{2} \right] \right\}^{-1}$$

Since g is the prevalence of the prognostic factor in the study population, it should be equal to h except for chance imbalance. Some insight into the effect of $g \neq h$ is given by examining the case $g = 1 - h$. This leads to considerable simplification when $n_1 = n_2$.

Then

$$RE = 4g(1 - g)$$

A few calculations are instructive. Table 1 shows some relative efficiencies in this special case.

Imbalance is more likely for small sample sizes than for large. In cancer chemotherapy, sample sizes of 25–50 are not uncommon. When $n = 25$, imbalance as much as $g = 0.7$, $h = 0.3$ or greater could occur with some resulting loss of efficiency. For sample sizes of 50, however, large deviations are unlikely, which implies that randomization can be trusted to prevent large losses in efficiency.

McHugh and Matts (1983) derived the variance of a treatment contrast in the prestratification and poststratification cases using the permutation model induced by the act of randomization. They found the theoretical

Table 1. Relative Efficiency When $g = 1 - h$

g, h	RE
0.6, 0.4	0.96
0.7, 0.3	0.84
0.8, 0.2	0.64
0.9, 0.1	0.36

variance of a contrast for the post stratified estimate to be

Variance (contrast) poststratified

= Variance (contrast) prestratified + adjustment for poststratification

The adjustment for stratification is $O(1/n_k^2)$, where n_k is the sample size in the kth stratum.

Table 2, which has been abstracted from Table 1 of McHugh and Matts (1983), shows the relative efficiencies for small and moderate sample sizes within strata and small and moderate number of strata. Thus it seems clear that in the sample size and number of strata most likely to be found in practice, the Peto et al. (1976) assessment of the average loss in efficiency due to poststratification is approximately correct.

Mantel (1984) gives an example of 20 strata after poststratification in which each has 20 observations. His informal analysis of the loss of precision due to poststratification is that it is equivalent to a sample of 380 rather than 400. More precise assessment of the situation by McHugh and Matts (1984) yields an effective sample size of 360. Thus two extra observations per stratum would be required to compensate for the loss in efficiency due to poststratification in this rather extreme case.

Cox (1982) derived an inflation factor for the variance between two treatment means when covariance type of adjustment is used. The expression is

$$(1 + W_p/d_w)$$

where W_p is a χ^2 random variable with p degrees of freedom and d_w is the degrees of freedom within treatment. Since the expected value of W_p is p and d_w is approximately n, he finds that the inflation factor is approximately $1 + p/n$, which shows clearly the trade-off between n and p for large samples. Cox points out, however, that in any specific case the effects

Table 2. Relative Efficiency

Stratum Sample Size	Number of Strata 4	8	20
40	0.930	0.845[a]	0.602
80	0.964	0.920	0.803
160	—	0.998	0.894

[a] Interpolated value.

of imbalance could be considerably different from this expected value and a much safer requirement can be obtained by taking into account the variability in W_p that might be encountered by requiring $c^*_{i\varepsilon} < \frac{1}{10}n$, where $c^*_{p_\varepsilon}$ is the upper point of χ^2 with p degrees of freedom. For example when $\varepsilon = 0.01$ and $p = 1, 5, 10, 50, 108$, the minimum values of n are 67, 151, 232, 726, and 1358.

Further evidence supporting a small power loss was provided by Palta (1985), who investigated maximum power loss in survival studies with nonstratified randomization. Palta studied the log rank test and assumed a constant hazard ratio both between and within strata. The results obtained make it possible to derive the loss of power in large samples and to calculate the increase in sample size necessary to compensate for the loss of power due to poststratification. This loss can be compensated by adding 6–12 subjects to each treatment group depending on the specific design characteristics.

The experimental situation in which a binary response is analyzed by logistic regression was investigated by Begg and Kalish (1984). They compared a design that minimized the variance of the estimated treatment difference ($D_s - L$ optimal) to complete randomization, permuted blocks, and Begg–Inglewicz optimal. The results of several clinical studies were used to design simulations that should be similar to those found in practice. The efficiency was found to be quite high when there was a small number of covariates, but it tended to decline when the number of covariates was increased or when the number of covariates was small but had large effects. For sample sizes of 100 or more, however, the efficiency of the randomized design relative to $D_s - L$ optimal was generally above 90%. This loss of efficiency, measured by variance, is somewhat larger than found in the other work summarized here. It is due possibly to the fact that the variance depends on the probability of a positive response and that the difference in probability is taken into account in the optimum assignment rules. Complete randomization would tend to equalize the number assigned to each treatment, which could cause very different variances of the treatment effects from the optimum even when the sample sizes are equal.

Statisticians who are concerned about a large number of strata in a study sometimes adopt other strategies to control variance. A method that makes good use of previously collected data is based on deriving a functional relationship, usually by multiple regression, between the response and important predictors of the study endpoint. The trial subjects are then randomized to treatment within levels of that predicted value. Alternatively, postrandomization strata could be constructed by this method.

To cite an example, the Lipid Research Clinics Program conducted a clinical trial on the effects of cholesterol lowering in men who had high

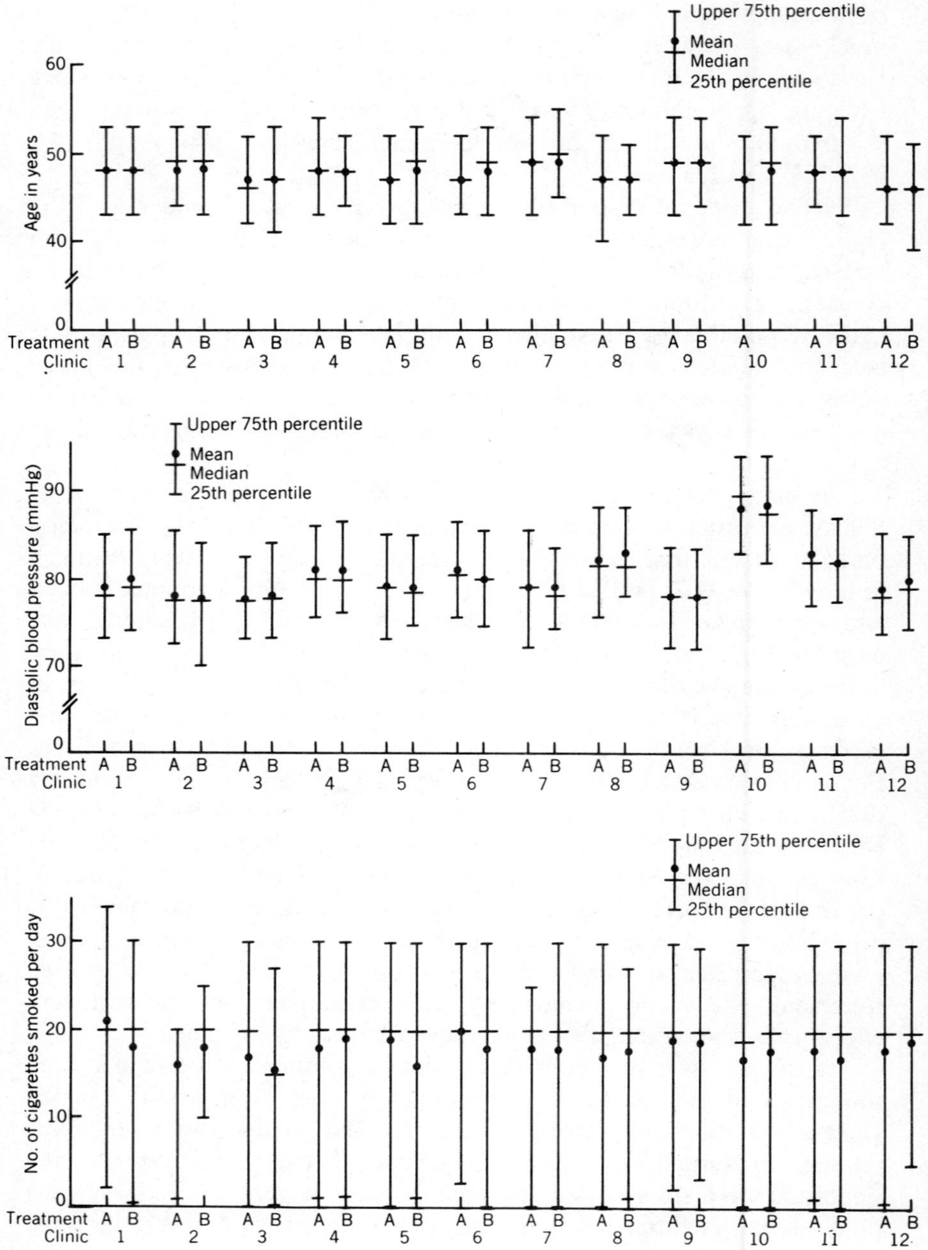
Upper 75th percentile
Mean
Median
25th percentile
Age in years
60
50
40
0
Treatment A B A B A B A B A B A B A B A B A B A A B A B
Clinic 1 2 3 4 5 6 7 8 9 10 11 12
Upper 75th percentile
Mean
Median
25th percentile
Diastolic blood pressure (mmHg)
90
80
70
0
Treatment A B A B A B A B A B A B A B A B A B A B A B A B
Clinic 1 2 3 4 5 6 7 8 9 10 11 12
Upper 75th percentile
Mean
Median
25th percentile
No. of cigarettes smoked per day
30
20
10
0
Treatment A B A B A B A B A B A B A B A B A B A B A B A B
Clinic 1 2 3 4 5 6 7 8 9 10 11 12

Figure 1.

cholesterol but no clinical heart disease. The stratification was by cholesterol level, exercise stress test, clinic, and a risk function that related age, systolic blood pressure, and smoking history by a logistic function that was derived from Framingham data to the probability of developing coronary heart disease.

As expected, blocking on the risk function resulted in balance among treatments for that factor. Also, as shown in Figure 1, randomization resulted in reasonable balance for each component of the risk function within each clinic. Thus components of the risk function then could be used as poststratification variables or in an analysis of covariance to investigate effects of the individual components of the risk function on survival if desired. This approach enables the analyst to keep the number of strata small while at the same time preserving the opportunity to do poststratification analysis that should have high efficiency.

3. CONCLUDING REMARKS

Armitage (1985) notes that many of the plans that are interesting theoretically are little used in practice. He cites a number of reasons why this may be so. Although he refers specifically to sequential analysis, his remarks may apply to complex designs that have been developed to increase the overall efficiency of a clinical trial.

Complex or unusual design tend to be avoided because of concern about acceptability of the results by the clinical community. Most large clinical studies will never be replicated, because of problems of cost and ethics. Health policy affecting millions of persons may be based on the outcome of such trials. These considerations cause the planners of such studies to be interested in conservative designs that are easily understood and widely accepted. Thus it is likely that innovations in design of these large cooperative studies will come in small increments that parallel the increase in statistical sophistication in the medical community rather than in the statistical community.

This appears to leave the statistician on the horns of a dilemma when deciding on whether to stratify or not to stratify. It seems clear that in large studies postrandomization stratification can be done without significant decrease in efficiency and that this is the preferred way to proceed when numerous strata would make a study difficult to administer or the results hard to interpret. Very small studies have a significant risk of loss of efficiency because of chance imbalance when postrandomization stratification is done. It is equally clear that the characteristics to be used for postrandomization stratification should be specified in the design stage and that they should not be too numerous. When a poststratification analysis of the analysis of covariance type is contemplated, it is important to have knowledge about the functional form of the relationship between the response and the poststratification variables. It should be kept in mind that in the poststratification analysis the analyst is attempting to accomplish same control by modeling that is done purposefully by design when the strata are defined in advance and that control at the design stage is less model dependent than poststratification, especially when adjustments of the analysis of covariance type are made in the analysis. Thus the experienced statistician is still faced with a judgment call that requires intimate knowledge of the area of application and the audience for whom the analytical results are intended.

REFERENCES

Armitage, P. (1985). The search for optimumality in clinical trials, *Int. Statist. Rev.* **53**, 15–24.

Begg, C. B. and Kalish, L. A. (1984). Treatment allocation for nonlinear models in clinical trials. *Biometrics* **40**, 409–420.

Cox, D. R. (1982). Randomization and concomitant variables in the design of experiments. In *Statistics and Probability: Essays in Honor of C. R. Rao,* (G. Kallianpur, P. R. Krishnaiah, and J. K. Ghosh, Eds.), Amsterdam: North-Holland, pp. 197–202.

Grizzle, J. E. (1982). A note on stratifying versus complete random assignment in clinical trials. *Control. Clin. Trials* **3**, 365–368.

Mantel, N. (1984). Pre-stratification or post stratification, (letter to the editor). *Biometrics* **40**, 256–257.

McHugh, R. and Matts, J. (1983). Post stratification in the randomized clinical trial. *Biometrics* **39**, 217–225.

McHugh, R. and Matts, J. (1984). Response to Mantel's letter. *Biometrics* **40**, 258.

Palta, M. (1985). Investigating maximum power losses in survival studies with non-stratified randomization. *Biometrics* **41**, 497–504.

Peto, R., Pike, M. C., Armitage, P., Breslow, W. E., Cox, D. R., Howard, S. V., Mantel, N., McPherson, K., Peto, J., and Smith, P. G. (1976). Design and analysis of randomized clinical trials requiring prolonged observation of each patient. I. Introduction and design. *Br. J. Cancer.* **34**, 585–612.

Using Systematic Plans for Factorial Experiments in Industry

Peter W. M. John
Department of Mathematics
The University of Texas
Austin, Texas

SUMMARY

In engineering experiments it is usual for the various points of a planned experimental design to be run in sequence. This differs importantly from agricultural experimentation, where the varieties are sown simultaneously and are assigned to the plots at random. In this chapter we argue that one should not arrange the points of the engineering experiment in random order. Instead one should take advantage of the sequential nature of the experiment by making the runs in a systematic order.

Three reasons are advanced for this:

1. Systematic designs can protect the experimenter against linear and quadratic time trends.
2. They can reduce the number of times that one has to change the levels of the factors, which can bring a saving in cost.
3. They can provide protection against the risk that the experiment may be stopped before all the points have been run.

Unfortunately, these three goals may conflict. The same design will not be the best for every criterion, and some compromises must be made.

The practical problems will be illustrated by examples of designs for factorial experiments with four factors at two levels each.

1. INTRODUCTION

The use of statistically designed experiments has increased steadily, but slowly, in industry since the end of World War II. Lately there has been an increase in interest because of the worldwide concern for quality in production. After years in which quality control meant only the passive approach of control charts and acceptance sampling, followed when necessary by some form of correction or rectification, we are now focusing our attention more on finding out how our processes really work and how their yields might be improved. Particular interest has centered upon the work in quality control in Japan (Taguchi and Wu, 1985).

The emphasis of this chapter is on what is sometimes called off-line quality control. In particular it is concerned with screening experiments in which the experimenter, or engineer, screens various factors to find out which of them are important in improving production. The main interest lies in finding one or two major factors among several candidates, rather than with a careful measurement of precision. We shall confine ourselves only to the simplest and, at the same time, most common case of the factorial experiments in which the factors appear at only two levels. Our development depends on a crucial difference between the traditional agricultural experiment and industrial experiments.

In the usual agricultural experiment the varieties are sown in their plots at the same time at the beginning of the growing season. They grow side by side until harvest time. Most industrial experiments, particularly engineering experiments, differ from the agronomist's experiments in one crucial point. They are made on the same equipment or pilot plant and must be carried out in sequence, one after another. The industrial experimenter is thus subject to three complications that do not concern the agronomist:

1. Since the engineer's experimental runs are made in sequence over time he or she may have to take into account a time trend, linear or quadratic; the catalyst may lose its potency as the experiment proceeds, or deposits may build up in an experimental motor.
2. Large costs may be incurred by all the changes in factor levels that are required. It can be expensive to change the catalyst in a pilot plant. It may take several hours for a system to settle down after the temperature has been changed. Can the number of these changes be minimized?
3. The budget of time or money may run out before the whole experiment as originally planned is finished, or a vital piece of equipment may fail with no replacement available. If this happens, it would be

preferable if the part that had been completed made up a good fraction.

The first of these problems has been investigated by, among others, Box (1952), Cox (1952), Hill (1960), Daniel and Wilcoxon (1966), and John (1985); the second by Dickinson (1974); and the third by John (1971) and Smith and Schmoyer (1982).

These are all practical problems that face the industrial experimenter. We shall illustrate them with plans for the 2^4 factorial experiment—four factors each at two levels with 16 points.

2. LINEAR AND QUADRATIC TRENDS

The main reference for this section is the paper by Daniel and Wilcoxon (1966). Cuthbert Daniel is a well-known industrial consultant. At that time Frank Wilcoxon had retired to a professorship at Florida State University after a distinguished career in the chemical industry with American Cyanamid Company. They were considering sequences in which the points of a factorial design might be run in order to eliminate the effect of linear or quadratic time trends. There is no way in which all the interactions as well as the main effects can be made free of trends. We shall discuss how one can arrange for the main effects and some of the two factor interactions to be uncorrelated with a linear trend.

Daniel and Wilcoxon noticed that if the points in the factorial were run in the standard order, beginning with the base point (1), main effects and two factor interactions were correlated with time trends. So they introduced four new factors by equating them to higher-order interactions. These transition factors were $F = ABD$, $G = ACD$, $H = BCD$, and $J = -ABCD$. Then they wrote a, b, c, d for f, g, h, j and obtained the following experimental plan:

Plan 1. (Daniel and Wilcoxon)

(1)	*abd*	*acd*	*bc*	*bcd*	*ac*	*ab*	*d*
abcd	*c*	*b*	*ad*	*a*	*bd*	*cd*	*abc*

In this plan the estimates of the main effects are uncorrelated with both the linear and the quadratic trends. Each of the two factor interactions has some degree of correlation with one, but not both, of the trends.

If J is defined by $J = ABC$ instead of $J = -ABCD$, one gets the design that had been quoted earlier by Hill (1960), another industrial statistician working for the Tennessee Eastman Company. In Hill's design the main effects are uncorrelated with linear and quadratic trends; all the two-factor interactions are uncorrelated with linear trends:

Plan 2. (Hill)

(1)	*bcd*	*acd*	*ab*	*abc*	*ad*	*bd*	*c*
abd	*ac*	*bc*	*d*	*cd*	*b*	*a*	*abcd*

In both papers the authors have assumed that the runs are independent (uncorrelated), and we shall continue with that assumption. When there is serial correlation between the runs, the situation is more complicated. Another potential source of complication would be the presence of periodic time trends.

The experimental plan proposed by Daniel and Wilcoxon is an ordered foldover design (John, 1985). The ninth point is obtained from the first point by changing the levels of all four factors. The tenth point is obtained in the same way from the second point, and so on. If each factor appears four times at its high level and four times at its low level in the first eight runs, the ordered foldover design will always have the main effects orthogonal to a linear trend.

The general result may be stated as follows. Let D be a plan with N points for n factors each at two levels and let D' be another plan, also with n points constructed in the following way. The ith point of D' is the complement of the ith point of N, that is, the levels of every factor are reversed. Then the combined plan $D^* = D + D'$ has the property that the main effect of each factor is orthogonal to a linear trend.

The first-degree orthogonal polynomial takes the values $-(2N - 1), -(2N - 3), \ldots, -1, +1, \ldots, +(2N - 3), +(2N - 1)$. A main effect is orthogonal to the linear trend if the sum of the values of this polynomial at the places where the factor is at its high level is zero. This always happens when a factor is folded over. Consider the factor A; it appears at its high level in places 2, 3, 6, 7, 9, 11, 12, 16. The corresponding values of the first degree polynomial sum to $-13 - 11 - 5 - 3 + 1 + 7 + 9 + 15 = 0$.

This method produces resolution IV designs. If D is a fractional plan with N points for n factors, which has resolution III (the main effects are estimable if we suppress all interactions), $D^* = D + D'$ is a plan with $2N$ points that has resolution IV (the main effects are estimable whether or not the two-factor interactions exist).

The following plan is a sequence for a one-eighth replicate of a 2^7 factorial. It is of resolution IV and has all the main effects uncorrelated with a linear trend. The first eight points form the traditional 2^{7-4} fraction of resolution III:

Plan 3. (John)

(1)	*abde*	*acdf*	*bcef*	*bcdg*	*aceg*	*abfg*	*defg*
abcdefg	*cfg*	*beg*	*adg*	*aef*	*bdf*	*cde*	*abc*

We are not confined to plans with 16 points. Plan 4 is a resolution IV plan for five factors in 12 points. In this plan the estimates of the main effects have less than 100% efficiency.

Plan 4.

(1)	*abce*	*ade*	*abcd*	*cd*	*be*
abcde	*d*	*bc*	*e*	*abe*	*ace*

The experimental plan of Hill (Plan 2) is not a foldover design. Factors A, B, and D are each folded over as we go from the first eight points to the second eight. Factor C remains unchanged; such a factor is called *kept* because the levels are kept unchanged. It can be shown (John, 1985) that the main effect of a kept factor will be orthogonal to a linear trend if the sum of the places in the first N points at which it takes its high level is $N(N + 1)/2$. In our case $N = 8$. We note that C appears at its high level in runs 2, 3, 5, and 8 ($2 + 3 + 5 + 8 = 18$).

The case of two-factor interactions is more complicated. If A and B are both folded over, their product AB is kept. If A is folded over and D is kept, the AD interaction is folded over also. In Hill's plan all six of the two-factor interactions are also uncorrelated with linear trend.

The idea of using ordered foldover designs as plans in which the main effects are robust against linear trends has also been extended to factorial experiments with factors at three levels by John (1985). When Taguchi and Wu (1985) consider experiments with factors at three levels, they use plans that they call *lattices*. These are main effects plans that are called by the agricultural statisticians orthogonal arrays. The ideas that are developed in this chapter and in John (1985) are directly applicable to Taguchi's lattices. In the case of the lattice with 18 points, the report contains a nonisomorphic lattice whose points can be arranged in a sequence such that the main effects are uncorrelated with a linear trend.

3. CHANGING ONE FACTOR AT A TIME

When a sequence of points in a factorial experiment is run on a pilot plant the levels of some factors must be changed as we go from point to point.

Some of these changes of level can be complex and costly. Dickinson (1974) was interested in experiments in chemical plants. He considered the problem of reducing the cost of experimentation by minimizing the number of changes of level that have to be made.

Consider the following two sequences for the three factor experiment.

Plan 5a.

(1)	*a*	*ab*	*abc*	*bc*	*c*	*ac*	*b*

At each stage the level of only one factor has to be changed until we get to the transition from the seventh point to the eighth; then all three factors have to be changed. In the whole experiment the level of *A* is changed four times, the level of *B* is changed three times, and the level of *C* is changed twice. The total number of changes of level is nine.

Plan 5b.

(1)	*abc*	*a*	*bc*	*ac*	*b*	*c*	*ab*

This is a much less desirable plan because there are 18 changes of level, twice as many as in Plan 5a.

Dickinson found several sequences for both the four-factor and the five-factor experiments in which only one factor had to be changed at each stage. These minimal plans occur in switching theory, where they are known as Gray codes. Dickinson's plans do not have the property that the main effects are orthogonal to linear trends. Indeed, he suggests that the two aims are to a certain extent opposed to one another. The following example comes close to combining both properties.

Plan 6. (Dickinson)

(1)	*a*	*ad*	*acd*	*abcd*	*bcd*	*bc*	*b*
bd	*abd*	*ab*	*abc*	*ac*	*c*	*cd*	*d*

In this plan the main effects of *B* and *D* are uncorrelated with linear trends. The factors *B* and *D* are both folded. Factors *A* and *C* are kept, but the sums of their places at high level are not equal to 18.

The emphasis on the number of changes may be criticized because it regards the cost of changing the level of each factor as the same. This may not be reasonable. If the cost of changing *C* and *D* is negligible compared to the cost of changing A or B, it would be reasonable to look for sequences that minimize the numbers of changes only in A or B. Plan 7 has only two changes of A and four changes of B, and all four main effects are orthogonal to linear trends:

Plan 7.

(1)	*c*	*bd*	*bcd*	*abcd*	*abd*	*a*	*ac*
acd	*ad*	*abc*	*ab*	*b*	*bc*	*cd*	*d*

4. EARLY TERMINATION

A third problem with an industrial experiment in which the points are run in sequence is that the experiment may suffer early termination. The budget of money or of time may run out before all the points have been run. Another negative possibility is that a crucial piece of equipment like a pump may fail, causing the experiment to be abandoned.

Suppose, for example, that the experiment stops after only 8 of the 16 points have been run. It would be better for the experimenter to have chosen a sequence in which the first 8 points were either the half replicate defined by $I = +ABCD$:

(1) *ab* *ac* *ad* *bc* *bd* *cd* *abcd*

or the complementary half replicate defined by $I = -ABCD$:

a *b* *c* *d* *abc* *abd* *acd* *bcd*

In each of those two cases the shortened experiment would be a design with resolution IV (main effects clear of two-factor interactions). On the other hand a random choice of order for the 16 points is likely to produce a first half that is far from being an efficient subdesign.

After one has chosen the first 8 points to be a good half replicate it would be desirable to choose the next 4 points so that all 12 would form a design from which all main effects and two-factor interactions are estimable. This can be done by having them be a three-quarter replicate (John, 1962, 1971). In that case one would select one of the factors, say A, and choose the ninth through twelfth points to be a quarter replicate from the family defined by $I = \pm A = \pm BCD = \pm ABCD$.

Again we are faced with competing criteria. A plan that is robust against early termination is not necessarily robust against linear trends. Plan 8 goes some way towards meeting both these goals:

Plan 8. (John)

(1)	*bc*	*bd*	*cd*	*ad*	*abcd*	*ac*	*ab*
abc	*a*	*acd*	*abd*	*bcd*	*d*	*b*	*c*

If the complete plan is run, we have a design in which all four main effects are free of linear trend. Factors A, B, and C are folded; factor D is kept. In the worst case, in which only four points can be run, we have a half replicate of the three-factor design (B, C, D) that has resolution III. The addition of the next two points makes a plan with six points for all four factors that is of resolution III. The first eight points are a resolution IV plan, as we have already said.

Adding the next 4 points produces a resolution V design in 12 runs. The 12 points can be divided into three overlapping half replicates. Each of the main effects and two-factor interactions can be estimated from one or more half replicates. The main effects B, C, D are estimated from the first 8 points; AB, AC, AD are estimated from the half replicate defined by $I = -BCD$; BC, BD, CD are estimated from the half replicate defined by $I = A$; A is estimated from two half replicates, $I = -BCD$ and $I = ABCD$.

5. CONCLUSION

We have used the 2^4 factorial experiment to illustrate three considerations that arise in industrial experiments when the points in a factorial design are run in sequence: protection against linear time trends, economy in the cost of changing the levels of factors, and defense against early termination of the experiment.

Unfortunately these considerations can be in conflict. To make the main effects orthogonal to linear trends, we look for ordered foldover plans. But a foldover plan cannot be a plan that minimizes the number of changes of levels of factors, and a foldover design does not necessarily give the best protection against premature termination.

The experimenter has to make compromises. In a particular experimental situation one of the three criteria may be more important than the other two. The important things to realize are that we can and should exploit the idea of experimental runs made in sequence to help us and that the industrial experimenter should talk with the statistician about planning the experiment before the first point is run. It is not enough that the experimenter should blindly take a plan from a standard textbook that has probably been written for agricultural scientists. The experimenter and the statistician should think beforehand about the aims of the experiment, and plan together how they may best be achieved.

REFERENCES

Box, G. E. P. (1952). Multifactor designs of first order. *Biometrika* **39**, 49–57.

Cox, D. R. (1952). Some recent work on systematic experimental designs. *J. R. Statist. Soc. B* **14**, 211–219.

Daniel, C. and Wilcoxon, F. (1966). Factorial 2^{p-q} plans robust against linear and quadratic trends. *Technometrics* **8**, 259–278.

Dickinson, A. W. (1974). Some run orders requiring a minimum number of factor level changes for the 2^4 and 2^5 main effects plans. *Technometrics* **16**, 31–38.

Hill, H. H. (1960). Experimental designs to adjust for time trends. *Technometrics* **2**, 67–82.

John, P. W. M. (1962). Three quarter replicates of 2^n designs. *Biometrics* **18**, 172–184.

John, P. W. M. (1971). *Statistical Design and Analysis of Experiments*, New York: Macmillan.

John, P. W. M. (1985). Time trends and screening experiments. Technical Report No. 16, Center for Statistical Sciences, University of Texas, Austin, Texas.

Smith, D. E. and Schmoyer, D. D. (1982). First order "interruptible" plans. *Technometrics* **24**, 55–58.

Taguchi, G. and Wu, Y. (1985). *Introduction to Off-Line Quality Control*. Central Japan Quality Control Association.

Some Recent Advances in Data Analysis Practice

Ludovic Lebart
Centre National de la Recherche Scientifique
Paris, France

SUMMARY

In this chapter we deal with a personal review of contributions relating to "multivariate exploratory data analysis" during the past five years. It aims at detecting the main trends and the new tracks of investigation. Six sections are considered:

1. Innovative theories, methods, or algorithms: This section is devoted to the more academic works. Among a large number of papers, some apparently significant contributions are selected: valuable syntheses, noticeable graphical techniques, promising theoretical findings.
2. Structured or constrained data: How to use a priori information to perform further exploration. This question leads to a rich domain of research with practical interest.
3. The interface between inference and exploration: Users are now more demanding about validity, stability of patterns, significance of results.
4. The new fields of applications: Survey data processing, textual data processing, chronological or spatial data. Each domain stimulates specific methodology or software.
5. Computational aspects, software, and environment: There is intense activity in connection with the dramatic increase of availability of microcomputers.
6. General scientific environment: The state of the institutional environment of the investigation in data analysis.

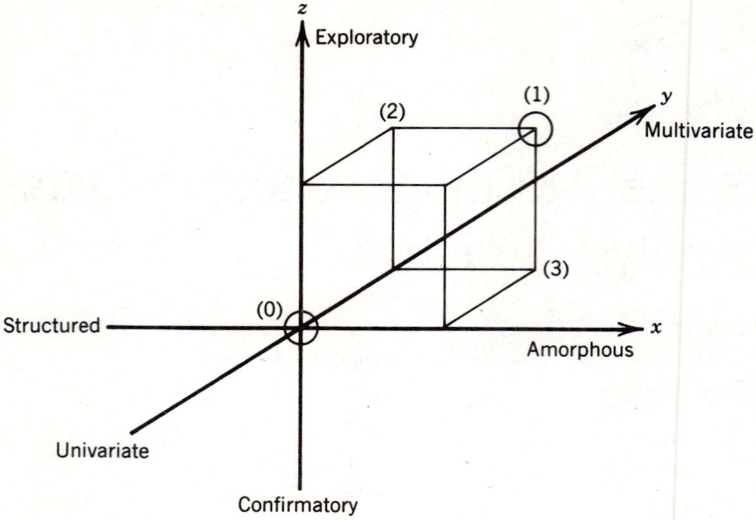

Figure 1.

This text presents a rather personal point of view on the work that has been done over the past five years in the area of multivariate exploratory data analysis. Thus, this is neither a synthesis nor a historical study, but an almost instantaneous picture—the time period under review is very short if we consider the lifespan of a discipline—of areas of interest, fields of application, and technical innovations. The extent of the field of study and the vague or conventional nature of the limits of data analysis make this study all the more arbitrary. Traditional inferential statistics, algorithmics, pattern recognition, econometrics, software invention; all these fields have common borders or overlap with data analysis, as we understand it in this chapter. Research in these overlapping areas is often active and fruitful.

How can we set limits in this field of study, which separates mathematics, informatics, and all experimental and observational sciences? In order to clearly situate our area of interest within a more general context, we make use of and extend a pattern put forward by Havranek and Chytil (1983).

Statisticians are faced with a situation that can be summarized as a point in a three-dimensional space (see Figure 1):

The first dimension (X) opposes structured data and amorphous data (about which there is no a priori information)

The second dimension (Y) opposes univariate and multivariate data

The third dimension (Z) involves the approach that can be either confirmatory (almost synonymous to inferential) or exploratory (i.e., of a more descriptive nature)

It thus becomes necessary to give a fuller interpretation of each of the two axes.

The first dimension (X), related to a priori information and to the monitoring of the phenomenon under study, often opposes experimental data (generally partially structured) to observational data.

The second dimension (Y) can more generally be described as a measurement of the complexity of data collection.

The two circled vertices in Figure 1 correspond to the two situations where the contrast is most obvious:

(0), the confirmatory approach on structured (or experimental) unidimensional data

(1), the exploratory approach on multivariate observations (or amorphous data)

Situation 1 is characteristic of the growth of data analysis in the 1970s as a "technique for describing data matrices" or as an "instrument for the observation of multidimensional reality": the larger the matrix and the more restricted the a priori information, the more justified and efficient the use of such instruments will be.

Section 1 will deal with contributions about observation instruments: the study of their characteristics, theoretical clarification, the development of alternatives, and so on. In practice, the situation is not as ideal as in (1). A priori information does exist, and we wish to use the information we already possess to the maximum in order to gain more information: data are partially structured and this structure must be borne in mind. The circle moves horizontally from vertex (1) to vertex (2) of the cube. Section 2 will deal with structured data, which involve a particular methodology or specific instruments. The user, on the other hand, is not necessarily satisfied with a simple description. He or she might desire information on the stability of the patterns obtained and might also wish to test a model, even a simple one. This situation corresponds to a vertical movement of the circle toward vertex (3). Section 3 will deal with research undertaken as a consequence of these new preoccupations.

A concrete situation will in fact correspond to a point situated within the triangle made up by the vertices (1), (2) and (3) of the cube in Figure 1.

Apart from these trends in research on methods, other factors significantly influence recent progress in data analysis: the new fields of application, studied in Section 4, are proof of the many instruments available and, in turn, encourage the development of methodology. The computational environment, dealt with in Section 5, is also undergoing rapid change and greatly interacts with description and analysis techniques.

Finally, Section 6 is a description of the institutional and scientific environment of data analysis.

1. GENERAL METHODS

The studies examined in this section are, perhaps, the most academic. Emphasis will be placed on synthesis, always welcome in new disciplines of research, the improvement or adaptation of existing methods, new methods, and new concepts.

We shall examine a few examples, without attempting to be exhaustive. "Multidimensional scaling" techniques, a very independent field of investigation, will be among those examples excluded.

Apart from very general surveys that review the entire field of exploratory statistical analysis, such as Mallows and Tukey (1982), syntheses according to theme or according to the method have been published over the past few years.

Along with articles of synthesis published in Barnett (1981), the classifications offered by Gower and Digby (1981) in relation to bidimensional representations constitutes one of the earliest attempts to bring together the wide variety of instruments, all of which aim at observing multidimensional data.

The different presentations and variants of correspondence analysis are listed in a very compact synthesis on this method compiled by Escoufier (1985), Tenenhaus and Young (1985) develop and compare the various approaches to multiple correspondence analysis, scattered throughout the literature and in space (see also Hayashi, 1984).

A number of recent contributions in English, such as Nishisato (1980), Greenacre (1984), Lebart et al. (1984), and Gifi (1981) help to spread information on this type of method. As far as classification goes, there are too many works on this subject for them all to be quoted here. We do, however, mention a recent synthesis on available algorithms by Murtagh (1985).

As a result of the proliferation of methods and of the ensuing confusion among users, many studies comparing methods on specific data sets are currently being published.

Thus, in solving a single problem, comparisons are made between log-linear models and correspondence analysis (Lauro and Decarli, 1982; Daudin and Trecourt, 1980), between correspondence analysis and ordination techniques (Ihm and van Groenewoud, 1984) and a logistic model (Leclerc et al., 1985), and principal components analysis in a number of special cases (Eppink, 1984). Numerous technical comparisons have been made in relation to more confirmatory methods such as discriminant analysis; see Nakache et al. (1980) as an example.

As far as tested methods are concerned, the improvement of computational techniques facilitates an easier or less costly application. As far as automatic classification goes, it is now possible to apply hierarchical clustering to several thousands of objects with such techniques as the reciprocal neighbors chain search algorithm (Benzecri, 1982; Juan, 1982) while the hybrid clustering method (Wong, 1982) allows for the processing of even larger groups. For principal-axes methods, large or sparse matrices can be processed with direct-reading algorithms (stochastic approximation type) (Lebart et al., 1984). Murtagh (1984) gives an interesting assessment of the complexities of hierarchic clustering algorithms. The following sections will deal largely with the improvement and diversification of methods, particularly when considering a priori structures. Among general improvements, "concept imports" should be mentioned: the *fuzzy set* notion has proved to be useful in classification (Bock, 1979b; Lerman et al., 1980) or simply through the introduction of "fuzzy scaling" (Gallego, 1982). This brings us to another successfully imported notion, that is, spline transformation from studies by Winsberg and Ramsay (1983) and Rijckevorsel (1982). These works fall within the context of nonlinear methods; so too do the studies by Leeuw (1982) and Gifi (1981).

Concerning classification, improvements have led to the publication of numerous studies. Several specialists such as Bock (1979a) and Govaert (1984) direct their attention to methods for the simultaneous clustering of the rows and columns of a matrix. Many other contributions are devoted to clarifying the optimization criterion (Diday et al., 1980; Marcotorchino and Michaud, 1982) or to determining the number of clusters (Matusita and Ohsumi, 1980; Milligan and Cooper, 1985). New concepts are being created, such as that of the "pyramids" (Bertrand and Diday, 1985), which extends the concept of "indexed hierarchy." It is very difficult, however, at this time to identify which of the new concepts will be decisive in the future. This will become evident after users have applied and compared methods for a long period of time.

To conclude this brief review, we may mention the interest shown in graphical methods. Interesting new techniques are worth being better known (see Wakimoto and Taguri, 1978, for example) especially since the increased use of microcomputers (see Section 5). Broersma and Molenaar (1985) have examined the graphical perception of the distributional aspects of data, which is increasingly used as a means of communication but is seldom studied. Thus, they have bridged the gap in the study of the following chain: "phenomenon or problem data, representations, interpretations, and conclusions."

2. STRUCTURED DATA

There are limits to the analogy between methods of data analysis and physical instruments of observation. When the multidimensional reality to be studied is totally unfamiliar to the statistician, methods of matrix description in effect function, we may say, like a microscope or an X-ray machine. This may be so for new fields of research such as, for example, certain cases in the collection of ecological data. The less a priori information the statistician has of the area under observation, the greater will be the benefit to him of the use of methods of data analysis. This is the ideal case represented by vertex (1) of the cube in Figure 1.

In practice, however, a priori information does exist due to the type of data available (partially experimental or controlled data, chronological data, spatial data etc.), or due to previous results obtained (the existence of a size effect, the obvious role of factors that are not easily controlled, etc.).

The emergence of trivial patterns at the end of computations, which are, nonetheless, costly in terms of the time spent by the specialist in doing them, is of no interest.

The problem, then, for the statistician is that he or she should know how to use all the information at hand in order to try to obtain more information.

This leads to a basic topic of research in data analysis: which is the methodology to be adopted in order to make the best use of a priori information or information gathered in the course of analysis? Should specific methods be created, or should one simply employ a sequence of recoding and processing? The broad range of studies dedicated to these choices is proof of the fact that there is no simple answer to the questions posed.

Among the structures most frequently used, we may mention multidimensional matrices, chronological data, spatial data, and heterogenous group matrices. The stacking of matrices giving, for each year, the pattern of consumption for 50 products in 21 regions, provides an example of a "multistructured matrix": chronological repetition introduces an additional dimension, which can be considered as a nominal variable with ordered categories and which one may wish to take into account. One may also wish to consider the spatial contiguity of each annual matrix. In relation to temporal data, structure is of such importance that the field of descriptive data analysis is often left behind in favor of that of procedures and models. There are, however, intermediate tools such as those proposed by Deville and Saporta (1980). Operator analyses, introduced by Escoufier and applied through the STATIS method (cf. Foucart, 1983), provide interesting information on series of matrices, although the methodology, the rules of interpretation, and the graphical results probably need to be tested and improved. Multiple factorial analysis (Escofier and Pages, 1984), allows, in a fairly similar manner, for the representation of several groups of variables.

Many spatially based data analyses are carried out without taking into account the relative positions of observations in space. One could, however, wish to create spatially connected clusters (cf. for example, Gordon, 1979; Ohsumi, 1984) or, otherwise, to question the local stability of links and relationships observed at an overall level (Aluja Banet and Lebart, 1984). Ripley (1981) does not fall within the realm of data analysis, but does present a review of methods likely to improve spatial description practices.

In the case of principal-component analysis, a context defined by a set of variables can be taken into account by analyzing partial correlation matrices, which leads back to an analysis of residuals, after eliminating the effect of instrumental variables (Rao, 1964).

The now classical dichotomy between active and supplementary variables (cf., for example, Cazes, 1982), allows for the data to be "photographed" from different viewpoints. The subjective principal-component method (Kohronen, 1984) is another interesting attempt to bring together the observed and the known.

3. INFERENCE–EXPLORATION INTERFACE

Users not only wish to make the best use of their a priori information, but they also often want, whenever possible, to have an idea of how much trust they may place in the observations and results obtained.

It would of course, generally, not be acceptable to discover a phenomenon and then to offer information on the plausibility of the said phenomenon by using the same data set. It would be necessary to carry out an a priori fragmentation of the data. The dichotomy between active and supplementary variables allows, precisely, for the application of tests on the position in relation to the origin of the supplementary variables, whether they be nominal or continuous in the space of the active variables (cf., for example, Lebart et al., 1984).

"Cross-validation" methods can also be useful. The "bootstrap" method has given rise to recent studies on correspondence analysis (Greenacre, 1984) and multiple correspondence analysis, a method known in Japan as Hayashi's type 3 quantification method (Tanaka, 1984).

At the same time, analytical studies are being done on the distribution of eigenvalues within the framework of multivariate analysis. In continuation of a series of studies synthesized by Muirhead (1982), several publications, particularly the *Journal of Multivariate Analysis*, attest to the progress made, although it has not yet had a really significant impact on data analysis practice.

The interfacing, itself, between exploratory analysis and the development of models is studied in the comparative articles cited in Section 1. This is an area of research in which the demands of users are often not met due, particularly, to a lack of software combining the two approaches.

4. NEW FIELDS OF APPLICATION

The wider use of data analysis has led to an increase in the number of its fields of application. These, in return, interact with methods and software.

In the field of teledetection, for example, the extraordinary volume of data calls for special computation and editing procedures. The same is true for pattern recognition. Mention will be made here of only one field, which is familiar to us, that of textual data. Automatic word processing requires processing algorithms specifically designed for sparse matrices, as well as equally specific input, coding, and printing software. Examples of processing may be found in Benzecri (1981, 1984) and Warnesson and Marcotorchino (1984), information on a software programme in Morineau (1984) and on the development of special tools in Salem (1984). The field of socioeconomic data has also experienced a major upsurge in the use of methods of data analysis, recognized for their effectiveness in the processing of official statistics (Deville and Malinvaud, 1983; EUROSTAT, 1984).

5. THE COMPUTER ENVIRONMENT

Data analysis owes its present development to the appearance of computers. It goes without saying that recent progress made has depended greatly on innovations in informatics.

The availability and suitability of software programs exert a considerable influence on the effective use of the various methods, and constantly highlight the "school of thought" effect and, thus, the compartmentalization of disciplines. Studies, such as those by Molenaar (1984), on the behavior of software users will become indispensable in the near future, as software is unlike other tools: its uses requires specific training and a high level of user involvement.

Bearing in mind the comments made by Chambers (1982, 1983), expert systems should allow statisticians to be informed and trained. The multiplicity of existing statistical software, as described by Francis (1983), and despite the omissions in relation to multidimensional exploratory data analysis, is proof that the task will not be an easy one for expert systems designers. (cf. also Schektman, 1982).

Interactivity is a new outlet that could lead to the development of new languages (Nelder, 1984), or facilitate access to existing libraries (Ralambondrainy, 1984). This interactivity is essential when dealing with microinformatics, a field that is expanding at a very quick pace. The survey of microcomputer software done by Woodward and Elliott (1983) gives an idea of the explosion underway. Unprecedented progress is also taking place in the area of graphical methods, with the almost systematic presence of graphical-display screens on most microcomputers. A synthesis of these graphical methods may be found in Boardman (1983).

Finally, another innovation, encouraging studies in relation to computation procedures, is parallelization. Sylwestrowicz (1982) has prepared an initial report, whereas developments related, more specifically, to data analysis have been published by Lafaye de Micheaux (1984).

A certain level of caution should, perhaps, be exercised vis-à-vis the increasing use of informatics and the diversification of hardware and software. The appearance of powerful means of computation has led to considerable changes, particularly in the attitude of statisticians towards data, models, and inference, but the present level of dispersion does not seem always to be free from danger—the danger of the undermining of the status of the statistician, who moves from being a creator of processing methods to a software consumer; the danger of seeing scientific congresses (exchanges of knowledge) turn into trade fairs (sale of products); the danger of dispersion in the face of an abundance of hardware and software. But these risks are the price that must be paid for a scientific and technological climate that has never before been so favorable to the profession of "data-analyst-statistician."

6. GENERAL SCIENTIFIC ENVIRONMENT

Is multidimensional data analysis becoming an autonomous discipline? The events of the past few years would seem to indicate that the reply is in the affirmative.

Proof of this may be found in the tremendous increase in the number of publications, journals, and congresses dedicated to the subject. Let us take the example of the journals. Articles on data analysis, as we understand it here, were formally included in reviews which focused on slightly different concerns (e.g., *Psychometrika*, *Applied Statistics*, *Biometrics*, *J.A.S.A.*), except for the journal *Les Cahiers de l'Analyse des Données* (Dunod), founded in 1976 by J. P. Benzecri.

Over the past three years, three new journals that focus on multidimensional data analysis have appeared:

Computational Statistics Quartely (Physica Verlag, Wien)
Computational Statistics and Data Analysis (North-Holland)
Applied Stochastics Models and Data Analysis (Wiley)

Other journals are about to be published, and still others are in the planning stages. This mushrooming of journals is quite characteristic of the creation of a new paradigm, according to the definition offered by T. S. Kuhn. The principles may be briefly described by the statement, in keeping with J. W. Tukey and J. P. Benzecri, that the computer processes data in a way that relegates models and classical inference to second place.

In accordance with each specific case, this processing will allow for an efficient reduction of the volume of information, clarifications or interactive testing, and simulation controls, all of which are tools that owe their use to the availability of powerful means of computation.

The series of processes will, generally, be complex, so the assessment of the final result will only be possible by the user.

A new and significant trend in studies on data analysis involves a comparison of approaches to the treatment of concrete problems in the presence of specialists from the fields concerned. This work is tedious, difficult to be explained, and not very much in keeping with normal academic standards (which give priority to isolated technical innovations). It is, however, undertaken in an increasing number of workshops.

Only by the test of concrete application will it be possible, using at times cost-effectiveness criteria, to select series of methods or algorithms.

ACKNOWLEDGMENT

The author wishes to express his gratitude to Catherine Davies and Anne Girardeau for their assistance in the translation of this paper.

REFERENCES

Aluja Banet, T. and Lebart, L. (1984). *Local and Partial Principal Component Analysis and Correspondence Analysis*, COMPSTAT, Vienna: Physica Verlag, pp. 113–118.

Barnett, V. (1981). *Interpreting Multivariate Data*. New York: Wiley.

Benzecri, J. P., et al. (1981). *Pratique de l'Analyse des Données*, Tome 3: Langue, Linguistique et Lexicologie, Paris: Dunod, pp. 566.

Benzecri, J. P. (1982). Construction d'une classification ascendante hiérarchique par la recherche en chaîne des voisins réciproques. *Cahiers Anal. Donn.* **7**, 209–218.

Benzecri, J. P. (1984). Description des textes et analyses documentaires. *Cahiers Anal. Donn.* **9**, 205–211.

Bertrand, P. and Diday, E. (1985). A visual representation of the compatibility between an order and a dissimilarity index: The pyramids. *Comp. Statist. Quart.* **1**, 31–42.

Boardman, T. J. (1983). Statistical graphics on smaller computer: The data analyst's new tool. *Bull. I.S.I.* **1**, 203–212.

Bock, H. H. (1979a). Simultaneous clustering of objects and variables. In *Analyse des Données et Informatiques, Cours C.E.E.* (R. Tomassone, Ed.), Le Chesnay: INRIA, pp. 187–204.

Bock, H. H. (1979b). Fuzzy clustering procedures. In *Analyse des Données et Informatiques, Cours C.E.E.*, Le Chesnay: INRIA, pp. 205–218.

Broersma, H. J. and Molenaar, I. W. (1985). Graphical perception of the distributional aspects of data. *Comp. Statist. Quart.* **2**, 53–72.

Cazes, P. (1982). Note sur les éléments supplémentaires en analyse des correspondances. *Cahiers Anal. Donn.* **1**, 9–23; **2**, 133–154.

Chambers, J. M. (1982). *Analytical Computing: Its Nature and Needs*, COMPSTAT, Vienna: Physica Verlag, pp. 22–29.

Chambers, J. M. (1983). The new future of data analysis. *Bull I.S.I.* **1**, 97–103.

Daudin, J. J. and Trecourt, P. (1980). Analyse des correspondances et modèle log-linéaire: Comparaison des deux méthodes sur un exemple. *Rev. Statist. Appl.* **1**, 5–24.

Diday, E., et al. (1980). *Optimisation en Classification automatique et Reconnaissance des Formes*, Le Chesnay: INRIA.

Deville, J. C. and Saporta, G. (1980). Analyse Harmonique Qualitative. In *Data Analysis and Informatics*, Amsterdam: North-Holland, pp. 375–389.

Deville, J. C. and Malinvaud, E. (1983). Data analysis in official socio-economic statistics. *J. R. Statist. Soc. A*, **146**, part 3.

Eppink, T. W. A. (1984). Correspondence analysis versus principal component analysis for highly skewed distributed variables. *Comp. Statist. Quart.* **1**, 61–76.

Escofier, B. and Pages, J (1984). L'analyse factorielle multiple: une méthode de comparaison de groupes de variables. In *Data Analysis and Informatics III*. Amsterdam: North-Holland, pp. 41–55.

Escoufier, Y. (1985). L'analyse des correspondances: ses propriétés et ses extensions (invited paper). *Bull. I.S.I.* **4**, 28—2.

EUROSTAT (1984). Developpements Récents dans l'Analyse des grands Ensembles de Données, Séminaire de Luxembourg, 1983 (Présentation de E. Malinvaud) (Office des Publ. Offic. des Comm. Européennes).

Foucart, T. (1983). Une nouvelle approche de la méthode STATIS. *Rev. Statist. Appl.* **31**, 61–75.

Francis, I. (1983). A survey of statistical software. *Comp. Statist. Data Anal.* **1**, 17–27.

Gallego, F. J. (1982). Codage flou en analyse des correspondances. *Cahiers Anal. Donn.* **7**, 413–430.

Gifi, A. (1981). Nonlinear Multivariate Analysis. Department of Data Theory, University of Leiden.

Gordon, A. D. (1979). Methods of constrained classification. In *Analyse des Données et Informatiques, Cours C.E.E.*, Le Chesnay: INRIA, pp. 161–172.

Govaert, G. (1984). Classification simultanée de tableaux binaires. In *Data Analysis and Informatics, III*, Amsterdam: North-Holland, pp. 223–236.

Gower, J. C. and Digby, P. G. N. (1981). Expressing complex relationship in two dimensions. In *Interpreting Multivariate Data* (V. Barnett, Ed.), New York: Wiley.

Greenacre, M. (1984). *Theory and Applications of Correspondence Analysis*. Orlando, Florida: Academic Press.

Havranek, T. and Chytil, M. (1983). Mechanizing hypothesis formation—A way for computerized exploratory data analysis? *Bull. I.S.I.* **1**, 104–121.

Hayashi, C. (1984). Recent development of multidimensional scaling and method of quantification in Japan. *Proc. 12th Int. Biometric Conf., Tokyo*, pp. 129–138.

Ihm P. and van Groenewoud, H. (1984). *Correspondence Analysis and Gaussian Ordination, Compstat Lectures* 3. Vienna: Physica Verlag, pp. 5–60.

Juan, J. (1982). Le Programme HIVOR de classification automatique selon les voisins réciproques et le critère de la variance. *Cahiers Anal. Donn.* **7**, 173–184.

Korhonen, P. J. (1984). Subjective principal component analysis. *Comp. Statist. Data Anal.* **2**, 243–255.

Lafaye de Micheaux, D. (1984). *Parallelization of Algorithms in the Practice of Statistical Data Analysis*, COMPSTAT, Vienna: Physica Verlag, pp. 293–300.

Lauro, N. C. and Decarli, A. (1982). Correspondence analysis and log-linear models in multiway contingency tables. *Study–Metron*, **1–2**, 213–234.

Lebart, L., Morineau, A. and Warwick, K. (1984). *Multivariate Descriptive Statistical Analysis*, New York: Wiley.

Lebart, L., et al. (1985). SPAD (Système Portable pour l'Analyse des Données). Paris: CESIA.

Leclerc, A., Chevalier, A., Luce D. and Blanc M. (1985). Analyse des correspondances et modèle logistique: Possibilités et intérêt d'approches complémentaires. *Rev. Statist. Appl.* **1**, 25–40.

Leeuw, J. de (1983). *Nonlinear Principal Component Analysis*, Vienna: *COMPSTAT*, Physica Verlag, pp. 77–85.

Lerman, I. C., Hardouin, M. and Chantrel, T. (1980). Analyse de la situation relative entre deux classifications floues. In *Data Analysis and Informatics*, Amsterdam: North-Holland, pp. 523–552.

Mallows, C. L. and Tukey, J. W. (1982). An overviews of techniques of data analysis emphasizing its exploratory aspects. In *Some Recent Advances in Statistics* (J. Tiago de Oliveira et al., Eds.), Orlando, Florida: Academic Press, pp. 111–172.

Marcotorchino, F. and Michaud, P. (1982). Agregation de similarité en classification automatique. *Rev. Statist. Appl.* **30**, 21–44.

Matusita, K. and Ohsumi, N. (1980). A criterion for choosing the number of clusters in cluster analysis. In *Recent Developments in Statistical Inference and Data Analysis*, Amsterdam: North-Holland, pp. 203–213.

Milligan, G. W. and Cooper, M. C. (1985). An examination of procedures for determining the number of clusters in a data set. *Psychometrika* **50**, 159–179.

Molenaar, I. W. (1984). Behavioral studies of the software user. *Comp. Statist. Data Anal.* **2**, 1–12.

Morineau, A. (1984). *Computational and Statistical Methods for Exploratory Analysis of Textual Data*, COMPSTAT, Vienna: Physica Verlag, p. 372–377.

Muirhead, R. J. (1982). *Aspects of Multivariate Statistical Theory*, New York: Wiley.

Murtagh, F. (1984). Complexities of hierarchic clustering algorithms: State of the art. *Comp. Statist. Quart.*, **1**, 101–113.

Murtagh, F. (1985). *Multidimensional Clustering Algorithms*, COMPSTAT Lectures 4, Vienna: Physica Verlag.

Nakache, J. P., Morice, V. and Golmard, J. L. (1980). Discrimination sur variables binaires basée sur l'estimation des distributions de probabilité par groupe. In *Data Analysis and Informatics*. Amsterdam: North-Holland, pp. 305–316.

Nelder, J. A. (1984). Do statisticians need special interactive languages? In *Data Analysis and Informatics III*, Amsterdam: North-Holland, pp. 511–516.

Nishisato, S. (1980). *Analysis of Categorical Data: Dual Scaling and Its Application*. Toronto: Univ. of Toronto Press.

Ohsumi, N. (1984). Practical Techniques for Areal Clustering. In *Data Analysis and Informatics*, Amsterdam: North-Holland, pp. 247–258.

Ralambondrainy, H. (1984). Le Système SICLA. In *Data Analysis and Informatics*, Amsterdam: North-Holland, pp. 529–537.

Rao, C. R. (1964). The use and interpretation of principal component in analysis in applied research. *Sankhya A*. **26**, 329–357.

Rijckevorsel, J. van (1982). *Canonical Analysis with B. Spline*, COMPSTAT, Vienna: Physica Verlag, pp. 393–398.

Ripley, B. D. (1981). *Spatial Statistics*. New York: Wiley.

Salem, A. (1984). La typologie des segments répétés dans un corpus, fondée sur l'analyse d'un tableau croisant mots et textes. *Cahiers Anal. Donné.*, **9**, 489–500.

Schektman, Y., Ibrahim, A., Jockin, J., Pastor, J. and Vielle, D. (1982). *Computer Science as a tool improving Data Analysis Research and Uses*, COMPSTAT, Part I, Vienna: Physica Verlag, pp. 103–112.

Sylwestrowicz, J. D. (1982). *Parallel Processing in Statistics*, COMPSTAT, Vienna: Physica Verlag, pp. 131–136.

Tanaka, Y. (1984). Sensitivity Analysis in Hayashi's Third Method of Quantification. *Behaviormetrika* **16**, 31–44.

Tenenhaus, M. and Young, F. W. (1985). An analysis and synthesis of multiple correspondence analysis, optimal scaling, dual scaling, homogeneity analysis and other methods for quantifying categorical multivariate data. *Psychometrika* **50**, 91–119.

Wakimoto, K. and Taguri, M. (1978). Constellation graphical method for representing multidimensional data. *Ann. Inst. Statist. Math*. **30**, 97–104.

Warnesson, I. and Marcotorchino, F. (1984). Agrégation de similarités et dictionnaire de synonymes. In *Data Analysis and Informatics*, Amsterdam: North-Holland, pp. 277–291.

Winsberg, S. and Ramsay, J. O. (1983). Monotone spline transformations for dimension reduction. *Psychometrika* **48**, 575–595.

Wong, M. A. (1982). A hybrid clustering method for identifying high density clusters. *J. Amer. Statist. Assoc.* **77**, 841–847.

Woodward, W. A. and Elliott, A. C. (1983). A survey of statistical packages on micro-computers. *Comp. Statist. Data Anal.* **3**, 191–200.

Analysis of Categorical Data from Sample Surveys

J. N. K. Rao
Department of Mathematics and Statistics
Carleton University
Ottawa, Ontario, Canada

SUMMARY

A review of some recent work on methods for analyzing categorical data arising from sample surveys is provided. The topics covered include: (a) The impact of survey design on standard χ^2 tests (or likelihood ratio tests G^2) for a multiway table, under log-linear models. (b) Simple corrections to χ^2 or G^2 taking account of the survey design but requiring the knowledge of only the design effects for the cells and/or the marginals. (c) Improved corrections to χ^2 (or G^2) or alternative tests requiring the knowledge of full estimated covariance matrix of cell estimates. (d) Standard errors of parameter estimates and of "smoothed" cell estimates, and residual analysis taking the design into account. (e) Small sample properties of test statistics for the special case of a simple goodness-of-fit test. (f) Simple corrections to χ^2 (or G^2) under a logistic regression (logit) model involving a binary response variable, in terms of deffs of response proportions for each factor combination. (g) Standard errors of parameter estimates and of "smoothed" estimates of response proportions, under a logit model.

This work was supported by a research grant from the Natural Sciences and Engineering Research Council of Canada.

1. INTRODUCTION

Standard statistical methods for analyzing data summarized in the form of cross-classified tables include χ^2 (or likelihood ratio) tests for the evaluation and selection of parsimonious models and residual analysis to detect model deviations. In particular, methods based on hierarchical (or nested) log-linear models and related logistic regression models are being extensively used because of their ability systematically to provide test statistics of various hypotheses in a multiway table. These techniques, however, are based on the assumption of multinomial sampling and as a result do not take into account the clustering and stratification employed in the survey design. Recent work on the impact of survey design on standard χ^2 tests has shown that clustering can have a substantial impact on the significance levels in the sense that the actual type I error rate can be much larger than the nominal level leading to models that are unnecessarily complex. Some adjustment to standard tests or the use of alternative tests, therefore, is necessary.

In this chapter we provide a review of some recent work on methods for analyzing categorical survey data. In particular, the following topics are covered: (a) The impact of survey design on standard χ^2 tests (or likelihood ratio tests G^2) for a multiway table, under log-linear models. (b) Simple corrections to χ^2 or G^2 that take account of the survey design but requiring the knowledge of only cell variances or design effects* (deffs) and/or deffs for marginal totals. (c) Improved corrections to χ^2 (or G^2) or alternative tests requiring the knowledge of full estimated covariance matrix of cell estimates. (d) Standard errors of parameter estimates and of "smoothed" cell estimates, and residual analysis taking the design into account. (e) Small sample properties of test statistics for the special case of a simple goodness-of-fit test. (f) Simple corrections to χ^2 (or G^2) under a logistic regression model involving a binary response variable, in terms of deffs or response proportions for each factor combination. (g) Standard errors of parameter estimates and of "smoothed" estimates of response proportions, under a logistic regression model.

2. LOG-LINEAR MODELS

Suppose a sample of n ultimate units is drawn according to a specified survey design and a consistent estimator $\hat{\mathbf{p}} = (\hat{p}_1, \ldots, \hat{p}_T)'$ of the vector of population cell proportions $\boldsymbol{\pi} = (\pi_1, \ldots, \pi_T)'$ in a multiway table is ob-

**Design effect* is defined as the ratio of the variance under the given design to the variance under simple random sampling (multinomial sampling).

tained ($\Sigma\pi_t = 1, t = 1, \ldots, T$). Typically, $\hat{p}_t$ is a ratio estimator of π_t depending on survey design weights. We assume that $\sqrt{n}(\hat{\mathbf{p}} - \boldsymbol{\pi})$ converges in distribution to a $N_T(\mathbf{0}, V)$ random vector as $n \to \infty$, where V is a positive semi-definite matrix of rank $T - 1$. In the case of multinomial sampling (simple random sampling with replacement), V reduces to $P = D_\pi - \boldsymbol{\pi}\boldsymbol{\pi}'$, where $D_\pi = \text{diag}(\pi_1, \ldots, \pi_T)$.

A general log-linear model on the population cell proportions, $\boldsymbol{\pi}$, is given by

$$\boldsymbol{\mu} = \tilde{u}(\boldsymbol{\theta})\mathbf{1} + X\boldsymbol{\theta} \tag{2.1}$$

where $\boldsymbol{\mu} = (\mu_1, \ldots, \mu_T)'$, with $\mu_t = \ln \pi_t$, X is a known $T \times r$ matrix of full rank r ($\leq T - 1$) and $X'\mathbf{1} = \mathbf{0}$, $\mathbf{1}$ is a T vector of 1's, $\mathbf{0}$ is an r vector of 0's, and $\boldsymbol{\theta}$ is an r vector of parameters. The factor $\tilde{u}(\boldsymbol{\theta})$ in (2.1) is a normalizing factor that ensures that the π_t defined by (2.1) satisfy the constraint $\Sigma\,\pi_t = 1$. If $r = T - 1$ in (2.1), then we get the saturated model.

We shall be mainly concerned with design-based inferences directed at the finite population parameters π. Such inferences, however, may also be considered as relating to inferences about infinite superpopulation parameters if the finite population size is large, which ensures that the finite population parameters π are very close to their superpopulation model analogs (Pfeffermann and Nathan, 1985).

Under multinomial sampling, the maximum likelihood estimates (m.l.e.) $\hat{\boldsymbol{\pi}}$ are obtained from the likelihood equations

$$X'\hat{\boldsymbol{\pi}} = X'(\mathbf{n}/n) \tag{2.2}$$

where $\mathbf{n}$ is the T vector of observed counts, n_t, in the sample ($\Sigma\,n_t = n$). For general survey designs, however, we do not have m.l.e. of $\boldsymbol{\pi}$, because of difficulties in getting appropriate likelihood functions. Hence, it is a common practice to employ "pseudo-m.l.e." $\hat{\boldsymbol{\pi}}$ obtained from (2.2) by replacing $\mathbf{n}/n$ by the survey estimate $\hat{\mathbf{p}}$. This is analogous to using ordinary least-squares estimates in regression analysis when the errors are not independent and identically distributed (i.i.d.). The consistency of $\hat{\mathbf{p}}$ ensures that of the "smoothed" estimates of $\hat{\boldsymbol{\pi}}$, under standard regularity conditions. The method of iterative proportional fitting provides $\hat{\boldsymbol{\pi}}$, the solution of (2.2), directly, while the Newton–Raphson method gives both $\hat{\boldsymbol{\theta}}$, the pseudo–m.l.e. of $\boldsymbol{\theta}$, and $\hat{\boldsymbol{\pi}} = \boldsymbol{\pi}(\hat{\boldsymbol{\theta}})$.

2.1. Standard Errors

Under the model (2.1), the asymptotic covariance matrices of $\hat{\boldsymbol{\theta}}$ and $\hat{\boldsymbol{\pi}}$ are given by

$$D(\hat{\boldsymbol{\theta}}) = n^{-1}(X'PX)^{-1}(X'VX)(X'PX)^{-1} \tag{2.3}$$

and

$$D(\hat{\boldsymbol{\pi}}) = PXD(\hat{\boldsymbol{\theta}})X'P \tag{2.4}$$

respectively, where $P = \text{diag}(\boldsymbol{\pi}) - \boldsymbol{\pi}\boldsymbol{\pi}'$ (Rao and Scott, 1982, 1984). The asymptotic covariance matrix of the residual vector $\hat{\mathbf{r}} = \hat{\mathbf{p}} - \hat{\boldsymbol{\pi}}$ is given by

$$D(\hat{\mathbf{r}}) = n^{-1}\left[I - PX(X'PX)^{-1}X'\right]V\left[I - X(X'PX)^{-1}X'P\right] = B \quad \text{(say)} \tag{2.5}$$

If a consistent estimator $\hat{V}$ of V is available, then (2.3)–(2.5) can be consistently estimated by replacing V by $\hat{V}$ and $\boldsymbol{\pi}$ in P by $\hat{\boldsymbol{\pi}}$ or $\hat{\mathbf{p}}$. In the multinomial case, $V = P$ and (2.3)–(2.5) reduce to $D(\hat{\boldsymbol{\theta}}) = n^{-1}(X'PX)^{-1}$, $D(\hat{\boldsymbol{\pi}}) = n^{-1}PX(X'PX)^{-1}X'P$ and $D(\hat{\mathbf{r}}) = n^{-1}[P - PX(X'PX)^{-1}X'P]$.

Using consistent estimates $\hat{\sigma}_{tt,r}$ of the diagonal elements of $D(\hat{\mathbf{r}})$, one can compute the standardized residuals $\hat{e}_t = \hat{r}_t/\hat{\sigma}_{tt,r}^{1/2}$, which are approximately $N(0,1)$ under the model (2.1). Deviations from the model are indicated by cells with large values of $|\hat{e}_t|$. Hidiroglou and Rao (1985) illustrate the use of standardized residuals in the context of Canada Health Survey, 1978–1979.

Ignoring the survey design, and hence using $V = P$, could lead to serious understatement of standard errors of $\hat{\theta}_i$, $\hat{\pi}_t$, and $\hat{r}_t$ and to inflated values of $|\hat{e}_t|$.

2.2. Goodness of Fit

The standard χ^2 statistic for testing the goodness-of-fit of the model (2.1) is given by

$$\chi^2 = n \sum_{t=1}^{T} \frac{(\hat{p}_t - \hat{\pi}_t)^2}{\hat{\pi}_t} \tag{2.6}$$

Alternatively, the likelihood ratio statistic is given by

$$G^2 = 2n \sum_{t=1}^{T} \hat{p}_t \ln\left(\frac{\hat{p}_t}{\hat{\pi}_t}\right) \tag{2.7}$$

Effect of the Design

The statistics χ^2 and G^2 are asymptotically equivalent under the model, and their asymptotic distribution is a weighted sum $\sum \delta_i W_i$ of $T - r - 1$ independent χ_1^2 variables W_i, where the weights δ_i are the eigenvalues of the "generalized design effect matrix" $V_{0\phi}^{-1}V_{\phi}$ (Rao and Scott, 1982, 1984). Here $n^{-1}V_{0\phi} = n^{-1}(C'D_{\pi}^{-1}C)$ and $n^{-1}V_{\phi} = n^{-1}(C'D_{\pi}^{-1}VD_{\pi}^{-1}C)$ are the asymptotic covariance matrices of $\hat{\boldsymbol{\phi}} = C'\hat{\boldsymbol{\mu}}$ under multinomial sampling and under the survey design respectively, where $D_{\pi} = \text{diag}(\boldsymbol{\pi})$, $\hat{\boldsymbol{\mu}} = (\hat{\mu}_1, \ldots, \hat{\mu}_T)'$ with $\hat{\mu}_t = \ln \hat{p}_t$ and C is any $T \times (T - r - 1)$ full-rank matrix such that $C'X = 0$ and $C'\mathbf{1} = \mathbf{0}$. In particular, under the standard parametrization of a log-linear model, C may be chosen as the matrix that complements X to form the model matrix of the saturated model. For example, in a 2×2 table the saturated model for (i, j)th cell may be written $\mu_{ij} = \tilde{u} + u_{1(i)} + u_{2(j)} + u_{12(ij)}$ with $\Sigma_i u_{1(i)} = \Sigma_j u_{2(j)} = 0$ and $\Sigma_i u_{12(ij)} = 0$, $\Sigma_j u_{12(ij)} = 0$, and the usual independence model is equivalent to setting $u_{12(ij)} = 0$ for all (i, j) in the saturated model, that is, $\boldsymbol{\theta} = (u_{1(1)}, u_{2(1)})$,

$$X' = \begin{bmatrix} 1 & 1 & -1 & -1 \\ 1 & -1 & 1 & -1 \end{bmatrix}, \qquad C = [1 \quad -1 \quad -1 \quad 1]'$$

provided $\boldsymbol{\mu}$ is ordered as $(\mu_{11}, \mu_{12}, \mu_{21}, \mu_{22})'$. In the multinomial case, the generalized deffs δ_i are all equal to 1 and hence the standard result that χ^2 (or G^2) is asymptotically distributed as $(\approx)\chi^2_{T-r-1}$, a χ^2 variable with $T - r - 1$ degrees of freedom (d.f.), follows as a special case.

The general result that $\chi^2 \approx \Sigma_i \delta_i W_i$ shows that the survey design can have a substantial impact on the significance levels, that is, $P[\chi^2 > \chi^2_{T-r-1}(\alpha)]$ can be much larger than α under the model, where $\chi^2_{T-r-1}(\alpha)$ is the upper α point of χ^2_{T-r-1}. For instance, Brier (1980) and Rao and Scott (1981) studied constant deff clustering models for which $V = \lambda P$ with $\lambda > 1$. In this case, $\delta_i = \lambda$ for all i and $P[\chi^2 > \chi^2_{T-r-1}(\alpha)] \doteq P[\chi^2_{T-r-1} > \lambda^{-1}\chi^2_{T-r-1}(\alpha)]$, which increases with λ. Although the constant deff model may not be realistic in practice, it illustrates clearly the effect of clustering on the significance level of standard tests based on χ^2 (or G^2).

First-Order Correction to χ^2

A first-order correction to χ^2 refers $\chi^2_c = \chi^2/\hat{\delta}_.$ to χ^2_{T-r-1}, where $\hat{\delta}_.$ is a consistent estimator of the mean of eigenvalues δ_i, that is,

$$(T - r - 1)\hat{\delta}_. = \text{tr}\left(C'D_{\hat{\pi}}^{-1}C\right)^{-1}\left(C'D_{\hat{\pi}}^{-1}\hat{V}D_{\hat{\pi}}^{-1}C\right) \tag{2.8}$$

where tr denotes the trace operator. One could also use $D_{\hat{\mathbf{p}}}$, instead of $D_{\hat{\pi}}$, in (2.8). The correction χ^2_c is asymptotically valid in the case of constant deffs ($V = \lambda P$), but, in general, it should perform well when the coefficient of variation of the δ_i is small. It may be noted that $\hat{\delta}_.$ can be simply obtained from (2.8) without evaluating the individual eigenvalues $\hat{\delta}_i$. The first-order correction, in general, requires the knowledge of estimated covariance matrix, $n^{-1}\hat{V}$, of the survey estimates $\hat{\mathbf{p}}$. However, in the case of a model admitting direct estimates $\hat{\pi}$ of the form

$$\hat{\pi}_\theta = \prod_i \hat{p}_{\theta_i} \Big/ \prod_j \hat{p}_{\phi_j} \tag{2.9}$$

Rao and Scott (1982, 1984) have shown that

$$(T - r - 1)\hat{\delta}_. = \sum_\theta \frac{\hat{p}_\theta}{\hat{\pi}_\theta}(1 - \hat{p}_\theta)\hat{d}_\theta$$

$$- \sum_i \left\{ \sum_{\theta_i} (1 - \hat{p}_{\theta_i})\hat{d}_{\theta_i} \right\} + \sum_j \left\{ \sum_{\phi_j} (1 - \hat{p}_{\phi_j})\hat{d}_{\phi_j} \right\} \tag{2.10}$$

(see also Bedrick, 1983). In (2.10), θ denotes the set of subscripts for an arbitrary cell, θ_i is a subset of subscripts in θ, ϕ_j is the set of subscripts common to θ_i and θ_l for some i and l, and $\hat{d}_\theta = [\text{est var}(\hat{p}_\theta)] [\hat{p}_\theta(1 - \hat{p}_\theta)/n]^{-1}$ is the estimated deff of the cell estimate $\hat{p}_\theta$ irrespective of the model, $\hat{d}_{\theta_i} = [\text{est var}(\hat{p}_{\theta_i})][\hat{p}_{\theta_i}(1 - \hat{p}_{\theta_i})/n]^{-1}$ is the estimated deff of marginal total $\hat{p}_{\theta_i}$, and a similar expression for $\hat{d}_{\phi_j}$. It follows from (2.10) that the first-order correction to χ^2 can be computed from published tables providing the estimated cell deffs and the estimated deffs of marginals. For example, under the independence model in an $I \times J$ table, (2.10) reduces to

$$(I - 1)(J - 1)\hat{\delta}_. = \sum_{i=1}^{I} \sum_{j=1}^{J} \frac{\hat{p}_{ij}}{\hat{p}_{i+}\hat{p}_{+j}} \hat{d}_{ij} - \sum_{i=1}^{I} (1 - \hat{p}_{i+})\,\hat{d}_i(r)$$

$$- \sum_{j=1}^{J} (1 - \hat{p}_{+j})\,\hat{d}_j(c) \tag{2.11}$$

where $\hat{d}_{ij}$ is the estimated deff of the cell estimate $\hat{p}_{ij}$ and $\hat{d}_i(r)$, $\hat{d}_j(c)$ are the estimated deffs of the corresponding row and column marginals $\hat{p}_{i+}$ and $\hat{p}_{+j}$, respectively. In the case of a three-way table, the models of complete independence, conditional independence, and independence of one variable from the remaining two jointly all admit direct estimates $\hat{\pi}$, but the important model of no three-factor interaction does not permit a direct solution.

In the case of one set of fixed margins, Roberts (1985) has developed an algorithm to express $\delta_.$ in the closed-form case [viz., (2.10)] in terms of deffs within populations corresponding to fixed margins. The latter deffs are usually reported in published tables.

In the case of a model (2.1) not admitting direct estimates, Rao and Scott (1982, 1984) suggested the use of $\chi^2/\hat{\delta}_.(1)$, where $\hat{\delta}_.(1)$ is the average estimated deff of the form (2.10) corresponding to a nested model permitting direct estimates and "close" to the given model in the sense that its model matrix is $T \times s$ of rank s and $r - s$ is small. Rao and Scott (1985) suggested an improved correction:

$$\tilde{\chi}_c^2 = \chi^2 \left[\frac{T - s - 1}{T - r - 1} \hat{\delta}_.(1) \right]^{-1} \qquad (2.12)$$

The correction (2.12) is conservative compared to using $\chi^2/\hat{\delta}_.(1)$. Note that $\tilde{\chi}_c^2$ depends only on the estimated deffs of cells and marginals, unlike χ_c^2.

Rao and Scott (1985) also obtained an exact upper bound on $\hat{\delta}_.$:

$$\hat{\delta}_. \le (T - r - 1)^{-1} \sum_{i=1}^{T-r-1} \hat{\lambda}_i \qquad (2.13)$$

where $\hat{\lambda}_1 \ge \hat{\lambda}_2 \ge \cdots \ge \hat{\lambda}_{T-1} > 0$ are nonzero eigenvalues of $D_{\hat{p}}^{-1}\hat{V}$. Using the upper bound (2.13), one can obtain a conservative test compared to $\chi^2/\hat{\delta}_.$ if the published tables also report all the estimated eigenvalues $\hat{\lambda}_i$. This should be feasible since the $\hat{\lambda}_i$ do not depend on any model, unlike $\hat{\delta}_.$. If the $\hat{\lambda}_i$ are not available, a simple upper bound on $\hat{\delta}_.$ depending only on the estimated cell deffs is given by

$$(T - r - 1)\hat{\delta}_. \le (T - 1)\hat{\lambda}_. = \sum_{t=1}^{T} (1 - \hat{p}_t)\, \hat{d}_t \qquad (2.14)$$

where the $\hat{d}_t$ are the estimated deffs of the cell estimates $\hat{p}_t$ and $\hat{\lambda}_.$ is the mean of the $T - 1$ eigenvalues $\hat{\lambda}_i$. The corresponding test $[(T - r - 1)/(T - 1)](\chi^2/\hat{\lambda}_.)$, however, is likely to be overly conservative unless r is small compared to T.

We now turn to methods that take account of the variation in the generalized deffs δ_i, unlike χ_c^2 or its variations. These methods, however, require access to original data or an estimate of the complete covariance matrix, $n^{-1}\hat{V}$.

Satterthwaite Approximation

Rao and Scott (1982, 1984) obtained the following second-order correction to χ^2, using the well-known Satterthwaite approximation to the distribution of a weighted sum of independent χ^2 variables:

$$\chi_S^2 = \left[\hat{\delta}_{\cdot}(1 + \hat{a}^2)\right]^{-1} \chi^2 = (1 + \hat{a}^2)^{-1} \chi_c^2 \tag{2.15}$$

where $\hat{a}^2 = [(T - r - 1)^{-1} \Sigma_i (\hat{\delta}_i - \hat{\delta}_{\cdot})^2] \hat{\delta}_{\cdot}^{-2}$ is the square of the coefficient of variation (c.v.) of the $\hat{\delta}_i$. The statistic χ_S^2 is referred to χ_ν^2 where $\nu = (T - r - 1)/(1 + \hat{a}^2)$. The value of $\hat{a}$ can be computed without evaluating the individual $\hat{\delta}_i$, noting that

$$(T - r - 1)\hat{\delta}_{\cdot} = \sum_{t=1}^{T} \frac{\hat{b}_{tt}}{\hat{\pi}_t} \quad \text{and} \quad \sum \hat{\delta}_i^2 = \sum_{t=1}^{T} \sum_{l=1}^{T} \frac{\hat{b}_{tl}^2}{(\hat{\pi}_t \hat{\pi}_l)}, \tag{2.16}$$

where $\hat{B} = (\hat{b}_{tl})$ is the estimated covariance matrix of the residuals $\hat{\mathbf{r}}$, obtained from (2.5). The Satterthwaite correction, χ_S^2, requires the knowledge of $\hat{V}$ even in the case of a model admitting direct estimates, $\hat{\boldsymbol{\pi}}$, but it controls the type I error rate much better than χ_c^2 when the variation in the δ_i is substantial. For two-way and three-way tables, Hidiroglou and Rao (1985) provide simplified formula for $\hat{B}$ for stratified multistage designs with primary clusters sampled with replacement with probability proportional to size (p.p.s.).

Wald Statistic

If the estimated covariance matrix $\hat{V}$ is known, a Wald test statistic based on $\hat{\boldsymbol{\phi}}_p = C'\hat{\boldsymbol{\mu}}_p$, the vector of contrasts of log probabilities $\ln \hat{p}_t$, is given by (Koch et al., 1975)

$$\chi_W^2 = \hat{\boldsymbol{\phi}}_p' \left[C' D_{\hat{\mathbf{p}}}^{-1} \hat{V} D_{\hat{\mathbf{p}}}^{-1} C\right]^{-1} \hat{\boldsymbol{\phi}}_p \tag{2.17}$$

(see also Nathan, 1975; Shuster and Downing, 1976). Under the model (2.1), χ_W^2 is asymptotically distributed as a χ_{T-r-1}^2, that is, the test based on treating χ_W^2 as χ_{T-r-1}^2 is asymptotically correct. As explained below, however, the Wald statistic may not perform well in finite samples. It may be noted that χ_W^2 is independent of the choice C.

Jackknifed χ^2 Test

Fay (1985) has developed a jackknifed χ^2 test applicable to survey designs permitting the use of a replication method, such as the jackknife. For example, the jackknife or the balanced half-sample replication (McCarthy, 1969) can be applied to stratified multistage designs with primary clusters sampled with replacement (or without replacement provided the sampling fractions at the first stage of sampling are negligible, as in the case of the

Canada Health Survey). Like χ_S^2 or χ_W^2, the jackknife statistic essentially requires access to the microdata file or at least to the cell estimates at the primary cluster level.

Let $\hat{\mathbf{N}}$ be the T vector of survey estimates of the population cell counts, N_t, and $\hat{\mathbf{N}}^{(a,b)} = \hat{\mathbf{N}} + \mathbf{W}^{(a,b)}$ be the corresponding T vector of estimates based on the (pseudo-)replicates (a, b), $a = 1,\ldots, A$; $b = 1,\ldots, B_a$, such that $\Sigma_b \mathbf{W}^{(a,b)} = \mathbf{0}$. An estimate of the covariance matrix of $\hat{\mathbf{N}}$ is of the form $\Sigma_a g_a \Sigma_b \mathbf{W}^{(a,b)}\mathbf{W}^{(a,b)\prime}$, where the g_a are fixed constants. For example, in the case of a multistage design with n_a primary clusters sampled with replacement from ath stratum, the jackknife method leads to $g_a = (n_a - 1)/n_a$. The jackknifed statistic for testing the goodness of fit of the model (2.1) is given by

$$\chi_J = \left[\chi^2(\hat{\mathbf{N}}) - (K^+)^{1/2}\right]\left[V/(8\chi^2(\hat{N}))\right]^{-1/2} \tag{2.18}$$

where $\chi^2(\hat{\mathbf{N}}) = \Sigma(\hat{N}_t - \tilde{N}_t)^2/\tilde{N}_t$ with $\tilde{N}_t$ denoting the fitted cell counts, under the model, obtained from the survey estimates $\hat{N}_t$,

$$K = \sum_a g_a \sum_b Q_{ab}, \qquad V = \sum_a g_a \sum_b Q_{ab}^2 \tag{2.19}$$

$$Q_{ab} = \chi^2(\hat{\mathbf{N}}^{(a,b)}) - \chi^2(\hat{\mathbf{N}}) \tag{2.20}$$

and $K^+ = \max(K, 0)$. The G^2 version, G_J, is similarly defined by replacing $\chi^2(\hat{\mathbf{N}})$ by $G^2(\hat{\mathbf{N}}) = 2\Sigma_t \hat{N}_t \ln(\hat{N}_t/\tilde{N}_t)$. Note that the fitted cell counts $\tilde{N}_t^{(a,b)}$ used in $\chi^2(\hat{\mathbf{N}}^{(a,b)}) = \Sigma_t(\hat{N}_t^{(a,b)} - \tilde{N}_t^{(a,b)})^2/\tilde{N}_t^{(a,b)}$ or $G^2(\hat{\mathbf{N}}^{(a,b)})$ are obtained from the (pseudo-)replicate survey estimates $\hat{N}_t^{(a,b)}$, that is, by refitting the given log-linear model (2.1) to the replicate estimates $\hat{N}_t^{(a,b)}$ for each (a, b), thus involving $\Sigma_a B_a$ refittings of the model. In the special case of constant deffs, Fay (1985) has shown that χ_J is asymptotically distributed as $\tilde{\chi}_J = 2^{1/2}\{(\chi_{T-r-1}^2)^{1/2} - (T - r - 1)^{1/2}\}$, and also provided the upper 5% and 1% points of χ_J. The test based on treating χ_J as $\tilde{\chi}_J$, however, is not asymptotically correct in the general case of nonconstant deffs, but simulation results indicate that it is generally asymptotically conservative. Fay has developed a computer program, CPLX, to implement the jackknifed tests.

Like χ_S^2, the test χ_J is not affected, in finite samples, by the lack of precision in the estimated covariance matrix (as the number of cells increases and the number of sampled primary clusters decreases), whereas the Wald statistic χ_W^2 could lead to inflated type I error rates due to instability in the inverse of the estimated covariance matrix of the contrast vector $\hat{\boldsymbol{\phi}}_p$.

2.3. Nested Models

Denoting $X\boldsymbol{\theta} = X_1\boldsymbol{\theta}_1 + X_2\boldsymbol{\theta}_2$ in the model (2.1), where X_1 is $T \times s$ of rank s and X_2 is $T \times u$ of rank u $(s + u = r)$, we are often interested in testing the hypothesis H: $\boldsymbol{\theta}_2 = 0$, given the model. The standard test statistics χ^2 and G^2 in this case are given by

$$\chi^2(2|1) = n \sum_{t=1}^{T} \frac{(\hat{\pi}_t - \hat{\hat{\pi}}_t)^2}{\hat{\hat{\pi}}_t} \tag{2.21}$$

$$\begin{aligned} G^2(2|1) &= 2n \sum_{t=1}^{T} \hat{\pi}_t \ln\left(\frac{\hat{\pi}_t}{\hat{\hat{\pi}}_t}\right) \\ &= 2n \sum_{t=1}^{T} \hat{p}_t \ln\left(\frac{\hat{p}_t}{\hat{\hat{\pi}}_t}\right) - 2n \sum_{t=1}^{T} \hat{p}_t \ln\left(\frac{\hat{p}_t}{\hat{\pi}_t}\right) \\ &= G^2(2) - G^2(1) \quad \text{(say)} \end{aligned} \tag{2.22}$$

where the $\hat{\hat{\pi}}_t$ are the pseudo-m.l.e. of the π_t under the reduced model. The first-order correction refers $\chi_c^2(2|1) = \chi^2(2|1)/\hat{\delta}_.(2|1)$ or $G_c^2(2|1) = G^2(2|1)/\hat{\delta}_.(2|1)$ to χ_u^2, where $\hat{\delta}_.(2|1)$ is the estimate of $\delta_.(2|1)$ given by

$$u\delta_.(2|1) = \text{tr}\left[(X'PX)^{-1}(X'VX)\right] - \text{tr}\left[(X_1'PX_1)^{-1}(X_1'VX_1)\right] \tag{2.23}$$

(Rao and Scott, 1982, 1984). In the case of constant deffs, $V = \lambda P$, $\delta_.(2|1)$ reduces to λ.

In general, the first-order correction requires the knowledge of $\hat{V}$. If *both* $\hat{\boldsymbol{\pi}}$ and $\hat{\hat{\boldsymbol{\pi}}}$ admit direct estimates of the form (2.9), however, then $u\hat{\delta}_.(2|1) = (T - r - 1)\hat{\delta}_. - (T - s - 1)\hat{\delta}_.(1)$, where $\hat{\delta}_.$ is given by (2.10) and $\hat{\delta}_.(1)$ is also given by a similar expression corresponding to the model $\boldsymbol{\mu} = \tilde{u}\begin{pmatrix}\boldsymbol{\theta}_1 \\ \mathbf{0}\end{pmatrix}\mathbf{1} + X_1\boldsymbol{\theta}_1$, and depends only on the cell deffs and the marginal deffs. Otherwise, an upper bound on $\hat{\delta}_.(2|1)$ is given by (Rao and Scott, 1985)

$$\hat{\delta}_.(2|1) \le u^{-1} \sum_{i=1}^{u} \hat{\lambda}_i \tag{2.24}$$

where the eigenvalues $\hat{\lambda}_1 \ge \cdots \ge \hat{\lambda}_u$ do not depend on any hypothesis, as in (2.13). The bound (2.24), however, is conservative if u is small. If the individual $\hat{\lambda}_i$ are not available, a simple upper bound on $\hat{\delta}_.(2|1)$, depending only on the estimated cell deffs $\hat{d}_t$, is given by $u^{-1}\Sigma_t(1 - \hat{\pi}_t)\,\hat{d}_t = u^{-1}$

$(T-1)\hat{\lambda}_.$, but the resulting test will be over conservative since u is usually small relative to T. The F statistic

$$F = \left[u^{-1}G^2(2|1)\right]\left[(T-r-1)^{-1}G^2(1)\right]^{-1} \tag{2.25}$$

is used in the GLIM (see McCullagh and Nelder, 1983) to test the nested hypothesis H: $\boldsymbol{\theta}_2 = \mathbf{0}$. Under H, F is treated as a F variable with u and $T-r-1$ degrees of freedom, respectively. Note that the F statistic does not require the knowledge of any deffs. This method might work well relative to $\chi_c^2(2|1)$, if $\delta_.(2|1) \doteq \delta_.$ and $T-r-1$ is large (Rao and Scott, 1985).

Rao and Scott (1982, 1984) have given the Satterthwaite correction to $\chi^2(2|1)$ or $G^2(2|1)$, requiring the knowledge of $\hat{V}$. Details of this correction are somewhat involved and hence omitted. A Wald statistic is given by

$$\chi_W^2(2|1) = \hat{\boldsymbol{\phi}}'\left[C_2'D_{\hat{\boldsymbol{\pi}}}^{-1}\hat{D}(\hat{\boldsymbol{\pi}})D_{\hat{\boldsymbol{\pi}}}^{-1}C_2\right]^{-1}\hat{\boldsymbol{\phi}} \tag{2.26}$$

where $\hat{D}(\hat{\boldsymbol{\pi}})$ is the estimate of $D(\hat{\boldsymbol{\pi}})$ given by (2.4), $\hat{\boldsymbol{\phi}} = \mathbf{c}_2'\hat{\boldsymbol{\mu}}$, $\hat{\boldsymbol{\mu}}$ is the T vector of log probabilities $\ln \hat{p}_t$, and C_2 is any $T \times u$ matrix of rank u such that $C_2'x_1 = 0$, $C_2'\mathbf{1} = \mathbf{0}$, and $C_2'x_2$ is nonsingular. In particular, if $X_1'X_2 = 0$ as in the standard parametrization of a log-linear model, then $C_2 = X_2$ is a convenient choice. Under the nested hypothesis H: $\boldsymbol{\theta}_2 = \mathbf{0}$, the Wald statistic $\chi_W^2(2|1)$ is asymptotically distributed as a χ_u^2. Koch et al. (1975) proposed a Wald statistic, for testing H: $\boldsymbol{\theta}_2 = \mathbf{0}$, based on a weighted least-squares estimator of $\boldsymbol{\theta}$, requiring the knowledge of $\hat{V}$ or the cell estimates at the primary cluster level. The jackknifed statistic $G_J(2|1)$ is simply obtained from (2.18) if we replace $X^2(\hat{\mathbf{N}})$ in (2.18) by $G_2^2(\hat{\mathbf{N}}) - G_1^2(\hat{\mathbf{N}})$, where $G_2^2(\hat{\mathbf{N}}) = 2\Sigma_t \hat{N}_t \ln(\hat{N}_t/\tilde{\tilde{N}}_t)$ and $G_1^2(\hat{\mathbf{N}}) = 2\Sigma_t \hat{N}_t \ln(\hat{N}_t/\tilde{N}_t)$, and the $\tilde{\tilde{N}}_t$ denote the fitted cell counts under the reduced model. The statistic $G_J(2|1)$ is treated as $\tilde{\chi}_J(2|1) = 2^{1/2}\{(\chi_u^2)^{1/2} - u^{1/2}\}$ under the nested hypothesis H: $\boldsymbol{\theta}_2 = \mathbf{0}$.

2.4. Small Sample Properties

Thomas and Rao (1984) employed simulated cluster sampling to examine the type I error performances of test statistics for the special case of a simple goodness-of-fit test: $\boldsymbol{\pi} = \boldsymbol{\pi}_0$, where $\boldsymbol{\pi}_0$ is completely specified. A range of cluster sampling situations, including the realistic case of nonconstant deffs, was modeled using mixtures of Dirichlet-multinomial distributions. Their simulation study led to the following important results: (a) The standard χ^2 or G^2 can lead to unacceptably high significance levels compared to the nominal level α. (b) The asymptotically valid Wald statistic χ_W^2 also leads to high levels as the number of sample clusters, l, decreases and the number of cells, T, increases ($l \geq T$). However, a modified χ_W^2 (Fellegi, 1980; Hidiroglou et al., 1980) given by

$$F_W = \frac{(l - T + 1)}{(T - 1)(l - 1)} \chi_W^2 \tag{2.27}$$

when referred to an F distribution with $T - 1$ and $l - T + 1$ d.f., respectively, performs well when $\boldsymbol{\pi}_0$ is the uniform probability vector with elements T^{-1}, but is sensitive to skewness in $\boldsymbol{\pi}_0$. (c) An F-based version of the first-order correction χ_c^2, given by

$$F\chi_c^2 = (T - 1)^{-1} \chi_c^2 \tag{2.28}$$

when referred to an F distribution with $T - 1$ and $(l - 1)(T - 1)$ d.f., respectively, performs better than χ_c^2 as l decreases. Note that $F\chi_c^2$ depends only on the cell deffs as in the case of χ_c^2. The statistic $F\chi_c^2$ performs well provided the coefficient of variation, a, of the generalized deffs, λ_i, is small, but tends to lead to higher significance levels than α when a is not small. (d) In general, the second-order Satterthwaite correction χ_S^2 and Fay's jackknifed test, χ_J, perform well even when a is not small.

Thomas and Rao (1985) have also examined the power performances of test statistics for the simple goodness-of-fit problem, again using simulated cluster sampling under the mixture Dirichlet-multinomial model.

3. LOGIT MODELS

Logistic regression (logit) models are commonly used to analyze the variation in the estimated proportions associated with a binary response variable. Suppose that the population of interest consists of I subpopulations (domains) corresponding to the levels of one or more factors. Let $\hat{N}_i$ and $\hat{N}_{i1}$ be the survey estimates of the ith domain size, N_i, and the ith domain total, N_{i1}, of a binary (0, 1) response variable ($\Sigma_i N_i = N$). The population proportion $F_i = N_{i1}/N_i$ is estimated by $\hat{P}_i = \hat{N}_{i1}/\hat{N}_i$.

A logit model on the response proportions F_i is given by

$$\ln\{F_i/(1 - F_i)\} = \mathbf{z}_i'\boldsymbol{\beta} \tag{3.1}$$

where $\mathbf{z}_i$ is an m vector of known constants and $\boldsymbol{\beta}$ is an m vector of parameters. The pseudo-m.l.e. $\hat{\boldsymbol{\beta}}$ and $\hat{F}_i$ of β and F_i are obtained through iterative calculations from the likelihood equations under independent binomial sampling by replacing n_i/n by w_i and n_{i1}/n_i by the ratio estimate $\hat{P}_i$, where n_i and n_{i1} are the sample size and the sample total of the binary response variable from the ith domain ($\Sigma_i n_i = n$):

$$Z'D_w\hat{\mathbf{F}} = Z'D_w\hat{\mathbf{P}} \tag{3.2}$$

where $Z' = (\mathbf{z}_1, \ldots, \mathbf{z}_I)$, $D_w = \text{diag}(w_1, \ldots, w_I)$ with $w_i = \hat{N}_i/\hat{N}$ and $\Sigma_i \hat{N}_i = \hat{N}$, $\hat{\mathbf{F}} = (\hat{F}_1, \ldots, \hat{F}_I)'$ and $\hat{\mathbf{P}} = (\hat{P}_1, \ldots, \hat{P}_I)'$.

3.1. Standard Errors

The asymptotic covariance matrices of $\hat{\boldsymbol{\beta}}$ and $\hat{\mathbf{F}}$ are given by

$$D(\hat{\boldsymbol{\beta}}) = n^{-1}(Z'\Delta Z)^{-1}(Z'D_W\Sigma_P D_W Z)(Z'\Delta Z)^{-1} \tag{3.3}$$

and

$$D(\hat{\mathbf{F}}) = D_W^{-1}\Delta Z D(\hat{\boldsymbol{\beta}}) Z'\Delta D_W^{-1} \tag{3.4}$$

where $\Delta = \text{diag}(W_1F_1(1 - F_1), \ldots, W_IF_I(1 - F_I))$, $n^{-1}\Sigma_P$ is the asymptotic covariance matrix of $\hat{\mathbf{P}}$, and $D_W = \text{diag}(W_1, \ldots, W_I)$ with $W_i = N_i/N$ (Roberts et al., 1985). The "smoothed" estimates $\hat{\mathbf{F}}$ can be considerably more efficient than the survey estimates $\hat{\mathbf{P}}$, especially for domains with small samples, if the model (3.1) provides an adequate fit to $\hat{\mathbf{P}}$. The asymptotic covariance matrix of the residual vector $\hat{\mathbf{R}} = \hat{\mathbf{P}} - \hat{\mathbf{F}}$ is given by

$$D(\hat{R}) = n^{-1}A\Sigma_P A' = \Sigma_R \quad \text{(say)} \tag{3.5}$$

where

$$A = I - D_W^{-1}\Delta Z(Z'\Delta Z)^{-1}Z'D_W \tag{3.6}$$

If a consistent estimator $\hat{\Sigma}_P$ of Σ_P is available, then (3.3)–(3.6) can be obtained by replacing Σ_P by $\hat{\Sigma}_P$, $\mathbf{W}$ by $\mathbf{w}$ and $\mathbf{F}$ by $\hat{\mathbf{P}}$ or $\hat{\mathbf{F}}$. Using consistent estimates $\hat{\sigma}_{ii,R}$ of the diagonal elements of $D(\hat{\mathbf{R}})$, the standardized residuals $\hat{E}_i = \hat{R}_i/\hat{\sigma}_{ii,R}^{1/2}$ can be computed to detect deviations from the model, noting that the $\hat{E}_i$ are approximately $N(0, 1)$ under the model (3.1).

If a consistent estimator of Σ_P is not available, A. J. Scott (1986) found that an approximation to $D(\hat{\boldsymbol{\beta}})$ can be obtained by considering the binomial case, namely $D(\hat{\boldsymbol{\beta}}) = n^{-1}(Z'\Delta Z)^{-1}$, and replacing nW_i by the "effective" domain sample size nW_i/D_i, where D_i is the ith domain design effect, $D_i = [\text{Var}(\hat{P}_i)][(nW_i)^{-1}P_i(1 - P_i)]^{-1}$. The approximation requires the knowledge of only the domain deffs and seems to be accurate when applied to survey data.

3.2. Goodness of Fit

The standard χ^2 and likelihood ratio statistics for testing the goodness of fit of the model (3.1) are given by

$$\chi^2 = n \sum_{i=1}^{I} \frac{(\hat{P}_i - \hat{F}_i)^2 w_i}{[\hat{F}_i(1 - \hat{F}_i)]} \tag{3.7}$$

and

$$G^2 = 2n \sum_{i=1}^{I} w_i \left\{ \hat{P}_i \ln\left(\frac{\hat{P}_i}{\hat{F}_i}\right) + (1 - \hat{P}_i)\ln\left[\frac{(1 - \hat{P}_i)}{(1 - \hat{F}_i)}\right]\right\}. \tag{3.8}$$

A first-order correction to χ^2 refers $\chi_c^2 = \chi^2/\hat{\delta}_.$ to χ^2_{I-m}, where

$$(I - m)\hat{\delta}_. = n \sum_{i=1}^{I} \frac{w_i \hat{\sigma}_{ii,R}}{[\hat{F}_i(1 - \hat{F}_i)]} \tag{3.9}$$

(Roberts et al., 1985). This correction, in general, requires the knowledge of the estimated covariance matrix, $n^{-1}\hat{\Sigma}_P$, of the response proportions $\hat{P}_i$. Rao and Scott (1985), however, have obtained a simple upper bound on $\hat{\delta}_.$, depending only on the estimated domain deffs $\hat{D}_i$:

$$(I - m)\hat{\delta}_. \leq I\hat{D}_., \tag{3.10}$$

where $\hat{D}_. = I^{-1}\Sigma_i \hat{D}_i$ and $\hat{D}_i = [\text{est var}(\hat{P}_i)][(nw_i)^{-1}\hat{P}_i(1 - \hat{P}_i)]^{-1}$.

The logit model (3.1) can be viewed as a special case of a log-linear model containing the saturated submodel for all the factors. Hence, the bound (2.14), namely, $\hat{\delta}_. \leqq [(T - 1)/(T - r - 1)]\hat{\lambda}_.$, in terms of the cell deffs, $\hat{d}_t$, in the extended table for log-linear model analysis, is applicable, where $T = 2I$. The bound (3.10), namely, $\hat{\delta}_. \leqq [I/(I - m)]\hat{D}_.$, in terms of the domain deffs $\hat{D}_i$, leads to much less conservative tests compared to those based on the bound (2.14) since r will be large (because of the inclusion of the saturated submodel for all the factors) and m is usually small in the context of logit models. It follows that an excellent approximation to χ_c^2 can be obtained from published tables for logit analysis providing the estimated response proportions, $\hat{P}_i$, and their deffs.

Roberts et al. (1985) have obtained a second-order correction to χ^2, based on the Satterthwaite approximation, and a Wald statistic, both requiring the knowledge of $\hat{\Sigma}_P$. They have also provided results for testing nested hypotheses, and diagnostic procedures to detect any outlying domain proportions and influential points in the factor space, taking account of the survey design.

REFERENCES

Bedrick, E. J. (1983). Adjusted goodness-of-fit tests for survey data. *Biometrika* **70**, 591–595.

Brier, S. E. (1980). Analysis of contingency tables under cluster sampling. *Biometrika* **67**, 591–596.

Fay, R. E. (1985). A jackknifed chi-squared test for complex samples. *J. Amer. Statist. Associ.* **80**, 148–157.

Fellegi, I. P. (1980). Approximate tests of independence and goodness of fit based on stratified multistage samples. *J. Amer. Statist. Assoc.* **71**, 665–670.

Hidiroglou, M. A., Fuller, W. A. and Hickman, R. D. (1980). MINI CARP: A program for stratified and cluster samples. Ames, Iowa: Survey Section, Iowa State University.

Hidiroglou, M. A. and Rao, J. N. K. (1985). Chi-squared tests for the analysis of categorical data from the Canada Health Survey. Technical Report, Statistics Canada, Ottawa.

Koch, G. G., Freeman, D. H. Jr. and Freeman, J. L. (1975). Strategies in the multivariate analysis of data from complex surveys. *Int. Statist. Rev.* **43**, 59–78.

McCarthy, P. J. (1969). Pseudo-replication: half samples. *Rev. Int. Statist. Inst.* **37**, 239–264.

McCullagh, P. and Nelder, J. A. (1983). *Generalized Linear Models*. London: Chapman and Hall.

Nathan, G. (1975). Tests of independence in contingency tables from stratified proportional samples. *Sankhyā C* **37**, 77–87.

Pfefferman, D. and Nathan, G. (1985). Problems in model identification based on data from complex sample surveys. *Bull. Int. Statist. Inst. 51* Book 2, Topic 12.

Rao, J. N. K. and Scott, A. J. (1981). The analysis of categorical data from complex sample surveys: chi-squared tests for goodness of fit and independence in two-way tables. *J. Amer. Statist. Assoc.* **76**, 221–230.

Rao, J. N. K. and Scott, A. J. (1982). On chi-square tests for multi-way contingency tables with cell proportions estimated from survey data. Paper presented at Israel Statist. Assoc. Intl. Meetings on Analysis of Sample Survey Data and on Sequential Analysis, Jerusalem, June 14–18, 1982.

Rao, J. N. K. and Scott, A. J. (1984). On chi-squared tests for multiway contingency tables with cell proportions estimated from survey data. *Ann. Statist.* **12**, 46–60.

Rao, J. N. K. and Scott, A. J. (1985). On simple adjustments to chi-square tests with sample survey data. In *Analysis of Categorical Data from Sample Surveys: A Collection of Five Papers*, Technical Report No. 66, Laboratory for Research in Statistics and Probability, Carleton University, Ottawa.

Roberts, G. (1985). Correction to the chi-squared statistic for a survey design imposing fixed marginal totals in multi-way contingency tables. In *Analysis of Categorical Data from Sample Surveys: A Collection of Five Papers*, Technical Report No. 66, Laboratory for Research in Statistics and Probability, Carleton University, Ottawa.

Roberts, G., Rao, J. N. K. and Kumar, S. (1985). Logistic regression analysis of sample survey data. In *Analysis of Categorical Data from Sample Surveys: A Collection of Five Papers*, Technical Report No. 66, Laboratory for Research in Statistics and Probability, Carleton University, Ottawa.

Shuster, J. J. and Downing, D. J. (1976). Two-way contingency tables for complex sampling schemes. *Biometrika* **63**, 271–276.

Scott, A. J. (1986). Fitting logistic models to survey data. *Proc. Amer. Statist. Assoc.*, Section on Survey Research Methods, in press.

Thomas, D. R. and Rao, J. N. K. (1984). A Monte Carlo study of exact levels of goodness-of-fit statistics under cluster sampling. *Proc. Amer. Statist. Assoc.*, Section on Survey Research Methods, 207–211.

Thomas, D. R. and Rao, J. N. K. (1985). On the power of some goodness-of-fit tests under cluster sampling. *Proc. Amer. Statist. Assoc.*, Section on Survey Research Methods, 291–296.

PART 2

Time-Series Analysis

Growth Curves Analysis for Patterned Covariance Matrices

Adelchi Azzalini
Dipartimento di Scienze Statistiche
Università di Padova
Padua, Italy

SUMMARY

The likelihood analysis of growth curves is considered for the case of patterned covariance matrices of the error term in a normal theory context. Both fixed-effect and random-effect models are treated. An illustrative example is presented.

1. INTRODUCTION

Much work on growth curves analysis (more generally, on longitudinal data) deals with the model

$$Y = X\gamma F + \xi \tag{1.1}$$

where Y is an $n \times T$ matrix of response variables, X is an $n \times p$ matrix of covariates, γ is a $p \times q$ matrix of unknown parameters, F is a $q \times T$ matrix of known constants that we may think of as containing the values of q functions of time at T time points, ξ is an $n \times T$ matrix of random

Much of this work was done while the author was visiting the Department of Statistics, Stanford University, with financial support from the Consiglio Nazionale delle Ricerche, Italy.

variables; each row of ξ is distributed as $N_T(0, \Sigma)$, different rows of ξ being independent. Model (1.1) can be derived from the following argument. Denote by y_i the vector of T observations taken at T time points from the ith individual of a group of n, that is, y_i represents the growth curve of the ith individual ($i = 1, \ldots, n$). Suppose that

$$E(y_i) = F'\beta_i \tag{1.2}$$

where β_i is a $q \times 1$ vector of coefficients associated to the ith individual and represents the vector of coefficients of the q functions of time contained in F. Suppose further that β_i can be expressed as a linear combination of p covariates, x_i say, associated to the ith individual, that is,

$$\beta_i = \gamma' x_i \tag{1.3}$$

for some unknown $p \times q$ matrix γ. Combining (1.2), (1.3), and the assumptions on ξ, we obtain (1.1), having indicated with X the matrix containing the n row vectors x_i', and Y the matrix containing the row vectors y_i' for $i = 1, \ldots, n$.

The most common approach to model (1.1) is by multivariate analysis techniques, regarding the variance matrix Σ as any positive-definite matrix. Reviews are given by Timm (1980), Ware (1983), and Geisser (1980). Treating Σ as unpatterned does not incorporate into the analysis the known fact that the observations are taken in a specified order (often equally spaced), so that successive observations can be supposed to be more correlated than observations far apart. Moreover, if Σ is unpatterned, the number of unknown parameters increases rapidly with T.

In many situations, it is reasonable to suppose that Σ can be adequately represented by some patterned matrix depending on a parameter whose dimensionality does not increase with T. In this way, it is likely that more efficient estimates of the parameter of interest γ can be produced. In particular, it is sensible to consider variance matrices associated to stationary stochastic processes, such as ARMA models.

Of course, there is no claim that this issue is being considered for the first time, but no systematic study seems to have been attempted. Rao (1967) and Krishnaiah (1975) consider the case of Σ corresponding to a k-order autoregressive error structure and F with columns of the form $(1, t, t^2, \ldots, t^q)'$ for $t = 1, \ldots, T$. Besides being limited to a specific choice of F, their results are obtained regarding the first k observations of each row of y as fixed, which is equivalent to dealing with the likelihood conditional on the first k observations; this conditioning leads to loss of efficiency in the parameter estimation, if the errors are stationary (Azzalini,

1981). Crowder and Tredger (1981) consider observations unequally spaced in time, with the error structure of a (continuous-time) normal Markov process. Glasbey (1979) seeks a variance stabilizing transformation for logistic trend plus autoregressive errors.

In Section 2, we deal with model (1.1) when Σ is of the form $\sigma^2\Omega$, where $\Omega = \Omega(\alpha)$ is some scale-invariant patterned matrix depending on a parameter α of low dimension. In Section 3, we add a random effect term on the right-hand side of (3). In Section 4, we consider some hypothesis-testing problems. In Section 5, we present a numerical example. The development is in the spirit of a previous paper of the author (Azzalini, 1984).

2. FIXED-EFFECT MODEL

We consider model (1.1) for the case in which $\Sigma = \sigma^2\Omega$, where $\Omega = \Omega(\alpha)$ is a nonsingular matrix depending on a parameter α of low dimension. Although in the situation we have in mind $\sigma^2\Omega$ is the covariance matrix of a stationary stochastic process with innovation variance σ^2, the approach is valid for any patterned matrix Ω, provided we avoid overparametrization due to the presence of σ^2. For reasons that will be clear soon, it is convenient that $\Omega(\alpha)$ have an explicit expression for the inverse and the determinant. This is indeed the case if the rows of ξ in (1.1) are stationary AR(k) processes (see, e.g., Anderson and Mentz, 1980) or ARMA(1, 1) processes (Tiao and Ali, 1971).

Twice the log-likelihood of model (1.1) is

$$2l(\gamma, \sigma^2, \alpha) = -nT \log \sigma^2 - n \log|\Omega(\alpha)| - \operatorname{tr}(\sigma^{-2}Q)$$

with

$$Q = Y\Omega^{-1}Y' - 2X\gamma F\Omega^{-1}Y' + X\gamma F\Omega^{-1}(X\gamma F)' \tag{2.1}$$

We note that, if Ω refers to an AR(k) process, then reduction to nontrivial sufficient statistics is possible. In fact, considering the case $k = 1$ for the sake of simplicity, for which

$$\Omega(\alpha)^{-1} = \begin{bmatrix} 1 & -\alpha & 0 & \cdots & 0 & 0 \\ -\alpha & 1+\alpha^2 & -\alpha & \cdots & 0 & 0 \\ 0 & -\alpha & 1+\alpha^2 & \cdots & 0 & 0 \\ \vdots & \vdots & \vdots & & \vdots & \vdots \\ 0 & 0 & 0 & \cdots & -\alpha & 1 \end{bmatrix}$$

then the generic entry of $F\Omega^{-1}Y'X$ is

$$[-1 \quad \alpha]\begin{bmatrix} \sum_{t=1}^{T} f_t z_t & \sum_{t=2}^{T} f_t z_{t-1} \\ \sum_{t=2}^{T} f_{t-1} z_t & \sum_{t=2}^{T-1} f_t z_t \end{bmatrix}\begin{bmatrix} -1 \\ \alpha \end{bmatrix}$$

where $(f_1, \ldots, f_T)$ is the generic row of F and $(z_1, \ldots, z_T)'$ is the generic column of $Z = Y'X$. Therefore

$$\mathrm{tr}(\gamma F\Omega^{-1}Y'X) = \mathrm{tr}(\gamma D_{11}) - \alpha\,\mathrm{tr}(\gamma D_{12}) + \alpha^2 \mathrm{tr}(\gamma D_{22})$$

for suitable $q \times p$ matrices D_{ij}. A similar fact holds for $\mathrm{tr}(Y\Omega^{-1}Y')$. By the factorization theorem, a $(3pq + 3)$-dimensional sufficient statistic exists. For specific choices of F and X, further reduction of dimensionality may be possible. We shall not pursue this point further.

For maximizing the log-likelihood, we first regard α as fixed and find the corresponding maximum likelihood estimate (m.l.e.) of the other parameters, $\hat{\gamma}(\alpha)$ and $\hat{\sigma}^2(\alpha)$ say. Then, substituting these values in the log-likelihood, we obtain the profile log-likelihood that has to be maximized numerically. Since α will usually be of very low dimensionality, this final maximization should usually be easy to perform.

Differentiating $\mathrm{tr}(Q)$ with respect to γ, we obtain

$$\hat{\gamma}(\alpha) = (X'X)^{-1}X'Y\Omega^{-1}F'(F\Omega^{-1}F')^{-1} \tag{2.2}$$

having assumed that X and F have rank p and q, respectively. Substituting $\hat{\gamma}(\alpha)$ in Q and maximizing l with respect to σ^2, we obtain

$$\hat{\sigma}^2(\alpha) = \mathrm{tr}\{Y\Omega^{-1}Y' - X\hat{\gamma}(\alpha)F\Omega^{-1}Y'\}/(nT)$$

and the corresponding value of twice the log-likelihood is

$$2l^*(\alpha) = -nT \log \hat{\sigma}^2(\alpha) - n \log|\Omega(\alpha)|$$

When the value $\hat{\alpha}$ that maximizes $l^*(\alpha)$ has been found (usually by numerical maximization), the corresponding $\hat{\gamma}(\hat{\alpha})$ and $\hat{\sigma}^2(\hat{\alpha})$ are the m.l.e. of γ and σ^2. Since $\hat{\gamma}(\alpha)$ and $\hat{\sigma}^2(\alpha)$ need to be computed for several values of α, the computational burden is much reduced if $\Omega(\alpha)^{-1}$ and $|\Omega(\alpha)|$ are known explicity as functions of α.

3. RANDOM-EFFECT MODEL

Experience with real data growth curves suggests there is often strong variability between individuals, which cannot be captured completely by the covariates x_i. For this reason, it will usually be necessary to replace (1.3) by the more general

$$\beta_i = \gamma' x_i + \delta_i \tag{3.1}$$

where $\delta_1, \ldots, \delta_n$ are independent $N_q(0, \Delta)$ for some positive-semidefinite matrix Δ. Here δ_i represents the discrepancy between the coefficients β_i of the ith individual and the values obtained averaging over the population with given covariates x_i. Substitution of (3.1) in (1.3) changes the previous expression of Σ to $\Sigma = \sigma^2\Omega + F'\Delta F$; of course, when Δ shrinks to the null matrix, we are back to the previous case. It would be interesting to parametrize Δ by a few parameters, but in the following we shall impose only the condition that Δ is positive-semidefinite.

To expand Σ^{-1}, we shall make use of the binomial inverse theorem, which states

$$(A + BCD)^{-1} = A^{-1} - A^{-1}B\left(C^{-1} + DA^{-1}B\right)^{-1}DA^{-1}$$

when the matrices are conformable and the necessary inverses exist. Then

$$\begin{aligned}
\Sigma^{-1} &= \left\{\Omega\left(\sigma^2 I_T + \Omega^{-1}F'\Delta F\right)\right\}^{-1} \\
&= \sigma^{-2}\Omega^{-1} - \sigma^{-2}\Omega^{-1}F'\left(\sigma^2 I_q + \Delta\tau\right)^{-1}\Delta F\Omega^{-1} \\
&= \sigma^{-2}\Omega^{-1} - \sigma^{-2}\Omega^{-1}F'\tau^{-1/2}\Psi^{-1}\left(\Psi - \sigma^2 I_q\right)\left(\tau^{-1/2}\right)'F\Omega^{-1} \\
&= \sigma^{-2}\Omega^{-1} - \sigma^{-2}\Omega^{-1}F'\tau^{-1}F\Omega^{-1} + \Omega^{-1}F'\tau^{-1/2}\Psi^{-1}\left(\tau^{-1/2}\right)'F\Omega^{-1}
\end{aligned}$$

having written

$$\begin{aligned}
\tau = \tau(\alpha) &= \left(\tau^{1/2}\right)'\tau^{1/2} = F\Omega^{-1}F' \\
\Psi &= \sigma^2 I_q + \tau^{1/2}\Delta\left(\tau^{1/2}\right)'
\end{aligned}$$

Moreover

$$\begin{aligned}
|\Sigma| &= |\sigma^2\Omega| \cdot |I_T + (\sigma^2\Omega)^{-1}F'\Delta F| \\
&= (\sigma^2)^{T-q}|\Omega| \cdot |\sigma^2 I_q + \Delta\tau| \\
&= (\sigma^2)^{T-q}|\Omega| \cdot |\Psi|
\end{aligned}$$

Then twice the log-likelihood is

$$2l(\gamma, \sigma^2, \alpha, \Psi) = -n\{(T-q)\log \sigma^2 + \log|\Omega| + \log|\Psi|\} - \operatorname{tr}(Q) \quad (3.2)$$

where Q is defined as in (2.1), except that Ω^{-1} is replaced by Σ^{-1}. The corresponding expression of $\hat{\gamma}(\alpha)$ is analogous to (2.2), with Ω^{-1} replaced by Σ^{-1}. Note, however, that

$$(F\Sigma^{-1}F')^{-1} = \{(\tau^{1/2})'\Psi^{-1}\tau^{1/2}\}^{-1} = \sigma^2\tau^{-1} + \Delta$$

Substituting the last relationship and the expression of Σ^{-1} in (2.2) and simplifying, we find that (2.2) gives the m.l.e. for fixed α also for the random-effect model. This last statement could also be deduced from Lemma 4 of Rao (1968). Of course, the variance of $\hat{\gamma}(\alpha)$ is changed. In fact

$$\begin{aligned}\operatorname{var}(\operatorname{vec}\hat{\gamma}(\alpha)) &= \operatorname{var}\{(\tau^{-1}F\Omega^{-1} \otimes (X'X)^{-1}X')\operatorname{vec} Y\}\\ &= (\tau^{-1}F\Omega^{-1} \otimes (X'X)^{-1}X')(\Sigma \otimes I_n)\\ &\quad \times(\Omega^{-1}F'\tau^{-1} \otimes X(X'X)^{-1})\end{aligned}$$

which depends on Δ; here vec is the operator that stacks the columns of a matrix. At $\alpha = \alpha_0$, the true value of the parameter, the above quantity simplifies to

$$\operatorname{var}(\operatorname{vec}\hat{\gamma}(\alpha_0)) = (\sigma^2\tau^{-1} + \Delta) \otimes (X'X)^{-1}$$

We now come to estimating σ^2 and Ψ. The reparametrization from Δ to Ψ is introduced because (a) estimation of σ^2 and Δ does not allow explicit solution of the likelihood equations, while this is possible with σ^2 and Ψ, and (b) the corresponding solutions $\tilde{\sigma}^2(\alpha)$ and $\tilde{\Psi}(\alpha)$ are independent random variables at $\alpha = \alpha_0$. These points will be clarified in the following. Substituting $\hat{\gamma}(\alpha)$ in Q gives

$$\begin{aligned}Q_{|\gamma=\hat{\gamma}(\alpha)} &= \sigma^{-2}Y\Omega^{-1}Y' - \sigma^{-2}U\tau^{-1}U' + U\tau^{-1/2}\Psi^{-1}(\tau^{-1/2})'U'\\ &\quad +\hat{\gamma}(\alpha)(\tau^{1/2})'\Psi^{-1}(\tau^{-1/2})'U'X\end{aligned}$$

where $U = Y\Omega^{-1}F'$. Substituting this quantity in (3.2) and differentiating the resulting expression with respect to σ^2 and Ψ^{-1}, we obtain the solutions

$$\tilde{\sigma}^2(\alpha) = \mathrm{tr}(Y\Omega^{-1}Y' - U\tau^{-1}U')/\{n(T-q)\} \tag{3.3}$$

$$\tilde{\Psi}(\alpha) = n^{-1}(\tau^{-1/2})'U'\{I_n - X(X'X)^{-1}X'\}U\tau^{-1/2} \tag{3.4}$$

To see that $\tilde{\sigma}^2(\alpha_0)$ and $\tilde{\Psi}(\alpha_0)$ are independent, write $Y = (X\gamma + \delta)F + \eta$, where δ is the $n \times q$ matrix with generic row δ_i' and η is an $n \times T$ matrix whose generic row is $N_T(0, \sigma^2\Omega)$, different rows being independent and independent of δ. Substituting the preceding expression of Y in (3.3) and (3.4), and expanding, one sees that $\tilde{\sigma}^2(\alpha)$ is a function of η but not of δ, while $\tilde{\Psi}(\alpha)$ depends on δ and some components of η that at $\alpha = \alpha_0$ are orthogonal to those that enter into $\tilde{\sigma}^2(\alpha)$. This computation also shows that $n(T-q)\tilde{\sigma}^2(\alpha_0)/\sigma^2$ is $\chi^2_{n(T-q)}$.

One must keep in mind, however, that $\Psi - \sigma^2 I_q$ is non-negative definite, while this property may not hold for $\tilde{\Psi}(\alpha) - \tilde{\sigma}^2(\alpha)I_q$ for certain values of α. Therefore, having computed (3.3) and (3.4), one must check that these quantities lie in the feasible area; in the case they do not, then they must be rejected as m.l.e.'s. In this event, the m.l.e.'s of σ^2 and Ψ, for fixed α, are to be found on the boundary of the parameter space; since $\sigma^2 = 0$ gives value $-\infty$ to the log-likelihood, one has to consider the set $S = \{(\sigma^2, \Psi): \sigma^2 > 0, \Psi = \Psi', \Psi - \sigma^2 I_q \geq 0\}$. It appears that such a constrained maximization problem has no explicit solution, and one has to resort to numerical methods. In order to avoid overparametrization problems, one needs a parsimonious representation of the set S. It turns out that the set of feasible matrices Ψ can be obtained as $\Psi = \sigma^2 I_q + LL'$, where L is a lower triangular matrix whose diagonal elements are non-negative and such that their product is 0. Then the function to be minimized numerically with respect to σ^2 and L is

$$(T-q)\log\sigma^2 + \log|\Psi| + \sigma^{-2}(T-q)\tilde{\sigma}^2(\alpha) + \mathrm{tr}(\Psi^{-1}\tilde{\Psi}(\alpha)) \tag{3.5}$$

where $\Psi = \sigma^2 I_q + LL'$.

Denote by $\hat{\sigma}^2(\alpha)$ and $\hat{\Psi}(\alpha)$ the m.l.e.'s of σ^2 and Ψ, for fixed α, both for the case that they are given by (3.3) and (3.4) or have been obtained from (3.5). Then the profile log-likelihood is

$$l^*(\alpha) = -\tfrac{1}{2}n\{(T-q)\log\hat{\sigma}^2(\alpha) + \log|\Omega(\alpha)| + \log|\hat{\Psi}(\alpha)| + R\} \tag{3.6}$$

where

$$R = \begin{cases} T & \text{if } \tilde{\Psi}(\alpha) - \tilde{\sigma}^2(\alpha) I_q \text{ is non-negative-definite} \\ \hat{\sigma}^{-2}(\alpha)\tilde{\sigma}^2(\alpha)(T-q) + \operatorname{tr}\left(\hat{\Psi}(\alpha)^{-1}\tilde{\Psi}(\alpha)\right) & \text{otherwise} \end{cases}$$

Function (3.6) has to be maximized numerically. It must be remarked that optimization of (3.5) involves a search of dimensionality $1 + q(q+1)/2$, which becomes lengthy if q is greater than, say, 3.

4. HYPOTHESES TESTING

Consider first the problem of testing the fixed-effect model versus the random-effect one, that is, testing H_0: $\Delta = 0$ versus H_1: $\Delta > 0$ or $\Delta \geq 0$. Standard results on the asymptotic distribution of the likelihood ratio test are not useful here, because the alternative hypothesis is one-sided.

Alternatively, observe that testing $\Delta = 0$ is equivalent to testing $\Psi = \sigma^2 I_q$. To this end, consider the distribution of $n\tilde{\Psi}(\alpha_0)$. Replacing Y by $Y - X\gamma F$ in the definition of $\tilde{\Psi}(\alpha)$ does not change this quantity; therefore we can argue as if $E(U) = 0$. Since each row of U has variance $F\Omega(\alpha)^{-1}\Sigma\Omega(\alpha)^{-1}F'$, which is equal to $(\tau^{1/2})'\Psi\tau^{1/2}$ when $\alpha = \alpha_0$, then, by the Cochran theorem, we have that $n\tilde{\Psi}(\alpha_0)$ is distributed as a Wishart random variable $W_q(\Psi, n-p)$.

Various possible tests based on the eigenvalues of $\tilde{\Psi}(\alpha_0)$ could be constructed now, but the simplest one is

$$\operatorname{tr}\{n\tilde{\Psi}(\alpha_0)\}/\{q(n-p)\tilde{\sigma}^2(\alpha_0)$$

which, under the null hypothesis, is distributed as $F(q(n-p), n(T-q))$, and is stochastically larger under the alternative. In practice, one will replace α_0 by some estimate, such as the m.l.e., so the preceding statement holds approximately. The effect of replacing α by $\hat{\alpha}$ is currently under study.

Consider now the problem of testing H_0: $A\gamma B = C$ versus H_1: $A\gamma B \neq C$, for given matrices $A(r \times p)$, $B(q \times s)$, $C(r \times s)$, with A of rank r and B of rank s. The development treats the random-effect case, the fixed-effect case being included. With computations similar to Tubbs et al. (1975), the constrained m.l.e. is given by

$$\tilde{\gamma}(\alpha) = \hat{\gamma}(\alpha) + (X'X)^{-1}A'\{A(X'X)^{-1}A'\}^{-1}$$
$$\times(C - A\hat{\gamma}(\alpha)B)(B'\tau_1^{-1}B)^{-1}B'\tau_1^{-1}$$

where $\tau_1 = F\Sigma^{-1}F' = (\sigma^2\tau^{-1} + \Delta)^{-1}$. The corresponding increment of Q in (3.2) is

$$Q_0 = (C - A\hat{\gamma}(\alpha)B)'\{A(X'X)^{-1}A'\}^{-1}(C - A\hat{\gamma}(\alpha)B)(B'\tau_1^{-1}B)^{-1}$$

whose trace can be used for testing the null hypothesis. Since

$$\text{var}(\text{vec}\,A\hat{\gamma}(\alpha_0)B) = (B'\tau_1^{-1}B) \otimes (A(X'X)^{-1}A')$$

by a computation similar to the one for var(vec $\hat{\gamma}(\alpha_0)$), then

$$W = (A(X'X)^{-1}A')^{-1/2}(C - A\hat{\gamma}(\alpha_0)B)(B'\tau_1^{-1}B)^{-1/2}$$

is an $r \times s$ matrix of independent $N(0,1)$ variates. Therefore $\text{tr}(Q_0) = \text{tr}(W'W)$ is distributed under the null hypothesis as χ^2_{rs}. Since in practice τ_1^{-1} must be replaced by an estimate, the χ^2 distribution will be approximate.

5. AN EXAMPLE

To illustrate the methods presented, we reanalyze the data of Table 1 of Grizzle and Allen (1969), which contains $T = 7$ observations equally spaced in time on each of $n = 36$ subjects divided into $p = 4$ groups of size 9, 10, 8, and 9. Matrix X is defined as containing dummy variables indicating the group to which each subject belongs; the rows of the 4×7 matrix F contain the values of orthogonal polynomials of degree up to 3. Hence X is defined as A' and F as B' at pages 358–360 of Grizzle and Allen's paper. Matrix Ω has been chosen to be the one of a stationary ARMA(1,1) process, that is, a process of the form

$$\eta_t - \alpha_1\eta_{t-1} = \varepsilon_t - \alpha_2\varepsilon_{t-1} \qquad (t = 2,\ldots,T)$$

where $|\alpha_1| < 1$, $|\alpha_2| < 1$, and the initial value η_1 is taken from the stationary distribution (see, for instance, Box and Jenkins, 1976, p. 76). This

Table 1. m.l.e.'s of γ and Δ for Data of Grizzle and Allen (1969)

$\hat{\gamma}$				s.e. of $\hat{\gamma}$			
4.617	0.163	−0.041	−0.188	0.168	0.036	0.015	0.049
3.553	−0.015	−0.009	0.014	0.159	0.034	0.014	0.046
4.153	0.120	−0.045	−0.051	0.178	0.038	0.016	0.052
3.941	0.060	−0.019	0.009	0.168	0.036	0.015	0.049
$\hat{\Delta}$							
0.204	0.023	−0.001	−0.016				
0.023	0.003	0.000	−0.002				
−0.001	0.000	0.000	0.000				
−0.016	−0.002	0.000	0.002				
				$\hat{\sigma}^2 = 0.1104$			

choice of Ω is believed to give sufficient flexibility to the autocorrelation structure, without introducing too many parameters. The inverse and the determinant of Ω have been computed using the formulae of Tiao and Ali (1971).

Maximization of (3.6) has led to $\hat{\alpha} = (\hat{\alpha}_1, \hat{\alpha}_2)' = (0.292, -0.337)'$ with associated log-likelihood 116.48, which is significantly different from 104.42 obtained at $\alpha = (0, 0)'$, when we compare twice the difference with the χ_2^2 percentage points. The m.l.e. of α lies in the region where (3.3) and (3.4) are unfeasible and maximization of (3.6) has involved several maximizations of (3.5) for various values of α.

The m.l.e.'s of γ and Δ are given in Table 1. They are obtained at $\alpha = \hat{\alpha}$, but, for this example, very similar values would be obtained in a fairly large region of α values, including the origin. In particular, the m.l.e.'s of γ are close to the estimates obtained by Grizzle and Allen, both with their weighted and unweighted analysis. The test of Section 4 for nullity of Δ gives 2.18 on (128, 108) degrees of freedom, which is highly significant. The test for nullity of last column of γ gives 15.96 on 4 degrees of freedom, which again is significantly large.

The computer program used for the preceding computations is available from the author on request.

REFERENCES

Anderson, T. W. and Mentz, R. P. (1980). On the structure of the likelihood function of autoregressive and moving average models. *J. Time Ser. Anal.* **1**, 83–94.

Azzalini, A. (1981). Replicated observations of low order autoregressive models. *J. Time Ser. Anal.* **2**, 63–70.

Azzalini, A. (1984). Estimation and hypothesis testing for collections of autoregressive time series. *Biometrika* **71**, 85–90.

Box, G. E. P. and Jenkins, G. M. (1976). *Time Series Analysis Forecasting and Control*, San Francisco: Holden Day.

Crowder, M. J. and Tredger, J. A. (1981). The use of exponentially damped polynomials for biological recovery data. *Appl. Statist.* **30**, 147–152.

Geisser, S. (1980). Growth curve analysis. In *Handbook of Statistics*, Vol. 1: Analysis of Variance (P. R. Krishnaiah, Ed.) Amsterdam: North-Holland, pp. 89–115.

Glasbey, C. A. (1979). Correlated residuals in non linear regression applied to growth data. *Appl. Statist.* **28**, 251–9.

Grizzle, J. E. and Allen, D. M. (1969). Analysis of growth and dose response curves. *Biometrics* **25**, 357–81.

Krishnaiah, P. R. (1975). Simultaneous tests for multiple comparisons of growth curves when errors are autocorrelated. *Proc. 40th Session Int. Stat. Inst.* **4**, 62–66.

Rao, C. R. (1967). Least squares theory using an estimated dispersion matrix and its applications to measurements of signals. *Proc. 5th Berkeley Symp.* **1**, 355–72.

Rao, C. R. (1968). A note on a previous lemma in the theory of least squares and some further results. *Sankhyā A* **30**, 259–266.

Tiao, G. C. and Ali, M. M. (1971). Analysis of correlated random effects: linear models with two components. *Biometrika* **58**, 37–51.

Timm, N. H. (1980). Multivariate analysis of variance of repeated measurements. In *Handbook of Statistics*, Vol. 1: Analysis of Variance, Amsterdam: North-Holland, pp. 41–87.

Tubbs, J. D., Lewis, T. O. and Duran, B. S. (1975). A note on the analysis of the MANOVA model and its application to growth curves. *Comm. Statist.* **4**, 643–653.

Ware, J. H. (1983). Growth curves. In *Encyclopedia of Statistical Sciences*, Vol. 3 (N. L. Johnson and S. Kotz, Eds.), New York: Wiley, pp. 539–542.

ARIMA Modeling of the "New" U.S. Monetary Aggregates

Giorgio Canarella and Stephen K. Pollard
Department of Economics and Statistics
California State University, Los Angeles
Los Angeles, California

SUMMARY

Univariate autoregressive-integrated-moving-average models are frequently used to analyze the stochastic properties of time series. In order to estimate such time-series models a decision must be made on the dimensions of the autoregressive-moving-average structure and the order of differencing required to achieve stationarity. In this study we investigate the optimal specification of the ARIMA models for the new U.S. monetary aggregates using the procedures introduced by Akaike (1973), Schwarz (1978), and Hannan (1980). The models selected are evaluated on the basis of their forecasting ability. The results indicate that the Akaike order criterion selects models that meet inferential and diagnostic checking, but do not forecast as well as those chosen by the Schwarz and other criterion.

1. INTRODUCTION

Autoregressive-integrated-moving-average (ARIMA) models, until recently, have not been much employed in economic research. Indeed, for most of the past two decades, while the relative forecasting performance of univariate ARIMA and econometric models has been the subject of extensive research (Leuthold et. al., 1970; Cooper, 1972; Nelson, 1972; Christ, 1975), the relationship between the two approaches has been viewed, in the economics circles, from an antagonistic perspective. This reflected, on the one hand, the role of theory in model parametrization in classical econometric models, and on the other hand, the reliance on the structure of the data and the atheoretical foundations of the process of identification of ARIMA models. Recent theoretical developments in economic analysis, however, have shown an interface between the two approaches, which has shifted the original perspective into one of complementarity (Zellner and Palm, 1974; Sims, 1980; Wallis, 1980).

Applied monetary economics, and in particular the analysis of the behavior of monetary authorities, is an area in which the link between ARIMA and econometric models has been strengthening in the past few years and research on ARIMA modeling has been developed (Nelson, 1973; Granger and Newbold, 1977; Neftci, 1982; Harvey and Todd, 1983). The purpose of our research is to add to this growing body of work by exploring the stochastic properties, by means of ARIMA modeling, of the U.S. monetary aggregates as redefined by the Federal Reserve Board in 1980. Berkman (1980), Simpson (1980), and Hafer (1980) provide a detailed critical appraisal of the new definitions. There are, at least, two reasons for following this approach.

First, there is no well-developed theory of money supply growth and reaction functions of the monetary authorities. Conceptually, such functions are the output of a dynamic constrained maximization process, where the constraints are the reduced-form equations that characterize the economy, and the objective function is the set of preferences held by the monetary authorities. However, the various studies that have estimated models of monetary policy (for example, Froyen, 1974; Barro, 1978; Niskanen, 1978; Levy, 1981) cannot escape what Sims (1980) has called the "dubious exclusion restrictions," that is, the identifying restrictions that economic theory cannot rationalize. This can be overcome through ARIMA modeling of the money supply.

Second, we have the purpose of estimation of money supply growth. The current debate on policy neutrality that has emerged from the theory of rational expectations has emphasized the decomposition of observed monetary growth into what is defined to be its "anticipated" and "unanticipated" components, and has stressed the asymmetric impacts of these components on unemployment, real output, and the price level (Barro, 1977, 1978, 1979, 1980; Mishkin, 1982). Within this context, the ARIMA modeling procedures are ideally suited to aid in finding such a decomposition (Illmakunnas and Tsurumi, 1985).

The rest of this chapter is organized as follows. Some questions related to the ARIMA methodology, in particular the order selection problem, are discussed in Section 2. The estimates of the ARIMA models for the new U.S. monetary aggregates under alternative order determination rules are presented in Section 3. Also included in that section is a comparison of the relative forecasting ability of the models selected. Finally, we briefly summarize the results and comment on the avenues opened by this research in Section 4.

2. THEORETICAL CONSIDERATIONS OF ARIMA MODEL SELECTION

ARIMA modeling of a nonstationary time series, as it is the case with the U.S. monetary aggregates, consists of two stages. The first stage is to determine the appropriate transformations and differencing of the original series to induce stationarity. The second stage is the modeling and estimation of the resulting stationary time series. For our purposes, let T be the appropriate transformation that induces stationarity on the measure of a monetary aggregate M_t. Then the process $X_t = T(M_t)$ can be represented by the model

$$\left(1 - \phi_1 B - \cdots - \phi_p B^p\right) X_t = \delta + \left(1 - \theta_1 B - \cdots - \theta_q B^q\right) a_t \quad (2.1)$$

where B is the backshift operator, such that $B^k X_t = X_{t-k}$; a_t is a zero-mean white-noise process, so that $E(a_t a_s)$ is σ_a^2 for $t = s$ and zero for $t \neq s$. The parameter space $(\phi_1, \ldots, \phi_p, \theta_1, \ldots, \theta_q)$ is constrained by the stationary condition that the roots of the polynomial equation

$$\left(1 - \phi_1 B - \cdots - \phi_p B^p\right) = 0 \quad (2.2)$$

exceed unity in modulus, and by the invertibility condition that the roots of the polynomial equation

$$\left(1 - \theta_1 B - \cdots - \theta_q B^q\right) = 0 \quad (2.3)$$

are not less than unity in modulus.

The determination of the (p, q) order of the model is a problem that has attracted the attention of many researchers, such as Akaike (1974), Box and Jenkins (1970), Hannan (1970), and Quenouille (1947), to mention a few. The traditional method of choosing the "best" model has been the likelihood ratio test statistic. The test of the null hypothesis that the order is (p_0, q_0) against the alternative that it is (p_1, q_1) is suitable, however, only when (p_0, q_0) and (p_1, q_1) have been specified a priori. If these are unknown, as usually is the case, then the testing procedure has to be applied repeatedly for different values of (p_0, q_0) and (p_1, q_1) (Hannan, 1970; Potscher, 1982) with the consequent difficulty of determining the appropriate level of significance (Akaike, 1978). Box and Jenkins (1970) have suggested an informative graphical procedure based on the examination of the sample autocorrelation function. However, while an informal inspection of the sample autocorrelation and partial autocorrelation function plays an important part in the analysis, it cannot ordinarily stand by itself, since no standards of comparison are provided against which the observed discrepancies can be measured (Newbold and Granger, 1974; Chatfield and Prothero, 1973; Bhansali, 1983).

Although the problem of order selection has not been completely solved, the recent work of Akaike (1973, 1974), Schwarz (1978), Hannan and Quinn (1979), and Quinn (1980) on automatic selection rules for autoregressive models has stimulated extensions to ARIMA models (Howrey, 1978; Kitagawa, 1977; Ozaki, 1977; Hannan and Rissanen, 1982; Hannan, 1980). Hannan (1980) in particular has considered a generalized order criterion of the form:

$$GIC = \log \hat{\sigma}^2_{(p,q)} + (p+q)[f(N)/N] \tag{2.4}$$

where $\hat{\sigma}^2_{(p,q)}$ is the maximum likelihood estimate of the residual variance for a given p, q and N is the number of observations, and has shown that the (p, q) that minimizes GIC provides a weakly consistent estimate of the true (p, q) if $f(N)$ is an increasing function so that $f(N) \to \infty$ and $f(N)/N \to 0$ as $N \to \infty$. Furthermore, Hannan has shown that if $f(N) = C \log\log N$, then the minimizing (p, q) will be a strongly consistent estimate of (p, q) for $C > 2$.

Diagnostic tests have also been suggested for selecting the best among several competing models for the same process. These tests are based on comparing the response of the process with that of the model for a number of tests. The model is considered suitable if the two responses are "sufficiently" close. This closeness is quantified in terms of the sequence of residuals, defined as the difference between the actual sequence and the predicted output for the model. A model is considered satisfactory if the residuals form a white-noise sequence with zero mean and as small a variance as possible. The Box–Pierce Q statistic is used to test for a white-noise process of the residuals.

3. STATISTICAL RESULTS

Four measures of monthly, seasonally adjusted, monetary aggregates were chosen for analysis over the period January 1959–December 1981: M1, M2, M3, and liquidity (L).

Each series was first examined in terms of different transformations in order to induce stationarity. The transformations $(1 - B)M_t$ and $(1 - B)^2 M_t$ did not yield stationary behavior for any of the monetary measures. However the autocorrelation structure of $(1 - B)\log M_t$ revealed a pattern consistent with the notion of stationarity for all four measures. The relevance of this transformation is well established in the economic literature (Barro 1977, 1978, 1979, 1980; Mishkin, 1982). Then, using the first 12 observations on the logarithmic first differences of each series as "presample" values, the optimal (p, q) was searched over the range $0 \leq p \leq 12$ and $0 \leq q \leq 12$ using an approximate maximum likelihood procedure.

Five different order selection criteria were then employed. The first is with $f(N) = 2$. This is the AIC criterion proposed by Akaike (1973, 1974), and although it does not satisfy the Hannan conditions, it is included because of its widespread use. The second was with $f(N) = \log N$. This is the BIC criterion proposed by Schwarz (1978). The third is with $f(N) = 2 \log\log N$. This is the criterion proposed by Hannan and Quinn (1979) and Quinn (1980), which Paulsen (1984) calls the LIL criterion, because it is based on the law of the iterated logarithm. The fourth is with $f(N) = 3 \log\log N$, which satisfies the Hannan conditions for strong consistency. Finally, the last criterion considered is based on a family of goodness-of-fit procedures. This criterion chooses the model that minimizes the residual variance in the class of models that satisfy diagnostic checking.

The results of the search procedures can be summarized as follows. For M1, the optimal (p, q) according to the AIC criterion was found to be an ARMA(4, 4):

$$\begin{aligned}(1 - B)(1 - &\underset{(0.152)}{0.522}B + \underset{(0.153)}{0.405}B^2 - \underset{(0.140)}{0.513}B^3 - \underset{(0.139)}{0.380}B^4)\log M1_t \\ &= \underset{(0.0009)}{-0.033} + (1 - \underset{(0.126)}{0.234}B + \underset{(0.116)}{0.310}B^2 - \underset{(0.120)}{0.316}B^3 - \underset{(0.118)}{0.690}B^4)a_t\end{aligned}$$

$$Q(10) = 13.4, \qquad Q(20) = 26.4, \qquad Q(30) = 33.4$$

while the BIC criterion yielded an ARMA(0, 1):

$$(1 - B)\log M1_t = \underset{(0.0003)}{-0.004} + (1 - \underset{(0.059)}{0.3019}B)a_t$$

$$Q(10) = 26.3, \qquad Q(20) = 52.1, \qquad Q(30) = 60.2$$

The Hannan criterion, for $C = 2$, coincides with the AIC, while for $C = 3$ the Hannan criterion coincides with the BIC. The criterion that relies on a family of goodness-of-fit procedures also agreed with AIC. For M2, the AIC yielded an ARMA(3, 4):

$$
\begin{aligned}
&(1 - B)(1 - \underset{(0.023)}{0.039}B + \underset{(0.023)}{0.021}B^2 - \underset{(0.022)}{0.926}B^3)\log M2_t \\
&\quad = \underset{(0.001)}{-0.008} + (1 + \underset{(0.061)}{0.696}B + \underset{(0.067)}{0.391}B^2 - \underset{(0.067)}{0.560}B^3 - \underset{(0.059)}{0.419}B^4)a_t
\end{aligned}
$$

$$Q(10) = 3.85, \qquad Q(20) = 11.6, \qquad Q(30) = 21.2$$

as did all the other criteria with the exception of the goodness-of-fit criterion, which selected an ARMA(4, 4):

$$
\begin{aligned}
&(1 - B)(1 - \underset{(0.144)}{0.237}B + \underset{(0.024)}{0.021}B^2 - \underset{(0.024)}{0.969}B^3 + \underset{(0.141)}{0.196}B^4)\log M2_t \\
&\quad = \underset{(0.001)}{-0.001} + (1 + \underset{(0.135)}{0.252}B + \underset{(0.083)}{0.279}B^2 - \underset{(0.090)}{0.679}B^3 + \underset{(0.079)}{0.373}B^4)a_t
\end{aligned}
$$

$$Q(10) = 2.27, \qquad Q(20) = 9.91, \qquad Q(30) = 20.9$$

For M3, the AIC, goodness-of-fit, and the Hannan ($C = 2$) criteria chose an ARMA(6, 5):

$$
\begin{aligned}
&(1 - B)(1 + \underset{(0.099)}{0.318}B - \underset{(0.067)}{0.039}B^2 - \underset{(0.113)}{0.339}B^3 - \underset{(0.105)}{0.293}B^4 + \underset{(0.067)}{0.059}B^5 \\
&\quad \underset{(0.106)}{-0.617}B^6)\log M3_t = \underset{(0.001)}{-0.0008} + (1 + \underset{(0.083)}{1.052}B + \underset{(0.058)}{0.661}B^2 + \underset{(0.125)}{0.262}B^3 \\
&\qquad \underset{(0.182)}{+0.089}B^4 + \underset{(0.133)}{0.483}B^5)a_t
\end{aligned}
$$

$$Q(10) = 4.20, \qquad Q(20) = 10.8, \qquad Q(30) = 15.4$$

while the BIC and the Hannan ($C = 3$) criteria chose an ARMA(6, 2):

$$
\begin{aligned}
&(1 - B)(1 + \underset{(0.066)}{0.232}B + \underset{(0.064)}{0.311}B^2 - \underset{(0.065)}{0.729}B^3 - \underset{(0.065)}{0.206}B^4 - \underset{(0.063)}{0.289}B^5 \\
&\quad \underset{(0.064)}{-0.236}B^6)\log M3_t = \underset{(0.001)}{-0.002} + (1 + \underset{(0.030)}{0.924}B + \underset{(0.029)}{0.938}B^2)a_t
\end{aligned}
$$

$$Q(10) = 8.72, \qquad Q(20) = 12.7, \qquad Q(30) = 20.1$$

Finally, for L, the AIC, goodness-of-fit, and the Hannan ($C = 2$) criteria selected an ARMA(3, 5):

$$\begin{aligned}(1 - B)(1 + &\underset{(0.014)}{0.009}B - \underset{(0.014)}{0.003}B^2 - \underset{(0.014)}{1.004}B^3)\log L_t \\ = \underset{(0.0009)}{-0.0005} &+ (1 + \underset{(0.065)}{0.547}B + \underset{(0.072)}{0.238}B^2 + \underset{(0.063)}{0.667}B^3 - \underset{(0.071)}{0.243}B^4 + \underset{(0.064)}{0.056}B^5)a_t\end{aligned}$$

$$Q(10) = 2.36, \qquad Q(20) = 7.04, \qquad Q(30) = 12.8$$

while the BIC and the Hannan ($C = 3$) criteria chose an ARMA(3, 4):

$$\begin{aligned}(1 - B)(1 + &\underset{(0.057)}{0.024}B - \underset{(0.055)}{0.057}B^2 - \underset{(0.053)}{0.965}B^3)\log L_t \\ &= \underset{(0.0009)}{-0.0005} + (1 + \underset{(0.085)}{0.554}B + \underset{(0.083)}{0.181}B^2 - \underset{(0.079)}{0.617}B^3 - 0.230B^4)a_t\end{aligned}$$

$$Q(10) = 0.795, \qquad Q(20) = 5.57, \qquad Q(30) = 11.4$$

The different criteria appear to choose values of p and q that are fairly close, the only exception being M1, where the discrepancy between the AIC and the BIC specifications is quite extensive. The standard errors of the estimated autoregressive parameters can be used in diagnostic checking of the appropriateness of the estimated models as noted by Galbraith and Galbraith (1974). According to Galbraith and Galbraith, the standard errors of the autoregressive parameters should be approximately the same for each coefficient. This pattern holds for all models except the ARMA(4, 4) for M2 and the ARMA(6, 5) for M3, where only subsets of the standard errors are equal. All the estimated coefficients are significantly different from zero at the 5% level for both of the M1 models and the M3 ARMA(6, 2) model. The remaining models have at least one insignificant autoregressive or moving-average coefficient. Additional diagnostic checking was undertaken by examining whether the residuals of each model followed a white-noise sequence. This was confirmed by use of the Box–Pierce Q statistic for all models except the M1 ARMA(0, 1) model.

The above-chosen models were then evaluated on the basis of their postsample forecasting ability. Forecasts 1–6 months ahead were generated, in logarithmic first differences, for each model for the period January 1982–July 1983 (March 1983 for L). Each of these forecasted series was then compared to the actual series over this period and the summary statistics, the root-mean-square error (RMSE), and the mean absolute error (MAE), were calculated to judge the efficiency of the forecasts. (The complete set of the forecast summary statistics can be obtained from the authors.) For M3 and L, respectively, there is little difference between the forecasting efficiency of the different models chosen by the selection criteria. The 1-month-ahead forecast yields the smallest RMSE and MAE, and these measures increase as the forecast horizon increases for both M3 and L. A different picture emerges when the forecasts for M1 and M2 are analyzed. For M1, the ARMA(0, 1) 3-month-ahead forecast gives the smallest RMSE (0.0062) and MAE (0.00509), while for the ARMA(4, 4) the 1-month-ahead forecast gives the smallest RMSE (0.00992) and MAE (0.00805). Furthermore, the RMSE and MAE for all the forecasts of the ARMA(0, 1) are smaller than the corresponding forecasts generated by the ARMA(4, 4). This suggests that the ARMA(0, 1) provides "better" forecasts for M1 than the ARMA(4, 4), with the smallest RMSE and MAE associated with the 3-month-ahead forecast. For M2, the ARMA(4, 4) yields RMSE and MAE that are smaller for all six horizons than those of the ARMA(3, 4). The "best" forecast based on the RMSE (0.0105) and MAE (0.00846) is the 1-month-ahead forecast generated by the ARMA(4, 4).

While the growth rate of the monetary aggregates is a more meaningful measure of monetary policy analysis, it is, however, interesting to compare how the alternative specifications forecast the level of each aggregate. For M2, M3, and L, the results found for the growth rates also hold for the levels. However, a different pattern emerges for M1. The ARMA(4, 4) produces lower RMSE (4.973) and MAE (4.051) for the levels of M1 than the ARMA(0, 1) [RMSE (9.132), MAE (8.527)]. Furthermore, the 1-month-ahead RMSE and MAE are less than the corresponding values for the 3-month-ahead.

In terms of the models selected by the various criteria, there was no reported difference in the forecasting ability, for either growth rates or levels, of the models chosen for M3 and L. In the case of M1, the AIC, Hannan ($C = 2$), and the goodness-of-fit-criteria-chosen model gave better forecasts of M1, while the BIC and Hannan ($C = 3$) gave better forecasts of the growth rate of M1. The model ARMA(4, 4) chosen by the goodness-of-fit criteria gave the best forecasts for the level and growth rate of M2.

4. CONCLUSIONS

In this paper we analyzed the stochastic properties of the "new" monetary aggregates within the context of ARIMA modeling and order selection criteria. The models selected by the various criteria employed gave significantly more complicated structures than those currently found in the empirical literature. In all cases, the AIC and Hannan, $C = 2$, gave autoregressive and moving-average structures equal to or greater than those given by the BIC and Hannan ($C = 3$). The "best" specification for the individual aggregates in terms of inferential and diagnostic checking tend to be given by the AIC and Hannan ($C = 2$), while the "best" specification in terms of the forecasting efficiency is given by the BIC and Hannan ($C = 3$) and/or the goodness-of-fit criterion. These results have significance in the econometric analysis on the ongoing research into the application of the rational expectations hypothesis to macroeconomic theory, which, according to this theory, states that only monetary innovations have an impact on real economic activity. The few studies that have empirically examined this problem have either employed ad hoc autoregressive rules (Mishkin, 1982) or have strongly relied on Akaike's order selection criterion (Illmakunnas and Tsurumi, 1985) to model monetary innovations. Our results show that the selection of models to decompose monetary growth is a more complex issue and that use of the models selected by the BIC and Hannan ($C = 3$) criterion should be used because of their marked superiority based on their forecasting ability. The final relevance of this study will be reflected in our ongoing research on the macroeconometrics of rational expectations models. In particular, the analysis of the restrictions that rational expectations macroeconomics impose on the structure of an ARMA specification of monetary aggregates and the impact of monetary innovations on real economic activity.

REFERENCES

Akaike, H. (1973). Information theory and an extension of the maximum likelihood principle. In *2nd International Symposium on Information Theory*, B. N. Petruv and F. Csaki, Eds.), Budapest: Akademiai Kiado, pp. 267–281.

Akaike, H. (1974). Markovian representation of stochastic processes and its application to the analysis of autoregressive moving average processes. *Ann. Inst. Statist. Math.* **26**, 363–387.

Akaike, H. (1978). Comments on "On Model Structure Testing in System Identification. *Int. J. Cont.* **27**, 323–324.

Barro, Robert J. (1977). Unanticipated money growth and unemployment in the United States. *Amer. Econ. Rev.* **67**, 101–115.

Barro, Robert J. (1978). Unanticipated money, output and the price level in the United States. *J. Pol. Econ.* **86**, 549–580.

Barro, Robert J. (1979). Unanticipated money growth and unemployment in the United States: Reply. *Amer. Econ. Rev.* **69**, 1004–1009.

Barro, Robert J. (1980). A capital market in an equilibrium business cycle model. *Econometrica* **48**, 1393–1418.

Berkman, Neil G. (1980). The new monetary aggregates: A critical appraisal. *J. Money, Credit, Bank* 135–155.

Bhansali, R. J. (1983). Estimation of the order of a moving average model from autoregressive and window estimates of the inverse correlation function. *J. Time Ser. Anal.* **4**, 137–162.

Box, G. E. P. and Jenkins, G. M. (1970). *Time Series Analysis, Forecasting and Control*, San Francisco: Holden-Day.

Box, G. E. P. and Pierce, D. A. (1970). Distribution of residual autocorrelations in autocorrelated moving average models. *J. Amer. Statist. Assoc.* **65**, 1509–1526.

Chatfield, C. and Prothero, D. L. (1973). Box–Jenkins seasonal forecasting: Problems in a case-study. *J. R. Statist. Soc. A* **136**, 295–315.

Christ, C. F. (1975). Judging the performance of econometric models of the U.S. economy. *Int. Econ. Rev.* **16**, 54–74.

Cooper, R. L. (1972). The predictive performance of quarterly econometric models of the United States. In *Econometric Models of Cyclical Behavior*, Vol. 2 (B. G. Hickman, Ed.), New York: Columbia University Press, pp. 813–926.

Froyen, Richard (1974). A test of the endogenity of monetary policy. *J. Econometrics* **8**, 175–188.

Galbraith, R. F. and Galbraith, J. I. (1974). On the inverses of some patterned matrices arising in the theory of stationary time series. *J. Appl. Prob.* **11**, 63–71.

Granger, C. W. J. and Newbold, P. (1977). *Forecasting Economic Time Series*, New York: Academic Press.

Hafer, R. W. (1980). The new monetary aggregates. *Review*, **62**, 25–32.

Hannan, E. J. (1970). *Multiple Time Series*, New York: Wiley.

Hannan, E. J. (1980). The estimation of the order of an ARMA process. *Ann. Statist.* **8**, 1071–1081.

Hannan, E. J. and Quinn, B. G. (1979). The determination of the order of an autoregression. *J. R. Statist. Soc. B* **41**, 190–195.

Hannan, E. J. and Rissanen, J. (1982). Recursive estimation of mixed autoregressive-moving average order. *Biometrika* **69**, 81–94.

Harvey, A. C. and Todd, P. H. J. (1983). Forecasting economic time series with structural and Box–Jenkins models: A case study. *J. Bus. Econ. Statist.* **1**, 299–306.

Howrey, E. P. (1978). Comments on "Estimating Structural Models of Seasonality" by Robert F. Engle. In *Seasonal Analysis of Economic Time Series*, (A. Zellner, Ed.), Washington, D.C.: Government Printing Office, pp. 298–302.

Illmakunnas, Pekka and Tsurumi, Hiroki (1985). Testing the Lucas hypothesis on output-inflation trade-offs. *J. Bus. Econ. Statist.* **3**, 43–53.

Kitagawa, G. (1977). On a search procedure for the optimal AR-MA order. *Ann. Inst. Statist. Math.* **29**, 319–332.

Leuthold, R. M., MacCormick, A. J. A., Schmitz, A. and Watts, D. G. (1970). Forecasting daily hog prices and quantities: A study of alternative forecasting techniques. *J. Amer. Statist. Assoc.* **65**, 90–107.

Levy, Mickey D. (1981). Factors affecting monetary policy in an era of inflation. *J. Monetary Econ.* **8**, 351–373.

Lutkepohl, Helmult (1984). Forecasting contemporaneously aggregated vector ARMA processes. *J. Bus. Econ. Statist.* **2**, 201–214.

Meese, Richard and Geweke, John (1984). A comparison of autoregressive univariate forecasting procedures for macroeconomic time series. *J. Bus. Econ. Statist.* **2**, 191–200.

Mishkin, Fredric S. (1982). Does anticipated monetary policy matter? An econometric investigation. *J. Pol. Econ.* **90**, 22–51.

Neftci, Salih (1982). Specification of economic time series models using Akaike's criterion. *J. Amer. Statist. Assoc.* **77**, 537–540.

Nelson, C. R. (1972). The prediction performance of the FRB-MIT-Penn model of the U.S. economy. *Amer. Econ. Rev.* **62**, 902–917.

Nelson, C. R. (1973). *Applied Time Series Analysis*, San Francisco: Holden-Day.

Newbold, P. and Granger, C. W. (1974). Experience with forecasting univariate time series and the combination of forecasts. *J. R. Statist. Soc. A* **137**, 131–146.

Niskanen, William (1978). Deficits, government spending and inflation: What is the evidence? *J. Monetary Econ.* **4**, 591–602.

Ozaki, T. (1977). On the order determination of ARIMA models. *Appl. Statist.* **26**, 290–301.

Palm, F. (1976). Testing the dynamic specification of an econometric model with an application to Belgian data. *Euro. Econ. Rev.* **8**, 269–289.

Palm, F. (1977). On univariate time series methods and simultaneous equation econometric models. *J. Econometrics* **5**, 379–388.

Paulsen, Joslein (1984). Order determination of multivariate autoregressive time series with unit roots. *J. Time Ser. Anal.* **5**, 115–127.

Potscher, B. M. (1982). Order estimation of ARMA systems by the Lagrange multiplier test. Report No. 5, Institut für Okonometrie und Operations Research, Technische Universität Wien.

Prothero, D. L. and Wallis, K. F. (1976). Modeling Macroeconomic Time Series. *J. R. Statist. Assoc. A* **139**, 468–486.

Quenouille, M. H. (1947). A large sample test for the goodness of fit of autoregressive schemes. *J. R. Statist. Soc.* **110**: 123–129.

Quinn, B. G. (1980). Order determination for multivariate autoregression. *J. R. Statist. Soc. B* **42**, 182–185.

Schwarz, G. (1978). Estimating the dimension of a model of a model. *Ann. Statist.* **6**, 461–464.

Simpson, Thomas D. (1980). The redefined monetary aggregates. *Federal Reserve Bulletin* Feb: 100.

Sims, C. A. (1980). Macroeconomics and reality. *Econometrica* **48**, 1–48.

Trivedi, P. K. (1975). Time series analysis versus structural models: A case study of Canadian manufacturing behavior. *Int. Econ. Rev.* **16**, 587–608.

Wallis, K. F. (1977). Multiple time series analysis and the final form of econometric models. *Econometrica* **45**, 1481–1497.

Wallis, K. F. (1980). Econometric implications of the rational expectations hypothesis. *Econometrica* **48**, 49–73.

Zellner, A. and Palm, F. (1974). Time series analysis and simultaneous equation econometric models (paper presented at the Oslo Econometric Society Meeting, 1972). *J. Econometrics* **2**, 17–54.

Zellner, A. and Palm, F. (1975). Time series and structural analysis of monetary models of the U.S. economy. *Sankhyā C* **37**, 12–56.

Locally Asymptotically Optimal Tests for Randomness

Marc Hallin
Institut de Statistique
Université Libre de Bruxelles
Brussels, Belgium
Madan L. Puri
Department of Mathematics
Indiana University
Bloomington, Indiana

SUMMARY

A class of linear serial statistics is introduced for the problem of testing white noise against autocorrelated alternatives. The asymptotic distribution of these statistics is given, both under the null hypothesis as well as under contiguous alternatives of ARMA dependence. It is shown that the asymptotically most powerful test statistic against a specified contiguous ARMA alternative belongs to the class considered. A rank-based version of this optimal test is also provided. Quadratic serial statistics, leading to χ^2-type tests, are also introduced for the problem of testing white noise against unspecified ARMA alternatives. The asymptotically maximin most powerful test statistic is given, and is shown to constitute a generalization of the

Research of Madan L. Puri is supported by the Office of Naval Research Contract N00014-85-K-0648.

classical Box–Pierce and Ljung–Box *portmanteau* statistics. A rank-based version of these generalized portmanteau statistics is also provided.

1. INTRODUCTION

The problem of testing for randomness—more specifically, the problem of testing for randomness against serial dependence—is certainly one of the oldest and possibly the most fundamental one in statistical inference, since independence assumption underlies the vast majority of classical inference procedures. Moreover, this independence assumption is often a crucial one, in the sense that, most of the time, whenever it is violated, classical methods cannot be applied safely any longer; much of the development of time-series methods was motivated by the fact that classical linear model methods do not work in the presence of autocorrelated error terms.

Not surprisingly, this problem of testing for randomness has been extensively investigated, from a parametric point of view (procedures based on sample autocorrelations, including, e.g., Durbin–Watson's, von Neumann's ratio, or Box–Pierce's tests) as well as from a nonparametric point of view (the run test, turning point test, serial rank correlation test, etc.).

However, in spite of the recognized importance of this problem, and in spite of its long history [the rank-based approach, for example, can be traced back as far as Hotelling and Pabst (1936) and Wald and Wolfowitz (1943)], very little is known about the power of the many existing procedures—and hence about their optimality properties.

The reason for this lack of power results is that nonindependence is rather a vast and diversified alternative: whereas there exists only one way for identically distributed observations to be independent, they can yield infinitely many distinct serial dependence structures, which makes it particularly difficult to obtain overall efficient testing procedures. Actually, it can be shown [from Hallin and Puri (1985, Proposition 4.5)] that, in the problem of testing for randomness against the whole alternative of contiguous serial dependence, the envelope power function converges to the level α as the sample size n tends to infinity: the trivial test $\phi(\mathbf{X}) \equiv \alpha$ therefore provides an asymptotically (locally) maximin most powerful α-level test for the problem.

Thus it appears hopeless to look for optimal tests against completely unspecified serial dependencies: if efficient testing procedures are to be obtained, more restricted alternatives have to be considered. A convenient way of restricting the alternative consists in introducing some appropriate, finite-dimensional, parametrization of serial dependence; the one we adopt here is provided by the class of ARMA models, the wide applicability of which is now well established.

Our objective is to provide asymptotically locally optimal tests for randomness against two types of ARMA alternatives:

1. *Specified* ARMA *alternatives* (under which the observations are generated by some ARMA model with specified coefficients, driven by a white noise with specified density), against which we shall give asymptotically locally most powerful tests.
2. *Unspecified* ARMA *alternatives* (under which the observations are generated by some ARMA (p_1, p_2) model, with $\max(p_1, p_2) \leq p$, p fixed, unspecified coefficients but specified white-noise density), for which asymptotically locally maximin most powerful tests are given.

In both cases, we shall provide *parametric and nonparametric* (rank-based) versions of the optimal tests; the word *locally* refers to the fact that *contiguous* alternatives are considered.

2. BASIC NOTATION AND TECHNICAL ASSUMPTIONS

Denote by $\{\varepsilon_t | t \in \mathbb{Z}\}$ an absolutely continuous (with respect to the Lebesgue measure) stationary white noise, that is, a sequence of independent and identically distributed (i.i.d.) random variables with density function $f_\sigma(x) = \sigma^{-1} f_1(x/\sigma)$, $\sigma > 0$. Assume that $E(\varepsilon_t) = 0$, $E(\varepsilon_t^2) = \sigma^2$, and $E(\varepsilon_t^4) < \infty$, $t \in \mathbb{Z}$ ($\mathbb{Z} = \{0, \pm 1, \pm 2, \ldots\}$). We make the following assumptions on $f_\sigma(x)$:

1. $\dot{f}_\sigma(x) = df_\sigma(x)/dx$ exists (a.e.); put $\phi_\sigma(x) = -\dot{f}_\sigma(x)/f_\sigma(x) = \sigma^{-1}\phi_1(x/\sigma)$.
2. $f_\sigma(x)$ has finite Fisher's information: $0 < I(f_\sigma) = E[\phi_\sigma^2(\varepsilon_t)] = \sigma^{-2} I(f_1) < \infty$.
3. $\dot{\phi}_\sigma(x) = d\phi_\sigma(x)/dx$ exists (a.e.) and satisfies (a.e.) a Lipschitz condition of the form $|\dot{\phi}_\sigma(x) - \dot{\phi}_\sigma(y)| \leq A_\sigma |x - y|$.
4. $f_\sigma(x)$ is *strongly unimodal* (cf. Hájek and Šidák, 1967).

We say that ε_t has a specified *density type* $f_.(x)$ when its density belongs to the scale parameter family $\{f_\sigma | \sigma > 0\}$ without σ being specified.

Consider a sequence of stochastic difference equations ($n = 1, 2, \dots$)

$$X_t^{(n)} - n^{-1/2} \sum_{i=1}^{p_1} a_i X_{t-i}^{(n)} = \varepsilon_t + n^{-1/2} \sum_{i=1}^{p_2} b_i \varepsilon_{t-i}, \qquad t \in \mathbb{Z} \tag{2.1}$$

If n is sufficiently large (so that the *stationarity condition* is satisfied), and $a_{p_1} \neq 0 \neq b_{p_2}$, then (2.1) characterizes an ARMA(p_1, p_2) process $\{X_t^{(n)} | t \in \mathbb{Z}\}$, namely, the unique stationary solution of (2.1).

Denote by $\mathbf{X}^{(n)} = (X_1^{(n)}, \dots, X_t^{(n)}, \dots, X_n^{(n)})$ an n-tuple of observations, and by $\mathbf{R}^{(n)} = (R_1^{(n)}, \dots, R_t^{(n)}, \dots, R_n^{(n)})$ the corresponding n-tuple of ranks.

3. ASYMPTOTICALLY MOST POWERFUL TESTS AGAINST SPECIFIED ALTERNATIVES

3.1. Autocorrelation Coefficients Associated with Density Types

The null hypothesis under which the observation $\mathbf{X}^{(n)} = (X_1^{(n)}, \dots, X_n^{(n)})$ constitutes a finite realization of a white-noise process with density type $f_\cdot(x)$ is denoted by $H_{0;f}^{(n)}$ whenever $f_\cdot(x)$ is specified and by $H_0^{(n)}$ otherwise.

The alternative under which $\mathbf{X}^{(n)}$ is a finite realization of some ARMA process characterized by (2.1) with specified parameters

$$\mathbf{a} = \left(a_1, a_2, \dots, a_{p_1}, \dots, a_{\max(p_1, p_2)}\right),$$

$$\mathbf{b} = \left(b_1, b_2, \dots, b_{p_2}, \dots, b_{\max(p_1, p_2)}\right), \qquad \begin{pmatrix} a_i = 0,\ i > p_1 \text{ and} \\ b_j = 0,\ j > p_2 \end{pmatrix}$$

and specified density type $f_\cdot(x)$ is denoted by $H_{\mathbf{a},\mathbf{b};f}^{(n)}$.

Denote by $(S^{(n)})^2 = (n^{-1}\sum_{t=1}^n X_t^2)$ the sample variance [note that $E(X_t^{(n)}) = 0$ both under $H_0^{(n)}$ and $H_{\mathbf{a},\mathbf{b};f}^{(n)}$], and let (for $i < n$)

$$r_{i;f}^{(n)} = (n-i)^{-1} \sum_{t=i+1}^{n} \phi_\sigma\left(X_t^{(n)}\right) X_{t-i}^{(n)} \left[\sigma^2 I(f_\sigma)\right]^{-1/2} \tag{3.1}$$

$$\hat{r}_{i;f}^{(n)} = (n-i)^{-1} \sum_{t=i+1}^{n} \phi_1\left(\frac{X_t^{(n)}}{S^{(n)}}\right) \frac{X_{t-i}^{(n)}}{S^{(n)}} \left[I(f_1)\right]^{-1/2} \tag{3.2}$$

and

$$\tilde{r}_{i;f}^{(n)} = \left[(n-i)^{-1} \sum_{t=i+1}^{n} \phi_1\left(F_1^{-1}\left(\frac{R_t^{(n)}}{n+1}\right)\right) F_1^{-1}\left(\frac{R_{t-i}^{(n)}}{n+1}\right) - m^{(n)}\right]\left[\sigma^{(n)}\right]^{-1} \tag{3.3}$$

where

$$m^{(n)} = [n(n-1)]^{-1} \sum\sum_{1 \le t_1 \ne t_2 \le n} \phi_1\left(F_1^{-1}\left(\frac{t_1}{n+1} \right)\right) F_1^{-1}\left(\frac{t_2}{n+1} \right) \quad (3.4)$$

and

$$\begin{aligned}
(\sigma^{(n)})^2 = {} & \frac{1}{n(n-1)} \sum\sum_{1 \le t_1 \ne t_2 \le n} \left[\phi_1\left(F_1^{-1}\left(\frac{t_1}{n+1} \right)\right) F_1^{-1}\left(\frac{t_2}{n+1} \right) \right]^2 \\
& + 2\frac{(n-2i)}{(n-i)} \frac{1}{n(n-1)(n-2)} \\
& \times \sum\sum\sum_{1 \le t_1 \ne t_2 \ne t_3 \le n} \left[\phi_1\left(F_1^{-1}\left(\frac{t_1}{n+1} \right)\right) \phi_1\left(F_1^{-1}\left(\frac{t_2}{n+1} \right)\right) \right. \\
& \qquad \left. \times F_1^{-1}\left(\frac{t_2}{n+1} \right) F_1^{-1}\left(\frac{t_3}{n+1} \right) \right] \\
& + \frac{n^2 - n(2i+3) + i^2 + 5i}{n(n-1)(n-2)(n-3)(n-i)} \\
& \times \sum\sum\sum\sum_{1 \le t_1 \ne t_2 \ne t_3 \ne t_4 \le n} \left[\phi_1\left(F_1^{-1}\left(\frac{t_1}{n+1} \right)\right) \phi_1\left(F_1^{-1}\left(\frac{t_2}{n+1} \right)\right) \right. \\
& \qquad \left. \times F_1^{-1}\left(\frac{t_3}{n+1} \right) F_1^{-1}\left(\frac{t_4}{n+1} \right) \right] \\
& - (n-i)(m^{(n)})^2
\end{aligned}$$

[$F_\sigma(x)$ stands for the distribution function associated with $f_\sigma(x)$.]

As we shall see (Proposition 3.1), (3.1), (3.2), and (3.3) are essentially equivalent measures of dependence (of order i). (3.1), however, is of little practical use, since it depends on σ^2, an unknown quantity in most problems; (3.2) is obtained from (3.1) by substituting $S^{(n)}$ for σ (indeed, $\sigma^2 I(f_\sigma) = I(f_1)$ and $\phi_1(x/\sigma)y/\sigma = \phi_\sigma(x)y$, $\sigma > 0$); as for (3.3), it constitutes a rank-based (hence distribution-free under $H_0^{(n)}$) version of (3.1). Note that $(n-i)^{1/2} r_{i;f}^{(n)}$ and $(n-i)^{1/2} \tilde{r}_{i;f}^{(n)}$ are exactly standardized random variables (under $H_{0;f}^{(n)}$ and $H_0^{(n)}$ respectively) [the exact variance of $\tilde{r}_{i;f}^{(n)}$ follows from Hallin et al. (1985a, Appendix 3)].

In the gaussian case, $\sigma^2 I(f_\sigma) = I(f_1) = 1$, and $\phi_1(x) = x$; $\hat{r}_{i;f}^{(n)}$ (write it $\hat{r}_{i;\mathcal{N}}^{(n)}$) thus reduces to the *sample autocorrelation of order* i: $r_i^{(n)} = \sum_{t=i+1}^{n}(X_t^{(n)}X_{t-i}^{(n)})/\Sigma(X_t^{(n)})^2$. It is well known that optimal inference procedures for gaussian series can be based [at least asymptotically, cf. Anderson (1971) or Hallin et al. (1985a, Section 5)] on sample autocorrelations $r_i^{(n)} = \hat{r}_{i;\mathcal{N}}^{(n)} \simeq \tilde{r}_{i;\mathcal{N}}^{(n)}$, and remain asymptotically valid for nongaussian densities, in which case, however, their optimality properties are lost. Our objective is to show that optimality can be maintained, provided that $r_{i;f}^{(n)}$, $\hat{r}_{i;f}^{(n)}$, or $\tilde{r}_{i;f}^{(n)}$ are substituted for the usual sample autocorrelations $r_i^{(n)}$.

The prominent role ascribed to sample autocorrelations in testing for randomness (as well as in most of time-series analysis) thus actually belongs to the $\hat{r}_{i;f}^{(n)}$, to which we therefore shall refer as the *autocorrelation coefficient of order i associated with the density type* $f_{\cdot}(x)$—shortly, the *f-autocorrelation of order i*. Accordingly, we call $\tilde{r}_{i;f}^{(n)}$ the *rank autocorrelation coefficient of order i associated with the density type* $f_{\cdot}(x)$—shortly, the *f-rank autocorrelation of order i*. Note that *f*-rank autocorrelations never reduce to rank autocorrelations in the usual (Spearman) sense. Also note that $\hat{r}_{i;f}^{(n)}$ and $\tilde{r}_{i;f}^{(n)}$ are not *stricto sensu* correlation or autocorrelation coefficients (except for gaussian densities). To be precise, they are standardized [up to $(n-i)^{1/2}$ factor] statistics asymptotically equivalent with the *cross*-correlation coefficient of order i between $\phi_\sigma(X_t^{(n)})$ and $X_t^{(n)}$, and they reduce to an autocorrelation coefficient if and only if f is gaussian. The term *f-autocorrelation* coefficient is adopted because of its intuitive appeal.

3.2. Asymptotic Properties of *f*-Autocorrelations

We say that two random variables X_n and Y_n are asymptotically equivalent (as $n \to \infty$) if their difference converges in probability to zero.

Proposition 3.1. The statistics $n^{1/2}r_{i;f}^{(n)}$, $n^{1/2}\hat{r}_{i;f}^{(n)}$ and $n^{1/2}\tilde{r}_{i;f}^{(n)}$

1. are asymptotically equivalent (under both $H_{0;f}^{(n)}$ and $H_{\mathbf{a},\mathbf{b};f}^{(n)}$),
2. are asymptotically normal, with mean 0 and variance 1 under $H_0^{(n)}$, and with mean

$$\begin{cases} (\mathbf{a}+\mathbf{b})_i\sigma^2[I(f_\sigma)]^{1/2} & \text{for } i \le p \\ 0 & \text{for } i > p \end{cases}$$

   and variance 1 under $H_{\mathbf{a},\mathbf{b};f}^{(n)}$,
3. are asymptotically independent of $n^{1/2}r_{j;f}^{(n)}$, $n^{1/2}\hat{r}_{j;f}^{(n)}$ and $n^{1/2}\tilde{r}_{j;f}^{(n)}$, respectively $j \neq i$ under both $H_0^{(n)}$ and $H_{\mathbf{a},\mathbf{b};f}^{(n)}$.

Proof. 1. *Asymptotic distribution of* $n^{1/2} r_{i;f}^{(n)}$. Let $\mathbf{Y}_t^{(n)} = (X_t^{(n)} \cdots X_{t-q}^{(n)})'$, with $q = \max(p_1, p_2, i)$. Being q-dependent, $\{\mathbf{Y}_t^{(n)} | t \in \mathbb{Z}\}$ constitutes an *absolutely regular process* (cf., e.g., Ibragimov and Rozanov, 1978) under $H_0^{(n)}$. The log-likelihood ratio $\log L^{(n)}$ for $H_{0;f}^{(n)}$ against $H_{\mathbf{a},\mathbf{b};f}^{(n)}$ can be written (cf. Hallin et al., 1985a, Proposition 3.1) as [put $\|\mathbf{a} + \mathbf{b}\|^2 = \sum_{j=1}^{p} (\mathbf{a} + \mathbf{b})_j^2$]

$$\log L^{(n)} = n^{-1/2} \sum_{t=p+1}^{n} \phi_\sigma\left(X_t^{(n)}\right) \sum_{j=1}^{p} (\mathbf{a} + \mathbf{b})_j X_{t-j}^{(n)} - \tfrac{1}{2}\|\mathbf{a} + \mathbf{b}\|^2 + o_p(1)$$

$$= n^{-1/2} \mathscr{L}_{0;f}^{(n)} - \tfrac{1}{2}\|\mathbf{a} + \mathbf{b}\|^2 + o_p(1)$$

An arbitrary linear combination of $r_{i;f}^{(n)}$ and $\mathscr{L}_{0;f}^{(n)}$ therefore takes the form [up to $o_p(n^{-1/2})$ terms] of a differentiable von Mises's statistical functional (with respect to the observed series $(\mathbf{Y}_q^{(n)}, \ldots, \mathbf{Y}_n^{(n)})$. Applying Yoshihara (1976)'s central-limit theorem, we obtain that $n^{1/2} r_{i;f}^{(n)}$ and $\log L^{(n)}$ admit a joint asymptotic normal distribution (under $H_{0;f}^{(n)}$) with mean $(0, -\tfrac{1}{2}\|\mathbf{a} + \mathbf{b}\|^2 \sigma^2 I(f_\sigma))$ and covariance matrix

$$\begin{bmatrix} 1 & \begin{cases} (\mathbf{a} + \mathbf{b})_i \left[\sigma^2 I(f_\sigma)\right]^{1/2} & \text{if } i \le p \\ 0 & \text{if } i > p \end{cases} \\ \cdot & \|\mathbf{a} + \mathbf{b}\|^2 \sigma^2 I(f_\sigma) \end{bmatrix} \tag{3.5}$$

The desired result then follows from LeCam's third lemma.

2. *Joint asymptotic normality and asymptotic independence.* The joint asymptotic normality of $n^{1/2} r_{i;f}^{(n)}$ and $n^{1/2} r_{j;f}^{(n)}$, $i \neq j$, can be established as in 1, and their asymptotic covariance matrix can be shown to be diagonal. Note, however, that $r_{i;f}^{(n)}$ and $r_{j;f}^{(n)}$ are exactly uncorrelated under $H_{o;f}^{(n)}$ for any value of n, since $E[\phi_\sigma(X_{t_1}^{(n)}) X_{t_1-i}^{(n)} \phi_\sigma(X_{t_2}^{(n)}) X_{t_2-j}^{(n)}] = 0$ for every $n, t_1, t_2, 0 \neq i \neq j \neq 0$.

3. *Asymptotic equivalence of* $n^{1/2} r_{i;f}^{(n)}$ *and* $n^{1/2} \tilde{r}_{i;f}^{(n)}$. This equivalence straightforwardly follows from Hallin et al. (1985a, Section 4.1), under $H_{0;f}^{(n)}$.

4. *Asymptotic equivalence of* $n^{1/2} \hat{r}_{i;f}^{(n)}$ *and* $n^{1/2} r_{i;f}^{(n)}$. Put $T^{(n)}(\mathbf{X}^{(n)}; \theta) = (n-i)^{-1} \sum_{t=i+1}^{n} \phi_1(\theta X_t^{(n)}) X_{t-i}^{(n)}$. Then $n^{1/2} r_{i;f}^{(n)} = \sigma^{-1} n^{1/2} T^{(n)}(\mathbf{X}^{(n)}; \sigma^{-1})$ and $n^{1/2} \hat{r}_{i;f}^{(n)} = (S^{(n)})^{-1} T^{(n)}(\mathbf{X}^{(n)}; (S^{(n)})^{-1})$. Since $S^{(n)}$ is a consistent estimator of σ, we just have to show that $\Delta^{(n)} = n^{1/2}[T^{(n)}(\mathbf{X}^{(n)}; (S^{(n)})^{-1}) - T^{(n)}(\mathbf{X}^{(n)}; \sigma^{-1})]$ is $o_p(1)$. Since $T^{(n)}(\mathbf{X}^{(n)}; \theta)$ is a differentiably smooth function of θ at $\theta = \sigma^{-1}$, we have $\Delta^{(n)} = n^{1/2}((S^{(n)})^{-1} - \sigma^{-1}) \partial_\theta T^{(n)} (\mathbf{X}^{(n)}; \theta^*)$, where $\theta^* = \theta^*(\mathbf{X}^{(n)})$ is some (random) intermediate point be-

tween $(S^{(n)})^{-1}$ and σ^{-1}. Now

$$\partial_\theta T^{(n)}(\mathbf{X}^{(n)}; \theta^*) - \partial_\theta T^{(n)}(\mathbf{X}^{(n)}; \sigma^{-1})$$

$$= (n-i)^{-1} \sum_{t=i+1}^{n} \left[\left(\dot{\phi}_1\left(\theta^* X_t^{(n)}\right) - \dot{\phi}_1\left(\sigma^{-1} X_t^{(n)}\right)\right) X_t^{(n)} X_{t-i}^{(n)}\right] \quad (3.6)$$

Under $H_0^{(n)}$, the terms of this latter sum are mutually uncorrelated, and the variance of (3.6) is

$$(n-i)^{-1} E\left\{\left[\dot{\phi}_1\left(\theta^* X_2^{(n)}\right) - \dot{\phi}_1\left(\sigma^{-1} X_2^{(n)}\right)\right]^2 \left(X_2^{(n)}\right)^2 \left(X_1^{(n)}\right)^2\right\}$$

$$\leq (n-i)^{-1} A^2 E\left\{\left[(S^{(n)})^{-1} - \sigma^{-1}\right]^2 \cdot \left(X_2^{(n)}\right)^4 \left(X_1^{(n)}\right)^2\right\} = o_p(n^{-1})$$

Hence $\Delta^{(n)} = n^{1/2}((S^{(n)})^{-1} - \sigma^{-1}) \cdot (n-i)^{-1} \Sigma \dot{\phi}_1(\sigma^{-1} X_t^{(n)}) X_t^{(n)} X_{t-i}^{(n)} + o_p(1)$; it then follows from Yoshihara (1976)'s and Slutzky's theorems that $\Delta^{(n)}$ converges to zero in probability, which completes the proof.

3.3. Asymptotically Most Powerful Test

Denote by $\Phi(x)$ and $k_{1-\alpha}$, respectively, the distribution function and $(1-\alpha)$ quantile of the standard normal distribution. It then follows from Proposition 3.1 that the three tests

$$\text{reject } H_0^{(n)} \text{ if } \sum_{i=1}^{p} (n-i)^{1/2} \frac{(\mathbf{a}+\mathbf{b})_i}{\|\mathbf{a}+\mathbf{b}\|} r_{i;f}^{(n)} > k_{1-\alpha} \quad (3.7)$$

$$\text{reject } H_0^{(n)} \text{ if } \sum_{i=1}^{p} (n-i)^{1/2} \frac{(\mathbf{a}+\mathbf{b})_i}{\|\mathbf{a}+\mathbf{b}\|} \hat{r}_{i;f}^{(n)} > k_{1-\alpha} \quad (3.8)$$

$$\text{reject } H_0^{(n)} \text{ if } \sum_{i=1}^{p} (n-i)^{1/2} \frac{(\mathbf{a}+\mathbf{b})_i}{\|\mathbf{a}+\mathbf{b}\|} \tilde{r}_{i;f}^{(n)} > k_{1-\alpha} \quad (3.9)$$

have asymptotic probability level α.

Proposition 3.2. The tests (3.7), (3.8), and (3.9) are asymptotically most powerful against $H_{\mathbf{a},\mathbf{b};f}^{(n)}$ (within the class of all tests of asymptotic level α). Their asymptotic power is $1 - \Phi(k_{1-\alpha} - \|\mathbf{a}+\mathbf{b}\|(\sigma^2 I(f_\sigma))^{1/2})$.

Proof. Since the three test statistics are asymptotically equivalent, let us consider (3.7) only. Since under $H_{0;f}^{(n)}$, $\log L^{(n)}$ is asymptotically normal with mean $-\frac{1}{2}\|\mathbf{a}+\mathbf{b}\|^2\sigma^2 I(f_\sigma)$ and variance $\|\mathbf{a}+\mathbf{b}\|^2\sigma^2 I(f_\sigma)$, an asymptotically most powerful test against $H^{(n)}_{\mathbf{a},\mathbf{b};f}$ is

$$\text{reject } H_0^{(n)} \text{ if } \frac{\log L^{(n)}}{\|\mathbf{a}+\mathbf{b}\|\left(\sigma^2 I(f_\sigma)\right)^{1/2}} + \frac{1}{2} > k_{1-\alpha} \tag{3.10}$$

Now it follows from Proposition 3.1 (3) and the form of the autocovariance matrix (3.5) that

$$\sum_{i=1}^{p} (n-i)^{1/2} \frac{(\mathbf{a}+\mathbf{b})_i}{\|\mathbf{a}+\mathbf{b}\|} r_{i;f}^{(n)} - \frac{\log L^{(n)}}{\|\mathbf{a}+\mathbf{b}\|\left(\sigma^2 I(f_\sigma)\right)^{1/2}} - \frac{1}{2} = o_p(1)$$

(3.10), (3.7), (3.8), and (3.9) are thus asymptotically equivalent tests. Their common asymptotic power straightforwardly follows from Proposition 3.1.

REMARK 3.1. The tests (3.7), (3.8), and (3.9) actually are *uniformly* asymptotically most powerful against alternatives of the form

$$U\left\{H^{(n)}_{\boldsymbol{\alpha},\boldsymbol{\beta};f} \,\middle|\, \frac{\boldsymbol{\alpha}+\boldsymbol{\beta}}{\|\boldsymbol{\alpha}+\boldsymbol{\beta}\|} = \frac{\mathbf{a}+\mathbf{b}}{\|\mathbf{a}+\mathbf{b}\|}\right\}$$

Sequences of ARMA models (2.1) associated with coefficients $(\boldsymbol{\alpha}, \boldsymbol{\beta})$ and $(\mathbf{a}, \mathbf{b})$ such that $\boldsymbol{\alpha}+\boldsymbol{\beta} = \mathbf{a}+\mathbf{b}$ are, indeed, asymptotically observationally equivalent, in the sense that the corresponding likelihood functions are asymptotically equivalent, and the whole problem is scale-invariant. As a consequence, Proposition 3.2 provides uniformly asymptotically locally most powerful tests against unilateral alternatives of positive (negative) first-order dependence [AR(1), MA(1), or ARMA(1, 1)]. As another consequence, the sequences of ARMA models associated with coefficients $(\mathbf{a}, \mathbf{b})$ such that $\mathbf{a}+\mathbf{b} = \mathbf{0}$ are asymptotically observationally equivalent with white noise in which case the asymptotic power of the optimal test reduces to its size α.

REMARK 3.2. For n sufficiently large, (2.1) always constitutes a sequence of stationary invertible ARMA models. The tests in Proposition 3.2 thus should be interpreted as optimal against sequences of ARMA processes whose *stationary invertible representations* admit coefficients proportional to $n^{-1/2}(\mathbf{a}+\mathbf{b})$; their corresponding noninvertible or nonstationary representations—if any—*cannot* be of the form (2.1).

4. ASYMPTOTICALLY MAXIMIN MOST POWERFUL TESTS AGAINST UNSPECIFIED ALTERNATIVES

4.1. Asymptotically Maximin Most Powerful Tests and Asymptotically Sufficient Statistics

The asymptotically most powerful tests in Section 3 are perhaps of theoretical rather than practical interest, since the alternative in practice is seldom specified. A uniformly asymptotically most powerful test procedure being impossible (cf. Hallin et al., 1985b, c)—except for the case of unilateral first-order alternatives [such as positive AR(1) or positive MA(1) dependence]—a weaker form of asymptotic optimality has to be considered.

Before establishing the existence of asymptotically locally maximin most powerful tests and providing their explicit form, we first recall some basic definitions.

Denote by $\beta(\alpha; H^{(n)}, K^{(n)})$ the envelope power function for some null hypothesis $H^{(n)} = \{l_0^{(n)}\}$ against some alternative hypothesis $K^{(n)} = \{l_1^{(n)}\}$:

$$\beta(\alpha; H^{(n)}, K^{(n)}) = \sup_{\psi} \inf_{l_1^{(n)} \in K^{(n)}} E_{l_1^{(n)}}[\psi],$$

where the $\sup_{\psi}$ is taken with respect to all the α-level tests, $\alpha \in (0,1)$. A sequence of tests $\psi^{(n)*}$ is said to be *asymptotically maximin* for (the sequence of tests) $H^{(n)}$ against $K^{(n)}$ if

$$\begin{cases} \liminf_{n\to\infty} \left[E_{l_1^{(n)}}\left[\psi^{(n)*}\right] - \beta\left(\alpha; H^{(n)}, K^{(n)}\right)\right] \geq 0, & l_1^{(n)} \in K^{(n)} \\ \limsup_{n\to\infty} E_{l_0^{(n)}}\left[\psi^{(n)*}\right] \leq \alpha, & l_0^{(n)} \in H^{(n)} \end{cases}$$

Another concept that plays a fundamental role in the asymptotic theory of tests is that of asymptotically sufficient statistics, introduced by LeCam (1960) and studied by Hájek and Šidák (1967, Chapter 7). Consider again the problem of testing $H^{(n)}$ against $K^{(n)}$. Denote by $\mathbf{T}^{(n)}$ a sequence of statistics, and by $H_T^{(n)}$ and $K_T^{(n)}$, respectively, the hypotheses induced by $\mathbf{T}^{(n)}$ (i.e., the families of distributions $\{l_0^{(n)T}\}$ and $\{l_1^{(n)T}\}$ induced from $\{l_0^{(n)}\}$ and $\{l_1^{(n)}\}$ by $\mathbf{T}^{(n)}$). Let $\beta(\alpha; H_T^{(n)}, K_T^{(n)})$ stand for the corresponding envelope power function. If

$$\lim_{n\to\infty} \left[\beta\left(\alpha; H^{(n)}, K^{(n)}\right) - \beta\left(\alpha; H_T^{(n)}, K_T^{(n)}\right)\right] = 0, \qquad \alpha \in (0,1)$$

then $\mathbf{T}^{(n)}$ is said to be *asymptotically sufficient* for $H^{(n)}$ against $K^{(n)}$. This definition immediately implies that any asymptotically maximin most powerful test for $H_T^{(n)}$ against $K_T^{(n)}$ is also asymptotically maximin most powerful for $H^{(n)}$ against $K^{(n)}$. One can thus safely restrict to $\mathbf{T}^{(n)}$-measurable tests in the problem of obtaining asymptotically maximin most powerful tests for $H^{(n)}$ against $K^{(n)}$.

4.2. Unspecified ARMA Alternatives

An asymptotically maximin most powerful test for the null hypothesis of white noise (specified *or* unspecified density type $f.$) against, say, the whole family of ARMA(p_1, p_2) dependencies (2.1) [specified maximal order p such that $\max(p_1, p_2) \leq p$; specified density type $f.$ but unspecified coefficients a_i and b_i] is clearly the trivial test $\phi \equiv \alpha$. If we allow, indeed, arbitrarily small coefficients a_i and b_i in (2.1), the alternative (for fixed n) contains distributions that are arbitrarily close to the null hypothesis.

If asymptotically maximin most powerful tests are to be obtained, the alternative has thus to be bounded away from the null hypothesis. Let us call unspecified ARMA alternative of order p an alternative of the form

$$K^{(n)}(f.|p, d) = U\left\{ H_{\mathbf{a},\mathbf{b};f}^{(n)} | \mathbf{a}, \mathbf{b} \in \mathbb{R}^p, \|\mathbf{a} + \mathbf{b}\| \geq d \right\}$$

($\|\mathbf{a} + \mathbf{b}\|$ denotes the euclidean norm; d is an arbitrary strictly positive constant.) The following result then follows as a particular case of Hallin and Puri (1985, Proposition 3.2).

Proposition 4.1. Any of the three vectors of f-autocorrelations $(r_{1;f}^{(n)}, \ldots, r_{p;f}^{(n)})'$, $(\hat{r}_{1;f}^{(n)}, \ldots, \hat{r}_{p;f}^{(n)})'$ or $(\tilde{r}_{1;f}^{(n)}, \ldots, \tilde{r}_{p;f}^{(n)})$ constitutes an asymptotically sufficient statistic for $H_0^{(n)}$ against $K^{(n)}(f.|p, d)$, for any $d > 0$.

4.3. Optimal Portmanteau Tests

According to the definitions given in Section 4.1, asymptotically (locally) maximin most powerful tests can be based on f-autocorrelations. It follows easily from Proposition 3.1 (2) and (3), an invariance argument, and the Hunt–Stein theorem that this test can be based on either of the quadratic statistics $Q_{p;f}^{(n)*} = \Sigma_{i=1}^p (n-i)(r_{i;f}^{(n)})^2$, $\hat{Q}_{p;f}^{(n)*} = \Sigma_{i=1}^p (n-i)(\hat{r}_{i;f}^{(n)})^2$ or $\tilde{Q}_{p;f}^{(n)*} = \Sigma_{i=1}^p (n-i)(\tilde{r}_{i;f}^{(n)})^2$. Because of their close similarity with the usual Box–Pierce (1970) or Ljung–Box (1978) *portmanteau statistics*, we call

$Q_{p;f}^{(n)*}$ and $\hat{Q}_{p;f}^{(n)*}$ *f-portmanteau statistics*, and $\tilde{Q}_{p;f}^{(n)}$ an *f-rank portmanteau statistic* (of order p).

We then have the following result.

Proposition 4.2. The (asymptotically equivalent under $H_{0;f}^{(n)}$) portmanteau statistics $Q_{p;f}^{(n)*}$, $\hat{Q}_{p;f}^{(n)*}$, and $\tilde{Q}_{p;f}^{(n)}$

1. are asymptotically χ^2 with p degrees of freedom under $H_0^{(n)}$, and asymptotically noncentral χ^2, with p degrees of freedom and noncentrality parameter $\lambda_f^*(\mathbf{a}, \mathbf{b}) = (1/2)\|\mathbf{a} + \mathbf{b}\|^2 I(f_1)$ under $H_{\mathbf{a},\mathbf{b};f}^{(n)}$ (with $\mathbf{a}, \mathbf{b} \in \mathbb{R}^p$).
2. provide an asymptotically locally maximin most powerful test for randomness against the unspecified ARMA alternative $K^{(n)}(f.|p, d)$ (for any positive d):

$$\text{reject } H_0^{(n)} \quad \text{if } Q_{p;f}^{(n)*} \left(\text{or } \hat{Q}_{p;f}^{(n)*} \text{ or } \tilde{Q}_{p;f}^{(n)*}\right) > \chi_{1-\alpha}^2$$

[where $\chi_{1-\alpha}^2$ is the $(1 - \alpha)$-fractile of the χ^2 distribution with p degrees of freedom]. The ARE of these tests with respect to the usual Box–Pierce or Ljung–Box portmanteau tests, based on sample autocorrelations, is $I(f_1) \geq 1$ (cf. Hallin et al., 1985b).

ACKNOWLEDGMENT

We would like to thank the referee for a number of useful comments and suggestions.

REFERENCES

Anderson, T. W. (1971). *The Statistical Analysis of Time Series*. New York: Wiley.

Box, G. E. P. and Pierce, D. A. (1970). Distribution of residual autocorrelations in autoregressive-integrated moving average time series models. *J. Amer. Statist. Assoc.* **65**, 1509–1526.

Hájek, J. and Šidák, Z. (1967). *Theory of Rank Tests*. Orlando, Florida: Academic Press.

Hallin, M., Ingenbleek, J.-Fr. and Puri, M. L. (1985a). Linear serial rank tests for randomness against ARMA alternatives. *Ann. Statist* **13**, 1156–1181.

Hallin, M., Ingenbleek, J.-Fr. and Puri, M. L. (1985b). Tests de rangs quadratiques pour une hypothèse de bruit blanc. *C. R. Acad. Sc. Paris* Série I **301**, 935–938.

Hallin, M., Ingenbleek, J.-Fr. and Puri, M. L. (1985c). Linear and quadratic serial rank tests for randomness. *J. Time Ser. Anal.* (to appear).

Hallin, M. and Puri, M. L. (1985). Optimal rank-based procedures for time series analysis: testing an ARMA model against other ARMA models. Technical Report, Department of Mathematics, Indiana University.

Hotelling, H. and Pabst, M. R. (1936). Rank correlation and tests of significance involving no assumption of normality. *Ann. Math. Statist.* **7**, 29–43.

Ibragimov, I. and Rozanov, Y. (1978). *Gaussian Random Processes*. Berlin and New York: Springer-Verlag.

Le Cam, L. (1960). Locally asymptotically normal families of distributions. *Univ. Calif. Publ. Statist.* **3**, 27–90.

Ljung, G. and Box, G. E. P. (1978). On a measure of lack of fit in time series models. *Biometrika* **65**, 297–303.

Wald, A. and Wolfowitz, J. (1943). An exact test for randomness in the nonparametric case based on serial correlation. *Ann. Math. Statist.* **14**, 378–388.

Yoshihara, K. (1976). Limiting behavior of *U*-statistics for stationary absolutely regular processes. *Z. Wahrsch. Geb.* **35**, 237–252.

Estimation of Missing Values in Time Series Adjusted by Autoregressive Models

Luis J. López Martín
Departamento de Estadística y Econometría
Facultad Ciencias Económicas
Universidad de la Laguna
Canary Islands, Spain

Miguel Sánchez García
Departamento de Estadística e Investigación Operativa
Facultad de Matemáticas
Universidad de la Laguna
Canary Islands, Spain

SUMMARY

A sample of two temporal series lacking some data in at least one of them have been taken. We assume that both series adjust to autoregressive models of order *p* and *q*, respectively. Moreover, we assume the existence of correlation between both series. We pose and solve a model to estimate the missing values.

1. INTRODUCTION

The amount of information that an individual receives every day is enormous. This fact together with the development of the technical means available today, for storing and keeping that information, provokes the creation of data banks to file and preserve it. The future usefulness of that information is undeniable.

Nevertheless, it frequently happens that we do not have part of the data about the observed phenomena during a period of time. The reason for this lack of data lies either with the fact that they were not observed before or that the data disappear with the passing of time. One of the problems that is currently of great interest is the estimation of missing data in order to complete the series. These estimations can be done under the hypothesis that the missing observed data follow a particular model.

The calculation of the unknown values was usually done either by estimating the parameters of the model, and evaluating the missing data, or, in the case of data of temporal series, by estimating the spectral density function of the corresponding process. With regard to this, we can mention the papers of Jones (1962, 1971), Parzen (1963), Dunsmuir and Robinson (1981), and Bloomfield (1970). In the time domain, we will mention Andel (1976).

Brown and Kadiyala (1983) estimated data values adjusted to a linear model, considering the model as well as the correlation with other data. Afterwards, the authors generalized the result of Brown and Kadiyala (1983) (Sánchez G. and López M., 1984; López M. and Sánchez García, 1985).

In this paper, we develop a method to estimate the missing values of two temporal series that are adjusted by autoregressive models.

2. NOTATION

The notation we shall use is the following:

Y	the data of a temporal series
Z	the data of another series
σ	the weight to construct the objective function
β, γ	parameters of the autoregressive models
$\mathbf{E}$	except when something different is said, a column vector with all its elements equal to 1 (its dimension is appropriate to the specific case)

3. THE MODEL

Suppose that we have two unidimensional temporal series Y and Z that have been observed during a given period of time, denoted by the numbers

$1, 2, 3, \ldots, N$; also, assume that certain registers are unknown, though not simultaneously for both series. These assumptions will allow us to make a partition of the two data series in three groups (Y_1, Z_1), (Y_2, Z_2), (Y_3, Z_3), where Y_1, Z_1, Y_2 and Z_3 are known and Z_2, Y_3 are unknown.

If we suppose that Y and Z conform to an autoregressive model of order p and q, respectively, and that there exists correlation between both series, our problem can be stated in the following terms:

Find Z_2, Y_3, $\boldsymbol{\beta}$, $\boldsymbol{\gamma}$, a, and b in order to minimize the following objective function:

$$(1-\sigma)\left[\sum_{j=p+1}^{N}\left(y_j - \sum_{i=1}^{p}\beta_i y_{j-i}\right)^2 + \sum_{j=q+1}^{N}\left(z_j - \sum_{l=1}^{q}\gamma_l z_{j-l}\right)^2\right] + \sigma\sum_{j=1}^{N}(y_j - a - bz_j)^2 \tag{3.1}$$

The complete solution to this problem is quite difficult. In this paper, we try to solve it by iterative methods.

4. INITIAL ESTIMATIONS

Let us consider, separately, the three functions which enter the objective function:

$$\sum_{j=p+1}^{N}\left(y_j - \sum_{i=1}^{p}\beta_i y_{j-i}\right)^2 \tag{4.1}$$

$$\sum_{j=q+1}^{N}\left(z_j - \sum_{l=1}^{q}\gamma_l z_{j-l}\right)^2 \tag{4.2}$$

$$\sum_{j=1}^{N}(y_j - a - bz_j)^2 \tag{4.3}$$

and we are trying, as an initial solution, to find the values of Z_2, Y_3, $\boldsymbol{\beta}$, $\boldsymbol{\gamma}$, a, and b that minimize them. Obviously, there will be two values for Z_2 (Z_2^1 and Z_2^2) and two others for Y_3 (Y_3^1 and Y_3^2).

To find the solution to these problems, set:

$$J = \{j/y_j, y_{j-1}, \ldots, y_{j-p} \in \{Y_1 \cup Y_2\}\} = \{j_1, j_2, \ldots, j_{N_1}\}$$

$$L = \{l/z_l, z_{l-1}, \ldots, z_{l-q} \in \{Z_1 \cup Z_3\}\} = \{l_1, l_2, \ldots, l_{N_2}\}$$

$$K = \{k/y_k \in Y_1, z_k \in Z_1\} = \{k_1, k_2, \ldots, k_{N_3}\}$$

With this new notation, and supposing that $\mathbf{Y}_{0J}^{\mathrm{T}}\mathbf{Y}_{0J}$ is nonsingular and that $\mathbf{Z}_{0L}^{\mathrm{T}}\mathbf{Z}_{0L}$ is also nonsingular, we have the following results:

$$\hat{\boldsymbol{\beta}} = \left(\mathbf{Y}_{0J}^{\mathrm{T}}\mathbf{Y}_{0J}\right)^{-1}\mathbf{Y}_{0J}^{\mathrm{T}}\mathbf{Y}_J, \qquad \hat{\boldsymbol{\gamma}} = \left(\mathbf{Z}_{0L}^{\mathrm{T}}\mathbf{Z}_{0L}\right)^{-1}\mathbf{Z}_{0L}^{\mathrm{T}}\mathbf{Z}_L,$$

$$\bar{y} = \frac{1}{|K|}\sum_{k\in K} y_k \quad \bar{z} = \frac{1}{|K|}\sum_{k\in K} z_k,$$

$$\hat{b} = \sum_{k\in K}(y_k - \bar{y})(z_k - \bar{z}) \Big/ \sum_{k\in K}(z_k - \bar{z})^2, \qquad \hat{a} = \bar{y} - \hat{b}\bar{z}$$

where

$$\mathbf{Y}_j = \left(y_{j_1}, y_{j_2}, \ldots, y_{j_{N_1}}\right)^{\mathrm{T}}, \qquad \mathbf{Z}_L = \left(z_{l_1}, z_{l_2}, \ldots, z_{l_{N_2}}\right)^{\mathrm{T}}$$

$$\mathbf{Y}_{0J} = \left(y_{j_i-1}, y_{j_i-2}, \ldots, y_{j_i-p}\right), \qquad \text{matrix } N_1 \times p \text{ taking } i = 1, 2, \ldots, N_1$$

$$\mathbf{Z}_{0L} = \left(z_{l_i-1}, y_{l_i-2}, \ldots, z_{l_i-q}\right), \qquad \text{matrix } N_2 \times q \text{ taking } i = 1, 2, \ldots, N_2$$

The values of Y_3^2 and Z_2^2 are calculated through the expressions:

$$z_j^2 = \frac{1}{\hat{b}}(y_j - \hat{a}) \quad \text{and} \quad y_j^2 = \hat{a} + \hat{b}z_j$$

The values of Y_3^1 are calculated with the following algorithm:

Step 1. Choose $y_j \in Y_3$, such that $k \in J$ with $j = k + 1$ or $j = k - p + 1$. If $j = k + 1$, we take $y_j^1 = \sum_{i=1}^{p} \hat{\beta}_i y_{j-i}$; on the contrary, under the other assumption, we take

$$y_j^1 = \frac{1}{\hat{\beta}_p}\left(y_{j+p} - \sum_{i=1}^{p-1} \beta_i y_{j+p-i}\right)$$

Step 2. Place $Y_3 = Y_3 - \{y_j\}$; $J = J - \{j\}$. If $Y_3 = \{\phi\}$ we stop; in other case, we return to step 1.

With a similar algorithm to the previous one, we are able to calculate the values of Z_2^1.

Obviously $\hat{\beta}_p \neq 0$, since, in another case, the model will not be autoregressive of order p.

The previous data have the following properties:

Theorem 3.1

$$\forall \boldsymbol{\beta}, \forall Y_3 \sum_{j=p+1}^{N} \left(y_j - \sum_{i=1}^{p} \beta_i y_{j-i} \right)^2 \geq \sum_{j=p+1}^{N} \left(y_j^1 - \sum_{i=1}^{p} \hat{\beta}_i y_{j-i}^1 \right)^2 \tag{4.4}$$

$$\forall \boldsymbol{\gamma}, \forall Z_2 \sum_{j=q+1}^{N} \left(z_j - \sum_{i=1}^{q} \gamma_i z_{j-i} \right)^2 \geq \sum_{j=q+1}^{N} \left(z_j^1 - \sum_{i=1}^{q} \hat{\gamma}_i z_{j-i}^1 \right)^2 \tag{4.5}$$

$$\forall a, b, Z_2, Y_3 \sum_{k=1}^{N} (y_k - a - bz_k)^2 \geq \sum_{k=1}^{N} (y_k^2 - \hat{a} - \hat{b} z_k^2)^2 \tag{4.6}$$

Proof. We shall examine only (4.4) since the others are analogous:

$$\sum_{j=p+1}^{N} \left(y_j - \sum_{i=1}^{p} \beta_i y_{j-i} \right)^2 \geq \sum_{j \in J} \left(y_j - \sum_{i=1}^{p} \beta_i y_{j-1} \right)^2 \geq \sum_{j \in J} \left(y_j - \sum_{i=1}^{p} \hat{\beta}_i y_{j-i} \right)^2$$

$$= \sum_{j=p+1}^{N} \left(y_j^1 - \sum_{i=1}^{p} \hat{\beta}_i y_{j-i}^1 \right)^2$$

Taking $Z_2 = (1 - \sigma)Z_2^1 + \sigma Z_2^2$, $Y_3 = (1 - \sigma)Y_3^1 + \sigma Y_3^2$; we estimate, again, $\hat{\boldsymbol{\beta}}$, $\hat{\boldsymbol{\gamma}}$, $\hat{a}$, and $\hat{b}$ minimizing (4.1), (4.2) and (4.3); going afterwards to the iterative algorithm (Section 5).

5. ALGORITHM OF ESTIMATION

Let $\hat{\boldsymbol{\beta}}$, $\hat{\boldsymbol{\gamma}}$, $\hat{a}$, and $\hat{b}$ be the estimated parameters in Section 4. The objective function (3.1) can be arranged in matrix form:

$$(1 - \sigma)(\mathbf{Y}^{\mathrm{T}}\mathbf{B}^{\mathrm{T}}\mathbf{B}\mathbf{Y} + \mathbf{Z}^{\mathrm{T}}\mathbf{C}^{\mathrm{T}}\mathbf{C}\mathbf{Z}) + \sigma(\mathbf{Y} - a\mathbf{E} - b\mathbf{Z})^2 \tag{5.1}$$

where **Y** and **Z** are column vectors and

$$\mathbf{B} = \begin{bmatrix} \beta_p & \beta_{p-1} & \cdots & \beta_1 & -1 & 0 & 0 & 0 & \cdots & 0 & 0 \\ 0 & \beta_p & \cdots & \beta_2 & \beta_1 & -1 & 0 & 0 & \cdots & 0 & 0 \\ \vdots & \vdots & & \vdots & \vdots & \vdots & \vdots & \vdots & & \vdots & \vdots \\ 0 & 0 & \cdots & 0 & 0 & 0 & 0 & \beta_p & \cdots & \beta_1 & -1 \end{bmatrix} \tag{5.2}$$

with the same meaning for **C** with respect to γ and q.

Step 1. Using $\hat{\boldsymbol{\beta}}$, $\hat{\boldsymbol{\gamma}}$, $\hat{a}$, and $\hat{b}$, we find $\mathbf{Y}_3$ and $\mathbf{Z}_2$ that minimize (5.1). Taking the derivative with respect to $\mathbf{Y}_3$ and $\mathbf{Z}_2$, we have

$$(1-\sigma)(2\mathbf{B}_{31}\mathbf{Y}_3^* + 2\mathbf{B}_{33}\mathbf{Y}_3) + 2\sigma(\mathbf{Y}_3 - \hat{a}\mathbf{E}_3 - \hat{b}\mathbf{Z}_3) = 0$$

$$(1-\sigma)(2\mathbf{C}_{21}\mathbf{Z}_2^* + 2\mathbf{C}_{22}\mathbf{Z}_2) - 2\hat{b}\sigma(\mathbf{Y}_2 - \hat{a}\mathbf{E}_2 - \hat{b}\mathbf{Z}_2) = 0 \quad (5.3)$$

where $\mathbf{B}_{33}$ is the submatrix of $\mathbf{B}$ corresponding to the variables of $\mathbf{Y}_3$; $\mathbf{Y}_3^*$ the observations that are not in $\mathbf{Y}_3$; with the same meaning for $\mathbf{Z}_2^*$ and $\mathbf{C}_{22}$. With the appropriate changes in the formulas (5.3) we have

$$\hat{\mathbf{Y}}_3 = \left((1-\sigma)\mathbf{B}_{33} + \sigma\mathbf{I}\right)^{-1}\left(\hat{b}\mathbf{Z}_3 + \hat{a}\mathbf{E}_3 - \mathbf{B}_{31}\mathbf{Y}_3^*\right)$$

$$\hat{\mathbf{Z}}_2 = \left((1-\sigma)\mathbf{C}_{22} + \hat{b}^2\mathbf{I}\right)^{-1}\left(\hat{b}\mathbf{Y}_2 - \hat{b}\hat{a}\mathbf{E}_2 - \mathbf{C}_{21}\mathbf{Z}_2^*\right)$$

Step 2. With these values we return to the objective function (3.1) and we calculate $\hat{\boldsymbol{\beta}}$, $\hat{\boldsymbol{\gamma}}$, $\hat{a}$, and $\hat{b}$ using the expressions

$$\hat{\boldsymbol{\beta}} = \left(\mathbf{Y}_0^{\mathrm{T}}\mathbf{Y}_0\right)^{-1}\mathbf{Y}_0^{\mathrm{T}}\mathbf{Y}, \qquad \hat{\boldsymbol{\gamma}} = \left(\mathbf{Z}_0^{\mathrm{T}}\mathbf{Z}_0\right)^{-1}\mathbf{Z}_0^{\mathrm{T}}\mathbf{Z},$$

$$\bar{y} = \frac{1}{N}\sum_{i=1}^{N} y_i \qquad \bar{z} = \frac{1}{N}\sum_{i=1}^{N} z_i,$$

$$\hat{a} = \bar{y} - \hat{b}\bar{z}, \qquad \hat{b} = \sum_{i=1}^{N}(y_i - \bar{y})(z_i - \bar{z}) \Big/ \sum_{i=1}^{N}(z_i - \bar{z})^2$$

where $\mathbf{Y} = (y_{p+1}, y_{p+2}, \ldots, y_N)^{\mathrm{T}}$, $\mathbf{Y}_0 = (y_j, y_{j-1}, \ldots, y_{j-p+1})$ a matrix $(N-p)\times p$, $j = p, \ldots, N-1$; with the same meaning for $\mathbf{Z}$ and $\mathbf{Z}_0$.

Obviously, the previous algorithm is convergent since the value of the objective function in each step (i.e., with the calculation of new coefficients) diminishes.

It is also obvious that if in Section 4 $Z_2^1 = Z_2^2$ and $Y_3^1 = Y_3^2$, then the values of Y_3 and Z_2 calculated in step 1 of this section verify $Y_3 = Y_3^1$ and $Z_2 = Z_2^1$ and, therefore, the algorithm is convergent to a unique step.

We can start with a value for σ, estimated as the correlation coefficient of the data (Y_1, Z_1). Once the convergency is reached, we can substitute the initial σ for a new correlation coefficient between (Y, Z) for the total data; and we can start again.

REFERENCES

Andel, J. (1976). *Multiple Autoregressive Processes*. Prague: Charles University.

Bloomfield, P. (1970). Spectral analysis with randomly missing observations. *J. R. Statist. Soc. B* **32**, 369–380.

Brown, K. C. and Kadiyala, K. R. (1983). Construction of economic index numbers with an incomplete set of data. *Rev. Econ. Statist.* **3**, 520–524.

Dunsmuir, W. and Robinson, P. M. (1981). Parametric estimators for stationary time series with missing observations. *Adv. Appl. Probab.* **13**, 129–146.

Harvey, A. C. and Pierse, R. G. (1984). Estimating missing observations in economic time series. *J. Amer. Statist. Assoc.* **79**, 125–131.

Jones, R. H. (1962). Spectral analysis with regularly missed observations. *Ann. Math. Statist.* **33**, 455–461.

Jones, R. H. (1971). Spectrum estimation with missing observations. *Ann. Inst. Statist. Math.* **23**, 387–398.

López M., L. J. and Sánchez García, M. (1985). Modelo para estimar valores desconocidos en series de datos. *Actas X Jornadas Hispano-Lusas de Matemáticas*, 278–284.

Parzen, E. (1963). On spectral analysis with missing observations and amplitude modulation. *Sankhya A* **25**, 383–392.

Sánchez G., M. and López M., L. J. (1985). Estimación de registros desconocidos en series de datos. *Trabajos Estadíst. Investigación Oper.* **36**, 259–268.

Measuring the Importance of Outliers in ARIMA Models

Daniel Peña
Departamento de Estadistica
E.T.S. Ingenieros Industriales
Universidad Politécnica de Madrid
Madrid, Spain

SUMMARY

In this chapter we present a measure for the influence of each observation on the estimated parameter values of an ARIMA model, that is based on the their standardized change when these parameters are estimated assuming that the observation is missing. It is argued that this measure represents a natural generalization to dependent data of influence measures previously suggested for independent observations.

1. INTRODUCTION

Real data almost always contain some kind of heterogeneity or contamination, and the study and detection of outliers has been a topic of steady statistical research (Barnett and Lewis, 1978). Only recently, however, has it been recognized that (1) some outliers have very small influence on the estimated parameters of the model, and (2) data points accepted as normal with the standard outlier tests can have strong influence on estimation. The class of influential observations, includes, therefore, not only the outliers that really matter, but other heterogeneous points as well.

The measures of influence observations so far developed really on the study of the change in the model when (1) one observation (or subset) is deleted and (2) one observation is assumed to have more variance (less weight) than the others. The first approach is usually called *global influence*, and the second, *local influence*. Although both aspects are closely related there are clear differences between them.

The deleted approach has been justified on several grounds: first, as sensible distance of the empirical sample influence curve (Cook and Weisberg, 1982); second, as a generalized likelihood distance (Cook and Weisberg, 1982; Cook et al., 1984); third, as Kullblack–Leibler distance between either the predictive distributions (Johnson and Geisser, 1982), or the posterior distributions (Pettit and Smith, 1983). However, the limitations of this approach is that it cannot easily be extended to dependent data, where deleting one observation change the structural pattern observed.

In this chapter we suggest the following general procedure to build global influence measures: treat each observation (or subset) as a missing value, compute $\hat{\boldsymbol{\theta}}$, the maximum likelihood estimator of the parameter using the whole data set, and $\hat{\boldsymbol{\theta}}_{(i)}$ the estimator assuming that the observation i is missing; then, the distance between $\hat{\boldsymbol{\theta}}$ and $\hat{\boldsymbol{\theta}}_{(i)}$ in some adequate metric will be a general measure of influence.

The missing-value treatment seems to be the sensible way to proceed when the observations are dependent: it leads to substitute the piece of data by its forecast using the rest of the sample. This same idea can be extended in a straightforward way to other statistical procedures, such as jackknife or cross-validation methods, that are based on the deleting approach.

2. SOME INFLUENCE MEASURES

We assume the linear statistical model for the observed series y_t:

$$y_t = \mu + \mathbf{b}'\mathbf{x}_t + \boldsymbol{\phi}'\mathbf{Y}_{t-1} + \boldsymbol{\theta}'\mathbf{a}_{t-1} + a_t \qquad (2.1)$$

where μ is a global level parameter, $\mathbf{b}$, $\boldsymbol{\phi}$, and $\boldsymbol{\theta}$ are vector parameters, $\mathbf{x}_t$ is a vector of explanatory variables, $\mathbf{Y}_{t-1}$ a vector of lagged values of the response, and $\mathbf{a}_{t-1}$ a vector of normal independent and identically distributed (i.i.d.) $(0, \sigma^2)$ variables. Model (2.1) includes as particular cases the linear model, autoregressive moving-average time-series models, and transfer function dynamic models. Furthermore, it can easily be generalized for y_t being a vector, but in what follows we assume it is a scalar.

Model (2.1) includes a level parameter μ, a scale parameter σ^2, and a vector of structural parameters $\boldsymbol{\beta}' = (\mathbf{b}'\boldsymbol{\phi}'\boldsymbol{\theta}')$.

Let $\hat{\mu}$, $\hat{\boldsymbol{\beta}}$ be the ML estimators of these parameters; $\hat{\sigma}_\mu^2$ and $\hat{\Sigma}_\beta$ the variance and covariance matrix of these estimators; $\hat{\mu}_{(i)}$, $\hat{\boldsymbol{\beta}}_{(i)}$ the ML estimators treating the ith observation as missing; $\hat{\sigma}^2$ the usual unbiased estimator of the variance σ^2; and $\hat{\sigma}_{(i)}^2$ this same estimator when the ith observation is missing. Then, we define

1. The influence on the global level

$$D_L(i) = \frac{(\hat{\mu} - \hat{\mu}_{(i)})^2}{\hat{\sigma}_\mu^2}$$

2. The influence on the structural parameters

$$D_S(i) = (\hat{\boldsymbol{\beta}} - \hat{\boldsymbol{\beta}}_{(i)})'\hat{\Sigma}_\beta(\hat{\boldsymbol{\beta}} - \hat{\boldsymbol{\beta}}_{(i)})$$

this statistic can be decomposed to separate the effects on the components of $\boldsymbol{\beta}$.

3. The influence on the scale, given by

$$D_V(i) = \frac{\hat{\sigma}^2 - \hat{\sigma}_{(i)}^2}{\hat{\sigma}_{(i)}^2}$$

It is easy to see (Peña, 1985) that the application of these definitions to regression models leads to statistics similar to those proposed previously. For instance, the statistic suggested by Cook (1977) is the addition of D_S and D_L, whereas DFFITS, advocated by Belsley et al. (1980), is a mixture of the three kinds of effects. Finally, the measure of influence on the scale parameters is equivalent to the likelihood ratio test for outliers in linear models (Barnett and Lewis, 1978).

In the next section we will apply these ideas to ARIMA time-series models.

3. TIME-SERIES MODELS

Suppose a time series y_t that follows a univariate ARIMA(p, d, q) process

$$\phi_p(B)\nabla^d(y_t - \mu) = \theta_q(B)a_t \tag{3.1}$$

where B is the backshift operator, $\nabla = 1 - B$, $\phi_p(B)$ and $\theta_q(B)$ are autoregressive and moving-average polynomials with roots outside the unit circle, and μ is some constant that will be the mean if the series is stationary ($d = 0$). The model can be written in the autoregression representation

$$\pi(B)y_t = \pi(1)\mu + a_t$$

where $\pi(B) = \phi(B)\nabla^d\theta_q^{-1}(B)$. Assuming that the process is nonstationary, $d > 1$, then $\pi(1) = 0$, and for being invertible, the π weights will become practically zero after some lag h.

To measure the influence of observation y_i on the estimated parameters, π, we face the problem that y_i is, at the same time, a dependent and independent variable.

We would need an estimator of the parameter vector, $\hat{\pi}_{(i)}$, computed free of any effect from the ith observation—that is, as if the y_i observation were missing—but without disrupting the dynamic structure of the series. It is clear that deleting the observation is not, in this case, a sensible way to proceed, and that we have to take into account the dependence among the data to obtain such estimate.

There are two basic approaches to compute the estimator, $\hat{\pi}_{(i)}$, in time-series data: the first is intervention analysis (Box and Tiao, 1975), that produces a "smoothed" value for the missing point, as well as an estimated parameter vector; the second, is to build the likelihood and use the Kalman filter to "smooth" the missing value (Jones, 1980; Harvey and Pierse, 1984; Ansley and Kohn, 1984). In Section 3.1 we shall compare both approaches.

3.1. Missing Values and Intervention Analysis

Intervention analysis is a procedure developped by Box and Tiao (1975) to evaluate the effect of exogeneous interventions on a given response variable with dynamic noise structure. Suppose a given series y_t with univariate ARIMA model

$$y_t = \Psi(B)a_t$$

Then the effect of an intervention at time $t = i$ can be estimated by the parameter w in the model

$$y_t = wI_t^{(i)} + \Psi(B)a_t$$

where $I_t^{(i)}$ is an impulse variable defined as

$$I_t^{(i)} = \begin{cases} 0, & t \neq i \\ 1, & t = i \end{cases}$$

and has the same role than a dummy variable in regression models.

To study this estimator, let us assume that the noise structure is autoregressive of order p. Then, calling $\phi(B)$ the AR polynomial, the model can be written

$$\phi(B)y_t = w_0\phi(B)I_t^{(i)} + a_t$$

That is, calling $\boldsymbol{\phi}' = (\phi_1, \ldots, \phi_p)$

$$y_t = \boldsymbol{\phi}'\mathbf{Y}_{t-1} + w\phi(B)\mathbf{I}_t^{(i)} + \mathbf{a}_t$$

where $\mathbf{Y}_{t-1} = (y_{t-1}, \ldots, y_{t-p})$. Then, the likelihood function for the observed series y_t, conditional to the first p values, is

$$L(\phi, \sigma^2, w) = -\frac{n-p}{2}\ln \sigma^2 - \sum_{s_1} \frac{(y_t - \boldsymbol{\phi}'\mathbf{Y}_{t-1})^2}{2\sigma^2}$$

$$-\sum_{0}^{p} \frac{(y_{i+l} + \phi_l w - \boldsymbol{\phi}'\mathbf{Y}_{i+l-1})^2}{2\sigma^2}$$

where $\phi_0 = -1$ and s_1 is the set $(p+1, \ldots, i-1, i+p+1, \ldots, n)$. The conditional maximum likelihood estimates of ϕ and w are

$$\hat{w} = y_i - \hat{y}_{i/n} \tag{3.2}$$

$$\hat{y}_{i/n} = \sum_{j=1}^{p} c_j(y_{i+j} + y_{i-j}) \tag{3.3}$$

where the coefficients c_j are given by

$$c_j = \frac{\hat{\phi}_j^{(i)} - \sum_1^p \hat{\phi}_l^{(i)} \hat{\phi}_{l+j}^{(i)}}{\sum_0^p \hat{\phi}_l(i)^2} \tag{3.4}$$

The estimator parameters are given by the Yule–Walker equations

$$\hat{\boldsymbol{\phi}}_{(i)} = \mathbf{R}_p^{-1}(i)\mathbf{A}_p(i)$$

where $\mathbf{R}_p(i)$ is the $p \times p$ matrix of autocorrelation coefficients and $\mathbf{A}_p(i)$ is the vector of autocorrelation coefficients from lag 1 to p, computed using $\hat{y}_{i/n}$ instead of y_i as value at point i.

These estimates can be interpreted as follows: (1) the intervention $\hat{w}$ is the difference between the observed value and its smoothed estimator using all the sample information: (2) the estimated parameters are the result of substituting the y_i value by its smoothed value $\hat{y}_{i/n}$.

The smoothed value, $\hat{y}_{i/n}$, summarizes all the information about the assumed missing value. For instance, suppose that $p = 1$, then two unbiased estimators of y_i are available: ϕy_{i-1} and $y_{i+1}\phi^{-1}$, with variances σ^2 and $\sigma^2\phi^{-2}$.

The minimum variance estimator using both estimators is

$$\hat{y}_{i/n} = \left(\sigma^{-2} + \phi^2\sigma^{-2}\right)^{-1}\left(\sigma^{-2}\phi y_{i-1} + \sigma^{-2}\phi y_{i+1}\right)$$

$$= \left(1 + \phi^2\right)^{-1}\phi\left(y_{i+1} + y_{i-1}\right)$$

that is, $\hat{y}_{i/n}$, the intervention analysis estimate. This same result can be easily obtained for any AR(p) process: the smoothed intervention analysis estimator is a linear combination of the $2p$ unbiased estimators that can be built, with weights equal to their relative precision. This analysis points out that intervention analysis is really carrying out the following estimation procedure: (1) treat observation y_i as missing, and substitute it by its estimation, $\hat{y}_{i/n}$, using some initial parameter values; (2) estimate the parameters by maximum likelihood with $y_i = \hat{y}_{i/n}$; (3) compute a new $\hat{y}_{i/n}$ value, with the new estimated parameters and iterate points (2) and (3) until convergence.

3.2. A Measure of Structural Influence

Let $\hat{\boldsymbol{\beta}}$ be the maximum likelihood estimator of the parameter vector of the ARIMA model. These ML estimators are asympotically normally distributed with covariance matrix equal to the inverse of the information matrix (see Fuller, 1976). Let $\hat{\boldsymbol{\beta}}_{(i)}$ be the estimator computed within the intervention model. Then

$$D_s(i) = \frac{\left(\hat{\boldsymbol{\beta}} - \hat{\boldsymbol{\beta}}_{(i)}\right)'\mathbf{I}(\hat{\beta})\left(\boldsymbol{\beta} - \hat{\boldsymbol{\beta}}_{(i)}\right)}{\sigma^2}$$

where $\sigma^{-2}\mathbf{I}(\hat{\boldsymbol{\beta}})$ is the observed Fisher information matrix. For a purely autoregressive process, this reduces to

$$D_s(i) = \frac{\|\hat{\mathbf{Y}}_t - \hat{\mathbf{Y}}_t\left(\hat{\phi}_{(i)}\right)\|^2}{\hat{\sigma}^2}$$

where $\hat{\mathbf{Y}}_t(\hat{\phi}_{(i)})$ is the one step ahead forecast of y_t using the parameter values $\phi_{(i)}$. Calling $\hat{\mathbf{Y}}_t(i)$ the forecast wth the intervention model and $\mathbf{1}' = (1, 1, \ldots, 1)$, then

$$\hat{\mathbf{Y}}_t(\hat{\phi}_{(i)}) = \hat{\mathbf{Y}}_{t(i)} - \mathbf{1}\hat{w}\hat{\phi}_{(i)}(B)I_t^{(i)} \tag{3.5}$$

Now, let $\hat{a}_{t(i)}$ be the intervention models residual and $\hat{a}_t$ the standard residual, then, calling $\phi_0 = -1$, $D_s(i)$ can be written

$$D_s(i) = \frac{\Sigma(\hat{a}_t - \hat{a}_{t(i)})^2}{\hat{\sigma}^2} + \hat{w}^2\frac{\Sigma_0^p\hat{\phi}_{l(i)}^2}{\hat{\sigma}^2} + 2\frac{\hat{w}}{\hat{\sigma}}\frac{\Sigma_0^p\hat{\phi}_l(\hat{a}_{i+l}(i) - \hat{a}_{i+l})}{\hat{\sigma}}$$

The second term is the square of the likelihood ratio test for observation y_i being an additive outlier (see Chang and Tiao, 1983). Calling $\lambda_a(i)$ this statistic and

$$e_t(i) = \hat{a}_t - \hat{a}_t(i)$$

and assuming that $e_t^2(i) \simeq 0$ for $t \neq i, i+1, \ldots, i+1$, then

$$D_s(i) = \lambda_a^2(i) + \sigma^{-2}\sum_{l=0}^{p} e_{i+l}(i)(e_{i+l}(i) - 2\hat{w}\hat{\phi}_l(i))$$

Therefore, to be influential the observation must be an additive outlier, having large $\lambda_a(i)$. However, observations with the same likelihood ratio value for additive outliers can have different influence, depending on the particular values of the observations after the point in which the additive outlier happened.

The computation of $D_s(i)$ requires the nonlinear estimation of the intervention. An approximation suggested by Peña (1984) is to use

$$D_s(i) = \frac{\hat{w}_0^2(i)}{\hat{\sigma}^2}\mathbf{E}'(i)(\mathbf{X}'\mathbf{X})^{-1}\mathbf{E}(i)$$

Where $\hat{w}_0(i)$ is computed for all the observations using in (3.2) the same parameters $\hat{\phi}$, $\mathbf{E}(i)$ is a vector of "pseudoresiduals" based on the p forward and backward residuals, and $\sigma^{-2}(\mathbf{X}'\mathbf{X})$ is the observed information matrix.

3.3. Use of the Kalman Filter to Compute the Estimators Assuming Missing Data

The previous analysis can also be carried out using the state space representation of the ARIMA model and the Kalman filter to estimate the missing values (see Priestley, 1981; Harvey, 1981).

There are some computational advantages in using the Kalman filter to compute the likelihood function when some observations are missing via the prediction error decomposition. A smoothed value of the missing observation can also be computed using a fixed internal algorithm, based on a set of backwards recursions starting from the end of the sample (see Harvey and Pierse, 1984).

For instance, for an AR(1) process the likelihood function computed by prediction error decomposition assuming that y_i is missing is

$$L(\sigma^2, \phi) = -\frac{n-2}{2}\ln\sigma^2 - \tfrac{1}{2}\ln(1+\phi^2) - \sum_{\substack{t\neq i\\ t\neq i+1}} \frac{(y_t - \phi y_{t-1})^2}{2\sigma^2}$$

$$-\left(2\sigma^2(1+\phi^2)\right)^{-1}\left(y_{i+1} - \phi^2 y_{i-1}\right)^2$$

and for the intervention analysis model, concentrating the parameter w, the likelihood is

$$L(\sigma^2, \phi) = -\frac{n-1}{2}\ln\sigma^2$$

$$-\sum_{\substack{t\neq i\\ t\neq i+1}} \frac{(y_t - \phi y_{t-1})^2}{2\sigma^2} - \frac{(\hat{y}_{i/n} - y_{i-1})^2}{2\sigma^2} - \frac{(y_{i+1} - \hat{y}_{i/n})^2}{2\sigma^2}$$

and using that

$$\hat{y}_{i/n} = \frac{\phi(y_{i+1} + y_{i-1})}{1+\phi^2}$$

both functions will be equal, apart from the term $\ln(1+\phi^2)$. If the sample is large, this term can be safely ignored, and the intervention analysis and missing value likelihoods, will be equivalent.

4. TRANSFER FUNCTION MODELS

The general principles outlined can be extended to transfer function models. Let

$$y_t = v(B)x_t + \psi(B)a_t$$

a transfer function or dynamic regression model. Let $\mathbf{v}$ be the parameters of the transfer function and $\boldsymbol{\psi}$ those of the noise structure, and write $\boldsymbol{\beta}' = (\mathbf{v}'\boldsymbol{\psi}')$. We may be now interested on the influence of y_i, x_i, or both. The influence of y_i will be obtained estimating the model

$$y_t = wI_t^{(i)} + v(B)x_t + \psi(B)a_t$$

calling $\boldsymbol{\beta}'_{(i)} = (\mathbf{v}'_{(i)}\boldsymbol{\psi}'_{(i)})$, the effect of observation i on the structural parameters is

$$D_s(i) = (\hat{\boldsymbol{\beta}} - \hat{\boldsymbol{\beta}}_{(i)})'\mathbf{I}(\boldsymbol{\beta})(\hat{\boldsymbol{\beta}} - \hat{\boldsymbol{\beta}}_{(i)})/\hat{\sigma}^2$$

where, as before, $\hat{\sigma}^{-2}\mathbf{I}(\hat{\boldsymbol{\beta}})$ will be the observed information matrix.

In many cases, the effects on the transfer function and on the noise structure would be very different, and it is interesting to separate both. The effect on the transfer function is

$$D_T(i) = (\hat{\mathbf{V}} - \hat{\mathbf{V}}_{(i)})'\mathbf{I}(\hat{V})(\hat{\mathbf{V}} - \hat{\mathbf{V}}_{(i)})/\hat{\sigma}^2$$

and on the noise structure

$$D_N(i) = (\hat{\boldsymbol{\Psi}} - \hat{\boldsymbol{\Psi}}_{(i)})'\mathbf{I}(\hat{\boldsymbol{\Psi}})(\hat{\boldsymbol{\Psi}} - \hat{\boldsymbol{\Psi}}_{(i)})'/\hat{\sigma}^2$$

Other measures of influence could be built following the same guidelines.

ACKNOWLEDGMENT

The research has been sponsored by the Comisión Asesora de Investigación Cientifica y Técnica, MEC, Spain. The author thanks Agustin Maravall for helpful comments.

REFERENCES

Ansley, C. F. and Kohn, R. (1984). Exact likelihood of vector autoregressive-moving average process with missing or aggregated data. *Biometrika* **70**, 275–278.

Barnett, V. and Lewis, T. (1978). *Outliers in Statistical Data*. New York: Wiley.

Belsley, D. A., Kuh, E. and Welsch, R. E. (1980). *Regression Diagnostics*. New York: Wiley.

Box, G. E. P. and Tiao, G. C. (1975). Intervention analysis with Applications to Economic and Environmental Problems. *J. Amer. Statist. Assoc.* **70**, 70–79.

Chang, I. and Tiao, G. C. (1983). Estimation of time series parameters in the presence of outliers. Technical Report No. 8. University of Chicago, Graduate School of Business.

Cook, R. D. (1977). Detection of influential observations in linear regression. *Technometrics* **19**, 15–18.

Cook, R. D. and Weisberg, S. (1982). *Residuals and influence in Regression*. London: Chapman and Hall.

Cook, R. D., Peña D., and Weisberg S. (1984). *The Likelihood displacement: a unifying principle for influence meausures*. Mathematics Research Center Technical Report No. 2751. University of Wisconsin-Madison.

Fuller, W. A. (1976). *Introduction to Statistical Time Series*. New York: Wiley.

Harvey, A. C. (1981). *Time Series Models*. London: Philip Allan.

Harvey, A. C. and Pierse, R. G. (1984). Estimating missing observations in economic time series. *J. Amer. Statist. Assoc.* **79**, 125–131.

Johnson, W. and Geisser, S. (1982). A predictive view of the detection and characterization of influential observation in regression analysis. *J. Amer. Statist. Assoc.* **78**, 137–143.

Jones, R. H. (1980). Maximum likelihood fitting of ARIMA models to time series with missing observations. *Technometrics* **22**, 339–395.

Peña, D. (1984). *Influential observations in time series*. Mathematics Research Center Technical Report No. 2718. University of Wisconsin-Madison.

Peña, D. (1985). Influential observations in statistical models: A generalized approach. *Proc. 25th Anniversary Conf. Dept. of Statistics*. University of Wisconsin-Madison.

Pettit, L. L. and Smith, A. F. (1983). Outliers and influential observations in linear models. *Second Valencia International Meeting on Bayesian Statistics* (in press).

Priestley, M. B. (1981). *Spectral Analysis and Time Series*, Vol. 2, London: Academic Press.

New Developments in Time-Series Analysis

M. B. Priestley
Department of Mathematics
The University of Manchester
Institute of Science and Technology
Manchester, England

SUMMARY

In this paper we describe some recent developments in the area of nonlinear time-series models, with a brief-summary of the basic theory of linear models.

1. INTRODUCTION

Traditional methods of time-series analysis rest heavily on the twin assumptions of stationarity and linearity. *Stationarity* means that the data are generated by a stochastic process whose probabilistic properties are invariant over time, whereas *linearity* means that the current observation can be expressed as a linear combination of present and past values of a white-noise process (a sequence of independent random variables). Within this framework the most popular class of models is that given by the so-called ARMA (mixed autoregressive-moving-average) processes, first introduced by Box and Jenkins (1970), and this type of model has provided the basis for much of the traditional model-fitting methodology. However, during the past five years or so interest has shifted to the relatively unexplored territory of nonlinear modeling, and a number of nonlinear models have been developed. Experience so far gained indicates that these nonlinear models offer the analyst much greater scope than the more conventional linear models, and can reproduce a rich variety of structures —including some typically associated with the study of nonlinear differential equations, such as "limit cycles" behavior.

2. BASIC THEORY OF LINEAR MODELS

The fundamental "building block" in all forms of time-series modeling is that of the white-noise process, which we denote by $\{\varepsilon_t\}$. In this context a white-noise process is defined as a sequence of zero-mean random variables, with constant variance, σ_ε^2, that is, $E\{\varepsilon_t\} = 0$, $E\{\varepsilon_t^2\} = \sigma_\varepsilon^2$, $E\{\varepsilon_s \varepsilon_t\} = 0$, $s \neq t$.

Given a general discrete parameter time series $\{X_t | t = 0, \pm 1, \pm 2, \ldots\}$, a "model" is specified by finding that function of future, present, and past observations that transforms the given series into a white-noise process. Thus, we seek a function h which is such that

$$h(\ldots, X_{t-2}, X_{t-1}, X_t, X_{t+1}, X_{t+2}, \ldots) = \varepsilon_t \tag{2.1}$$

In particular, the class of *linear* models corresponds to the case where the function h is *linear*, so that (2.1) then takes the form,

$$\sum_{u=-\infty}^{\infty} h_u X_{t-u} = \varepsilon_t \tag{2.2}$$

This is the general form of the nonlinear model, but it is "anticipative," that is, X_t is allowed to depend on future values as well as on past values. In practice we would usually assume that X_t depends only on past values, that is, we would set $h_u = 0$, $u < 0$, and rewrite (2.2) in the "one-sided" form as

$$\sum_{u=0}^{\infty} h_u X_{t-u} = \varepsilon_t \tag{2.3}$$

Introducing the backward shift operator B defined by $BX_t = X_{t-1}$, etc., (2.3) can now be expressed in the form,

$$H(B) X_t = \varepsilon_t \tag{2.4}$$

where

$$H(z) = \sum_{u=0}^{\infty} h_u z^u \tag{2.5}$$

Equation (2.4) may be "solved" to express X_t as a linear function of present and past values of ε_t. Thus, inverting (2.4) formally we may write

$$X_t = H^{-1}(B).\varepsilon_t \tag{2.6}$$

and if $H^{-1}(z)$ can be expanded as a convergent series for $|z| < 1$ [i.e., if $H(z)$ has no zeros inside the unit circle] we may write further

$$H^{-1}(z) = g_0 + g_1 z + g_2 z^2 + \cdots \tag{2.7}$$

We may now write X_t explicitly

$$X_t = \sum_{u=0}^{\infty} g_u \varepsilon_{t-u} \tag{2.8}$$

or

$$X_t = \Gamma(B).\varepsilon_t \tag{2.9}$$

where

$$\Gamma(z) = \sum_{u=0}^{\infty} g_u z^u \tag{2.10}$$

Note that $\Gamma(z) \equiv H^{-1}(z)$, or $\Gamma(z)H(z) \equiv 1$.

Equation (2.8) gives us an alternative formulation for the general (non-anticipative) linear model in which X_t is expressed as a linear combination of present and past values of a white noise process, $\{\varepsilon_t\}$.

2.1. AR, MA, and ARMA Models

The general linear model, as given by (2.3) or (2.8), involves an indefinite number of parameters and cannot therefore be fitted directly to data unless further assumptions are made. In the context of practical time-series model fitting we use a standard set of finite parameter models that arise as special cases of the general linear model (2.8). Thus, if we assume that $\Gamma^{-1}(z)$ $(\equiv H(z))$ may be approximated by a finite-order polynomial, say,

$$\Gamma^{-1}(z) = 1 + \alpha_1 z + \cdots + \alpha_k z^k \tag{2.11}$$

then (2.8) reduces to the AR(k) (*autoregressive model of order* k) form

$$X_t + \alpha_1 X_{t-1} + \cdots + \alpha_k X_{t-k} = \varepsilon_t \tag{2.12}$$

On the other hand, if we approximate to $\Gamma(z)$ by a finite-order polynomial, say,

$$\Gamma_1(z) = 1 + \beta_1 z + \cdots + \beta_l z^l \tag{2.13}$$

then (2.8) reduces to the MA(l) (*moving average of order* l) form

$$X_t = \varepsilon_t + \beta_1 \varepsilon_{t-1} + \cdots + \beta_l \varepsilon_{t-l} \tag{2.14}$$

If $\Gamma(z)$ and $\Gamma^{-1}(z)$ are "well-behaved" functions we can, to an arbitrary degree of accuracy, approximate either by a finite polynomial of sufficiently high degree. It follows, therefore, that we can, in general, approximate a general linear model by either an AR or MA model of sufficiently high order. However, we can obtain a more parsimonious finite-parameter representation of $\Gamma(z)$ by approximating it by a *rational function*. If we take $\Gamma(z)$ to have the form

$$\Gamma(z) = \frac{1 + \beta_1 z + \cdots + \beta_l z^l}{1 + \alpha_1 z + \cdots + \alpha_k z^k} \tag{2.15}$$

then (2.8) reduces to the ARMA(k, l) [*mixed autoregressive-moving-average of order* (k, l)] model

$$X_t + \alpha_1 X_{t-1} + \cdots + \alpha_k X_{t-k} = \varepsilon_t + \beta_1 \varepsilon_{t-1} + \cdots + \beta_l \varepsilon_{t-l} \tag{2.16}$$

This model can be written in operator form (with an obvious notation)

$$\alpha(B) X_t = \beta(B) \varepsilon_t \tag{2.17}$$

[The slightly more general ARIMA models which have been studied, in particular, by Box and Jenkins (1970), correspond to the case where the operator $\alpha(B)$ in (2.17) contains a factor of the form $(1 - B)^d$. In this case $\Gamma(z) = [\beta(z)/\alpha(z)]$ contains a dth-order pole at $z = 1$ and consequently $\{X_t\}$ is no longer a stationary process—although its dth difference, $\Delta^d X_t$, would, in general, be stationary.]

2.2. State-Space Representations

There is an interesting way of rewriting the general ARMA model (2.16) using the so-called *state-space*, (or *Markovian*) representation, which provides a compact description of any finite-parameter linear model. The basic idea rests simply on the well-known result that any finite-order linear differential or difference equation can be expressed as a vector first-order equation. For example, if we take the AR(2) model,

$$X_t + \alpha_1 X_{t-1} + \alpha_2 X_{t-2} = \varepsilon_t \tag{2.18}$$

and write $x_t^{(2)} = X_t$, $x_t^{(1)} = -\alpha_2 X_{t-1}(= -\alpha_2 X_{t-1}^{(2)})$, then (2.18) may be rewritten

$$\begin{bmatrix} x_t^{(1)} \\ x_t^{(2)} \end{bmatrix} = \begin{bmatrix} 0 & -\alpha_2 \\ 1 & -\alpha_1 \end{bmatrix} \begin{bmatrix} x_{t-1}^{(1)} \\ x_{t-1}^{(2)} \end{bmatrix} + \begin{bmatrix} 0 \\ 1 \end{bmatrix} \varepsilon_t \tag{2.19}$$

To recover X_t from the vector $(x_t^{(1)}, x_t^{(2)})$, we now write,

$$X_t = (0, 1) \begin{pmatrix} x_t^{(1)} \\ x_t^{(2)} \end{pmatrix} \tag{2.20}$$

The pair of equations (2.19) and (2.20) are completely equivalent to (2.18), but whereas (2.18) involves a two-stage dependence (so that X_t is non-Markovian), (2.19) involves only a one-stage dependence (so that $(x_t^{(1)}, x_t^{(2)})'$ is a vector Markov process). [The device of expressing (2.18) in the form (2.19) may be viewed alternatively as a special case of the technique of writing certain types of non-Markov processes involving only finite stage dependence as vector Markov processes.]

The same approach may be used to write the general ARMA model in the state-space form. Thus, assuming that $l \le k - 1$ in (2.16), we first rewrite (2.16) as

$$X_t + \alpha_1 X_{t-1} + \cdots + \alpha_k X_{t-k} = \varepsilon_t + \beta_1 \varepsilon_{t-1} + \cdots + \beta_{k-1} \varepsilon_{t-k+1} \tag{2.21}$$

(where $\beta_j = 0$, $j > l$), and it is then easy to verify that (2.21) can be rewritten in state space form

$$\mathbf{x}(t+1) = \boldsymbol{\theta}\mathbf{x}(t) + \boldsymbol{\gamma}\varepsilon_t \tag{2.22}$$

$$X_t = \mathbf{H}\mathbf{x}(t) \tag{2.23}$$

where the $k \times k$ matrix $\boldsymbol{\theta}$ is given by

$$\boldsymbol{\theta} = \begin{pmatrix} 0 & 0 & \cdots & 0 & -\alpha_k \\ 1 & 0 & \cdots & 0 & -\alpha_{k-1} \\ 0 & 1 & \cdots & 0 & -\alpha_{k-2} \\ \vdots & \vdots & \cdots & \vdots & \vdots \\ 0 & 0 & \cdots & 1 & -\alpha_1 \end{pmatrix}$$

the $k \times 1$ matrix γ by

$$\gamma' = [\beta_{k-1}, \quad \beta_{k-2}, \ldots, \beta_1, \quad 1]$$

the $1 \times k$ matrix $\mathbf{H}$ by

$$\mathbf{H} = (0, 0, \ldots 0, 1)$$

and the $k \times 1$ vector $x(t)$ is partitioned as

$$\mathbf{x}'(t) = (x_1(t), x_2(t), \ldots, x_k(t))$$

[In fact, $x_k(t)$ is simply the variable X_t in (2.21), and $x_{k-1}(t)$, $x_{k-2}(t), \ldots,$ are defined recursively in terms of $x_k(t)$.] In this formulation $\mathbf{x}(t)$ is called the *state vector*, $\boldsymbol{\theta}$ is called the *system matrix*, γ the *input matrix*, and $\mathbf{H}$ the *observation matrix*. [Note that the condition for (2.21) to represent a stationary process is that $\alpha(z)$ has no zeros inside or on the unit circle—in which case the matrix $\boldsymbol{\theta}$ is "stable", that is, all its eigenvalues have modulus less than 1, and consequently the process $\mathbf{x}(t)$ is stationary.]

Akaike (1974) has given a very general treatment of finite-order linear models, and has shown, using a geometrical approach that a *minimal* realization [i.e., a state-space representation with $\mathbf{x}(t)$ having the smallest possible dimension] can be derived by selecting the state vector $\mathbf{x}(t)$ as any basis of the "predictor space at time t," $R(t + | t -)$, namely, the space spanned by the linear least-squares predictors of $X_t, X_{t+1}, X_{t+2}, \ldots,$ given $\varepsilon_t, \varepsilon_{t-1}, \varepsilon_{t-2}, \ldots$. The crucial feature of this space is that, for models described by finite-order difference equations, it has *finite dimensions* k. A minimal realization can now be constructed by choosing the state vector (at time t) as *any* basis of $R(t + | t -)$. If we think of the model as a physical system with ε_t as the "input" and X_t as the "output," the preceding specification of the state vector is in complete accord with the physical interpretation of the "state" of the system at time t_0 as that set of quantities which, together with the input for all time points $\geq t_0$, uniquely determines the output for all time points $\geq t_0$. (When, e.g., the relationship between input and output is described by a finite-order difference or differential equation, the "state" at time t_0 is simply the set of "initial conditions" required to determine the solution of the difference or differential equation for all $t \geq t_0$.) Intuitively, therefore, we may think of the state at any time point as the totality of information on the future output contained within the past input, and this is precisely the information contained within the set of predictors, $\{X_t, X_{t+1|t}, X_{t+2|t}, \ldots\}$, or equivalently, within any set of basis variables of the space spanned by these predictors.

3. GENERAL NONLINEAR MODELS

We now return to the general model (2.1). If we impose the condition of "physical realisability" on (2.1) we obtain a model of the form

$$h(X_t, X_{t-1}, X_{t-2}, \ldots) = \varepsilon_t \tag{3.1}$$

and if we assume that this model is "invertible" [i.e., that (3.1) may be "solved" to yield an expression for X_t in terms of $\varepsilon_t, \varepsilon_{t-1}, \varepsilon_{t-2}, \ldots$], we then obtain, say,

$$X_t = f(\varepsilon_t, \varepsilon_{t-1}, \varepsilon_{t-2}; \ldots) \tag{3.2}$$

From the standpoint of statistical analysis (3.2) is just as intractable as (2.1), but we may attempt to parametrize (3.2) by assuming further that the function f is analytic, so that it can be expanded about some fixed point—say about the point $\mathbf{0} = (0, 0, 0, \ldots)$. This yields the formal expansion

$$X_t = \mu + \sum_{u=0}^{\infty} g_u \varepsilon_{t-u} + \sum_{u=0}^{\infty} \sum_{v=0}^{\infty} g_{uv} \varepsilon_{t-u} \varepsilon_{t-v} + \sum_{u=0}^{\infty} \sum_{v=0}^{\infty} \sum_{w=0}^{\infty} g_{uvw} \varepsilon_{t-u} \varepsilon_{t-v} \varepsilon_{t-w} + \cdots \tag{3.3}$$

where we have written $\mu = f(\mathbf{0})$, $g_u = [\partial f / \partial \varepsilon_{t-u}]_{\mathbf{0}}, \ldots$. This expansion is known as a *Volterra series*, and was used by Wiener (1958) in his pioneering study of nonlinear systems. Wiener's objective was to construct transformations of these series in which the successive terms are "orthogonal" (i.e., uncorrelated)—rather like the use of orthogonal polynomials in regression analysis. However, there is clearly no possibility of fitting such a Volterra series to a finite stretch of data since the first term contains an infinite sequence of parameters, the second term a doubly infinite sequence of parameters, and so on. To circumvent this difficulty we may assume either (a) that the Fourier transforms of the sequences $\{g_u\}, \{g_{uv}\}, \ldots$ possess certain "smoothness" properties, or (b) that each of these transforms has a known functional form involving only a finite number of unknown parameters. Approach (a) leads to "polyspectral analysis" [the Fourier analysis of the higher-order moments, see, e.g., Brillinger (1970)], but is feasible only when (3.3) is of homogeneous form with only one term present. On the other hand, (b) leads to the development of special forms of nonlinear schemes, three particularly important types being (1) *bilinear models*

(Granger and Andersen (1978), Subba Rao (1981)), (2) *threshold autoregressive models* (Tong and Lim (1980)), and (3) *exponential autoregressive models* (Haggan and Ozaki (1981)). For example, a bilinear model takes the form

$$X_t + a_1 X_{t-1} + \cdots + a_k X_{t-k} = \varepsilon_t + b_1 \varepsilon_{t-1} + \cdots + b_l \varepsilon_{t-l} + \sum_{i=1}^{m} \sum_{j=1}^{p} c_{ij} \varepsilon_{t-i} X_{t-j} \tag{3.4}$$

and is thus seen to be a generalization of the (linear) ARMA model (2.16) in which a bilinear form in $\{\varepsilon_{t-1}\}$, $\{X_{t-j}\}$ has been added to the right-hand side. Threshold AR models are constructed by starting with the (linear) AR model (2.12) and allowing each of the coefficients $a_1, \ldots, a_k$ to "switch" between two different values according as to whether some past observation X_{t-d} falls below or above a given "threshold" level. Exponential AR models are constructed by again starting with the AR model (2.12) and replacing each of the coefficients by exponential functions of X_{t-1}. Each of the special nonlinear schemes describes, in its own way, a specific type of nonlinearity; the threshold AR models can (with suitably chosen parameters) give rise to "limit cycle" behavior, while the exponential AR model can reproduce certain well-known features of nonlinear random vibrations theory, such as amplitude-dependent frequency and jump phenomena. In practice, however, it may be difficult to decide which, if any, of these special models is best suited to a given set of data. In Section 4 we describe a more general class of nonlinear models, called *state-dependent models*, which includes the bilinear, threshold autoregressive, and exponential autoregressive models as special cases, but allows much greater flexibility in the characterization of the nonlinear structure. Although these state-dependent models are of a general nature, they are nevertheless amenable to statistical analysis and may be fitted to data. This approach offers two major advantages, namely, (1) state-dependent models may be used directly in connection with, for example, the prediction of future values, and (2) since they can be fitted to data without any specific prior assumptions about the form of nonlinearity, they may be used to give us an "overview" of the character of the nonlinearity inherent in the data and thus indicate whether, e.g., a bilinear, threshold AR, or exponential AR model is appropriate.

4. STATE-DEPENDENT MODELS

In developing the approach outlined at the end of Section 3 the basic difficulty that we encounter is that of striking a judicious balance between generality and tractability. To highlight this point, we may note that the model (3.1) is completely general but totally intractable (as far as statistical analysis is concerned), whereas, for example, the bilinear model (3.4) is certainly tractable but constrains us to a particular form of nonlinear structure. Our approach is to start from the general model (3.1), and then impose further restrictions until we have reached the point at which statistical analysis may commence.

As it stands, the model (3.1) is "infinite dimensional," in the sense that it involves a relationship between infinitely many variables. We now reduce this to a "finite-dimensional" form by assuming that the relationship between X_t and the "past history" of the series can be described in terms of finitely many values of past X_t and ε_t, so that we may write, say,

$$X_t = h(X_{t-1}, \ldots, X_{t-k}, \varepsilon_{t-1}, \ldots, \varepsilon_{t-l}) + \varepsilon_t \tag{4.1}$$

In this formulation ε_t plays the role of the "innovations" process for x_t, and the function h describes the information on X_t contained in its past history. In the language of probability theory we may interpret h as the "projection" (i.e., conditional expectation) of X_t on $\mathscr{B}_{t-1}^{\varepsilon}$, the Borel field generated by $\varepsilon_t, \varepsilon_{t-1}, \varepsilon_{t-2}, f \ldots$. In postulating the model (4.1) we are saying, in effect, that *this "projection" can be expressed* in terms of a finite number of functions of $\varepsilon_t, \varepsilon_{t-1}, \varepsilon_{t-2}, \ldots$. (Recall that, for each u, X_{t-u} is itself a function of $\varepsilon_t, \varepsilon_{t-1}, \varepsilon_{t-2}, \ldots$.)

Assuming further that the function h in (4.1) is analytic, we may expand the right-hand side of (4.1) about an arbitrary but fixed time point t_0. Using only a first-order expansion we obtain

$$\begin{aligned} Y_t = h(\mathbf{x}_{t_0}) &+ \sum_{u=1}^{k} f_u(\mathbf{x}_{t-1})(X_{t-u} - X_{t_0-u}) \\ &+ \sum_{u=1}^{l} g_u(\mathbf{x}_{t-1})\varepsilon_{t-u} - \varepsilon_{t_0-u}) \qquad \text{(say)} \end{aligned} \tag{4.2}$$

where $\mathbf{x}_t$ denotes the "state vector"

$$\mathbf{x}_t = (\varepsilon_{t-l+1}, \ldots, \varepsilon_t, X_{t-k+1}, \ldots, X_t)' \tag{4.3}$$

and f_u, g_u depend on the first-order partial derivatives of h. We can now rewrite (3.2) in the form

$$X_t + \sum_{u=1}^{k} \phi_u(\mathbf{x}_{t-1}) X_{t-u} = \mu(\mathbf{x}_{t-1}) + \varepsilon_t + \sum_{u=1}^{l} \psi_u(\mathbf{x}_{t-1})\varepsilon_{t-u} \tag{4.4}$$

This is the basic model with which we now work, and we call (4.4) a *state-dependent model* (*SDM*) *of order* (k, l). It has a natural interpretation as a *locally linear* ARMA model in which the evolution of the series at time $(t-1)$ is governed by a set of AR coefficients $\{\phi_u\}$, a set of MA coefficients $\{\psi_u\}$, and a local mean μ, all of which depend on the "state" of the series at time $(t-1)$.

For each $\mathbf{x}_{t-1}$, (4.4) may be interpreted geometrically as the *tangent plane* to the surface generated by $h(\mathbf{x}_{t-1})$. In this sense we may regard the SDM model as being formed by "bending" the linear model around each point in the state space.

It is easily seen that (4.4) includes as special cases all the special nonlinear models described in Section 3. Thus, if we take μ and $\{\phi_u\}$ as constants and set

$$\psi_u(\mathbf{x}_{t-1}) = b_u + \sum_{i=1}^{p} c_{uj} X_{t-j}$$

then (4.4) reduces to the bilinear model (3.4) [with the value of l in (4.4) chosen as $\max(m, l)$, m and l being the corresponding parameters in (3.4)]. Similarly, if we take all the ψ_u to be zero and let μ and the ϕ_u take the form of step functions, each switching between two levels, then (4.4) reduces to the AR threshold model, while setting all the ψ_u to zero and taking each ϕ_u to be an exponential function of X_{t-1} gives the AR exponential model.

A formal state-space representation of the SDM is easily obtained by augmenting the vector (4.3) with the additional element unity in the leading position. Specifically, if we redefine $\mathbf{x}_t$ as

$$\mathbf{x}_t = (1, \varepsilon_{t-l+1}, \ldots, \varepsilon_t, X_{t-k+1}, \ldots, X_t)'$$

then it is not difficult to show that, with suitably defined matrices F, H, we can rewrite (4.4)

$$\mathbf{x}_{t+1} = \mathbf{F}(\mathbf{x}_t)\mathbf{x}_t + \varepsilon_{t+1},$$

$$X_t = \mathbf{H}\mathbf{x}_t \tag{4.5}$$

If μ, ϕ_u, μ_u are each analytic functions of $\mathbf{x}_{t-1}$, then (4.5) is a special case of the class of "linear analytic" systems studied in the control theory literature (see, e.g., Brockett, 1976).

5. IDENTIFICATION OF SDMs

We now consider the problem of estimating or "identifying" the functional forms of the parameters $\mu(\mathbf{x}_t)$, $\phi_u(\mathbf{x}_t)$, $\mu_u(\mathbf{x}_t)$ from observational data. The simplest assumption we can make about these parameters is that each is a linear function of the components of $\mathbf{x}_t$, so that, for example, we may write

$$\psi_u(\mathbf{x}_t) = \psi_u^{(0)} + \mathbf{x}_t'\boldsymbol{\beta}_u,$$

$$\phi_u(\mathbf{x}_t) = \phi_u^{(0)} + \mathbf{x}_t'\boldsymbol{\gamma}_u$$

This form would be appropriate to bilinear models (with $\boldsymbol{\gamma}_u = \mathbf{0}$ for all u), but it is unreasonable to suppose that all nonlinear models can be approximated with sufficient accuracy with ψ_u, ϕ_u set as linear functions of $\mathbf{x}_t$. If ψ_u, ϕ_u are "smooth" functions of $\mathbf{x}_t$, however, it would be reasonable to suppose that they may be represented *locally* as linear functions. This means that we can obtain a fairly general representation of these parameters by allowing both $\boldsymbol{\beta}_u$ and $\boldsymbol{\gamma}_u$ to be themselves state dependent. We are then faced with the problem of specifying the functional forms of $\boldsymbol{\beta}_u$, $\boldsymbol{\gamma}_u$, but we can avoid this difficulty by simply letting $\boldsymbol{\beta}_u$ and $\boldsymbol{\gamma}_u$ "wander" over time, that is, we allow $\boldsymbol{\beta}_u = \boldsymbol{\beta}_u^{(t)}$ and $\boldsymbol{\gamma}_u = \boldsymbol{\gamma}_u^{(t)}$ to depend purely on the time parameter. This gives the SDM considerable flexibility, and if we now allow $\boldsymbol{\beta}_u^{(t)}$ and $\boldsymbol{\gamma}_u^{(t)}$ to "wander" in the form of random walks we can construct an estimation procedure which, for each t, determines those values of $\boldsymbol{\beta}_u^{(t)}$, $\boldsymbol{\gamma}_u^{(t)}$, which, in effect, minimizes the discrepancy between X_{t+1} and its predictor $\tilde{X}_{t+1}$ computed from the model. The estimation procedure is thus based on a sequential type of algorithm, similar in nature to the *Kalman filter* algorithm.

The estimation algorithm involves some rather technical statistical features, and here we illustrate its operation simply by reference to the "pure AR" SDM type of model. [For a full description, see Haggan et al. (1984).] We consider the SDM model (4.4) in which $\psi_u = 0$ for all u, and rewrite this in a different state-space form in which the state vector no longer represents the state of the series but now represents the current state of all the relevant parameters. Specifically, we define the vector $\boldsymbol{\theta}_t$ by

$$\boldsymbol{\theta}_t = \left(\mu_{t-1}, \phi_{1,t-1}, \ldots, \phi_{k,t-1}, \boldsymbol{\alpha}'^{(t)}, \boldsymbol{\gamma}_1'^{(t)}, \ldots, \boldsymbol{\gamma}_k'^{(t)}\right)'$$

[where, for brevity, we have written $\phi_{1,t-1}$ in place of $\phi_1(\mathbf{x}_{t-1})$, and similarly for the other parameters]. Writing

$$\mathbf{H}_t^* = (1, -X_{t-1}, \ldots, -X_{t-k}, 0, 0, \ldots, 0),$$

we may now express (4.4) in the form

$$X_t = \mathbf{H}_t^*\boldsymbol{\theta}_t + \varepsilon_t \tag{4.6}$$

where $\boldsymbol{\theta}_t$ satisfies an evolution of the form

$$\boldsymbol{\theta}_t = \mathbf{F}^*_{t-1}\boldsymbol{\theta}_{t-1} + \mathbf{W}_t \tag{4.7}$$

Equations (4.6) and (4.7) are now of the standard form to which the (extended) Kalman filter can be applied. If, on the basis of observations up to time $(t-1)$, we have constructed an estimate $\hat{\boldsymbol{\theta}}_{t-1}$ of $\boldsymbol{\theta}_{t-1}$, then given the next observation X_t, the updating equation for $\hat{\boldsymbol{\theta}}_t$ takes the form

$$\hat{\boldsymbol{\theta}}_t = \mathbf{F}_{t-1}\hat{\boldsymbol{\theta}}_{t-1} + \mathbf{K}^*_t[X_t - \mathbf{H}^*_t\mathbf{F}^*_{t-1}\hat{\boldsymbol{\theta}}_{t-1}]$$

where $\mathbf{K}^*_t$, the "Kalman gain" matrix, may be evaluated from the standard recursive algorithm for the Kalman filter [see, e.g., Jazwinski (1970)]. Thus, at each time point we obtain a new estimate of each of the parameters, $\mu, \phi_1, \ldots, \phi_k$, and these estimates can then be plotted against the corresponding state vector $\mathbf{x}_t$. In this way we gradually build up a set of ordinates for each of the parameter functions, and these ordinates can then be transformed into smooth surfaces by applying one of the standard nonparametric function fitting routines (e.g., Priestley and Chao, 1972). Once these surfaces have been constructed, we can then examine their shapes to gain insight into the nonlinear structure of the model.

These techniques have been applied with considerable success to a wide variety of different data sets (including both simulated and physical data), and have produced some very interesting results. In particular, when applied to data simulated from known (nonlinear) models the SDM approach has generally reproduced parameter surfaces that clearly indicate the correct model structure. This technique, allied with extensive use of computer graphics, thus offers the time-series analyst a powerful method for studying nonlinear models without any prior commitment to a specific form of nonlinearity. A detailed account of these numerical studies is given in Haggan et al. (1984).

REFERENCES

Akaike, H. (1974). Markovian representation of stochastic processes and its application to the analysis of autoregressive moving average processes. *Ann. Inst. Math.* **26**, 363–387.

Box, G. E. P. and Jenkins, G. M. (1970). *Time Series Analysis, Forecasting and Control*, San Francisco: Holden-Day.

Brillinger, D. R. (1970). The identification of polynomial systems by means of higher order spectra. *J. Sound. Vib.* **12**, 301–313.

Brockett, R. W. (1976). Volterra series and geometric control theory. *Automatica* **12**, 167–172.

Granger, C. W. J. and Andersen, A. P. (1978). *An Introduction to Bilinear Time Series Models*, Göttingen: Vandenhoek and Ruprecht.

Haggan, V. and Ozaki, T. (1981). Modelling non-linear random vibrations using an amplitude-dependent autoregressive time series model. *Biometrika* **68**, 189–196.

Haggan, V. Heravi, S. M. and Priestley, M. B. (1984). A study of the application of state-dependent models in non-linear time series analysis. *J. Time Ser. Anal.* **5**, No. 2, 69–102.

Jazwinski, A. H. (1970). *Stochastic Processes and Filtering Theory*, New York: Academic Press.

Priestley, M. B. and Chao, M. T. (1972). Non-parametric function fitting. *J. R. Statist. Soc. B* **34**, 385–392.

Subba Rao, T. (1981). On the theory of bilinear time series models. *J. R. Statist. Soc. B* **43**, 244–255.

Tong, H. and Lim, K-S. (1980). Threshold autoregression, limit cycles and cyclical data. *J. R. Statist. Soc. B* **42**, 245–292.

Wiener, N. (1958). *Non-linear Problems in Random Theory*, Cambridge, Massachusetts: MIT Press.

PART 3

Nonparametric Theory

P-P Plots, Rank Processes, and Chernoff–Savage Theorems

Emad-Eldin A. A. Aly
Department of Statistics and Applied Probability
The University of Alberta
Edmonton, Alberta, Canada

Miklós Csörgő
Department of Mathematics and Statistics
Carleton University
Ottawa, Ontario, Canada

Lajos Horváth
Bolyai Institute
Szeged University
Szeged, Hungary

SUMMARY

P-P plot processes, empirical processes, and quantile rank processes are described, and weak approximations of them are established in terms of appropriate gaussian processes. The main tools are weighted approximations of the uniform empirical and quantile processes. The results are then used directly to prove Chernoff–Savage-type theorems.

1. INTRODUCTION

Let $X_1, \ldots, X_m (m \geq 1)$ and $Y_1, \ldots, Y_n (n \geq 1)$ be two independent random samples on the random variables (rv) X and Y with continuous distribution functions F and G, respectively, and let F^{-1} and G^{-1} be their corresponding quantile functions, that is,

$$F^{-1}(y) = \inf\{x \in \mathbb{R} \mid F(x) \geq y\}, \qquad 0 < y \leq 1, \qquad F^{-1}(0) = F^{-1}(0+) \tag{1.1}$$

and a similar definition for G^{-1}. Let $X_{1,m} \leq \cdots \leq X_{m,m}$ be the order statistics of the X sample, and let $Y_{1,n} \leq \cdots \leq Y_{n,n}$ be those of the Y sample. Let F_m be the empirical distribution function of the X sample, defined by

$$F_m(x) = \begin{cases} 0 & \text{if } x < X_{1,m} \\ k/m & \text{if } X_{k,m} \leq x < X_{k+1,m} \quad (k = 1, \ldots, m-1) \\ 1 & \text{if } X_{m,m} \leq x \end{cases} \tag{1.2}$$

and let G_n be the empirical distribution function of the Y sample, similarly defined in terms of the Y-sample order statistics. Let F_m^{-1} and G_n^{-1} be the sample quantile functions of the X sample and Y sample, respectively, that is,

$$\begin{aligned} F_m^{-1}(y) &= \inf\{x \in \mathbb{R} \mid F_m(x) \geq y\} \\ &= X_{k,m} \quad \text{if } (k-1)/m < y \leq k/m \quad (k = 1, \ldots, m) \end{aligned} \tag{1.3}$$

and $F_m^{-1}(0) = X_{1,m}$, and similarly for G_n^{-1} in terms of the Y-sample order statistics.

Let $N = m + n$, and denote by H_N the sample distribution function of the pooled random sample $X_1, \ldots, X_m, Y_1, \ldots, Y_n$. Then

$$H_N(x) = \lambda_N F_m(x) + (1 - \lambda_N) G_n(x), \qquad x \in \mathbb{R}, \quad \lambda_N = m/N \tag{1.4}$$

with the corresponding theoretical distribution function

$$H(x) = \lambda_N F(x) + (1 - \lambda_N) G(x), \qquad x \in \mathbb{R}, \quad \lambda_N = m/N \tag{1.5}$$

We shall assume throughout that for *all* N the inequalities

$$0 < \lambda_* \leq \lambda_N \leq 1 - \lambda_* < 1, \qquad \lambda_* \leq \tfrac{1}{2} \tag{1.6}$$

hold true. We shall *not* assume that λ_N tends to a limit $\lambda_0 \in (0, 1)$ as $m \wedge n \to \infty$, *unless* we write λ_0 instead of λ_N in any one of the statements in the sequel. Also, we will use $\wedge$ for minimum and $\vee$ for maximum.

When two samples are compared, frequently quantile–quantile (Q-Q) and procentile–procentile (P-P) as well as P-Q plots are used for statistical data analysis. For an impressive review and annotated bibliography of these and other graphical methods in nonparametric statistics, we refer to Fisher (1983). Studies on the asymptotic behavior of Q-Q plot processes were carried out by Aly (1983a, b) and by Aly and Bleuer (1986). One of the aims of this exposition is to study the large sample behavior of P-P plots and their empirical processes. In this chapter the P-P plot of G against F is defined to be

$$\left(GF^{-1}(k/m), G_nF_m^{-1}(k/m)\right) \qquad (k = 1, \ldots, m) \tag{1.7}$$

which in turn leads us to study the empirical P-P plot process

$$l_N(y) = N^{1/2}\left(G_nF_m^{-1}(y) - GF^{-1}(y)\right), \qquad 0 \le y \le 1 \tag{1.8}$$

Concerning (1.7) we note that under the null hypothesis H_0: $F = G$ it reduces to the P-P plot of G against F as it is used in practice

$$\left(k/m, G_nF_m^{-1}(k/m)\right) = \left(k/m, G_n(X_{k,m})\right) \qquad (k = 1, \ldots, m)$$

If now R_k stands for the rank of $X_{k,m}$ in the pooled sample, then $nG_n(X_{k,m}) = R_k - k$ $(k = 1, \ldots, m)$. This relationship shows that ranks provide an alternative way of looking at P-P plots. The latter connection and Parzen's (1983) inspiring expository paper on two sample statistics have led us to also study rank plots and rank processes here.

Letting

$$D_N(y) = H_NF_m^{-1}(y), \qquad 0 \le y \le 1 \tag{1.9}$$

it can be easily seen that

$$D_N(y) = R_j/N, \qquad (j-1)/m < y \le j/m \qquad (j = 1, \ldots, m) \tag{1.10}$$

The theoretical counterpart of D_N is

$$D(y) = HF^{-1}(y), \qquad 0 \le y \le 1 \tag{1.11}$$

and we define the quantile rank process by

$$d_N(y) = N^{1/2}\left(D_N(y) - D(y)\right), \qquad 0 \le y \le 1 \tag{1.12}$$

If we now take the inverse D_N^{-1} of D_N, we get

$$D_N^{-1}(y) = \begin{cases} 0 & \text{if } 0 \le y < R_1/N \\ j/m & \text{if } R_j/N \le y < R_{j+1}/N \qquad (j = 1, \ldots, m-1) \\ 1 & \text{if } R_m/N \le y \le 1 \end{cases} \tag{1.13}$$

and observe that D_N^{-1} is like an empirical distribution function of the relative ranks $\{R_j/N\}_{j=1}^m$, and D_N is like an empirical quantile function of them [cf. (1.2) and (1.3)]. We introduce the empirical rank process

$$r_N(y) = N^{1/2}\left(D_N^{-1}(y) - D^{-1}(y)\right), \qquad 0 \le y \le 1 \tag{1.14}$$

where D^{-1} is

$$D^{-1}(y) = FH^{-1}(y), \qquad 0 \le y \le 1 \tag{1.15}$$

Pyke and Shorack (1968) considered the process $N^{1/2}(F_m H_N^{-1}(y) - FH^{-1}(y))$ $(0 \le y \le 1)$, which is asymptotically equivalent to r_N of (1.14). They proved its weak convergence in weighted metrics to a gaussian process. Our approach to the rank process r_N is more direct and lends itself to bootstrapping immediately [cf. Section 17 in Csörgő et al. (1986a)]. We also consider asymptotics of l_N and d_N in weighted metrics and apply our results to prove a Chernoff–Savage-type theorem.

2. TECHNICAL BACKGROUND TOOLS

For further use in the sequel we introduce the uniform empirical and quantile processes of the X sample,

$$\alpha_m(y) = m^{1/2}\left(F_m F^{-1}(y) - y\right), \qquad 0 \le y \le 1 \tag{2.1}$$

and

$$u_m(y) = m^{1/2}\left(y - FF_m^{-1}(y)\right), \qquad 0 \le y \le 1 \tag{2.2}$$

and the uniform empirical process of the Y sample

$$\beta_n(y) = n^{1/2}\left(G_n G^{-1}(y) - y\right), \qquad 0 \le y \le 1 \tag{2.3}$$

As far as these processes are being concerned, we shall need their asymptotic behavior in weighted metrics.

Let q be any *positive* function on $(0,1)$, that is, $\inf_{\delta \le s \le 1-\delta} q(s) > 0$ for all $0 < \delta < \frac{1}{2}$, nondecreasing in a neighborhood of 0 and nonincreasing in a neighborhood of 1. Such a function q will be called a Chibisov–O'Reilly function if and only if

$$I(q,c) = \int_0^1 (t(1-t))^{-1} \exp\{-c(t(1-t))^{-1} q^2(t)\}\, dt < \infty \tag{2.4}$$

for all $c > 0$. The latter integral being finite for all $c > 0$ is equivalent to

$$\lim_{\varepsilon \downarrow 0} \Pr\Big\{ \sup_{0 \le t \le \varepsilon} |B(t)|/q(t) \ge \delta \Big\} = 0 \quad \text{for all } \delta > 0 \tag{2.5}$$

where $\{B(s) | 0 \le s \le 1\}$ is a standard Brownian bridge.

Theorem 2.1. The underlying probability space of the X and Y random samples can be extended in such a way that with two independent sequences of Brownian bridges $\{B_m^{(1)}(y) | 0 \le y \le 1\}$ and $\{B_n^{(2)}(y) | 0 \le y \le 1\}$ we have as $m \wedge n \to \infty$

$$\sup_{0 \le y \le 1} |\alpha_m(y) - B_m^{(1)}(y)|/q(y) = o_P(1) \tag{2.6}$$

$$\sup_{1/(m+1) \le y \le 1 - 1/(m+1)} |u_m(y) - B_m^{(1)}(y)|/q(y) = o_P(1) \tag{2.7}$$

and

$$\sup_{0 \le y \le 1} |\beta_n(y) - B_n^{(2)}(y)|/q(y) = o_P(1) \tag{2.8}$$

with any Chibisov–O'Reilly weight function q.

This weighted convergence of the empirical and quantile processes was first proved by Chibisov (1964) for functions q satisfying (2.4) and some further tail regularity conditions. O'Reilly (1974) proved (2.6) and (2.7) for continuous q satisfying (2.4). There have been many attempts in the literature to re-prove this theorem (cf. Shorack, 1979; Shorack and Wellner, 1982; Stute, 1982; and Csörgő et al., 1986a). In the generality of (2.6) and (2.7), the Chibisov–O'Reilly theorem is a consequence of a weighted approximation theorem for α_m and u_m in Csörgő et al. (1986b).

The following result of Wellner (1978) will be also needed later on.

Theorem 2.2. We have, as $m \to \infty$,

$$\sup_{1/(m+1) \le y \le 1 - 1/(m+1)} FF_m^{-1}(y)/y = O_P(1) \tag{2.9}$$

and

$$\sup_{1/(m+1)\leq y\leq 1-1/(m+1)} (1-FF_m^{-1}(y))/(1-y) = O_P(1) \tag{2.10}$$

3. WEAK CONVERGENCE OF RANK PROCESSES

Our first two results are concerned with the P-P process l_N of (1.8) and the quantile rank process d_N of (1.12). The weak approximation of these two processes will be proven under the following conditions:

C.1. f and g are continuous positive density functions on the open support of the distribution functions F and G, respectively, and

C.2. with some Chibisov–O'Reilly function q we have

$$\sup_{0\leq y\leq 1} \frac{gF^{-1}(y)}{fF^{-1}(y)} q(y) < \infty$$

Theorem 3.1. Assuming the conditions C.1, C.2, and (1.6) on the probability space of Theorem 2.1 we have, as $m \wedge n \to \infty$,

$$\sup_{0\leq y\leq 1} |l_N(y) - \Gamma_N(y)| = o_P(1) \tag{3.1}$$

where

$$\Gamma_N(y) = (1-\lambda_N)^{-1/2} B_n^{(2)}(GF^{-1}(y)) - \lambda_N^{-1/2} \frac{gF^{-1}(y)}{fF^{-1}(y)} B_m^{(1)}(y),$$

$$0 \leq y \leq 1$$

It is easily seen that Γ_N is a gaussian process with covariance structure $E\Gamma_N(y) = 0$ and

$$\begin{aligned} E\Gamma_N(x)\Gamma_N(y) &= (1-\lambda_N)^{-1}(GF^{-1}(x) \wedge GF^{-1}(y) - GF^{-1}(x)GF^{-1}(y)) \\ &\quad + \lambda_N^{-1} gF^{-1}(x) gF^{-1}(y)(x \wedge y - xy)/(fF^{-1}(x) fF^{-1}(y)) \end{aligned} \tag{3.2}$$

We note also that whenever $\lambda_N \to \lambda_0 \in (0,1)$ $(m \wedge n \to \infty)$, then l_N converges weakly to a gaussian process with mean zero and covariance as in (3.2) where λ_N is replaced by λ_0.

Our conditions for having (3.1) are optimal in the following sense: Let q^* be a positive function on $(0,1)$ that is nondecreasing in a neighborhood of 0 and nonincreasing in a neighborhood of 1, $I(q^*, c) = \infty$ with some $c > 0$ and assume that

$$\liminf_{y \downarrow 0} q^*(y) \frac{gF^{-1}(y)}{fF^{-1}(y)} > 0 \tag{3.3}$$

Then by (2.5)

$$\liminf_{y \downarrow 0} |\Gamma_N(y)| > K > 0$$

with probability 1 for each N. We shall see in the proof of this theorem that under C.2 as well as under (3.3) we have

$$l_N(0) \xrightarrow{P} 0, \qquad m \wedge n \to \infty$$

Hence under the condition (3.3), which says that $gF^{-1}(y)/fF^{-1}(y) \to \infty$ faster than any Chibisov–O'Reilly function q could go to 0 as $y \downarrow 0$, we cannot have (3.1).

A similar result is true for the quantile rank process d_N.

Theorem 3.2. Under the conditions C.1, C.2, and (1.6) on the probability space of Theorem 2.1 we have, as $m \wedge n \to \infty$,

$$\sup_{0 \le y \le 1} |d_N(y) - (1 - \lambda_N)\Gamma_N(y)| = o_P(1)$$

Our next result is the weak approximation of the empirical rank process r_N of (1.14).

Theorem 3.3. Under the conditions C.1 and (1.6) on the probability space of Theorem 2.1 we have, as $m \wedge n \to \infty$,

$$\sup_{0 \le y \le 1} |r_N(y) - \Delta_N(y)| = o_P(1) \tag{3.4}$$

where

$$\Delta_N(y)$$

$$= \frac{1 - \lambda_N}{\lambda_N fH^{-1}(y) + (1 - \lambda_N) gH^{-1}(y)} \left\{ \lambda_N^{-1/2} gH^{-1}(y) B_m^{(1)}(FH^{-1}(y)) \right.$$

$$\left. - (1 - \lambda_N)^{-1/2} fH^{-1}(y) B_n^{(2)}(GH^{-1}(y)) \right\}, \qquad 0 \le y \le 1$$

The process Δ_N is a gaussian process with $E\Delta_N(y) = 0$,

$$E\Delta_N(x)\Delta_N(y)$$

$$= \frac{(1-\lambda_N)^2}{\left(\lambda_N fH^{-1}(x) + (1-\lambda_N)gH^{-1}(x)\right)\left(\lambda_N fH^{-1}(y) + (1-\lambda_N)gH^{-1}(y)\right)}$$

$$\times\left\{\lambda_N^{-1} gH^{-1}(x)gH^{-1}(y)\left(FH^{-1}(x) \wedge FH^{-1}(y) - FH^{-1}(x)FH^{-1}(y)\right)\right.$$

$$+ (1-\lambda_N)^{-1} fH^{-1}(x)fH^{-1}(y)\left(GH^{-1}(x) \wedge GH^{-1}(y)\right.$$

$$\left.\left. - GH^{-1}(x)GH^{-1}(y)\right)\right\} \tag{3.5}$$

We note that if $F = G$, then

$$\{\Gamma_N(y)|0 \le y \le 1\} \overset{\mathscr{D}}{=} \left\{(\lambda_N(1-\lambda_N))^{-1/2} B(y)|0 \le y \le 1\right\}$$

and

$$\{\Delta_N(y)|0 \le y \le 1\} \overset{\mathscr{D}}{=} \left\{\left(\frac{1}{\lambda_N}(1-\lambda_N)\right)^{1/2} B(y)|0 \le y \le 1\right\}$$

for each N, where B is a standard Brownian bridge. The latter statements easily follow from (3.2) and (3.5).

4. PROOFS OF THE RESULTS IN SECTION 3

Proof of Theorem 3.1. We have

$$l_N(y) = (1-\lambda_N)^{-1/2}\beta_n\left(GF_m^{-1}(y)\right) - \lambda_N^{-1/2}\hat{l}_m(y) \tag{4.1}$$

where

$$\hat{l}_m(y) = m^{1/2}\left(GF^{-1}(y) - GF_m^{-1}(y)\right)$$

By Theorem 2.1 we get

$$\sup_{0 \le y \le 1} |\beta_n\left(GF_m^{-1}(y)\right) - B_n^{(2)}\left(GF_m^{-1}(y)\right)| = o_P(1) \tag{4.2}$$

and

$$\sup_{0 \le y \le 1} |FF_m^{-1}(y) - y| = o_P(1), \qquad m \wedge n \to \infty \tag{4.3}$$

Since $B_n^{(2)}(GF^{-1}(y))$ is almost surely continuous for each n, by (4.3) we obtain

$$\sup_{0 \le y \le 1} |B_n^{(2)}(GF_m^{-1}(y)) - B_n^{(2)}(GF^{-1}(y))| = o_P(1) \tag{4.4}$$

Thus (4.2) and (4.4) yield

$$\sup_{0 \le y \le 1} |(1-\lambda_N)^{-1/2}\beta_n(GF_m^{-1}(y)) - (1-\lambda_N)^{-1/2}B_n^{(2)}(GF^{-1}(y))| = o_P(1) \tag{4.5}$$

Now we look at $\hat{l}_m$ and consider

$$\sup_{0 \le y \le 1/(m+1)} |\hat{l}_m(y)| \le m^{1/2}GF_m^{-1}\left(\frac{1}{m+1}\right) + m^{1/2}GF^{-1}\left(\frac{1}{m+1}\right) \tag{4.6}$$

In order to show that

$$\sup_{0 \le y \le 1/(m+1)} |\hat{l}_m(y)| = o_P(1) \tag{4.7}$$

by (4.6) and Theorem 2.2 it suffices to show that

$$m^{1/2}GF^{-1}\left(\frac{c}{m+1}\right) = o(1) \quad \text{for all fixed } c > 0 \tag{4.8}$$

By condition C.2 we have

$$GF^{-1}(y) \le K\int_0^y \frac{1}{q(s)}\,ds \quad \text{with some constant } K > 0 \tag{4.9}$$

Since for all Chibisov–O'Reilly functions q we have $s^{-1/2}q(s) \to \infty$ as $s \downarrow 0$, by (4.9) we get (4.8).

Let $\varepsilon \in (0, \frac{1}{2})$, and using the mean value theorem we consider

$$\sup_{1/(m+1) \le y \le \varepsilon} |\hat{l}_m(y)| = \sup_{1/(m+1) \le y \le \varepsilon} |u_m(y)| \frac{gF^{-1}(\xi_m(y))}{fF^{-1}(\xi_m(y))} \tag{4.10}$$

where

$$FF_m^{-1}(y) \wedge y \le \xi_m(y) \le FF_m^{-1}(y) \vee y \tag{4.11}$$

Since $q(\lambda y)$ is a Chibisov–O'Reilly function for all $\lambda > 0$ whenever $q(y)$ is, by Theorem 2.1 and (2.5) combined we get

$$\lim_{\varepsilon \downarrow 0} \limsup_{m \to \infty} P\left\{ \sup_{1/(m+1) \le y \le \varepsilon} |u_m(y)|/q(\lambda y) > \delta \right\} = 0 \quad \text{for all } \delta > 0 \tag{4.12}$$

Now using (4.11), Theorem 2.2, condition C.2, and (4.12) we obtain

$$\lim_{\varepsilon \downarrow 0} \limsup_{m \to \infty} P\left\{ \sup_{1/(m+1) \le y \le \varepsilon} |\hat{l}_m(y)| > \delta \right\} = 0 \quad \text{for all } \delta > 0 \tag{4.13}$$

One shows in a similar fashion that

$$\sup_{1-1/(m+1) \le y \le 1} |\hat{l}_m(y)| = o_P(1) \tag{4.14}$$

and

$$\lim_{\varepsilon \downarrow 0} \limsup_{m \to \infty} P\left\{ \sup_{1-\varepsilon \le y \le 1-1/(m+1)} |\hat{l}_m(y)| > \delta \right\} = 0 \quad \text{for all } \delta > 0 \tag{4.15}$$

Next we note that analogs of (4.7), (4.13), (4.14), and (4.15) are immediate also for the approximating gaussian sequences. Namely, by condition C.2 and (2.5) we have

$$\lim_{\varepsilon \downarrow 0} P\left\{ \sup_{0 \le y \le \varepsilon} \left| \frac{gF^{-1}(y)}{fF^{-1}(y)} B_m^{(1)}(y) \right| > \delta \right\} = 0 \quad \text{for all } \delta > 0 \tag{4.16}$$

and

$$\lim_{\varepsilon \downarrow 0} P\left\{ \sup_{1-\varepsilon \le y \le 1} \left| \frac{gF^{-1}(y)}{fF^{-1}(y)} B_m^{(1)}(y) \right| > \delta \right\} = 0 \quad \text{for all } \delta > 0 \tag{4.17}$$

Again by a one-term Taylor expansion

$$\hat{l}_m(y) = u_m(y) \frac{gF^{-1}(\xi_m(y))}{fF^{-1}(\xi_m(y))} \tag{4.18}$$

where $\sup_{0 \le y \le 1} |\xi_m(y) - y| = o_P(1)$ [cf. (4.11)]. Using now the fact that gF^{-1}/fF^{-1} is a continuous function over $[\varepsilon, 1-\varepsilon]$, Theorem 2.1 implies

$$\sup_{\varepsilon \le y \le 1-\varepsilon} \left| \hat{l}_m(y) - \frac{gF^{-1}(y)}{fF^{-1}(y)} B_m^{(1)}(y) \right| = o_P(1) \quad \text{for all } \varepsilon \in (0, 1/2) \tag{4.19}$$

Collecting now (4.1), (4.5), (4.7), (4.13)–(4.17), and (4.19), we get Theorem 3.1. ■

Proof of Theorem 3.2. By definition $d_N(y) = (1 - \lambda_N) l_N(y) + O(N^{-1/2})$ with probability 1 for each N, and Theorem 3.1 yields the desired result. ■

In order to prove Theorem 3.3, we need a lemma. Let

$$s_N(y) = N^{1/2}\left(FH^{-1}H_N F_m^{-1}(y) - y\right), \qquad 0 \le y \le 1$$

and

$$\Delta_N^*(y) = \frac{fF^{-1}(y)}{\lambda_N fF^{-1}(y) + (1-\lambda_N) gF^{-1}(y)} \times \left\{ \lambda_N^{1/2} B_m^{(1)}(y) + (1-\lambda_N)^{1/2} B_n^{(2)}\left(GF^{-1}(y)\right)\right\} - \lambda_N^{-1/2} B_m^{(1)}(y),$$
$$0 \le y \le 1$$

Lemma 4.1. Given conditions C.1 and (1.6), we have

$$\sup_{0 \le y \le 1} |s_N(y) - \Delta_N^*(y)| = o_P(1), \qquad m \wedge n \to \infty$$

Proof. We have

$$s_N(y) = \gamma_N\left(F_m^{-1}(y)\right) - \lambda_N^{-1/2} u_m(y), \qquad 0 \le y \le 1 \tag{4.20}$$

where

$$\gamma_N(x) = N^{1/2}\left(FH^{-1}H_N(x) - F(x)\right), \qquad -\infty < x < \infty$$

By definition of H_N

$$N^{1/2}\left(H_N(x) - H(x)\right) = \lambda_N^{1/2} \alpha_m\left(F(x)\right) + (1-\lambda_N)^{1/2} \beta_n\left(G(x)\right)$$

Theorem 2.1, (2.5), and a one-term Taylor expansion yield

$$\lim_{a \to -\infty} \limsup_{N \to \infty} \Big\{ \sup_{-\infty < x \le a} |\gamma_N(x)| > \delta \Big\} = 0 \quad \text{for all } \delta > 0 \tag{4.21}$$

We introduce the stochastic process

$$\Gamma_N^*(x) = \frac{f(x)}{\lambda_N f(x) + (1 - \lambda_N) g(x)} \big\{ \lambda_N^{1/2} B_m^{(1)}(F(x)) + (1 - \lambda_N)^{1/2} B_n^{(2)}(G(x)) \big\}$$

which will be approximating γ_N, and for which we have

$$\lim_{a \to -\infty} P\Big\{ \sup_{-\infty < x \le a} |\Gamma_N^*(x)| > \delta \Big\} = 0 \quad \text{for all } \delta > 0 \tag{4.22}$$

by (2.5). Similarly we have

$$\lim_{b \to \infty} \limsup_{N \to \infty} P\Big\{ \sup_{b \le x < \infty} |\gamma_N(x)| > \delta \Big\} = 0 \quad \text{for all } \delta > 0 \tag{4.23}$$

and

$$\lim_{b \to \infty} P\Big\{ \sup_{b \le x < \infty} |\gamma_N^*(x)| > \delta \Big\} = 0 \quad \text{for all } \delta > 0 \tag{4.24}$$

Since the function $f(x)/(\lambda_N f(x) + (1 - \lambda_N) g(x))$ is continuous on $[a, b]$ for all $-\infty < a < b < \infty$, on using a one-term Taylor expansion we get

$$\sup_{0 \le y \le 1} |\gamma_N(F_m^{-1}(y)) - \Gamma_N^*(F_m^{-1}(y))| = o_P(1) \tag{4.25}$$

by (4.21)–(4.24) and Theorem 2.1. Now (4.3) and a.s. continuity of the gaussian process $\Gamma_N^*(F^{-1}(y))$ imply

$$\sup_{0 \le y \le 1} |\Gamma_N^*(F_m^{-1}(y)) - \Gamma_N^*(F^{-1}(y))| = o_P(1) \tag{4.26}$$

and therefore we have

$$\sup_{0 \le y \le 1} |\gamma_N(F_m^{-1}(y)) - \Gamma_N^*(F^{-1}(y))| = o_P(1) \tag{4.27}$$

by (4.25). Hence also the lemma by (4.20), (4.27), and Theorem 2.1. ■

Now we are in the position of giving the

Proof of Theorem 3.3. We have

$$
\begin{aligned}
N^{1/2}&\left(y - \left(FH^{-1}H_N F_m^{-1}\right)^{-1}(y)\right) \\
&= s_N\left(\left(FH^{-1}H_N F_m^{-1}\right)^{-1}(y)\right) \\
&\quad + N^{1/2}\left(y - FH^{-1}H_N F_m^{-1}\left(FH^{-1}H_N F_m^{-1}\right)^{-1}(y)\right), \\
&\qquad\qquad 0 \le y \le 1 \qquad (4.28)
\end{aligned}
$$

By Lemma in Horváth (1984)

$$\sup_{0\le y\le 1} \left|\left(FH^{-1}H_N F_m^{-1}\right)^{-1}(y) - y\right| \le \sup_{0\le y\le 1} \left|FH^{-1}H_N F_m^{-1}(y) - y\right|,$$

and hence by Lemma 4.1

$$\sup_{0\le y\le 1} \left|\left(FH^{-1}H_N F_m^{-1}\right)^{-1}(y) - y\right| = o_P(1)$$

Consequently, on using a.s. continuity of the stochastic process Δ_N^*, we arrive at

$$\sup_{0\le y\le 1} \left|s_N\left(\left(FH^{-1}H_N F_m^{-1}\right)^{-1}(y)\right) - \Delta_N^*(y)\right| = o_P(1)$$

By definition of inverses, the second term of the right-hand side of equality (4.28) is bounded above by the largest jump of $s_N(y)$ over [0, 1]. Hence

$$
\begin{aligned}
&\sup_{0\le y\le 1} N^{1/2}\left|y - FH^{-1}H_N F_m^{-1}\left(FH^{-1}H_N F_m^{-1}\right)^{-1}(y)\right| \\
&\qquad \le \sup_{0\le y\le 1-1/N}\ \sup_{0\le x\le 1/N} \left|s_N(y+x) - s_N(y)\right|
\end{aligned}
$$

Next by Lemma 4.1 it suffices to show that

$$\sup_{0\le y\le 1-1/N}\ \sup_{0\le x\le 1/N} \left|\Delta_N^*(y+x) - \Delta_N^*(y)\right| = o_P(1)$$

The latter is true by definition of Δ_N^* being in terms of Brownian bridges.

Thus we proved

$$\sup_{0 \le y \le 1} |N^{1/2}(y - D_N^{-1}D(y)) - \Delta_N^*(y)| = o_P(1)$$

Since [cf. (1.14)]

$$r_N(D(y)) = -N^{1/2}(y - D_N^{-1}D(y))$$

and

$$\Delta_N(D(y)) = -\Delta_N^*(y)$$

our Theorem 3.3 is also proved. ■

5. WEAK CONVERGENCE IN WEIGHTED METRICS AND CHERNOFF–SAVAGE THEOREMS

The following class of functions will be frequently referred to in the sequel. Let Q be the class of positive functions q on $(0,1)$ that are nondecreasing in a neighborhood of 0 and nonincreasing in a neighborhood of 1, and for which the integral

$$L(q) = \int_0^1 \left(\frac{1}{q(t)}\right)^2 dt$$

is finite.

Theorem 5.1. Assume the conditions C.1 and (1.6) and that

C.3. $gF^{-1}(y)/fF^{-1}(y)$ is nonincreasing in a neighborhood of 0 and nondecreasing in a neighborhood of 1

C.4. with $q \in Q$ we have also

$$\int_0^1 \left(\frac{gF^{-1}(ay)}{q(y)fF^{-1}(ay)}\right)^2 dy < \infty \quad \text{for all } 0 < a < 1$$

Then, on the probability space of Theorem 2.1 and with q in C.4 we have, as $m \wedge n \to \infty$,

$$\sup_{1/(N+1) \le y \le N/(N+1)} |d_N(y) - (1 - \lambda_N)\Gamma_N(y)|/q(y) = o_P(1) \quad (5.1)$$

Theorem 5.2. Assume the conditions C.1 and (1.6) and that $q \in Q$. Then, on the probability space of Theorem 2.1, as $m \wedge n \to \infty$,

$$\sup_{0 \le y \le 1} \frac{|r_N(y) - \Delta_N(y)|}{q(y)} = o_P(1) \tag{5.2}$$

REMARK 5.1. From the definition of Γ_N and the proof of Theorem 5.1 it is clear that

1. if $\sup_{0 \le y \le 1} gF^{-1}(y)/fF^{-1}(y) < \infty$, then conditions C.3 and C.4 can be replaced by simply requiring $q \in Q$ in Theorem 5.1, or
2. condition C.3 in Theorem 5.1 can be replaced by assuming that $gF^{-1}(y)/fF^{-1}(y)$ is regularly varying near zero and 1 with a nonzero tail index.

When testing H_0: $F = G$ versus H_1: $F \neq G$, statistics of the form

$$T_N = m^{-1} \sum_{j=1}^{N} E_{Nj} z_{Nj} \tag{5.3}$$

are often used. Here E_{Nj} are given numbers and

$$z_{Nj} = \begin{cases} 1 & \text{if the } j\text{th smallest observation in the combined sample is an } X \\ 0 & \text{otherwise.} \end{cases}$$

Let J_N be a sequence of left continuous real-valued functions on $(0,1)$. If J_N is such that $J_N(D_N(y)) = E_{Nj}$ for $(j-1)/N < y \le j/N$ $(j = 1, \ldots, N)$, a so-called score function, then we have

$$m^{-1} E_{Nj} z_{Nj} = J_N\left(\frac{R_j}{N}\right)\frac{1}{m}, \qquad 1 \le j \le m$$

and so we get

$$\begin{aligned} T_N &= \sum_{j=1}^{m} J_N\left(\frac{R_j}{n}\right)\frac{1}{m} \\ &= \int_{-\infty}^{\infty} J_N(H_N(x))\, dF_m(x) \end{aligned} \tag{5.4}$$

the Chernoff–Savage (1958) representation of a linear rank statistic T_N, which they take as their definition of T_N even if $J_N(R_j/N) \neq E_{Nj}$. Chernoff and Savage (1958) established a global limit theorem for T_N of (5.4) *directly*, covering many situations with $F \neq G$. Pyke and Shorack (1968) introduced the following representation for T_N:

$$\begin{aligned} T_N &= \int_0^1 F_m H_N^{-1}(y)\, d\nu_N(y) \\ &= \sum_{i=1}^{N} R_m H_N^{-1}\left(\frac{i}{N}\right)\left\{\nu_N\left(\frac{i}{M}\right) - \nu_N\left(\frac{i-1}{N}\right)\right\} \end{aligned} \tag{5.5}$$

with a suitable signed measure ν_N on $[0,1]$. Parzen (1983) argues that the Chernoff–Savage representation of T_N as in (5.4), which preserves the direct role of the score functions J_N in terms of the relative ranks $\{R_j/N\}$, is directly usable for doing exploratory data analysis and that it should be directly studied also for the sake of theorem proving. As a starting point he suggests the representation

$$T_N = \int_{-\infty}^{\infty} J_N(H_N(x))\, dF_m(x) = \int_0^1 J_N(D_N(y))\, dy = \int_0^1 J_N(t)\, d\Delta_N^{-1}(t) \tag{5.6}$$

which we now take as definition of T_N even if $J_N(R_j/N) \neq E_{Nj}$. Hence, in order to study the asymptotic distribution of T_N, we are led to considering that of the rv

$$\begin{aligned} t_N &= N^{1/2}\left\{T_N - \int_0^1 J(y)\, dD^{-1}(y)\right\} \\ &= N^{1/2}\left\{\int_0^1 J_N(y)\, dD_N^{-1}(y) - \int_0^1 J(y)\, dD^{-1}(y)\right\} \end{aligned} \tag{5.7}$$

where J is a real-valued left continuous function on $(0,1)$. We prove

Theorem 5.3. Assume the conditions C.1 and (1.6). Then, on the probability space of Theorem 2.1 we have, as $m \wedge n \to \infty$,

$$\left| t_N - \int_0^1 J(y)\, d\Delta_N(y) \right| = o_P(1), \tag{5.8}$$

provided that

1. J is a function on $(0, 1)$ whose derivative J' exists almost everywhere with respect to Lebesgue measure on $(0, 1)$,

and for some $q \in Q$ we have

2. $\lim_{y \downarrow 0} q(y)J(y) = \lim_{y \uparrow 1} q(y)J(y) = O(1)$

and

3. $\int_0^1 q(y)|J'(y)|\,dy < \infty$,

while the $\{J_N\}$ $(N = 1, 2, \dots)$ are such functions on $(0, 1)$ for which we have

4. $N^{1/2}\int_0^1 (J_N(y))\,dD_N^{-1}(y) = o_P(1)$.

REMARK 5.2. The limiting rv $\int_0^1 J(y)\,dD_N(y)$ of (5.8) is normal with mean zero and variance σ_N^2, calculated by Chernoff and Savage (1958). An advantage of stating and proving a Chernoff–Savage (1958) theorem as in (5.8), is, as compared to going about it via (5.5), that the results on bootstrapping empirical functions of Chapter 17 in Csörgő et al. (1986a) are directly applicable to produce immediate bootstrapped versions of Theorems 5.1, 5.2, and 5.3. The latter in turn enable one to simulate appropriate percentage points of the normal rv $\int_0^1 J(y)\,d\Delta_N(y)$ via bootstrap—resampling our given samples without worrying about estimating its complicated variance.

REMARK 5.3. Chernoff–Savage-type theorems can also be based on Theorem 2.1 for functionals like $\int_0^1 J_N(y)\,dD_N(y)$. The thus gained results can again be bootstrapped immediately along the lines as suggested in Remark 5.2.

REMARK 5.4. In Theorem 5.3 we only illustrated how one can prove a Chernoff–Savage-type theorem using approximations of stochastic processes. The presented method is an easy and direct approach to proving asymptotic normality of t_N. There are other ways to deduce asymptotic normality of t_N that require less and/or different regularity conditions. The theory of rank statistics has a huge literature. We shall not compare our results with others. It will suffice for us to refer only to the books of Hájek and Šidák (1967) and Serfling (1980), and to the review paper of Ghosh (1984).

6. PROOFS OF THE RESULTS IN SECTION 5

Proof of Theorem 5.1. We have by definition

$$d_N(y) = (1 - \lambda_N)^{1/2}\beta_n\big(GF_m^{-1}(y)\big) - (1 - \lambda_N)\lambda_N^{-1/2}m^{1/2}\big(GF^{-1}(y) - GF_m^{-1}(y)\big) + O(N^{-1/2}) \quad (6.1)$$

with probability one for each N. Hence it follows from the proof of Theorem 3.2 that, on assuming C.1 and (1.6), we have for any $0 < \varepsilon < \frac{1}{2}$

$$\sup_{\varepsilon \le y \le 1-\varepsilon} |d_N(y) - (1 - \lambda_N)\Gamma_N(y)|/q(y) = o_P(1)$$

Hence it suffices to show that for all $\delta > 0$

$$\lim_{\varepsilon \downarrow 0} \limsup_{N \to \infty} P\left\{ \sup_{1/(N+1) \le y \le \varepsilon} |d_N(y)|/q(y) \ge \delta \right\} = 0 \tag{6.2}$$

and

$$\lim_{\varepsilon \downarrow 0} \limsup_{N \to \infty} P\left\{ \sup_{1/(N+1) \le y \le \varepsilon} |\Gamma_N(y)|/q(y) \ge \delta \right\} = 0 \tag{6.3}$$

and similarly over the intervals $1 - \varepsilon \le y \le N/(N+1)$. We shall prove only (6.2) and (6.3). The proofs on the right end are similar.

From the definition of the gaussian process Γ_N, for the sake of verifying (6.3) it is enough to show that we have

$$\lim_{\varepsilon \downarrow 0} \limsup_{N \to \infty} P\left\{ \sup_{1/(N+1) \le y \le \varepsilon} |B(GF^{-1}(y))|/q(y) \ge \delta \right\} = 0 \tag{6.4}$$

and

$$\lim_{\varepsilon \downarrow 0} \limsup_{N \to \infty} P\left\{ \sup_{1/(N+1) \le y \le \varepsilon} \frac{gF^{-1}(y)}{fF^{-1}(y)} \frac{|B(y)|}{q(y)} \ge \delta \right\} = 0 \tag{6.5}$$

for all $\delta > 0$ and any Brownian bridge B. On making the change of variable $s = GF^{-1}(y)$ in (6.4), we want to have

$$\lim_{\varepsilon \downarrow 0} \limsup_{N \to \infty} P\left\{ \sup_{1/(N+1) \le FG^{-1}(s) \le \varepsilon} |B(s)|/q^*(s) \ge \delta \right\} = 0 \tag{6.6}$$

for all $\delta > 0$ with $q^*(s) = q(FG^{-1}(s))$. First we note that $q^* \in Q$, that is,

$$\int_0^1 \left(\frac{1}{q^*(s)} \right)^2 ds = \int_0^1 \left(\frac{1}{q(y)} \right)^2 \frac{gF^{-1}(y)}{fF^{-1}(y)} dy < \infty \tag{6.7}$$

The latter is true by condition C.4, on applying Schwarz's inequality to $\int_0^1 l(y)\, dr(y)$, with $l(y) = gF^{-1}(y)/fF^{-1}(y)$ and $r(y) = \int_0^y (1/q(t))^2\, dt$.

Hence q^* is also a Chibisov–O'Reilly function and (6.4) follows by (2.5). Let

$$q^0(y) = q(y)(fF^{-1}(y)/gF^{-1}(y)) \tag{6.8}$$

By conditions C.3 and C.4 $q^0(y) \in Q$, and hence we have also (6.5) by the same argument. This also completes the proof of (6.3).

In order to verify (6.2), we first note that for all Chibisov–O'Reilly functions q we have $s^{-1/2}q(s) \to \infty$ as $s \downarrow 0$. Hence for the $O(N^{-1/2})$ term of d_N [cf. (6.1)] a statement like that of (6.2) for d_N is true with any Chibisov–O'Reilly function q. Consequently, in order to verify (6.2), it suffices to show that for all $\delta > 0$ we have

$$\lim_{\varepsilon \downarrow 0} \limsup_{N \to \infty} P\left\{ \sup_{1/(N+1) \le y \le \varepsilon} |\beta_n(GF_m^{-1}(y))|/q(y) > \delta \right\} = 0 \tag{6.9}$$

and

$$\lim_{\varepsilon \downarrow 0} \limsup_{N \to \infty} P\left\{ \sup_{1/(N+1) \le y \le \varepsilon} m^{1/2}|GF^{-1}(y) - GF_m^{-1}(y)|/q(y) > \delta \right\} = 0 \tag{6.10}$$

On using the mean value theorem

$$\sup_{1/(N+1) \le y \le \varepsilon} \frac{m^{1/2}|GF^{-1}(y) - GF_m^{-1}(y)|}{q(y)} = \sup_{1/(N+1) \le y \le \varepsilon} \frac{gF^{-1}(\xi_m(y))}{fF^{-1}(\xi_m(y))} \frac{|u_m(y)|}{q(y)} \tag{6.11}$$

where

$$FF_m^{-1}(y) \wedge y \le \xi_m(y) \le FF_m^{-1}(y) \vee y$$

Also by Wellner (1978), as $m \to \infty$,

$$\sup_{1/(m+1) \le y \le m/(m+1)} \frac{y}{FF_m^{-1}(y)} \vee \frac{y}{F_m F^{-1}(y)} = O_P(1) \tag{6.12}$$

Hence, instead of (6.9), by (6.12) it is enough to show that we have

$$\lim_{\varepsilon \downarrow 0} \limsup_{N \to \infty} P\left\{ \sup_{1/(N+1) \le y \le \varepsilon} \frac{|\beta_n(GF^{-1}(ay))|}{q(y) > \delta} \right\} = 0 \tag{6.13}$$

for all $\delta > 0$ and $a > 1$, and instead of (6.10), by (6.11) and (6.12) it suffices to show that we have

$$\lim_{\varepsilon \downarrow 0} \limsup_{N\to\infty} P\left\{ \sup_{1/(N+1)\le y\le \varepsilon} \frac{gF^{-1}(ay)}{fF^{-1}(ay)} \frac{|u_m(y)|}{q(y)} > \delta \right\} = 0 \quad (6.14)$$

for all $\delta > 0$ and $0 < a < 1$. On making the change of variable $s = GF^{-1}(ay)$ in (6.13), an argument like that of verifying (6.4) combined with (2.8) proves that (6.13) is true. Also, an argument like that of verifying (6.5) combined with (2.7) shows that (6.14) is true. This also completes the proof of Theorem 5.1. ■

Proof of Theorem 5.2. For any $0 < \varepsilon < \frac{1}{2}$ we have by Theorem 3.3 that, as $m \wedge n \to \infty$,

$$\sup_{\varepsilon\le y\le 1-\varepsilon} \frac{|r_N(y) - \Delta_N(y)|}{q(y)} = o_P(1) \quad (6.15)$$

Hence it suffices to show that we have for all $\delta > 0$

$$\lim_{\varepsilon \downarrow 0} \limsup_{N\to\infty} P\left\{ \sup_{0\le y\le \varepsilon} \frac{|\Delta_N(y)|}{q(y)} \ge \delta \right\} = 0 \quad (6.16)$$

and

$$\lim_{\varepsilon \downarrow 0} \limsup_{N\to\infty} P\left\{ \sup_{0\le y\le \varepsilon} \frac{|\Delta_N(y)|}{q(y)} \ge \delta \right\} = 0 \quad (6.17)$$

and similarly over the interval $1 - \varepsilon \le y \le 1$. We shall indicate only proof of (6.16) and (6.17). The proofs on the right end are similar.

From the definition of the gaussian process Δ_N it follows that in order to verify (6.17), it is enough to show that

$$\lim_{\varepsilon \downarrow 0} P\left\{ \sup_{0\le y\le \varepsilon} \frac{|B(FH^{-1}(y))|}{q(y)} \ge \delta \right\} = 0 \quad (6.18)$$

and

$$\lim_{\varepsilon \downarrow 0} P\left\{ \sup_{0\le y\le \varepsilon} \frac{|B(GH^{-1}(y))|}{q(y)} \ge \delta \right\} = 0 \quad (6.19)$$

for all $\delta > 0$ and any Brownian bridge. The latter in turn are immediate by (2.5), for if $q(y) \in Q$, then so are also $q(HF^{-1}(t))$ and $q(HG^{-1}(t))$.

As to (6.16), we note only that its proof hinges on the fact that r_N has a similar representation in terms of $\alpha_m(FH_N^{-1}(y))$ and $\beta_n(GH_N^{-1}(y))$ as Δ_N does in terms $B_m^{(1)}(FH^{-1}(y))$ and $B_n^{(2)}(GH^{-1}(y))$. Now an argument like that of (6.18) and (6.19) combined with (2.6) and (2.8) yields

$$\lim_{\varepsilon \downarrow 0} \limsup_{m\to\infty} P\left\{ \sup_{0 \le y \le \varepsilon} \frac{|\alpha_m(FH^{-1}(y))|}{q(y)} > \delta \right\} = 0 \tag{6.20}$$

and

$$\lim_{\varepsilon \downarrow 0} \limsup_{n\to\infty} P\left\{ \sup_{0 \le y \le \varepsilon} \frac{|\beta_n(GH^{-1}(y))|}{q(y)} > \delta \right\} = 0 \tag{6.21}$$

for all $\delta > 0$, and similarly over $(1 - \varepsilon, 1)$. Hence

$$\sup_{-\infty < x < \infty} H_N(x) \frac{1 - H_N(x)}{H(x)(1 - H(x))} = O_P(1), \qquad n \to \infty$$

combined with (6.20) and (6.21) also completes the proof of Theorem 5.2. ■

Proof of Theorem 5.3. By condition 4 we get, as $m \wedge n \to \infty$,

$$t_N = \int_0^1 J(y)\, dr_N(y) + o_P(1)$$

and hence, in order to verify (5.8), it suffices to show that as $m \wedge n \to \infty$

$$\left| \int_0^1 J(y)\, d(r_N(y) - \Delta_N(y)) \right| = o_P(1)$$

The latter follows immediately by utilizing Theorem 5.2 and conditions 1–3 by integration by parts. ■

ACKNOWLEDGMENTS

Research of Emad-Eldin A. A. Aly and Miklós Csörgő is partially supported by NSERC Canada grants. Research of Lajos Horváth was done while at Carleton University, also supported by NSERC Canada grants of M. Csörgő and D. A. Dawson, and an EMR Canada grant of M. Csörgő. Part of this work was done while M. Csörgő was visiting the Department of Statistics of King Saud University, Riyadh, Saudi Arabia, and the Department of Mathematics and Computer Science of the University of Leiden in Holland. The authors also wish to thank Murray Burke for sending them a copy of Parzen (1983).

REFERENCES

Aly, E.-E. A. A. (1983a). Strong approximations of the Q-Q process. Technical Report Series of the Laboratory for Research in Statistics and Probability, Carleton University, University of Ottawa, No. 16.

Aly, E.-E. A. A. (1983b). Weak convergence of Q-Q processes in weighted sup-norm metrics. Technical Report Series of the Laboratory for Research in Statistics and Probability, Carleton University, University of Ottawa, No. 16.

Aly, E.-E. A. A. and Bleuer, S. (1986). Confidence bands for Quantile-Quantile plots. *Statistics & Decisions* **4**, 205–225.

Chernoff, H. and Savage, I. R. (1958). Asymptotic normality and efficiency of certain nonparametric test statistics. *Ann. Math. Statist.* **29**, 972–994.

Chibisov, D. (1964). Some theorems on the limiting behaviour of empirical distribution functions. *Selected Translations in Mathematical Statistics and Probability* **6**, 147–156.

Csörgő, M., Csörgő, S., and Horváth, L. (1986a). An Asymptotic Theory for Empirical Reliability and Concentration Processes. *Lecture Notes in Statistics*, Vol. 33, New York: Springer-Verlag.

Csörgő, M., Csörgő, S., Horváth, L. and Mason, D. M. (1986b). Weighted empirical and quantile processes. *Ann. Probab.* **14**, 31–85.

Fisher, N. I. (1983). Graphical methods in nonparametric statistics: A review and annotated bibliography. *Int. Statist. Rev.* **51**, 25–58.

Ghosh, M. (1984). Rank statistics and limit theorems. *Handbook of Statistics* **4**, 145–171.

Hájek, T. and Šidák, Z. (1967). *Theory of Rank Tests*. Orlando, Florida: Academic Press.

Horváth, L. (1984). Strong approximation of extended renewal processes. *Ann. Probab.* **12**, 1149–1166.

O'Reilly, N. (1974). On the weak convergence of empirical processes in sup-norm metrics. *Ann. Probab.* **2**, 642–651.

Parzen, E. (1983). Fun. stat quantile approach to two sample statistical data analysis. Texas A & M University, Technical Report No. A-21.

Pyke, R. and Shorack, G. R. (1968). Weak convergence of a two-sample empirical process and a new approach to Chernoff–Savage theorems. *Ann. Math. Statist.* **39**, 755–771.

Serfling, R. J. (1980). *Approximation Theorems of Mathematical Statistics*, New York: Wiley.

Shorack, G. R. (1979). Weak convergence of empirical and quantile processes in sup-norm metrics via KMT-construction. *Stoch. Proc. Appl.* **9**, 95–98.

Shorack, G. R. and Wellner, J. A. (1982). Limit theorems and inequalities for the uniform empirical process indexed by intervals. *Ann. Probab.* **10**, 639–652.

Stute, W. (1982). The oscillation behaviour of empirical processes. *Ann. Probab.* **10**, 86–107.

Wellner, J. A. (1978). Limit theorems for the ratio of the empirical distribution function to the true distribution function. *Z. Wahrsch. verw. Gebiete* **45**, 73–88.

Some Aspects of Edgeworth Expansions in Statistics and Probability

R. N. Bhattacharya
Department of Mathematics
Indiana University
Bloomington, Indiana

SUMMARY

The article has two parts. In the first part an absolutely convergent Edgeworth expansion is shown to exist for expectations of a class of smooth functions of the normalized sum of i.i.d. random vectors; also, Edgeworth expansions are shown to exist for certain linear and nonlinear autoregressive models. In the second part, the so-called empirical Edgeworth expansion is proposed as an alternative to Efron's bootstrap estimate of distributions of normalized (Studentized) statistics of the type considered earlier by Bhattacharya and Ghosh (1978). Asymptotic comparisons of these two methods extend earlier results of Bickel and Freedman (1980) and Babu and Singh (1984). The empirical Edgeworth expansion also immediately applies to many time-series models.

1. INTRODUCTION

This article surveys some recent developments in the theory of Edgeworth expansions and provides some new results.

Research supported by NSF Grant DMS 8503358.

Section 2 begins with the independent and identically distributed (i.i.d.) case. It is shown that in case the common distribution of the sequence of i.i.d. random vectors has a finite moment-generating function in a neighborhood of 0, expectations of certain functions of the normalized sum have convergent Edgeworth exphasions, converging to the true expectation for a fixed number n of summands (Theorem 2.1). These are functions f such that, for some integer s, $(1 + \|x\|^2)^{-s}f(x)$ is the Fourier transform of a finite signed measure. This extends an earlier result of Bhattacharya and Puri (1983). Although such classes of functions must of necessity be restricted, one expects that under the present assumption the numerical approximation of the distribution of the normalized sum may improve by the addition of several terms beyond the normal, even in those cases where no valid asymptotic expansion exists for probabilities beyond the normal term. Theorem 2.1, as well as Berry–Esseen-type bounds and Edgeworth expansions of expectations of smooth functions of the normalized sum, may also be derived in case of a strictly stationary sequence provided cumulants of the normalized sum decay rapidly enough with the number n of summands, and do not grow too fast, for fixed n, with the order of the cumulant. Unfortunately, the so-called *Markov-type regularity* conditions under which such bounds have been derived by Statulevičius (1974), and Zhurbenko (1972), turn out to be essentially vacuous by a result of Bradley (1980). However, Götze and Hipp (1983) have recently obtained asymptotic expansions for a reasonably wide class of dependent sequences. Two examples are given from time series: rth-order autoregressive models, one linear and the other nonlinear.

Section 3 proposes the so-called *empirical Edgeworth expansion* as an alternative to the *bootstrap* of Efron (1979). Such an alternative was considered earlier to a certain extent by Bickel and Freedman (1980). In the case of normalized sums of n i.i.d. random variables, whose common distribution satisfies *Cramér's condition*, Babu and Singh (1984) have strengthened the results of Bickel and Freedman (1980), by showing that the bootstrap estimates of probabilities differ from the $(s-1)$-term empirical Edgeworth expansion of the same probabilities by only $o(n^{-(s-2)/2})$ if moments of order s are finite. In this chapter we obtain a corresponding result for a large class of statistics of interest, namely, those that may be expressed or approximated by statistics of the form $H(\overline{X})$. Using results of Bhattacharya and Ghosh (1978) it is shown that under appropriate assumptions the asymptotic equivalence up to $o(n^{-(s-2)/2})$ of the bootstrap estimate and the $(s-1)$-term empirical Edgeworth expansion is valid for this class of statistics as well (Theorem 3.3). For such statistics, under appropriate assumptions, it is shown that the error of estimation of the true probability by the two-term Edgeworth expansion is $O_p(n^{-1})$ uniformly over all Borel sets, if the dispersion matrix of the limiting normal distribution does not involve unknown population parameters as is true, for example, for *Studentized statistics* (Theorem 3.2). Preliminary investigations, theoretical as well as numerical, by Qumsiyeh (1986) indicate that the three-term empirical Edgeworth expansion compares well with the bootstrap. An advantage of the empirical Edgeworth expansion is that it extends to stationary sequences of dependent random variables for which the Edgeworth expansion is valid.

2. EDGEWORTH EXPANSIONS UNDER INDEPENDENCE AND DEPENDENCE

In the classical Cramér–Edgeworth expansion one considers a sequence of k-dimensional i.i.d. random vectors $X_n = (X_n^{(1)}, X_n^{(2)}, \ldots, X_n^{(k)})$ $(n = 1, 2, \ldots)$ with mean vector μ, nonsingular dispersion matrix V, and a finite sth absolute moment $\rho_s = E\|X_1\|^s$ for some integer $s \geq 3$. Let Q_1 denote the distribution of $X_1 - \mu$ and Q_n that of the normalized sum $Z_n = n^{-1/2}\Sigma_{j=1}^n(X_j - \mu)$. Then the *characteristic function* (c.f.) $\hat{Q}_n$ of Z_n is $Q_n(\xi) = (Q_1(\xi|\sqrt{n}))^n$ and its logarithm (principal branch, defined in a neighborhood of the origin) may be expanded in a Taylor's series: Writing $\langle x, y\rangle = \Sigma x^{(i)}y^{(i)}$ for $x, y \in R^k$, one has

$$\log Q_n(\xi) = -\tfrac{1}{2}\langle \xi, V\xi\rangle + \sum_{m=3}^{s} n^{-(m-2)/2} \sum_{|\nu|=m} \frac{\lambda_\nu}{\nu!}(i\xi)^\nu$$

$$+o\left(n^{-(s-2)/2}\right) \qquad \text{as } n \to \infty \tag{2.1}$$

Here $\nu = (\nu^{(1)}, \nu^{(2)}, \ldots, \nu^{(k)})$ is a multi-index, that is, a vector whose coordinates are non-negative integers; $|\nu| = \nu^{(1)} + \nu^{(2)} + \cdots + \nu^{(k)}$, $\nu! = (\nu^{(1)}!)(\nu^{(2)}!) \cdots (\nu^{(k)}!)$, $(i\xi)^\nu = (i\xi^{(1)})^{\nu^{(1)}} \cdots (i\xi^{(k)})^{\nu^{(k)}}$; and λ_ν is the νth *cumulant* of X_1, that is, $i^{|\nu|}\lambda_\nu = (D^\nu \log \hat{Q}_1(\xi))_{\xi=0}$ with $D^\nu = (\partial/\partial\xi^{(1)})^{\nu^{(1)}} \cdots (\partial/\partial\xi^{(k)})^{\nu^{(k)}}$. Raising (2.1) to the exponent, expanding the exponential of the sum $\Sigma_{m=3}^s$ on the right-hand side, and collecting together powers of $n^{-r/2}$, one gets the *Edgeworth expansion* of the c.f. $\hat{Q}_n$:

$$\hat{Q}_n(\xi) = \exp\{-\tfrac{1}{2}\langle \xi, V\xi\rangle\}\left[1 + \sum_{r=1}^{s-2} n^{-r/2}P_r(i\xi : Q_1)\right] + o\left(n^{-(s-2)/2}\right) \tag{2.2}$$

Here P_r is the rth *Cramér–Edgeworth* polynomial whose coefficients are polynomials in the cumulants λ_ν. Inverting the expression on the right-hand side of (2.2), leaving out the remainder $o(n^{-(s-2)/2})$, one gets formally the Edgeworth expansion of the "density" of Q_n as

$$\psi_{s,n}(x) = \left[1 + \sum_{r=1}^{s-2} n^{-r/2} P_r(-D : Q_1)\right] \phi_V(x), \tag{2.3}$$

where ϕ_V is the density of the normal distribution having mean 0 and dispersion matrix V. Also, $D = (\partial/\partial x^{(1)}, \ldots, \partial/\partial x^{(k)})$.

If, in addition to the hypotheses on X_n introduced above, one assumes that the moment-generating function (m.g.f.) $E\exp\{\langle \xi, X_1\rangle\}$ is finite in a neighborhood of $\xi = 0$ then it may be shown (Bhattacharya and Puri, 1983) that the c.f. $\hat{Q}_n$ has an absolutely convergent expansion

$$\hat{Q}_n(\xi) = \exp\{-\tfrac{1}{2}\langle \xi, V\xi\rangle\}\left[1 + \sum_{r=1}^{\infty} n^{-r/2} P_r(i\xi : Q_1)\right] \tag{2.4}$$

in a region $\{\|\xi\| < \delta\sqrt{n}\}$ for some $\delta > 0$. One may immediately obtain Berry–Esseen bounds and Edgeworth expansions of expectations of smooth functions of Z_n from (2.4).

Theorem 2.1. Let X_n $(n = 1, 2, \ldots)$ be an i.i.d. sequence of k-dimensional random vectors with common mean vector μ and nonsingular dispersion matrix V. If the m.g.f. of X_1 is finite in a neighborhood of the origin, then for all f such that $(1 + \|x\|^2)^{-s} f$ is the Fourier transform of a finite signed measure with compact support for some integer $s \geq 0$, one has, for all sufficiently large n

$$\int_{\mathbb{R}^k} f\,dQ_n = \lim_{N\to\infty} \int f(x)\left[1 + \sum_{r=1}^{N} n^{-r/2} P_r(-D : Q_1)\right] \phi_V(x)\,dx. \tag{2.5}$$

Proof. Let $(1 + \|x\|^2)^{-s} f = \hat{\mu}$ be the Fourier transform of a finite signed measure μ with compact support: $|\mu|(\{\xi \in \mathbb{R}^k \mid \|\xi\| > C_1\}) = 0$, where $|\mu|$ denotes variation. Then by (2.4) and Parseval's relation one has, for

$n > (C_1/\delta)^2$,

$$\int_{\mathbb{R}^k} f\,dQ_n = \lim_{N\to\infty} \int_{\mathbb{R}^k} \hat{\mu}(x)(1 + \|x\|^2)^s Q_n(dx)$$

$$= \int_{\mathbb{R}^k} (1-\Delta)^s \hat{Q}_n(\xi)\mu(d\xi)$$

$$= \int_{\{\|\xi\| \le C_1\}} (1-\Delta)^s \Bigg[\exp\{-\tfrac{1}{2}\langle \xi, V\xi\rangle\}$$

$$\cdot\left(1 + \sum_{r=1}^{\infty} n^{-r/2} P_r(i\xi : Q_1)\right)\Bigg]\mu(d\xi) \qquad \left(\Delta = \sum \frac{\partial^2}{(\partial x^{(i)})^2}\right)$$

$$= \lim_{N\to\infty} \int_{\mathbb{R}^k} (1-\Delta)^s$$

$$\times\left[\exp\{-\tfrac{1}{2}\langle \xi, V\xi\rangle\}\left(1 + \sum_{r=1}^{N} n^{-r/2} P_r(i\xi : Q_1)\right)\right]\mu(d\xi)$$

$$= \lim_{N\to\infty} \int_{\mathbb{R}^k} \hat{\mu}(x)(1 + \|x\|^2)^s$$

$$\times\left[1 + \sum_{r=1}^{N} n^{-r/2} P_r(-D : Q_1)\right]\phi_V(x)\,dx$$

$$= \lim_{N\to\infty} \int_{\mathbb{R}^k} f(x)\left[1 + \sum_{r=1}^{N} n^{-r/2} P_r(-D : Q_1)\right]\phi_V(x)\,dx \quad \blacksquare \qquad (2.6)$$

REMARK 2.1.1. The hypothesis of finiteness of the m.g.f. of Q_1 in a neighborhood of 0 is rather strong and the restriction imposed on functions f for which (2.5) holds severe. However, it is quite unusual that the Edgeworth expansion steadily improves as the number of terms is increased indefinitely (and in the limit equals $\int f\,dQ_n$) for *fixed* n (for all $n > c_1^2/\delta^2$). This is an indication that *numerical approximation of probabilities* $Q_n(A)$ will probably improve by adding quite a few terms even for *moderate* n, under additional assumptions on the type of the distribution Q_1 and on the function $\varepsilon \to \Phi_V((\partial A)^\varepsilon)$ as $\varepsilon \downarrow 0$.

REMARK 2.1.2. As has been pointed out in Bhattacharya and Puri (1983, Section 3), (2.4) holds if the c.f. $\hat{Q}_n(\xi)$ of a statistic T_n (with distribution Q_n) is analytic as a function of ξ and $n^{-1/2}$ in a neighborhood of $\xi = 0$, and $n^{-1/2} = 0$, and its νth cumulant is of the order $O(n^{-(|\nu|-2)/2})$, $|\nu| \geq 2$. Suppose X_n $(n = 1, 2, \ldots)$ is a stationary sequence (of random vectors with values in $\mathbb{R}^k$) such that the νth cumulant $K_{\nu,n}$ of $Z_n = n^{-1/2}\sum_{j=1}^n(X_j - EX_j)$ satisfies

$$|K_{\nu,n}| \leq \frac{C_3 C_2^{|\nu|}|\nu|^{|\nu|}}{n^{(|\nu|-2)/2}} \qquad (|\nu| \geq 2) \tag{2.7}$$

form some positive constants C_2, C_3 (not depending on ν or n). If Z_n has a nonsingular limiting dispersion matrix V, then similar arguments as indicated above yield (2.4) for $\|\xi\| < \delta\sqrt{n}$ with $P_r(i\xi : Q_1)$ replaced by a polynomial $\bar{P}_{r,n}(i\xi)$ whose coefficients are bounded in n. Consequently, an analog of Theorem 2.1 holds in this case, together with (1) Berry–Esseen bounds for $Q_n(A) - \Phi_V(A)$ where $\Phi_V((\partial A)^{\varepsilon}) = O(\varepsilon)$ as $\varepsilon \downarrow 0$, and (2) Edgeworth expansions of $\int f\, dQ_n$ for smooth f (Bhattacharya, 1985; Götze and Hipp, 1983). The estimate (2.7) has been derived by Statulevičius (1969–1970) when X_n is bounded, Markovian, ergodic, and satisfies Doeblin's condition. Another condition, called the Markov-type *regularity condition* by Statulevičius (1974) and used to derive (2.7) by Statulevičius (1974) and Zhurbenko (1972), has been shown by Bradley (1980) to *imply that* X_n *are i.i.d.*

Götze and Hipp (1983) have obtained a useful asymptotic expansion for dependent sequences which may be applied to the following two examples.

Example 2.1 [Götze and Hipp, 1983, Example (1.12)]. Let X_n be a *linear autoregressive process of order r* $(r \geq 1)$:

$$X_{j+r} = \sum_{i=0}^{r-1} \beta_i X_{j+i} + \varepsilon_{j+r} \tag{2.8}$$

where ε_n $(n = 1, 2, \ldots)$ is an i.i.d. sequence of random variables satisfying *Cramér's condition*:

$$\overline{\lim_{|\xi| \to \infty}} |E \exp\{i\xi\varepsilon_1\}| < 1 \tag{2.9}$$

Also, assume that all the zeros of the polynomial below lie within the unit circle $\{z \in \mathbb{C} : |z| < 1\}$:

$$\beta_0 + \beta_1 z + \beta_2 z^2 + \cdots + \beta_{r-1} z^{r-1} + z^r \tag{2.10}$$

Note that the r-dimensional sequence $Y_n \equiv (X_n, X_{n+1}, \ldots, X_{n+r-1})$, $n \geq 1$, is Markovian and admits a unique invariant probability π on $\mathbb{R}^r$ when all the zeros of (2.10) are within the unit circle. In this case if Y_1 has the (initial) distribution π, then $Y_n (n \geq 1)$ is a stationary ergodic Markov process. Hence $X_n (n \geq 1)$ becomes a stationary ergodic sequence. If, in addition, $E|X_n|^{s+1} < \infty$ for some integer $s \geq 3$, then one has an Edgeworth expansion for the distribution Q_n of $Z_n = n^{-1/2} \sum_{j=1}^n (X_j - EX_j)$:

$$\sup_{A \in \mathscr{A}} \left| Q_n(A) - \int_A \left[1 + \sum_{r=1}^{s-2} n^{-r/2} \overline{P}_r\left(-\frac{d}{dx} \right) \right] \phi_{\sigma^2}(x)\, dx \right|$$
$$= o\left(n^{-(s-2)/2}\right) \quad \text{as } n \to \infty \tag{2.11}$$

for every class $\mathscr{A}$ of Borel sets of $\mathbb{R}^1$ satisfying

$$\sup_{A \in \mathscr{A}} \int_{(\partial A)^\varepsilon} \phi_{\sigma^2}(x)\, dx = O(\varepsilon^a) \quad \text{as } \varepsilon \downarrow 0 \tag{2.12}$$

for some $a > 0$. In (2.11) $\overline{P}_r$ is a polynomial whose coefficients do not depend on n, and ϕ_{σ^2} is the density of the one-dimensional normal distribution with mean 0 and *positive* variance σ^2 given by

$$\sigma^2 = \rho_{00} + 2 \sum_{j=1}^{\infty} \rho_{0j} \tag{2.13}$$

where $\rho_{0j} = \operatorname{cov}(X_1, X_{j+1})$ $(j \geq 0)$.

Example 2.2. Let ε_n $(n = 1, 2, \ldots)$ be an i.i.d. sequence of random variables and X_n a sequence of random variables defined recursively (given $X_1, X_2, \ldots, X_r$) by

$$X_{j+r} = h(X_j, X_{j+1}, \ldots, X_{j+r-1}) + \varepsilon_{j+r} \qquad (j \geq 1), \tag{2.14}$$

where h is a real-valued Borel measurable function on $\mathbb{R}^r$. Such a process X_n $(n \geq 1)$ is called a *nonlinear autoregressive process of order r* if h is nonlinear. Note that the sequence of r-dimensional random vectors $Y_n \equiv (X_n, X_{n+1}, \ldots, X_{n+r-1})$, $n \geq 1$, is a Markov process. Various sufficient conditions for the existence of a unique invariant probability distribution for this Markov process are given in Bhattacharya and Lee (1986). We consider here a simple special case for which *Doeblin's condition and ergodicity* are relatively simple to establish. Assume (1) h is bounded: $a \le h(x) \le b$ for all $x \in \mathbb{R}^r$, (2) ε_n has a density that is bounded away from zero on an interval whose length exceeds $b - a$. Under these conditions the transition probability $p(x, dy)$ of the r-dimensional Markov process Y_n satisfies the condition

$$p(x, B) \ge \mu(B) \tag{2.15}$$

for all $x \in \mathbb{R}^r$ and all Borel subsets B of $\mathbb{R}^r$; here μ is a nonzero absolutely continuous measure on $\mathbb{R}^r$. This immediately implies Doeblin's condition and a single ergodic class. Hence there exists a unique invariant probability π for Y_n. If Y_1 has this (initial) distribution π, then Y_n $(n \ge 1)$ is a stationary ergodic Markov process which is uniformly ϕ mixing, with a mixing coefficient going to 0 exponentially fast. The sequence X_n $(n \ge 1)$ is then also stationary and ergodic. If $E|\varepsilon_n|^{s+1} < \infty$ for some integer $s \ge 3$, then all the conditions in Götze and Hipp (1983, Theorem 2.8) are met and one has the expansion (2.11) for the distribution Q_n of $Z_n = n^{-1/2}\sum_{j=1}^{n}(X_j - EX_j)$, if the class of sets $\mathscr{A}$ satisfies (2.12).

3. THE BOOTSTRAP AND THE EMPIRICAL EDGEWORTH EXPANSION

Suppose that $T(P)$ is a functional, say real-valued, on some space $\mathscr{P}$ of probability measures P on an Euclidean space, and one wishes to obtain a confidence interval for $T(P)$ based on n i.i.d. observations $X_1, X_2, \ldots, X_n$ with common distribution P. For example, if $\mathscr{P}$ is a parametric family then $T(P)$ is a function of the parameter, and one may use the maximum likelihood estimate $\hat{\Theta}_n$ of $T(P)$ and an estimate s_n of its standard error σ_n (i.e., the standard deviation of its asymptotic normal distribution) to form a confidence interval using normal approximation. Under approximate additional assumptions one may do better than normal approximation of the studentized statistic $(\hat{\Theta}_n - T(P))/s_n$. In this section we consider two procedures for improvement over the normal approximation: (1) the *bootstrap* proposed by Efron (1979), and (2) the *empirical Edgeworth expansion* that has been considered by Bickel and Freedman (1980) and Babu and Singh (1984). The comparison of these two procedures is extended in the present article to the class of statistics considered by Bhattacharya and Ghosh (1978). These are statistics which may be expressed as or approximated by statistics of the form $H(\overline{X})$ and include maximum likelihood estimates under regularity conditions. Instead of actually requiring studentization, we shall consider those statistics whose normal approximation [around $H(\mu)$ or $T(P)$] does not have an unknown variance.

Consider m-dimensional i.i.d. observations Y_j $(j = 1, 2, \dots)$ with common distribution G; let $f^{(1)}, f^{(2)}, \dots, f^{(k)}$ be real-valued Borel measurable functions on $\mathbb{R}^m$. Write

$$\begin{aligned} X_j &= \left(f^{(1)}(Y_j), f^{(2)}(Y_j), \dots, f^{(k)}(Y_j)\right) \\ \mu &= EX_j = \left(Ef^{(1)}(Y_j), Ef^{(2)}(Y_j), \dots, Ef^{(k)}(Y_j)\right) \\ v_{rr'} &= E\left\{\left(f^{(r)}(Y_j) - Ef^{(r)}(Y_j)\right)\left(f^{(r')}(Y_j) - Ef^{(r')}(Y_j)\right)\right\} \\ V &= ((v_{rr'})) \end{aligned} \tag{3.1}$$

Let $H(x) = (H^{(1)}(x), H^{(2)}(x), \dots, H^{(p)}(x))$ be a measurable function defined on $\mathbb{R}^k$ (actually it is enough to define it on a neighborhood of μ) into $\mathbb{R}^p$. We consider the asymptotic distribution of $H(\overline{X})$, where

$$\overline{X} = (X_1 + \cdots + X_n)/n \tag{3.2}$$

We make the following assumptions. Below s is an integer, $s \geq 3$.

(A_1): $E\|X_1\|^s < \infty$.

(A_2): Derivatives of $H^{(i)}$ $(i = 1, 2, \dots, p)$ of orders up to and including $s - 1$ exist and are continuous in a neighborhood of μ.

(A_3): grad $H^{(i)}(\mu)$

$$\equiv \left(\partial/\partial x^{(1)} H^{(i)}(x), \partial/\partial x^{(2)} H^{(i)}(x), \dots, \partial/\partial x^{(k)} H^{(i)}(x)\right)_{x=\mu}, 1 \leq i \leq p,$$

are linearly independent elements of $\mathbb{R}^p$.

(A_4): there exists a nonempty open subset U of $\mathbb{R}^m$ with the following properties: (1) the distribution G of Y_j has a nonzero absolutely continuous component (with respect to Lebesgue measure on $\mathbb{R}^m$) with a positive density on U, (2) $f^{(1)}, f^{(2)}, \dots, f^{(k)}$ are continuously differentiable on U, (3) $1, f^{(1)}, f^{(2)}, \dots, f^{(k)}$ are linearly independent as elements of the vector space of real continuous functions on U.

Under these assumptions an Edgeworth expansion (called the *formal Edgeworth expansion*) was obtained in Bhattacharya and Ghosh (1978) for the probabilities $\text{Prob}(\sqrt{n}(H(\overline{X}) - H(\mu)) \in B)$, uniformly over the class of Borel sets B. To state this, let W_n' denote the $(s-1)$-term Taylor expansion of $\sqrt{n}(H(\overline{X}) - H(\mu))$ around μ. Write $K_{\nu,n}$ for the νth cumulant of W_n'. The formal expansion for $K_{\nu,n}$ (i.e., assuming all moments finite) is a polynomial in powers of $n^{-1/2}$, the smallest power being $n^{-(|\nu|-2)/2}$, $|\nu| \geq 2$. Let $K_{\nu,n}$ be obtained by deleting from the above expression all terms that involve higher powers of $n^{-1/2}$ than $n^{-(s-2)/2}$. Now write

$$\exp\left\{ \sum_{1 \leq |\nu| \leq s} \bar{K}_{\nu,n} \frac{(i\xi)^{\nu}}{\nu!} + \frac{1}{2}\langle \xi, \Sigma \xi \rangle \right\}$$

$$= 1 + \sum_{r=1}^{s-2} n^{-r/2} \pi_r(i\xi : F) + o(n^{-(s-2)/2}) \qquad (\xi \in \mathbb{R}^p) \quad (3.3)$$

where $\pi_r(i\xi : F)$ is a polynomial in $i\xi$ whose coefficients are themselves polynomials in the moments (or cumulants) of the common distribution F of X_j. Also, $\Sigma = ((\sigma_{ij}))$ is the $p \times p$ matrix defined by

$$\sigma_{ij} = \langle \text{grad } H^{(i)}(\mu), V \text{grad } H^{(j)}(\mu) \rangle \qquad (1 \leq i, j \leq p) \quad (3.4)$$

The density $\psi_{s,n}$ of the $(s-1)$-term formal Edgeworth expansion of the distribution of $\sqrt{n}(H(\bar{X}) - H(\mu))$ is then

$$\psi_{s,n}(x) = \left[1 + \sum_{r=1}^{s-2} n^{-r/2} \pi_r(-D : F) \right] \phi_{\Sigma}(x) \qquad (x \in \mathbb{R}^p) \quad (3.5)$$

The following theorem is a slightly improved version of a result in Bhattacharya and Ghosh (1978), as given in Bhattacharya (1985, Theorem 1).

Theorem 3.1. Under the assumptions (A_1)–(A_4), one has

$$\sup_{B \in \mathscr{B}^p} \left| \text{Prob}\left(\sqrt{n}(H(\bar{X}) - H(\mu)) \in B\right) - \int_B \psi_{s,n}(x)\, dx \right| = o(n^{-(s-2)/2}), \quad (3.6)$$

where $\mathscr{B}^p$ is the class of all Borel subsets of $\mathbb{R}^p$.

Let now G_n denote the *empirical distribution* of $\{Y_1, Y_2, \ldots, Y_n\}$ and F_n that of $\{X_1, X_2, \ldots, X_n\}$. Suppose one wants to *estimate* $\text{Prob}(\sqrt{n}(H(\bar{X}) - H(\mu)) \in B)$ for a suitable class of sets B (e.g., the class of all rectangles, the class of all measurable convex sets, or even $\mathscr{B}^p$). One obvious estimate is the $(s-1)$-term *empirical Edgeworth expansion*

$$\int_B \tilde{\psi}_{s,n}(x)\, dx \quad (3.7)$$

where $\tilde{\psi}_{s,n}$ is obtained by replacing population moments of X_j (i.e., moments of F) by the corresponding sample moments:

$$\tilde{\psi}_{s,n}(x) = \left[1 + \sum_{r=1}^{s-2} n^{-r/2}\pi_r(-D : F_n)\right]\phi_{\tilde{\Sigma}}(x) \tag{3.8}$$

and $\tilde{\Sigma} = ((s_{ij}))$ with s_{ij} obtained by replacing μ by $\overline{X}$ and V by the sample dispersion matrix of $X_1, \dots, X_n$ in the expression (3.4). If Σ does not involve unknown quantities, then (3.7) with $s = 3$ estimates $\text{Prob}(\sqrt{n}(H(\overline{X}) - H(\mu)) \in B)$ better than the normal approximation $\int_B \phi_{\Sigma}(x)\,dx$. The result below may be compared to Theorem (3.15) in Bickel and Freedman (1980).

Theorem 3.2. Assume that $E\|X_1\|^3 < \infty$ and that assumptions (A_2)–(A_4) hold, (A_2) with $s = 3$. Assume also that the matrix Σ does not involve unknown population parameters, that is, $\tilde{\Sigma} = \Sigma$. (1) Then

$$\sup_{B \in \mathscr{B}^p} \left| \text{Prob}\big(\sqrt{n}(H(\overline{X}) - H(\mu)) \in B\big) - \int_B \tilde{\psi}_{3,n}(x)\,dx \right| = o(n^{-1/2}) \quad \text{a.s.} \tag{3.9}$$

as $n \to \infty$. (2) If, in addition, $E\|X_1\|^6 < \infty$ and (A_2) holds with $s = 4$ then

$$\sup_{B \in \mathscr{B}^p} \left| \text{Prob}\big(\sqrt{n}(H(X) - H(\mu)) \in B\big) - \int_B \psi_{3,n}(x)\,dx \right| = O_p(n^{-1}) \tag{3.10}$$

Proof. (1) Follows from (3.6) and the strong law of large numbers applied to the sample moments appearing in the coefficient of $n^{-1/2}$ in $\tilde{\psi}_{3,n}$. (2) Under the given hypothesis it follows from (3.6) that

$$\sup_{B \in \mathscr{B}^p} \left| \text{Prob}\big(\sqrt{n}(H(\overline{X}) - H(\mu)) \in B\big) - \int_B \psi_{3,n}(x)\,dx \right| = O(n^{-1}) \tag{3.11}$$

Now write the coefficient of $n^{-1/2}$ in $\psi_{3,n}(x)$ as

$$\left(\sum_{0 \le |\nu| \le r} \alpha_\nu x^\nu\right)\phi_{\Sigma}(x), \tag{3.12}$$

where α_ν $(0 \le |\nu| \le r)$ are polynomial functions of moments of X_1 of orders 3 and less, and $|\alpha_\nu| \le cE\|X_1 - \mu\|^3$ for some positive constant c

(not depending on moments). If $\tilde{\alpha}_\nu$ are obtained by replacing the population moments in α_ν by sample moments, then the random vector $\{\sqrt{n}(\tilde{\alpha}_\nu - \alpha_\nu)|0 \le |\nu| \le r\}$ converges in distribution to a multivariate normal law. From this (3.10) immediately follows. ■

REMARK 3.2.1. The *asymptotic order* of the error of estimation $\text{Prob}(\sqrt{n}(H(\overline{X}) - H(\mu)) \in B) - \int_B \tilde{\psi}_{s,n}(x)\,dx$ cannot be improved in general by taking $s > 3$. However, *numerical approximation* often improves by including additional terms in the estimate, that is, by using $\tilde{\psi}_{s,n}$ with $s > 3$. This is especially the case when the tail probabilities of F decay rapidly. See Remark 2.1.1 in this connection.

In contrast to the above the *bootstrap estimate* of $\text{Prob}(\sqrt{n}(H(\overline{X}) - H(\mu) \in B)$ is

$$P^*\left(\sqrt{n}\left(H(\overline{X^*}) - H(\overline{X})\right) \in B\right), \tag{3.13}$$

where P^* denotes the probability distribution (conditionally given $Y_1, Y_2, \dots, Y_n$) of n i.i.d. observations $Y_1^*, Y_2^*, \dots, Y_n^*$ drawn from the empirical G_n, and

$$X_j^* = \left(f^{(1)}(Y_j^*), f^{(2)}(Y_j^*), \dots, f^{(k)}(Y_j^*)\right), \qquad 1 \le j \le n,$$

$$\overline{X^*} = \sum_{j=1}^{n} \frac{X_j^*}{n} \tag{3.14}$$

The following result shows the relation between the two methods of estiamtion (3.7) and (3.13). For the case $H(\overline{X}) = \overline{X}$, this result is due to Babu and Singh (1984), which improved upon an earlier result of Bickel and Freedman (1980).

Theorem 3.3. Under the assumptions (A_1)–(A_4) one has

$$\sup_{B\in\mathscr{B}} \left| P^*(\sqrt{n}(H(\overline{X}^*) - H(\overline{X})) \in B) - \int_B \tilde{\psi}_{s,n}(x)\,dx \right| = o(n^{-(s-2)/2})$$

$$\text{a.s. as } n \to \infty \tag{3.15}$$

for every class $\mathscr{B}$ of Borel sets satisfying, for some $a > 0$,

$$\sup_{B\in\mathscr{B}} \Phi_{\mathfrak{X}}((\partial B)^\varepsilon) = 0(\varepsilon^a) \quad \text{as } \varepsilon \downarrow 0 \tag{3.16}$$

Proof. Roughly, one applies Theorem 3.1 with G replaced by G_n (i.e., F replaced by F_n). By a result of Babu and Singh (1984, Lemma 2), corrected for an obvious oversight, there exists for each $c > 0$ positive numbers ε and δ such that

$$\overline{\lim_{n\to\infty}} \sup_{c \le \|\xi\| \le e^{\delta_n}} |\hat{F}_n(\xi)| \le 1 - \varepsilon \quad \text{a.s.} \tag{3.17}$$

where $\hat{F}_n$ is the c.f. of F_n. Since X_j^* $(1 \le j \le n)$ are i.i.d. (conditionally given $Y_1, \ldots, Y_n$) with common distribution F_n, one uses Theorem 9.12, Lemma 11.6, and a simple modification of the proof of Theorem 20.1 in Bhattacharya and Ranga Rao (1976) using (1) a smoothing kernel K_1 whose Fourier transform vanishes outside a compact subset of $\mathbb{R}^k$ and (2) applying Lemma 5 of Sweeting (1977), together with (3.17). One then has

$$\sup_{A \in \mathscr{A}} \left| P^*\big(\sqrt{n}\,(\bar{X}^* - \bar{X}) \in A\big) - \int_A \left[1 + \sum_{r=1}^{s-2} n^{-r/2} P_r(-D : F_n)\right] \phi_{\tilde{V}}(x)\, dx \right|$$

$$= o\big(n^{-(s-2)/2}\big) \quad \text{a.s. as } n \to \infty \tag{3.18}$$

for every class $\mathscr{A}$ of Borel subsets of $\mathbb{R}^k$ satisfying, for some $\theta > 0$,

$$\sup_{A \in \mathscr{A}} \Phi_V\big((\partial A)^{\varepsilon}\big) = O(\varepsilon^{\theta}) \quad \text{as} \quad \varepsilon \downarrow 0 \tag{3.19}$$

Reduction of the expansion in (3.18) to the formal Edgeworth expansion $\int_B \tilde{\psi}_{s,n}(x)\, dx$ $(B \in \mathscr{B})$ follows as in Bhattacharya and Ghosh (1978) or Bhattacharya (1985). ∎

REMARK 3.3.1. In view of (3.15), the bootstrap estimate (3.13) is closer to the empirical Edgeworth expansion (3.7) the larger s is. That is, one may think of the bootstrap to be asymptotically like an empirical Edgeworth expansion with many terms. The question remains: is it better to take so many terms than taking one or two terms beyond the normal, say? Theoretical, as well as numerical, work done by Qumsiyeh (1986) shows that neither method is always superior. Of course, in parametric models one may often obtain more efficient estimates of population moments than just sample moments to replace the former in approximating the theoretical Edgeworth expansion $\psi_{3,n}$. In these situations the empirical Edgeworth, using such estimates, will fare better than the bootstrap.

REMARK 3.3.2. The empirical Edgeworth expansion may also be used for time series models such as Examples 2.1 and 2.2, especially under parametric model assumptions.

REFERENCES

Babu, G. J. and Singh, K. (1984). On one term Edgeworth correction by Efron's bootstrap. *Sankhyā A* **46**, 219–232.

Bhattacharya, R. N. (1985). Some recent results on Cramér-Edgeworth expansions with applications. *Mult. Analysis-IV* (P. R. Krishnaiah, Ed.), New York: Elsevier, pp. 57–75.

Bhattacharya, R. N. and Ghosh, J. K. (1978). On the validity of the formal Edgeworth expansion. *Ann. Statist.* **6**, 434–451.

Bhattacharya, R. N. and Lee, Oesook (1986). The existence of unique invariant probabilities and central limit theorems for two classes of Markov processes which are not irreducible (to appear).

Bhattacharya, R. N. and Puri, M. L. (1983). On the order of magnitude of cumulatns of von Mises functionals and related statistics. *Ann. Probab.* **11**, 346–354.

Bhattacharya, R. N. and Ranga Rao, R. (1986). *Normal Approximation and Asymptotic Expansions*, Krieger, Melbourne (Florida).

Bickel, P. J. and Freedman, D. A. (1980). On Edgeworth expansions for the bootstrap. Unpublished manuscript.

Bradley, R. C. (1980). A note on a mixing condition. *Ann. Probab.* **8**, 636–638.

Efron, B. (1979). Bootstrap—another look at jackknife. *Ann. Statist.* **7**, 1–26.

Götze, F. and Hipp. C. (1983). Asymptotic expansions for sums of weekly dependent random vectors. *Z. Wahrsch. Verw. Gebiete* **64**, 211–239.

Qumsiyeh, M. (1986). Ph.D. Dissertation, Indiana University.

Statulevičius, V. (1969–1970). Limit theorems for sums of random variables related to a Markov chain. *Litovskii Mat. Sb.* **9**, 345–362, 635–672; **10**, 161–169 (in Russian).

Statulevičius, V. (1974). Limit theorems for dependent random variables under various regularity conditions. *Proc. Int. Cong. Math. Vancouver*, pp. 173–181.

Sweeting, T. J. (1977). Speeds of convergance for the multidimensional central limit theorem. *Ann. Probab.* **5**, 28–41.

Zhurbenko, I. G. (1972). On strong estimates of mixed semi-invariants of random processes. *Siberian Math. J.* **13**, 202–213.

Linear Rank Statistics, Bounded Operators, and Weak Convergence

Manfred Denker
Institut für Mathematische Stochastik
Universität Göttingen
Göttingen, Federal Republic of Germany
Gerhard Keller
Institut für Angewandte Mathematik
Universität Heidelberg
Heidelberg, Federal Republic of Germany
Madan L. Puri
Department of Mathematics
Indiana University
Bloomington, Indiana

SUMMARY

Simple linear rank statistics based on random variables $X_{1N}, \ldots, X_{NN}$, $N \geq 1$ (independent or not) are considered as linear operators acting between certain function spaces. This viewpoint leads to a variety of asymptotic distribution theory results, some of which are presented here, including a new (and simpler) proof of the Dupač–Hájek (1969) theorem on the asymptotic normality of simple linear rank statistics. Many dynamical systems may be regarded as processes being Hölder-continuous func-

Research by Madan L. Puri supported by the Office of Naval Research Contract N00014-85-K-0648.

tions of some mixing process; thus statistical methods can be applied in such situations. Here we prove two theorems concerning the two sample rank statistics and the signed rank statistics, and we discuss briefly some simulations for the Wilcoxon statistic. Finally, we prove a functional central limit theorem for signed rank statistics together with an extension of the Denker-Rösler (1985a) approximation theorem.

1. INTRODUCTION

Consider a simple linear rank statistic

$$V_N = N^{-1} \sum_{i=1}^{N} C_N(i) a_N(R_{iN}) \tag{1.1}$$

where $\{C_N(i) | 1 \le i \le N, N \ge 1\}$ are regression constants, $\{a_N(i) | 1 \le i \le N, N \ge 1\}$ are scores, and R_{iN} is the rank of X_{iN} among the random variables $X_{1N}, \dots, X_{NN}$. (We assume that the ranks are always well defined.)

We assume that the scores $a_N(i)$ are given by $a_N(i) = h(i/(N+1))$, where h: $(0,1) \to \mathbb{R}$ is measurable. Then, instead of (1.1), we may consider

$$T_N(h) = N^{1/2}\left[\int_{-\infty}^{\infty} h(\hat{F}_N(t))\, d\hat{H}_N(t) - \int_{-\infty}^{\infty} h(F_N(t))\, dH_N(t)\right] \tag{1.2}$$

where

$$\hat{H}_N(t) = \frac{1}{N}\sum_{i=1}^{N} C_N(i) I_{[X_{iN},\infty)}(t), \qquad \hat{F}_N(t) = \frac{1}{(N+1)}\sum_{i=1}^{N} I_{[X_{iN},\infty)}(t)$$

$$H_N = E\hat{H}_N, \qquad F_N = \left(\frac{N+1}{N}\right)E(\hat{F}_N)$$

and I denotes the indicator function.

Observe that $T_N(h)$ is linear in h, and it is translation invariant, and taking h to be right continuous does not change the asymptotic behavior of $T_N(h)$, if h is given by a signed measure. Thus, we may consider all functions h: $(0,1) \to \mathbb{R}$, right continuous, satisfying $h(\frac{1}{2}) = 0$, and having bounded total variation on every compact subset of $(0,1)$. Then, we may write $h = h_1 - h_2$ where h_i, $i = 1,2$, share the same properties as h, and are nondecreasing. For $\delta \ge 0$, define $\mathscr{H}_\delta$ to be the space of all score functions described above, satisfying in addition, the condition

$$\|h\|_\delta = \int_0^1 (|h_1(t)| + |h_2(t)|)(t(1-t))^{-(1+\delta)/(2+\delta)}\, dt < \infty \tag{1.3}$$

Note that $(\mathscr{H}_\delta, \|\ \|_\delta)$ is a nonseparable Banach space, and we may consider T_N acting as a linear operator on each $\mathscr{H}_\delta$.

If B denotes a Banach space with norm $\|\ \|$, we can define a new Banach space B^* with norm $\|\ \|^*$ by setting $B^* = \{(b_1, b_2, \dots) \in B^{\mathbb{N}} | \sup\|b_n\| < \infty\}$, and $\|b\|^* = \sup_n\|b_n\|$, where $b = (b_1, b_2, \dots) \in B^*$. Observe that B^* is in fact a Banach space. Let $\{L_n | n \geq 1\}$ be a sequence of bounded linear maps L_n: $E \to B$ defined on some other Banach space E. Then $L(h) = (L_1(h), L_2(h), \dots)$ defines a linear map $L: E \to B^N$, and L is a map into B^* if $\{L_n | n \geq 1\}$ is uniformly bounded. In this case L is continuous.

Continuous operators of this form are of importance for the asymptotic theory of sequences of random variables. To give a simple example, let $(\Omega, \mathscr{F}, P)$ be a probability space, and let E be the space L_2^0 of square integrable random variables with zero expectation. (The random variables may even take values in a separable type 2 Banach space.) For $X \in L_2^0$, define $L_n(X) = n^{-1/2}\sum_{i=1}^n X_i$, where X_i, $1 \leq i \leq n$ are independent and identically distributed (i.i.d.) random variables with the same distribution as X. Then L_n maps E into itself. Since $\|L_n(X)\|_2 \leq C\|X\|_2$ for some constant C, independent of n, we see that L maps E into E^* continuously. If we know in addition that $L_n(X)$ converges to some Gaussian law $G(X)$ for discrete random variables, and if the covariance operator of $G(X)$ depends continuously on X, then the boundedness of L implies the weak convergence of $L_n(X)$ for every $X \in E$.

This method of proof has indeed been used for type 2 Banach-space-valued random variables and for von Mises functionals

$$V_n(h) = n^{m/2} \int \cdots \int h(x_1, \dots, x_m) \prod_{i=1}^{m} (F_n - F)(dx_i) \quad \text{where } \|h\| < \infty$$

Here the norm $\|h\|$ of a measurable function h: $\mathbb{R}^m \to \mathbb{R}$ is defined to be the sum of all $L_2(F^{|Q|})$ norms of h_Q where $Q = \{Q_1, \dots, Q_S\}$ is a partition of $\{1, \dots, m\}$, and where $h_Q(y_1, \dots, y_S) = h(x_1, \dots, x_m)$ with $x_i = y_j$ if and only if $i \in Q_j$.

It is well known that for each n, V_n: $L^{(m)}(F) \to L_2(P)$ where $L^{(m)}(F) = \{h | \|h\| < \infty\}$ and $\|V\|^* < \infty$. Indeed, a stronger version can be proved. Denote by $L_n(h)$ the random function on $[0,1]$ defined by $L_n(h)(t) = n^{m/2}t^m[nt]^{-m/2}V_{[nt]}(h)$ where $[a]$ denotes the largest integer $\leq a$. Then L_n: $L^{(m)}(F) \to L_2(P, D([0,1]))$, the space of square integrable functions with values in $D([0,1])$. ($D([0,1])$ denotes the space of functions on $[0,1]$ with no discontinuities of the second kind together with the supremum norm.) It is known that in this case $\|L\|^* < \infty$.

In this chapter, we are interested in studying the estimates of the above kinds for simple linear rank statistics $T_N(h)$ defined in (1.2). So, let $T(h) = (T_1(h), T_2(h), \dots)$ and we look for conditions such that T maps H_δ into some L_p^* continuously.

Definition 1.1. We say that T satisfies the condition $C(\delta, p, L, \Lambda)$ where $\delta \geq 0$, $2 \leq p \leq 2 + \delta$, Λ, $L > 0$, if for every $N \geq 1$, $|C_N(i)| \leq L$, and

$$\|\hat{H}_N(t) - H_N(t)\|_p \leq L\Lambda N^{-1/2}[F_N(t)(1 - F_N(t))]^{1/(2+\delta)}$$
$$\|\hat{F}_N(t) - F_N(t)\|_p \leq \Lambda N^{-1/2}[F_N(t)(1 - F_N(t))]^{1/(2+\delta)} \tag{1.4}$$

hold.

Then, we will need the following result, the proof of which is given in Denker and Rösler (1985b).

Theorem 1.1. If T satisfies $C(\delta, p, L, \Lambda)$, then $\|T\|^* \leq 2L(2\Lambda + 3)$.

Observe that the modulus of continuity of T does not depend on the specific choice of the $C_N(i)$ or the distribution function F_{iN} of X_{iN}. Finally, if the X_{iN}, $1 \leq i \leq N$, $N \geq 1$, are independent, and F_{iN} are continuous, then $\|T\|$ is not affected.

We shall use the basic ideas in the foregoing discussion together with differentiability arguments to derive a variety of results. In Section 2, we prove the asymptotic normality of $T_N(h)$ when the $\{X_{iN}|1 \leq i \leq N, N \geq 1\}$ form an array of independent random variables, and h belongs to a class of (not necessarily absolutely continuous) score functions in $\mathscr{H}_0$. This result is originally due to Dupač and Hájek (1969), but our proof is different and much simpler. A similar result can be obtained for (regression-type) signed rank statistics; our result in Section 5 implies it. In Section 3, we relax the assumption of independence of the underlying random variables, and consider a special type of dependent situation. Simulations of one- and two-dimensional dynamics are very popular among physicists in order to examine different parameters of the system, like symmetry or fractal dimension. In many cases, it can be shown that the trajectories of the dynamics may be regarded as the outcome of some function of a mixing process, and this serves as a motivation to prove limit theorems for rank statistics (signed as well as unsigned) under mixing conditions. An application of these theorems in the case of Wilcoxon scores is given in Section 4, where we discuss simulations of variables $X_n = f_\lambda^n$, ($f_\lambda^n = f_\lambda \circ f_\lambda^{n-1}$, $f_\lambda^0(x) = x$) where f_λ: $[0, 1] \to [0, 1]$ is defined by

$$f_\lambda(x) = \begin{cases} \lambda(x-1) + 2 & \text{if } 0 \leq x \leq 1 - \lambda^{-1} \\ \lambda(1-x) & \text{if } 1 - \lambda^{-1} \leq x \leq 1 \end{cases} \qquad \sqrt{2} < \lambda \leq 2 \tag{1.5}$$

Though there are still some open questions, the simulations show good agreement with the theoretical behavior.

Finally, in Section 5, we extend Theorem 1.1 to the $L_2(P, D([0,1]))$ case discussed above, when the random variables $X_{1N}, \ldots, X_{(N/2)N}$ are independent and identically distributed, and T_N corresponds to the signed rank statistic (regression type), that is, $X_{kN} = -X_{k-N/2,N}$, $k > N$. From this result, we derive a functional central limit theorem for signed rank statistics. Our result extends and complements earlier results due to Schulze-Pillot (1978) and Antille (1979), among others.

2. ASYMPTOTIC NORMALITY OF SIMPLE LINEAR RANK STATISTICS

For every $N \geq 1$, let X_{iN}, $i = 1, \ldots, N$, be independent random variables with continuous distribution functions F_{iN}, $i = 1, \ldots, N$, respectively. If $|C_N(i)| \leq L$, then the condition $C(0, 2, L, 1)$ is satisfied.

Given the families $\{F_{iN} | 1 \leq i \leq N,\ N \geq 1\}$ and $\{C_N(i) | 1 \leq i \leq N, N \geq 1\}$ let us denote by $\tilde{\mathscr{H}}_0$ the subset of $\mathscr{H}_0$ consisting of those score functions h for which $(H_N \circ F_N^{-1})'(t)$ exists for almost every t with respect to the total variation measure $|dh|$ of dh, and is uniformly continuous (in N). Then

Theorem 2.1 (Dupač and Hájek, 1969). Let $h \in \tilde{\mathscr{H}}_0$. Then $T_N(h)$ has asymptotically a normal $N(0, \sigma_N^2(h))$ distribution, where

$$\sigma_N^2(h) = E\left(\int_0^1 N^{-1/2} \sum_{i=1}^{N} \left(C_N(i) - \left(H_N \circ H_N^{-1} \right)'(t) \right) \right.$$

$$\left. \times \left(I_{\{X_{iN} < F_N^{-1}(t)\}} - F_{iN}\left(F_N^{-1}(t)\right) \right) dh(t) \right)^2 \qquad (2.1)$$

Note that $\tilde{\mathscr{H}}_0$ contains all absolutely continuous score functions in $\mathscr{H}_0$. Thus Theorem 2.1 contains the corresponding results of Hájek (1968) and Pyke and Shorack (1968a) among others. In this case

$$\sigma_N^2(h) = \frac{2}{N} \sum_{i \leq N} \iint_{s<t} F_{iN}(s)(1 - F_{iN}(t)) \left(\frac{dH_N}{dF_N}(t) - C_N(i) \right)$$

$$\times \left(\frac{dH_N}{dF_N}(s) - C_N(i) \right) dh(F_N(s))\, dh(F_N(t)) \qquad (2.2)$$

The proof of this theorem is based on the following two lemmas. (We may note that if G is a distribution function always assumed to be right continuous, we denote by G^{-1} its left continuous inverse, that is, $G^{-1}(s) = \inf\{t \mid G(t) \geq s\}$).

Lemma 2.1. For every $h \in \mathscr{H}_0$,

$$N^{-1/2}T_N(h) = -\int_0^1 \Bigg\{ \hat{H}_N\big(\hat{F}_N^{-1}(t)\big) - H_N\big(\hat{F}_N^{-1}(t)\big) - \frac{H_N\big(\hat{F}_N^{-1}(t)\big) - H_N\big(F_N^{-1}(t)\big)}{F_N\big(\hat{F}_N^{-1}(t)\big) - t} \times \Big[\hat{F}_N\big(\hat{F}_N^{-1}(t)\big) - F_N\big(\hat{F}_N^{-1}(t)\big) - \hat{t} + t\Big] \Bigg\}\, dh(t) \quad \text{a.s.} \tag{2.3}$$

where $\hat{t} = (i+1)(N+1)^{-1}$ if $i(N+1)^{-1} < t \leq (i+1)(N+1)^{-1}$, and where the integrand is defined to be zero if $F_N(\hat{F}_N^{-1}(t)) = t$.

Proof. First note that

$$N^{-1/2}T_N(h) = \int_0^1 h(t)\, d\hat{H}_N\big(\hat{F}_N^{-1}(t)\big) - \int_0^1 h(t)\, dH_N\big(F_N^{-1}(t)\big)$$

This is so since $\hat{H}_N((\hat{F}_N^{-1}(b), \bar{t})) = 0$ where $\bar{t}$ is such that $\{t \mid \hat{F}_N(t) \leq b\} = (-\infty, \bar{t})$. Using integration by parts, the right continuity of h, and the left continuity of $\hat{F}_N^{-1}$, we have for small $a < \frac{1}{2}$ and large $b > \frac{1}{2}$,

$$\int_a^{1/2} h(t)\, d\hat{H}_N\big(\hat{F}_N^{-1}(t)\big) + \int_a^{1/2} \hat{H}_N\big(\hat{F}_N^{-1}(t)\big)\, dh(t) = h\big(\tfrac{1}{2}\big)\hat{H}_N\big(\hat{F}_N^{-1}\big(\tfrac{1}{2}^{+}\big)\big)$$

$$\int_a^{1/2} h(t)\, dH_N\big(F_N^{-1}(t)\big) + \int_a^{1/2} H_N\big(F_N^{-1}(t)\big)\, dh(t) = h\big(\tfrac{1}{2}\big)H_N\big(F_N^{-1}\big(\tfrac{1}{2}\big)\big) - h(a-)H_N\big(F_N^{-1}(a)\big)$$

and

$$\int_{1/2}^{b} h(t)\, dH_N\big(F_N^{-1}(t)\big) - \int_{1/2}^{b} \big(L^* - H_N\big(F_N^{-1}(t)\big)\big)\, dh(t) = h\big(\tfrac{1}{2}^{-}\big)\big(L^* - H_N\big(F_N^{-1}\big(\tfrac{1}{2}\big)\big)\big) - h(b)\big(L^* - H_N\big(F_N^{-1}(b)\big)\big)$$

where $L^* = N^{-1}\Sigma C_N(i)$, and where $\int_c^d$ means integration over the closed interval $[c, d]$. We may, of course, assume that h is continuous at $\frac{1}{2}$ (otherwise choose a suitable value $t_0 \in (0,1)$).

Thus

$$\int_a^b h(t)\, d\hat{H}_N\big(\hat{F}_N^{-1}(t)\big) - \int_a^b h(t)\, dH_N\big(F_N^{-1}(t)\big)$$

$$= -\int_a^b \big(\hat{H}_N\big(\hat{F}_N^{-1}(t)\big) - H_N\big(F_N^{-1}(t)\big)\big)\, dh(t)$$

$$-h(b)\big(L^* - H_N\big(F_N^{-1}(b)\big)\big) - h(a-)H_N\big(F_N^{-1}(a)\big)$$

By Proposition 1 in Denker and Rösler (1985a), and using $H_N(F_N^{-1}(a)) \le La$ and $L^* - H_N(F_N^{-1}(b)) = (1/N)\Sigma C_N(i)(1 - F_{iN}(F_N^{-1}(b))) \le L(1-b)$, we obtain

$$\lim_{b\to 1} h(b)\big(L^* - H_N\big(F_N^{-1}(b)\big)\big) = 0 \quad \text{and} \quad \lim_{a\to 0} h(a-)H_N\big(F_N^{-1}(a)\big) = 0$$

It follows (letting $a \to 0$ and $b \to 1$) that

$$N^{-1/2}T_N(h) = -\int_0^1 \big(\hat{H}_N\big(\hat{F}_N^{-1}(t)\big) - H_N\big(F_N^{-1}(t)\big)\big)\, dh(t)$$

and the proof follows from this equality. ■

Lemma 2.2. If $h \in \tilde{\mathscr{H}}_0$, then

$$\lim_{N\to\infty} \frac{H_N\big(\hat{F}_N^{-1}(t)\big) - H_N\big(F_N^{-1}(t)\big)}{F_N\big(\hat{F}_N^{-1}(t)\big) - t} - \big(H_N \circ F_N^{-1}\big)'(t) = 0 \quad \text{a.s.} \qquad (2.4)$$

for $|dh|$ a.e. t and this sequence is bounded by $2L$, uniformly in N, and $|dh|$ a.e. t.

Proof. We first prove the boundedness. Let $N \ge 1$ and $0 < t < 1$ such that $(H_N \circ F_N^{-1})'(t)$ exists. Then

$$\big(H_N \circ F_N^{-1}\big)'(t) = \lim_{\varepsilon\to 0} \varepsilon^{-1}\big(H_N\big(F_N^{-1}(t+\varepsilon)\big) - H_N\big(F_N^{-1}(t)\big)\big)$$

$$= \lim_{\varepsilon\to 0} \varepsilon^{-1}N^{-1}\sum C_N(i)\big(F_{iN}\big(F_N^{-1}(t+\varepsilon)\big) - F_{iN}\big(F_N^{-1}(t)\big)\big)$$

$$\le L$$

Moreover, by the same reasoning,

$$\frac{H_N(u) - H_N(v)}{F_N(u) - F_N(v)} \le L$$

The boundedness follows.

We now prove (2.4). Since $h \in \tilde{\mathscr{H}}_0$, we have for $|dh|$ a.e. t: given $\varepsilon > 0$, there exists an $\eta > 0$ such that for all N and all s with $|t - s| < \eta$, we have $|(H_N \circ F_N^{-1})'(s) - (H_N \circ F_N^{-1})'(t)| < \varepsilon$ as long as $(H_N \circ F_N^{-1})'(s)$ is well defined. By the absolute continuity of $H_N \circ F_N^{-1}$, we have uniformly in N,

$$\left| \frac{H_N(F_N^{-1}(s)) - H_N(F_N^{-1}(t))}{s - t} - (H_N \circ F_N^{-1})'(t) \right|$$

$$= \left| (s-t)^{-1} \int_t^s \left((H_N \circ F_N^{-1})'(u) - (H_N \circ F_N^{-1})'(t) \right) du \right|$$

$$< \varepsilon \quad \text{if } |s - t| < \eta$$

Now (2.4) is true on the set where $F_N(\hat{F}_N^{-1}(t)) \to t$ for all t, and this set has probability 1. The proof follows. ■

Proof of Theorem 2.1. We first consider the case in which dh corresponds to a finite signed measure, and use the representation (2.3) for $T_N(h)$. Since $\hat{t} - t \to 0$ as $N \to \infty$ a.s. for every t, by Lemma 2.2 and the Lebesgue dominated convergence theorem, it follows that a.s.

$$T_N(h) + N^{1/2} \int_0^1 \left\{ \hat{H}_N(\hat{F}_N^{-1}(t)) - H_N(\hat{F}_N^{-1}(t)) - (H_N \circ F_N^{-1})'(t) \right.$$

$$\left. \times \left[\hat{F}_N(\hat{F}_N^{-1}(t)) - F_N(\hat{F}_N^{-1}(t)) \right] \right\} dh(t) \to 0$$

and so, the distributions of both statistics have the same asymptotic behavior. Moreover, since a.s. $\hat{F}_N^{-1}(t) - F_N^{-1}(t) \to 0$ for every t, we can once more use the dominated convergence theorem to obtain

$$T_N(h) + N^{1/2} \int_0^1 \left\{ \hat{H}_N(F_N^{-1}(t)) - H_N(F_N^{-1}(t)) - (H_N \circ F_N^{-1})'(t) \right.$$

$$\left. \times \left[\hat{F}_N(F_N^{-1}(t)) - t \right] \right\} dh(t) \to 0 \quad \text{a.s.}$$

Thus, it suffices to determine the asymptotic distribution of

$$-N^{-1/2} \int_0^1 \left\{ \hat{H}_N(F_N^{-1}(t)) - H_N(F_N^{-1}(t)) - (H_N \circ F_N^{-1})'(t) \right.$$

$$\left. \times \left[\hat{F}_N(F_N^{-1}(t)) - t \right] \right\} dh(t)$$

that is, of

$$-N^{-1/2}\int_0^1 \sum_{i=1}^{N} \left(C_N(i) - \left(H_N \circ F_N^{-1}\right)'(t)\right)$$

$$\times\left[I_{\{X_{iN} \le F_N^{-1}(t)\}} - F_{iN}\left(F_N(t)\right)\right] dh(t)$$

Since this is a sum of independent L_∞ random variables, the Lindeberg theorem applies; consequently $T_N(h)$ has asymptotically $N(0, \sigma_N^2(h))$ distribution. This proves the theorem for the case in which h corresponds to a finite measure.

Now let $h \in \tilde{\mathscr{H}}_0$ be arbitrary. For $\varepsilon > 0$, define $h_\varepsilon(t) = h(1 - \varepsilon)$ if $t > 1 - \varepsilon$, $h_\varepsilon(t) = h(\varepsilon)$ if $t < \varepsilon$, and $h_\varepsilon(t) = h(t)$ if $t \in [\varepsilon, 1 - \varepsilon]$. Then $h_\varepsilon \in \tilde{\mathscr{H}}_0$ and h_ε corresponds to a finite signed measure. Moreover, $\|h - h_\varepsilon\| \to 0$ as $\varepsilon \to 0$, and hence by Theorem 1.1, $\|T(h) - T(h_\varepsilon)\|_2^{*2} \to 0$ as $\varepsilon \to 0$. It has been shown in Denker and Rösler (1985a, Theorem 5) that $\sigma_N^2(h)$ is (uniformly in N) continuous as a function of h. From the general discussion in the introduction and the case of finite signed measure, the conclusion of the theorem follows. ■

3. DEPENDENT SITUATIONS

Considerable effort has been made in recent years to relax the assumption of independence in the problems of the type discussed in Section 2. In the literature most authors have dealt with the standard two sample problem where two independent processes $\{X_n | n \ge 1\}$ and $\{Y_m | m \ge 1\}$ of strongly or uniformly mixing random variables X_n (and Y_m) are given (see, e.g., Ahmad and Lin, 1980). For such a case the asymptotic normality follows easily as a consequence of Theorem 1.1 (see Denker and Rösler, 1985a). Here we consider signed as well as unsigned rank statistics for a special case of functionals of absolutely regular processes, which turn out to be useful for the study of certain dynamical systems. [See Denker and Keller (1985) for more information about the background of the problem.]

Dynamical systems depending on some parameters may show quite different asymptotic behavior ranging from convergence to a single stable fixed point over limit cycles to chaotic behavior. Most often *chaos* means that there are uncountably many orbits far from being periodic or the system is sensitive to initial conditions. Many authors claim randomness for these dynamical systems, and in fact it has been known in ergodic theory for quite some time before it became of broad interest to physicists that in many situations there is an underlying absolutely regular finite-state stationary process governing the dynamics of the system. The associated flow then may be considered as a Hölder-continuous function of this process. Consequently statistical questions arise naturally. Simulations are in good agreement with theoretical results. Here we give an example from Denker and Keller (1985). Consider the map $\hat{T}(x) = 1 - |2x - 1|$ for $x \in [0,1]$. Let u: $[0,1] \to [0,1]$ be a diffeomorphism with positive derivative u' being symmetric at $\frac{1}{2}$, and consider the family

$$T_\varepsilon(x) = u^{-1}(\hat{T}(u(x))) + \varepsilon G\big(u^{-1}(\hat{T}(u(x)))\big) \qquad (x \in [0,1])$$

where G is symmetric at $\frac{1}{2}$, analytic, and $0 \le y + G(y) \le 1$. If ε is small, the invariant density is analytic in ε and for $\varepsilon = 0$, it is symmetric at $\frac{1}{2}$. Based on randomly chosen initial values, x_0 creates the sequence x_0, $x_1 = T_\varepsilon(x_0), \dots, x_n = T_\varepsilon^n(x_{n-1})$. Testing these "observations" for symmetry using the mean ($\bar{x} = \sum_{i=1}^n x_i/n$) test or the Wilcoxon signed rank test, we arrive at the following result: if $G(x) = x(1-x)$ and $u' \equiv 1$, the asymptotic efficiency of the Wilcoxon test with respect to the mean $\bar{x}$ test is 1. If $u'(x) = \pi^{-1}(x(1-x))^{-1/2}$, then the $\bar{x}$ test is less efficient. The same is true if $u'(x) = \alpha + \beta(x - \frac{1}{2})^{2m}$, $0 \le \alpha \le 1$, $m \ge 1$, $\beta = (1-\alpha)(2m+1)$, and m is small. However, if m is large, the $\bar{x}$ test is asymptotically more efficient than the Wilcoxon test. The results of these types have been obtained by using U statistics (see Denker and Keller, 1985). Here we obtain the asymptotic normality of the general rank scores (signed as well as unsigned) statistics. An application of the results is given in Section 4.

A stationary process $(Z_n | n \ge 1)$ is called *absolutely regular* if

$$\beta(n) := E \sup\big\{|P(A|\mathcal{F}_1^s) - P(A)| \,\big|\, s \ge 1,\ A \in \mathcal{F}_{s+n}^\infty\big\} \tag{3.1}$$

decreases to zero as $n \to \infty$, where $\mathcal{F}_r^{\,t}$ denotes the σ-algebra generated by $Z_r, Z_{r+1}, \dots, Z_t$ $(1 \le r \le t \le \infty)$. The $\beta(n)$'s are called the mixing coefficients. A second stationary process $(X_n | n \ge 1)$ is said to be a functional of $(Z_n | n \ge 1)$ if there exists a Hölder-continuous function ϕ, defined on the infinite product space of the state space of Z_1 (which in the sequel is assumed to be a compact metric space) such that

$$X_n = \phi(Z_n, Z_{n+1}, \dots) \qquad (n \ge 1) \tag{3.2}$$

Now let $(X_n | n \ge 1)$ and $(Y_n | n \ge 1)$ be two functionals of $(Z_n | n \ge 1)$ with continuous distribution functions F and G, respectively. Denote by F_n and G_m the right continuous empirical distribution functions of the two sequences of sizes n and m, respectively, and define $\hat{H}_N = [n/(N+1)]F_n +$

$[m/(N+1)]G_m$, $H_N = (n/N)F + (m/N)G$, where $N = n + m$. For a measurable function h: $(0,1) \to \mathbb{R}$ denote by

$$T_N(h) = \sqrt{N}\left(\int h(\hat{H}_N(t))\, dF_n(t) - \int h(H_N(t))\, dF(t)\right) \tag{3.3}$$

the two sample linear rank statistic. If $N = 2n = 2m$ and if $Y_k = -X_k$ $(k \geq 1)$, then (3.3) reduces essentially to a signed rank statistic and will be denoted by $S_n(h)$. These definitions agree with the previous ones, up to a change in notation.

Theorem 3.1. Let $(X_n | n \geq 1)$ and $(Y_n | n \geq 1)$ be two functionals of the absolutely regular process $(Z_n | n \geq 1)$ with mixing coefficients

$$\beta(n) = O(n^{-2-\eta}) \quad \text{for some } \eta > 0. \tag{3.4}$$

Assume that the two processes are independent of each other and that

$$\max\{F(b) - F(a); G(b) - G(a)\} \leq C(b-a)^r \tag{3.5}$$

for $a < b$ where $C, r > 0$ denote some constants. If $n, m \to \infty$ in such a way that $nN^{-1} \to \lambda \in (0,1)$, then $T_N(h)$ is asymptotically normal $N(0, \sigma^2(h))$ provided h is absolutely continuous and $h \in \mathscr{H}_\delta$ for some $\delta > 2/(1+\eta)$. The variance $\sigma^2(h)$ is given by

$$\sigma^2(h) = \lambda\left(Ef(X_1)^2 + 2\sum_{t>1} Ef(X_1)f(X_t)\right)$$

$$+\lambda^2(1-\lambda)\left(Eg(Y_1)^2 + 2\sum_{t>1} Eg(Y_1)y(Y_t)\right)$$

where

$$f(u) = f_0(u) - \int_{-\infty}^{\infty} f_0(v)F(dv),$$

$$f_0(u) = h(H(u)) + \lambda\int_u^{\infty} h'(H(v))F(dv)$$

$$g(u) = g_0(u) - \int_{-\infty}^{\infty} g_0(v)F(dv), \qquad g_0(u) = \int_u^{\infty} h'(H(v))F(dv)$$

and

$$H = \lambda F + (1-\lambda)G$$

The proof of this theorem will be given below. In an analogous way one can obtain the following result:

Theorem 3.2. Let $(X_n|n \geq 1)$ be a functional of the absolutely regular process $(Z_n|n \geq 1)$ satisfying (3.4). Assume that F satisfies (3.5). Then as $n \to \infty$, $S_n(h)$ is asymptotically normal $N(0, \tau^2(h))$ provided h is absolutely continuous and $h \in \mathscr{H}_\delta$ for some $\delta > 1/(1+\eta)$. Here the variance $\tau^2(h)$ is given by

$$\tau^2(h) = \tfrac{1}{2}\Big(Ea(X_1)^2 + 2 \sum_{t>1} Ea(X_1)a(X_t)\Big)$$

where

$$a(u) = a_0(u) - \int_{-\infty}^{\infty} a_0(v)F(dv),$$

$$a_0(u) = h(H(u)) + \tfrac{1}{2}\int_u^{\infty} h'(H(v))F(dv) + \tfrac{1}{2}\int_{-u}^{\infty} h'(H(v))F(dv)$$

and

$$H(v) = \frac{1 + F(v) - F(-v)}{2}$$

Proof of Theorem 3.1. Choose $\delta > 0$ such that

$$\sum_n \beta(n)^{\delta(2+\delta)^{-1}} < \infty \quad \text{where } \delta > 2(1+\eta)^{-1} \tag{3.6}$$

We shall first show that the processes satisfy the condition $C(\delta, 2, 1, \Lambda)$, equivalently

$$\begin{aligned} nE(F_n(t) - F(t))^2 &\leq \Lambda^2[F(t)(1 - F(t))]^{2(2+\delta)^{-1}} \\ nE(G_n(t) - G(t))^2 &\leq \Lambda^2[G(t)(1 - G(t))]^{2(2+\delta)^{-1}} \end{aligned} \tag{3.7}$$

for $n \geq 1$, where Λ denotes some constant.

We prove (3.7) for F.

Denote by $\mathscr{M}$ the compact metric state space of Z_n, so that ϕ is defined on $\mathscr{M}^{\mathbb{N}}$ which may be equipped with the metric $d((z_n),(z_n')) = \sum 2^{-n}d(z_n, z_n')$. Since ϕ is Hölder-continuous (with exponent s, say) we have $|E(X_1|\mathscr{F}_1^n)(x) - X_1(x)| \leq C2^{-ns}$ for some constant $C > 0$. Choose measurable functions ϕ^l: $\mathscr{M}^l \to \mathbb{R}$ such that for $x_1, \ldots, x_l \in \mathscr{M}$,

$$\phi^l(x_1, \ldots, x_l) = \phi(x_1, \ldots, x_l, x_{l+1}, \ldots)$$

for some choice of $x_{l+1}, x_{l+2}, \ldots,$ and let

$$X_n^l = \phi^l(Z_n, Z_{n+1}, \ldots, Z_{n+l-1}) \qquad (n \geq 1)$$

Then $(X_n^l | n \geq 1)$ is absolutely regular for each l. Denote by F^l the distribution function of X_1^l. It is easy to verify that

$$E|I_{\{X_1 \leq t\}} - I_{\{X_1^l \leq t\}}|^2 \leq F(t + C2^{-ls}) - F(t - C2^{-ls}) = O(2^{-lrs}) \quad (3.8)$$

$$\operatorname{Var}\left(I_{\{X_1 \leq t\}} - I_{\{X_1^l \leq t\}}\right) = O(2^{-lrs}) \quad (3.9)$$

and

$$|F(t) - F^l(t)| = O(2^{-lrs}) \quad (3.10)$$

In order to derive (3.8)–(3.10) one uses (3.5), because it implies that

$$\sup_{\varepsilon > 0} \varepsilon^{-r} \int \sup\{|h(y) - h(y')| \mid \|y - x\|, \|y' - x\| < \varepsilon\} \, dF^m(x) < \infty \quad (3.11)$$

for functions $h: \mathbb{R}^m \to \mathbb{R}$ assuming only finitely many values and having points of discontinuity only in a finite union of hyperplanes (here $h(x) = I_{\{x < t\}}$, $m = 1$).

Using Davidov's (1970) inequality for strongly mixing sequences it follows that (K denotes some constant)

$$\begin{aligned}
&nE(F_n(t) - F(t))^2 \\
&= \operatorname{Var}\left(1_{\{X_1 \leq t\}}\right) + 2\sum_{k=2}^{n}\left(1 - \frac{k-1}{n}\right)\operatorname{cov}\left(I_{\{X_1 \leq t\}}, I_{\{X_k \leq t\}}\right) \\
&\leq F(t)(1 - F(t)) + K\sum_{k=1}^{\infty}\min\Big\{2^{-krs} + \beta(k)^{\delta(2+\delta)^{-1}} \\
&\qquad \times\left[F^l(t)(1 - F^l(t))\right]^{2(2+\delta)^{-1}}, F(t)(1 - F(t))\Big\} \\
&\leq F(t)(1 - F(t)) + 2K\sum_{k=1}^{\infty}\min\{2^{-krs}; F(t)(1 - F(t))\} \\
&\quad + K\sum_{k=1}^{\infty}\beta(k)^{\delta(2+\delta)^{-1}}\left[F(t)(1 - F(t))\right]^{2(2+\delta)^{-1}} \\
&= O\left(\left[F(t)(1 - F(t))\right]^{2(2+\delta)^{-1}}\right)
\end{aligned}$$

since

$$\min\{F(t)(1-F(t)); 2^{-krs}\} \le (2^{-krs})^{\delta(2+\delta)^{-1}}[F(t)(1-F(t))]^{2(2+\delta)^{-1}}$$

Now from Theorem 1.1 it follows that $ET_N(h)^2 \le C\|h\|_\delta^2$ for every $n, m \in \mathbb{N}$, where C denotes some constant. This inequality reduces now the proof of Theorem 3.1 to score functions h that have a bounded second derivative and to the continuity of $h \to \sigma^2(h)$ in $\| \|_\delta$ norm, as explained in Section 1.

In order to prove the continuity of $\sigma^2(h)$, consider

$$U_N(h) = \frac{n}{\sqrt{N}}\int h(H_N(t))\, d(F_n - F)(t) + \frac{n}{\sqrt{N}}\int h'(H_N(t))(\hat{H}_N(t) - H_N(t))\, dF(t) \tag{3.12}$$

We have

$$\begin{aligned}
\sigma^2(h) &= EU_N(h)^2 \\
&= N^{-1}\int\int\Big\{\frac{n^2m^2}{N^2}E[(F_n(t)-F(t))(F_n(s)-F(s))]\frac{dG}{dH_N}(t)\frac{dG}{dH_N}(s) \\
&\quad + \frac{n^2m^2}{N^2}E[(G_m(t)-G(t)) \\
&\quad \times (G_m(s)-G(s))]\frac{dF}{dH_N}(t)\frac{dF}{dH_N}(s)\Big\}\, dh(H_N(s))\, dh(H_N(t)) \\
&\le \text{const}\left[\int [H_N(t)(1-H_N(t))]^{(2+\delta)^{-1}}\, dh(H_N(t))\right]^2 \le \text{const}\|h\|_\delta^2
\end{aligned}$$

where we have used again (3.7).

It remains now to prove Theorem 3.1 for C^2 scores h. Using Taylor's theorem, we obtain $T_N(h) = U_N(h) + A + B + C$, where

$$A = \sqrt{N}\int h'(H_N(t))(\hat{H}_N(t) - H_N(t))\, d(F_n - F)(t)$$

$$B = \sqrt{N}(\hat{H}_N(t) - H_N(t))^2$$

$$C = \sqrt{N}\int(\hat{H}_N(t) - H_N(t))^2\, d(F_n - F)(t)$$

According to Theorem 18.6.2 in Ibragimov and Linnik (1971), $U_N(h)$ converges to $N(0, \sigma^2(h))$, so that it remains to show that A, B, and $C \to 0$ in probability. B tends to 0 because of condition (3.7). We show that $A \to 0$ in probability. The proof for the C term follows analogously.

Let $w\colon \mathbb{R}^m \to \mathbb{R}$ be a measurable function and $\Delta \subset \mathbb{N}^m$ be either $\{(t_1,\dots,t_m)|t_i \neq t_j,\ i \neq j\}$ or $\{(t_1,\dots,t_m)|t_i \neq t_j$ for $i < j \leq m_1$ or $i > j > m_1\}$ where $m_1 < m$ is fixed. Let $p \geq 1$. A process $(X_n|n \geq 1)$ satisfies the (A_p) condition with coefficients $\psi(n)\downarrow 0$ and $\beta(n)\downarrow 0$ if there exists an $\varepsilon > 0$, an absolutely regular process $(Z_n|n \geq 1)$ with mixing coefficients $\beta(n)$ and measurable functions ϕ, ϕ^l $(l \geq 1)$ such that

$$X_n = \phi(Z_n, Z_{n+1},\dots) \tag{3.13}$$

If $X_n^l = \phi^l(Z_n, Z_{n+1},\dots, Z_{n+l-1})$, then for $t = (t_1,\dots,t_m) \in \Delta$,

$$\left(E|w\left(X_{t_1},\dots,X_{t_m}\right) - w\left(X_{t_1}^l,\dots,X_{t_m}^l\right)|^p\right)^{1/p}$$
$$\leq \psi\left(\max\left\{1, \min_{1\neq j}|t_i - t_j| - l\right\}\right) + l^{-1-\varepsilon} \tag{3.14}$$

The function w is degenerate if $Ew(a_1,\dots,a_{t-1}, X_1, a_{t+1},\dots,a_m) = 0$ for all $1 \leq t \leq m$ and $a_1,\dots,a_m \in \mathbb{R}$. From Denker and Keller (1985) we deduce the following.

Lemma 3.1. Let $m \geq 2$ and let $(X_n|n \geq 1)$ satisfy the condition (A_p) with $\beta(n)$ satisfying (3.4), and $\psi(n)$ satisfying $\sum\psi(n) < \infty$. If w is degenerate and bounded, then for $q = \min(p, 2)$, $\|R_N(w)\|_q = o(N^{m-1/2})$, where

$$R_N(w) = \sum_{(t_1,\dots,t_m)\in\Delta_N} w\left(X_{t_1},\dots,X_{t_m}\right),$$

$$\Delta_N = \Delta \cap \{(t_1,\dots,t_m)|t_i \leq N\}$$

in the one-sample case, and $\Delta_N = \Delta \cap \{(t_1,\dots,t_m);\ t_i \leq N_1\ i \leq m_1,\ t_i \leq N_2\ i > m_1\}$ in the two-sample case $(N = (N_1, N_2))$.

Now A can be written

$$\sqrt{N}\int\left\{\frac{n}{N}\int\int g_{H_N^{-1}(z)}(s,t)\,d(F_n - F)(s)\,d(F_n - F)(t)\right.$$
$$\left.+\frac{m}{N}\int\int g_{H_N^{-1}(z)}(s,t)\,d(G_m - G)(s)\,d(F_n - F)(t)\right\}dh'(z)$$
$$+O(N^{-1/2}) \tag{3.15}$$

where $g_z(s,t) = 1_{\{\max(z,s)\leq t\}}$, $0 < z < 1$.

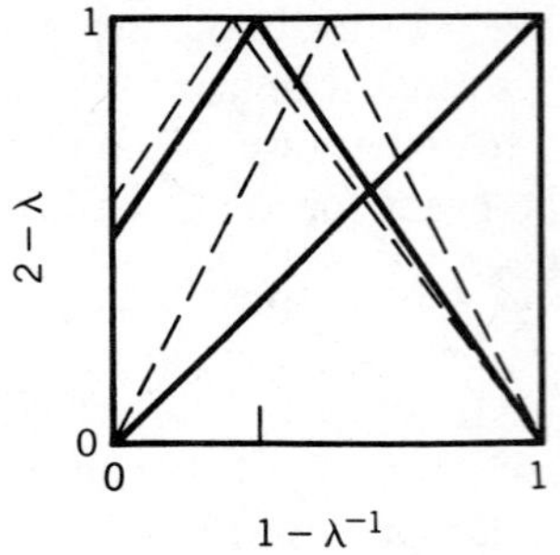

Figure 1.

It follows from (3.11) that each g_z and $((X_n, Y_n)|n \geq 1)$ satisfy the condition (A_1). Consequently, the lemma applies to the one-sample $(m = 2)$ and two-sample $(m = 2,\ m_1 = 1)$ cases involved in (3.15), and therefore the first moment of A is $o(1)$. The proof is completed. ■

4. A SIMULATION FOR THE WILCOXON STATISTIC

In this section, we give an application of Theorem 3.1 in the Wilcoxon case.

Consider the one-parameter family $\{f_\lambda\}$ of transformations f_λ: $[0,1] \to [0,1]$ defined by

$$f_\lambda(x) = \begin{cases} \lambda(x-1)+2 & \text{if } 0 \leq x \leq 1-\lambda^{-1} \\ \lambda(1-x) & \text{if } 1-\lambda^{-1} \leq x \leq 1 \end{cases} \tag{4.1}$$

In order to make f_λ well defined, we take $\lambda \in [1,2]$. The graph of f_λ is shown in Figure 1. We shall need the following theorem, due to Bowen (1978) and Hofbauer and Keller (1982).

Theorem 4.1. Let T: $[0,1] \to [0,1]$ be a piecewise C^2-transformation of the unit interval. If there are only two subintervals of monotonicity for T and if $\inf_x |T'(x)| > \sqrt{2}$, then, with respect to the unique absolutely continuous T-invariant measure, the label process of T is absolutely regular with an exponential decay of the mixing coefficients and $\{T^n|n \geq 0\}$ is a functional of this label process.

In the present situation f_λ restricted to each of the intervals $I_1 = (0, 1-\lambda^{-1})$ and $I_2 = (1-\lambda^{-1}, 1)$ is a straight line and $|f_\lambda'(x)| > \sqrt{2}$ if

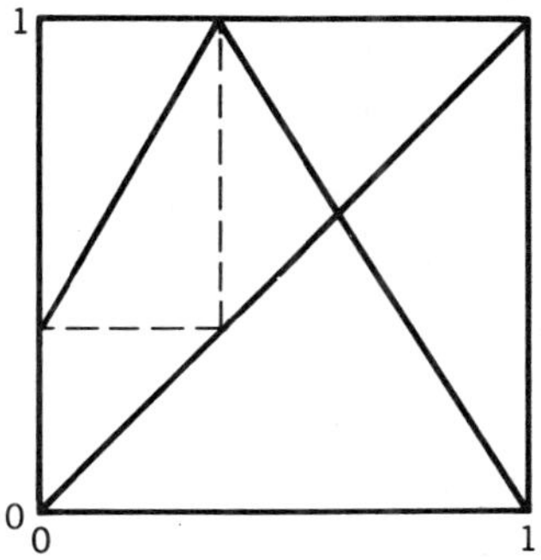

Figure 2.

$\lambda > \sqrt{2}$. The label process

$$Z_n(x) = \begin{cases} 1 & \text{if } T^n x \in I_1 \\ 2 & \text{if } T^n x \in I_2 \end{cases}$$

is a.e. well defined and absolutely regular for $\lambda > \sqrt{2}$. Since the T-invariant measure m on $[0, 1]$ has a bounded density with respect to the Lebesgue measure, (3.5) is satisfied for $m = F$, the distribution of $X_1 = f_\lambda$ (here $X_n = f_\lambda^n!$). Consequently, Theorems 3.1 and 4.1 apply.

For $\lambda = 2$, m becomes Lebesgue measure and the label process is independent. If $\lambda = 1/2(1 + \sqrt{5})$, then $f_\lambda^3(0) = 0$, $f_\lambda^2(0) = 1$, and $f_\lambda(0) = 2 - \lambda = 1 - \lambda^{-1}$ (cf. Figure 2), and we see that f_λ maps I_1 onto I_2, and I_2 onto $[0, 1]$. This means that the invariant density must be constant on each of the intervals, and in fact an easy computation shows that this density is given by

$$h_\lambda(x) = \begin{cases} (3 - \lambda)^{-1} & \text{if } 0 \le x \le 1 - \lambda^{-1} \\ [(3 - \lambda)(\lambda - 1)]^{-1} & \text{if } 1 - \lambda^{-1} \le x \le 1 \end{cases} \tag{4.2}$$

It follows that $\theta = EX_1 \simeq 0.553$, $\sigma^2 = \operatorname{Var} X_1 \simeq 0.076$, and the median of $m = F$ is $M = MX_1 \simeq 0.573$. Based on 500 iterations with a randomly chosen starting value, we estimated $\hat{\theta} = 0.556$, $\hat{\sigma}^2 = 0.078$, and $\hat{M} = 0.572$. The fit of the empirical distribution function with F is shown in Figure 3. Figure 4 shows the empirical distribution function for 100 iterations. It should be noted that the numbers 100 or 500 are small compared to the usual number of iterations taken for simulations.

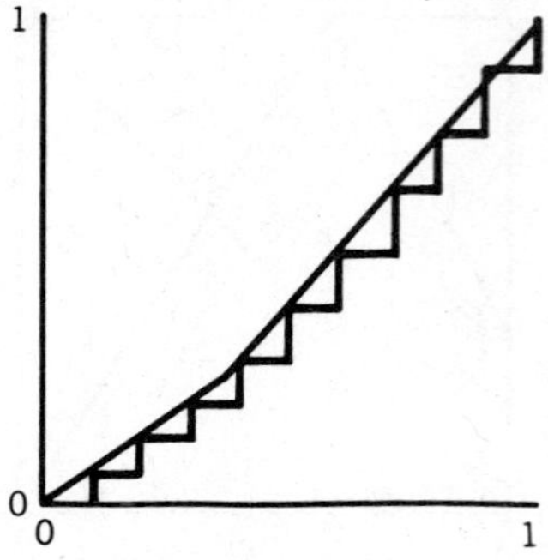

Figure 3.

The next case will be given for that value of λ when 0 is a periodic point of period 4 (Figure 5): $f_\lambda(0) = 2 - \lambda$, $f_\lambda^2 = 1 - \lambda^{-1}$, $f_\lambda^3(0) = 1$, $f_\lambda^4(0) = 0$. Equivalently λ is a solution of

$$\lambda^3 - \lambda^2 - \lambda - 1 = 0 \tag{4.3}$$

($\lambda \simeq 1.8392$). Here the invariant density is

$$h_\lambda(x) = \begin{cases} w\lambda^{-1} & \text{if } 0 \le x \le 2 - \lambda \\ w(\lambda^{-1} + \lambda^{-2}) & \text{if } 2 - \lambda \le x \le 1 - \lambda^{-1} \\ w & \text{if } 1 - \lambda^{-1} \le x \le 1 \end{cases} \tag{4.4}$$

where $w = \lambda^3[3\lambda^2 - 2\lambda - 1]^{-1}$.

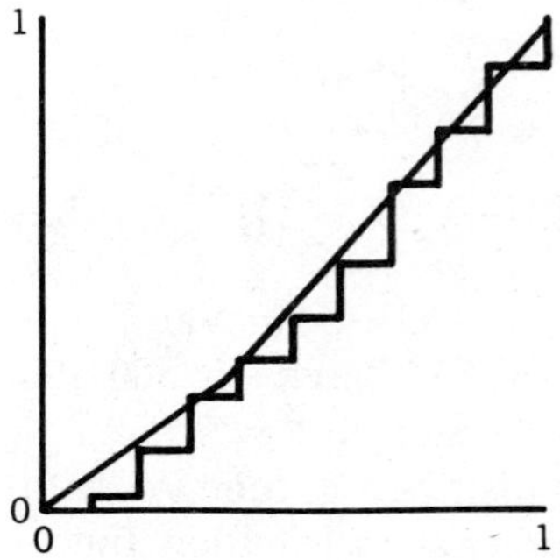

Figure 4.

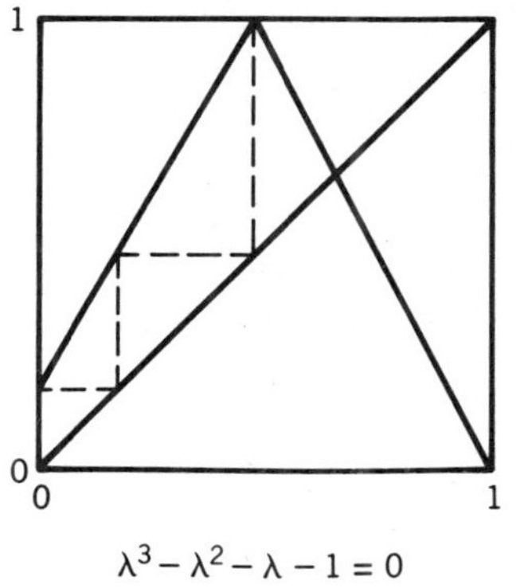

Figure 5.

We have $\theta = EX_1 \simeq 0.545$, $\sigma^2 = \operatorname{Var} X_1 \simeq 0.075$, and $M = MX_1 \simeq 0.560$. The estimates based on 500 iterations were 0.542, 0.077, and 0.569 in this case. Figure 6 shows the fit of the empirical distribution function again.

We have chosen this family $\{f_\lambda\}$ for various reasons. First, the true distributions can be computed in special cases, and according to Keller's result (1982) for every $\varepsilon > 0$, there exists a constant C_ε such that

$$\int |h_\lambda(x) - h_{\lambda'}(x)|\, dx \le C_\varepsilon |(\lambda - \lambda') \ln|\lambda - \lambda'|\,| \tag{4.5}$$

for all $\lambda, \lambda' \in [\sqrt{2} + \varepsilon, 2]$. Consequently, the distribution functions do not differ very much. Moreover, the asymptotic variance $\sigma^2(h)$ in Theorem 3.1

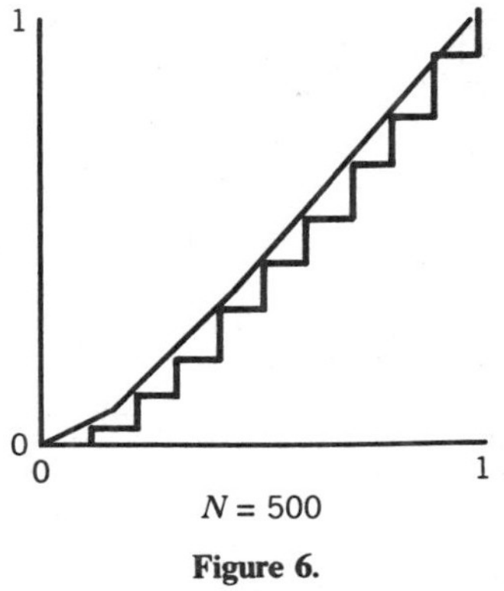

Figure 6.

for the Wilcoxon scores will be about 25% of the variance in the independent case. The formula for $\sigma^2(h)$ gives a consistent estimator using U statistics; however, because of the smallness of $\sigma^2(h)$ and the oscillating behavior of the covariances $Ef(X_1)f(X_t)$, it is still an open question to obtain reasonable estimates for $\sigma^2(h)$.

Taking two different values λ_1 and λ_2 and setting $X_n = f^n_{\lambda_1}$ and $Y_n = f^n_{\lambda_2}$, Theorem 3.1 gives a method of testing the hypothesis $F = G$ against $F \neq G$ using the Wilcoxon scores. In our simulations we estimated $\sigma^2(h)$ conservatively, but the results were satisfactory in some sense: if λ_1 is close to the lower bound and λ_2 is close to the upper bound, the test works well even at the 1% level. If λ_1 and λ_2 get closer, the test gets worse because of the theoretical behavior. It should be pointed out that in general the test works well if F and G are not too close, for example, if F corresponds to f_2 and if G is the invariant density of $x \to 1.4x \bmod 1$.

5. A FUNCTIONAL CENTRAL LIMIT THEOREM FOR SIGNED RANK STATISTICS

In this section, we consider signed rank statistics

$$L_n = \sum_{i=1}^{n} a_n(R_{in}^{+}) \operatorname{sgn} X_i \tag{5.1}$$

where R_{in}^{+} is the rank of X_i among $(|X_1|, \dots, |X_n|)$, and

$$a_n(i) = h\left(\frac{1}{2} + \frac{i}{2(n+1)}\right), \qquad i = 1, \dots, n \tag{5.2}$$

where h is some score function with $h(\frac{1}{2}) = 0$. We assume that $X_1, \dots, X_n$ are independent and identically distributed random variables with common continuous distribution function F. We prove a functional central limit theorem (FCLT) for L_n, and to do so, we represent L_n by an asymptotically equivalent two-sample linear rank statistic.

For $n \geq 1$, consider the random variables $X_1, \dots, X_n, -X_1, \dots, -X_n$ and the two-sample linear rank statistics $T_{2n}(g)$ given by

$$T_{2n}(g) = (2n)^{1/2}\left[\int_{-\infty}^{\infty} g(\hat{F}_{2n}(t))\, d\hat{H}_{2n}(t) - \int_{\infty}^{\infty} g(F_{2n}(t))\, dH_{2n}(t)\right] \tag{5.3}$$

where

$$\hat{H}_{2n}(t) = \frac{1}{2n}\sum_{i=1}^{n} I_{[X_i \le t]},$$

$$\hat{F}_{2n}(t) = \frac{n}{2n+1} + \frac{2n}{2n+1}\left(\hat{H}_{2n}(t) - \hat{H}_{2n}(-t)\right)$$

$$F_{2n} = \frac{(2n+1)}{2n} E(\hat{F}_{2n}), \qquad H_{2n} = E(\hat{H}_{2n}), \tag{5.4}$$

$$g(t) = \begin{cases} h(t) & \text{if } t \ge \frac{1}{2} \\ -h(1-t) & \text{if } t < \frac{1}{2} \end{cases} \qquad \text{[see (1.2)]}$$

Lemma 5.1. We have for $h \in \mathscr{H}_0$,

$$(2n)^{-1/2} \sup_{k \le n} \left(L_k - 2k \int_{-\infty}^{\infty} g(\hat{F}_{2k}(t))\, d\hat{H}_{2k}(t) \right) \quad \text{a.s.} \tag{5.5}$$

and $g \in \mathscr{H}_0$.

Proof. Let us first assume that h is increasing. Setting

$$g_k(t) = \begin{cases} h\left(\dfrac{2k+1}{2k+2}t + \dfrac{1}{2k+2}\right) & t \ge \frac{1}{2} \\ -h\left(1 - \dfrac{2k+1}{2k+2}t\right) & t < \frac{1}{2} \end{cases}$$

then $L_k = \Sigma_{i=1}^{2k} C_{2k}(i) g_k[R_i/(2k+1)]$ where

$$C_{2k}(i) = \begin{cases} 1 & i \le k \\ 0 & i > k \end{cases}$$

and where R_i denotes the rank of X_i among $\pm X_1, \ldots, \pm X_k$ for $i \le k$ and the rank of $-X_{i-k}$ among $\pm X_1, \ldots, \pm X_k$ for $i > k$. This follows immediately from

$$R_i^+ = \begin{cases} R_i - k & \text{if } X_i > 0 \\ k + 1 - R_i & \text{if } X_i < 0 \end{cases}$$

Consequently we can estimate

$$\left|(2n)^{-1/2} \sup_{k \le n}\left(L_k - 2k\int_{-\infty}^{\infty} g(\hat{F}_{2k}(t))\, d\hat{H}_{2k}(t)\right)\right|$$

$$= \left|(2n)^{-1/2} \sup_{k\le n} \sum_{\substack{i \le k\\ R_i > k}} \left(h\left(\frac{2k+1}{2k+2}\frac{R_i}{2k+1} + \frac{1}{2k+2}\right) - h\left(\frac{R_i}{2k+1}\right)\right)\right.$$

$$\left. + \sum_{i\le k}\left(-h\left(1 - \frac{2k+1}{2k+2}\frac{R_i}{2k+1}\right) + h\left(1 - \frac{R_i}{2k+1}\right)\right)\right|$$

$$\le (2n)^{-1/2} \sup_{k \le n}\left(\sum_{i \le k} h\left(1 - \frac{i}{2k+2}\right) - h\left(1 - \frac{i}{2k+1}\right)\right.$$

$$\left. + \sum_{i>k} h\left(\frac{i+1}{2k+2}\right) - h\left(\frac{i}{2k+1}\right)\right)$$

$$\le 2(2n)^{-1/2} \sup_{k\le n}\left(h\left(1 - \frac{1}{2k+2}\right) - h\left(1 - \frac{k}{2k+1}\right)\right)$$

Clearly

$$n^{-1/2} \sup_{k\le n} h\left(1 - \frac{k}{2k+1}\right) \to 0$$

and by Proposition 1(i) in Denker and Rösler (1985a)

$$n^{-1/2} \sup_{k\le n} h\left(1 - \frac{1}{2k+2}\right) \to 0$$

because $h \in \mathscr{H}_0$. This proves (5.5) if h is increasing. For general $h \in \mathscr{H}_0$, write $h = h_1 - h_2$ with h_i increasing and (5.5) follows from the previous case. ■

In order to study the limit behavior of $u \to (2n^{-1/2})L_{[un]}$, it suffices by Lemma 5.1 to consider the map

$$S_n\colon \mathscr{H}_0 \to L_2(P, D([0,1]))$$

defined by

$$S_n(h)(u) = (n)^{-1/2}([un])^{1/2} T_{2[un]}(g) \qquad (0 \leq u \leq 1) \tag{5.6}$$

where g is defined by (5.4). Recall that $[a]$ denotes the largest integer not exceeding a. In the sequel we shall also use the notation $a \ll b$ meaning that ab^{-1} is bounded (instead of the "0" notation).

We first establish the appropriate extension of Theorem 1.1. Note that the condition $C(0, 2, 1, \Lambda)$ holds in a more general form under our general assumptions in this section. For (1.4), replace the $L_2(P, \mathbb{R})$ norm on the left-hand side by the $L_2(P, D([0,1]))$ norm of the random function $u \to 2\hat{H}_{2[un]}(t) - F(t)$. This means that we have for each t

$$\begin{aligned} E \sup_{0 \leq u \leq 1} \frac{[un]^2}{4n^2} \left(2\hat{H}_{2[un]}(t) - F(t)\right)^2 \\ \leq 4\|2\hat{H}_{2n}(t) - F(t)\|^2_{L_2(P, \mathbf{R})} \ll n^{-1}F(t)(1 - F(t)) \end{aligned} \tag{5.7}$$

(use Doob's inequality for submartingales).

Theorem 5.1. $S = (S_1, S_2, \dots)$ is a continuous linear map from $\mathscr{H}_0$ into $L_2(P, D([0,1]))^*$.

Proof. It suffices to show that

$$\sup_n E\|S_n(g)\|^2_{L_2(P, D([0,1]))} \leq C\|g\|_0^2$$

for some constant C, independent of g, and for an increasing $g \in \mathscr{H}_0$ because of the definition of the norm $\|g\|_0$. We first estimate

$$\begin{aligned} \left(E \sup_{0 \leq u \leq 1} |S_n(g)(u)|^2\right)^{1/2} \leq \Bigg\{ & E \sup_{0 \leq u \leq 1} 2n^{-1}[un]^2 \\ & \times \left(\int_{-\infty}^{\infty} \left(g(\hat{F}_{2[un]}(t)) - g(F_{2[un]}(t))\right) d\hat{H}_{2[un]}(t)\right)^2 \Bigg\}^{1/2} \\ & + \Bigg\{ E \sup_{0 \leq u \leq 1} 2n^{-1}[un]^2 \\ & \times \left(\int_{-\infty}^{\infty} g(F_{2[un]}(t))\, d\left(\hat{H}_{2[un]} - H_{2[un]}\right)(t)\right)^2 \Bigg\}^{1/2} \end{aligned}$$

and the theorem follows from the two lemmas below. ∎

Lemma 5.2

$$E \sup_{0 \le u \le 1} [un]^2 \left(\int_{-\infty}^{\infty} \left(g\left(\hat{F}_{2[un]}(t) \right) - g\left(F_{2[un]}(t) \right) \right) d\hat{H}_{2[un]}(t) \right)^2 \ll n \|g\|_0^2$$

Proof. The proof is partly very much like the proof of Proposition 2(a) in Denker and Rösler (1985a). For completeness and since there are some significant differences, we give it in detail. Recall the inverse of a distribution function is always taken to be left continuous. Using

$$g(z) = \begin{cases} \int I_{(1/2, z]}(y)\, dg(y) & (z > \frac{1}{2}) \\ -\int I_{(z, 1/2)}(y)\, dg(y) & (z < \frac{1}{2}) \end{cases}$$

$$= \int_{1/2}^{z} dg(y)$$

we first observe that

$$\int_{-\infty}^{\infty} \left(g\left(\hat{F}_{2[un]}(t) \right) - g\left(F_{2[un]}(t) \right) \right) d\hat{H}_{2[un]}(t)$$

$$= \int_{-\infty}^{\infty} \left\{ \int_{1/2}^{\hat{F}_{2[un]}(t)} dg(z) - \int_{1/2}^{F_{2[un]}(t)} dg(z) \right\} d\hat{H}_{2[un]}(t)$$

$$= \int_{-\infty}^{\infty} I_{\{F_{2[un]}(t) \le \hat{F}_{2[un]}(t)\}} \int_{F_{2[un]}(t)}^{\hat{F}_{2[un]}(t)} dg(z)\, d\hat{H}_{2[un]}(t)$$

$$- \int_{-\infty}^{\infty} I_{\{\hat{F}_{2[un]}(t) < F_{2[un]}(t)\}} \int_{\hat{F}_{2[un]}(t)}^{F_{2[un]}(t)} dg(z)\, d\hat{H}_{2[un]}(t)$$

$$= \int_0^1 \left(\int_{-\infty}^{\infty} I_{\{(t,z) | F_{2[un]}(t) \le \hat{F}_{2[un]}(t),\, \hat{F}_{2[un]}^{-1}(z) \le t < F_{2[un]}^{-1}(z)\}} \right.$$

$$\left. - I_{\{(t,z) | \hat{F}_{2[un]}(t) < F_{2[un]}(t),\, F_{2[un]}^{-1}(z) \le t < \hat{F}_{2[un]}^{-1}(z)\}} \right) d\hat{H}_{2[un]}(t)\, dg(z)$$

$$= I_1 + I_2$$

where I_1 denotes the above integral with integration over $0 < z \le 2n/(2n + 1)$ and where I_2 denotes the above integral with integration over $2n/(2n + 1) < z < 1$.

We first estimate I_2. Since $z > 2n/(2n+1)$ we have

$$\hat{F}_{2[un]}(t) \le \frac{2[un]}{2[un]+1} \le \frac{2n}{2n+1} < z$$

and hence

$$I_{\{(t,z)|\hat{F}^{-1}_{2[un]}(z) \le t < F^{-1}_{2[un]}(z)\}} = 0$$

Consequently

$$[un]|I_2| \ll \int_{2n/(2n+1)}^{1} \int_{-\infty}^{\infty} I_{\{F^{-1}_{2[un]}(z) \le t\}} [un] \hat{F}_{2[un]}(t)\, dg(z)$$

$$= [un] \int_{2n/(2n+1)}^{1} \left(1 - \hat{F}_{2[un]}\left(F^{-1}_{2[un]}(z) -\right)\right) dg(z)$$

$$\ll [un] n^{-1/2} \int_{2n/(2n+1)}^{1} \sqrt{z(1-z)}\, dg(z)$$

$$+ ([un]+1) \int_{2n/(2n+1)}^{1} \left| \hat{F}_{2[un]}\left(F^{-1}_{2[un]}(z) -\right) - \frac{2[un]}{2[un]+1} z \right| dg(z)$$

since for $z \ge 2n(2n+1)^{-1}$ $1 - z \le (2n)^{-1}\sqrt{1-z}$.

For I_1 we proceed as follows. Let

$$\phi_u(t,z) = \begin{cases} 1 & \text{if } F_{2[un]}(t) < z \le \hat{F}_{2[un]}(t) \leftrightarrow \hat{F}^{-1}_{2[un]}(z) \le t < F^{-1}_{2[un]}(z) \\ -1 & \text{if } \hat{F}_{2[un]}(t) < z \le F_{2[un]}(t) \leftrightarrow F^{-1}_{2[un]}(z) \le t < \hat{F}^{-1}_{2[un]}(z) \\ 0 & \text{otherwise} \end{cases}$$

for $0 \le u \le 1$, $t \in \mathbb{R}$, and $z \le 2n/(2n+1)$. Then we have

$$\left| \int_{-\infty}^{\infty} \phi_u(t,z)\, d\hat{H}_{2[un]}(t) \right|$$

$$\ll \int_{-\infty}^{\infty} |\phi_u(t,z)|\, d\hat{F}_{2[un]}(t)$$

$$= \left| \hat{F}_{2[un]}\left(\hat{F}^{-1}_{2[un]}(z) -\right) - \hat{F}_{2[un]}\left(F^{-1}_{2[un]}(z) -\right) \right|$$

$$\ll \min\left\{z, (2[un]+1)^{-1}\right\} + \left| \hat{F}_{2[un]}\left(F^{-1}_{2[un]}(z) -\right) - \frac{2[un]}{2[un]+1} z \right|$$

and

$$[un]|I_1| = [un]\left|\int_0^{2n/(2n+1)}\int_{-\infty}^{\infty}\phi_u(t,z)\,d\hat{H}_{2[un]}(t)\,dg(z)\right|$$

$$\le [un]\int_0^{2n/(2n+1)}\int_{-\infty}^{\infty}|\phi_u(t,z)|\,d\hat{H}_{2[un]}(t)\,dg(z)$$

$$\ll \sqrt{n}\,\|g\|_0 + [un]\int_0^{2n/(2n+1)}\left|\hat{F}_{2[nu]}\left(F_{2[un]}^{-1}(z)-\right) - \frac{2[un]}{2[un]+1}z\right|\,dg(z)$$

since

$$n^{-1/2}\int_{1/(2n+1)}^{2n/(2n+1)} dg(z) = n^{-1/2}g\left(\frac{2n}{2n+1}\right) \ll \|g\|_0$$

and

$$\int_0^{1/(2n+1)} z\,dg(z) \ll n^{1/2}\|g\|_0$$

by Proposition 1 in Denker and Rösler (1985a).

Using

$$\hat{F}_{2[nu]}\left(F_{2[un]}^{-1}(z)-\right) - \frac{2[un]}{2[un]+1}z$$

$$= \hat{F}_{2[un]}\left(F_{2[un]}^{-1}(z)-\right) - E\hat{F}_{2[un]}\left(F_{2[un]}^{-1}(z)\right)$$

$$= \frac{1}{2[un]+1}\sum_{i\le[un]}\Big\{I_{\{X_i<F_{2[un]}^{-1}(z)\}} - F\left(F_{2[un]}^{-1}(z)\right) + I_{\{-X_i<F_{2[un]}^{-1}(z)\}} - \left(1-F\left(-F_{2[un]}^{-1}(z)\right)\right)\Big\}$$

$$= \frac{2[un]}{2[un]+1}\Big\{\hat{H}_{2[un]}\left(F_{2[un]}^{-1}(z)-\right) - H_{2[un]}\left(F_{2[un]}^{-1}(z)\right) - \hat{H}_{2[un]}\left(-F_{2[un]}^{-1}(z)\right) + H_{2[un]}\left(-F_{2[un]}^{-1}(z)\right)\Big\}$$

and

$$F_{2[un]}(t) = \frac{1}{2[un]} \sum_{i \le [un]} \left(EI_{\{X_i \le t\}} + EI_{\{-X_i \le t\}} \right)$$
$$= 1/2(1 + F(t) - F(-t)) =: H(t)$$

we obtain

$$\left([un] \int \left(g\left(\hat{F}_{2[un]}(t) \right) - g\left(F_{2[un]}(t) \right) \right) d\hat{H}_{2[un]}(t) \right)^2$$
$$\ll n\|g\|_0^2 + \left(\int_0^1 [un] |\hat{H}_{2[un]}(H^{-1}(z) -) - H_{2[un]}(H^{-1}(z))| \right.$$
$$\left. + [un] |\hat{H}_{2[un]}(-H^{-1}(z)) - H_{2[un]}(-H^{-1}(z))| \, dg(z) \right)^2$$

Finally it follows from (5.7) and Proposition 1(ii) in Denker and Rösler (1985a) that

$$E \sup_{0 \le u \le 1} [un]^2 \left(\int_{-\infty}^{\infty} \left(g\left(\hat{F}_{2[un]}(t) \right) - g\left(F_{2[un]}(t) \right) \right) d\hat{H}_{2[un]}(t) \right)^2$$
$$\ll n\|g\|^2 + E \sup_{0 \le u \le 1} [un]^2 \left(\int_0^1 |\hat{H}_{2[un]}(H^{-1}(z) -) \right.$$
$$\left. - H_{2[un]}(H^{-1}(z))| \, dg(z) \right)^2$$
$$+ E \sup_{0 \le u \le 1} [un]^2 \left(\int_0^1 |\hat{H}_{2[un]}(-H^{-1}(z)) - H_{2[un]}(-H^{-1}(z))| \, dg(z) \right)^2$$
$$\le n\|g\|^2 + \left(\int_0^1 \left(E \left(\sup_{0 \le u \le 1} [un] |\hat{H}_{2[un]}(H^{-1}(z) -) \right. \right. \right.$$
$$\left. \left. \left. - H_{2[un]}(H^{-1}(z))| \right)^2 \right)^{1/2} dg(z) \right)^2$$
$$+ \left(\int_0^1 \left(E \left(\sup_{0 \le u \le 1} [un] |\hat{H}_{2[un]}(-H^{-1}(z)) \right. \right. \right.$$
$$\left. \left. \left. - H_{2[un]}(-H^{-1}(z))| \right)^2 \right)^{1/2} dg(z) \right)^2$$

$$\ll n\|g\|_0^2 + \left(\int_0^1 \sqrt{n}\sqrt{F(H^{-1}(z))(1 - F(H^{-1}(z)))}\, dg(z)\right)^2$$

$$+\left(\int_0^1 \sqrt{n}\sqrt{F(-H^{-1}(z))(1 - F(-H^{-1}(z)))}\, dg(z)\right)^2$$

$$\ll n\|g\|_0^2 + n\left(\int_0^1 (z(1-z))^{1/2}\, dg(z)\right)^2 \ll n\|g_0\|^2$$

where we have also used that

$$F(t)(1 - F(t)) \leq (1 + F(t) - F(-t))(1 + F(-t) - F(t))$$

$$= 4H(t)(1 - H(t))$$

and $H(-t) = 1 - H(t)$. ■

Lemma 5.3

$$E \sup_{0 \leq u \leq 1} [un]^2 \left(\int_{-\infty}^{\infty} g\left(F_{2[un]}(t)\right) d\left(\hat{H}_{2[un]} - H_{2[un]}\right)(t)\right)^2 \ll n\|g\|_0^2$$

Proof. Since $F_{2[un]}(t) = H(t)$ is independent of u, the lemma follows from

$$E \max_{k \leq n} \left(\sum_{i=1}^{k} (g(H(X_i)) - Eg(H(X_1)))\right)^2$$

$$\leq 4E\left(\sum_{i=1}^{n} (g(H(X_i)) - Eg(H(X_1)))\right)^2$$

$$\leq 4n \int g^2(H(t))\, dF(t) \ll n\|g\|_0^2$$

where we used Proposition 1(iii) in Denker and Rösler (1985a) and $F(t) - F(s) \leq 2(H(t) - H(s))$ for $s \leq t$.

We can now proceed in a manner similar to that of Theorem 2.1. We shall say that a pair (h, F) satisfies condition (A) if $F \circ H^{-1}$ and $F \circ (-H^{-1})$ are $|dh|$ a.e. differentiable and $h \in \mathscr{H}_0$. Let K be a Kiefer process with covariance structure given by

$$EK(u, v) = 0$$
$$EK(u, v)K(u', v') = \min(u, u')F(v)(1 - F(v')) \tag{5.8}$$

for $0 \leq u,\ u' \leq 1$, and $v \leq v'$.

Theorem 5.2. Let (h, F) satisfy condition (A). Then $S_n(g)$ converges weakly in $D([0,1])$ to the random function

$$u \to \frac{1}{4\sqrt{2}} \int_0^1 \left[-(F \circ (-H^{-1}))'(t) K(u, H^{-1}(t)) \right.$$

$$\left. + (F \circ H^{-1})'(t) \cdot K(u, -H^{-1}(t)) \right] dg(t) \qquad (5.9)$$

Proof. Let us first show the theorem when g corresponds to a finite signed measure. Using Lemma 2.1 in a slightly different form we obtain

$$(2n)^{1/2} S_n(g)(u) = -[un] \int_0^1 \Bigg\{ \hat{H}_{2[un]}\left(\hat{F}^{-1}_{2[un]}(\hat{t})\right) - \tfrac{1}{2} F\left(\hat{F}^{-1}_{2[un]}(\hat{t})\right)$$

$$- \frac{F\left(\hat{F}^{-1}_{2[un]}(\hat{t})\right) - F\left(H^{-1}(t)\right)}{H\left(\hat{F}^{-1}_{2[un]}(\hat{t})\right) - t} \left[\hat{F}_{2[un]}\left(\hat{F}^{-1}_{2[un]}(\hat{t})\right) \right.$$

$$\left. - H\left(\hat{F}^{-1}_{2[un]}(\hat{t})\right) + \hat{t} - t \right] \Bigg\} dg(t)$$

$$= -[un] \int_0^1 - \frac{1}{2} \frac{F\left(-\hat{F}^{-1}_{2[un]}(\hat{t})\right) - F\left(-H^{-1}(t)\right)}{H\left(\hat{F}^{-1}_{2[un]}(\hat{t})\right) - t}$$

$$\times \left(\hat{H}_{2[un]}\left(\hat{F}^{-1}_{2[un]}(\hat{t})\right) - \tfrac{1}{2} F\left(\hat{F}^{-1}_{2[un]}(\hat{t})\right) \right)$$

$$+ \frac{1}{2} \frac{F\left(\hat{F}^{-1}_{2[un]}(\hat{t})\right) - F\left(H^{-1}(t)\right)}{H\left(\hat{F}^{-1}_{2[un]}(\hat{t})\right) - t} \left(\hat{H}_{2[un]}\left(-\left(\hat{F}^{-1}_{2[un]}(\hat{t}) -\right)\right) \right.$$

$$\left. - \tfrac{1}{2} F\left(-\left(\hat{F}^{-1}_{2[un]}(\hat{t}) -\right)\right) \right)$$

$$+ \frac{1}{2} \frac{F\left(\hat{F}^{-1}_{2[un]}(\hat{t})\right) - F\left(H^{-1}(t)\right)}{H\left(\hat{F}^{-1}_{2[un]}(\hat{t})\right) - t}$$

$$\times \left(\frac{1}{2[un]} \hat{F}_{2[un]}\left(\hat{F}^{-1}_{2[un]}(\hat{t})\right) + t - \hat{t} \right) dg(t) \qquad (5.10)$$

By the assumption [condition (A)] we have as in Lemma 2.2 that

$$\frac{F\left(\hat{F}_{2[un]}^{-1}(\hat{t})\right) - F\left(H^{-1}(t)\right)}{H\left(\hat{F}_{2[un]}^{-1}(\hat{t})\right) - t} - (F \circ H^{-1})'(t) \to 0 \quad \text{a.s.} \tag{5.11}$$

and

$$\frac{F\left(-\hat{F}_{2[un]}^{-1}(\hat{t})\right) - F\left(-H^{-1}(t)\right)}{H\left(\hat{F}_{2[un]}^{-1}(\hat{t})\right) - t} - \left(F \circ (-H^{-1})\right)'(t) \to 0 \quad \text{a.s.} \tag{5.12}$$

as $n \to \infty$ and uniformly in u with $2[un] \geq n^{1/2}$ for $|dh|$ a.e. t. Moreover the expressions in (5.11) and (5.12) are uniformly bounded for all u.

Considering $\sup_u |S_n(h)(u)|$ separately for $2[un] < \sqrt{n}$ and $2[un] \geq \sqrt{n}$, it follows from the representation (5.10) together with (5.11) and (5.12), Lebesgue's dominated convergence theorem and e.g. Doob's inequality that

$$\begin{aligned} \sup_{0 \leq u \leq 1} \Bigg| S_n(g)(u) & \\ &+ \frac{[un]}{\sqrt{2n}} \int_0^1 \Big\{ -\tfrac{1}{2}\left(F \circ (-H^{-1})\right)'(t)\Big(\hat{H}_{2[un]}\left(\hat{F}_{2[un]}^{-1}(\hat{t})\right) \\ &- \tfrac{1}{2}F\left(\hat{F}_{2[un]}^{-1}(\hat{t})\right)\Big) + \tfrac{1}{2}(F \circ H^{-1})'(t)\Big(\hat{H}_{2[un]}\left(-\hat{F}_{2[un]}^{-1}(\hat{t}) -\right) \\ &- \tfrac{1}{2}F\left(-\hat{F}_{2[un]}^{-1}(\hat{t})\right)\Big)\Big\}\, dg(t) \Bigg| \to 0 \end{aligned} \tag{5.13}$$

in probability.

We may assume by Skorohod's theorem that

$$\frac{[un]}{\sqrt{n}}\left[2\hat{H}_{2[un]}(t) - F(t)\right] \to K(u, t) \quad \text{a.s.}$$

for suitable versions. Since $\hat{\hat{F}}_{2[un]} \to H$ a.s. uniformly in t and since K has uniformly continuous paths,

$$\frac{[un]}{\sqrt{n}}\left(2\hat{H}_{2[un]}\left(\pm\hat{F}_{2[un]}^{-1}(\hat{t})\right) - F\left(\pm\hat{F}_{2[un]}^{-1}(\hat{t})\right)\right) - K\left(u, \pm H^{-1}(t)\right) \tag{5.14}$$

is small with large probability uniformly in t and u with $2[un] \geq n^{1/4}$, say.

Inserting (5.14) in (5.13) shows that

$$\sup_{0 \le u \le 1} \Bigg| S_N(g)(u) + \frac{1}{4\sqrt{2}} \int_0^1 \{ (F \circ H^{-1})'(t) K(u, -H^{-1}(t)) \\ - (F \circ (-H^{-1}))'(t) K(u, H^{-1}(t)) \} \, dg(t) \Bigg|$$

is small with large probability. This proves the theorem for finite signed measures dh.

In view of the general discussion in the introduction of our method and in view of Theorem 5.1 it, remains to show that

$$Z(g)(u) = \int_0^1 (-F \circ (-H^{-1}))'(t) K(u, H^{-1}(t)) \\ + (F \circ H^{-1})'(t) K(u, -H^{-1}(t)) \, dg(t) \qquad (5.15)$$

depends continuously on g, satisfying the condition (A).

Since for fixed t, $\{ K(u, t)^2 | 0 \le u \le 1 \}$ is a non-negative submartingale (in fact a Wiener process) we have

$$E \sup_{0 \le u \le 1} K(u, t)^2 \ll EK(1, t)^2 = F(t)(1 - F(t)).$$

In the proof of Lemma 5.2 we used already that $F(t)(1 - F(t)) \ll H(t)(1 - H(t))$ and $H(-t) = 1 - H(t)$. Thus it follows that

$$E \sup_{0 \le u \le 1} K(u, H^{-1}(t))^2 \ll F(H^{-1}(t))(1 - F(H^{-1}(t))) \\ \ll t(1 - t)$$

and

$$E \sup_{0 \le u \le 1} K(u, -H^{-1}(t))^2 \ll F(-H^{-1}(t))(1 - F(-H^{-1}(t))) \\ \ll t(1 - t)$$

Since by Lemma 2.2 $(F \circ H^{-1})'$ and $(F \circ (-H^{-1}))'$ are bounded functions in t, we conclude that

$$E \sup_{0 \le u \le 1} Z(g)(u)^2$$

$$\ll \left(\int_0^1 \left(-(F \circ (-H^{-1}))'(t) \left(E \sup_{0 \le u \le 1} K(u, H^{-1}(t))^2 \right)^{1/2} \right. \right.$$

$$\left. \left. + (F \circ H^{-1})'(t) \left(E \sup_{0 \le u \le 1} K(u, -H^{-1}(t))^2 \right)^{1/2} dg(t) \right) \right)^2$$

$$\ll \left(\int_0^1 (t(1-t))^{1/2} \, dg(t) \right)^2 \ll \|g\|_0^2$$

where we used Proposition 1(ii) in Denker and Rösler (1985a) again and that g satisfies condition (A).

General invariance principles for signed rank statistics have been obtained by Schulze-Pillot (1978) and Denker and Puri (1985), using an invariance principle in $D([0,1]^2)$ for the two-sample linear rank statistics. Here the condition on $h \in \mathscr{H}_0$ is such that, for every $0 < \lambda < 1$, $(F \circ H_\lambda^{-1})'$ exists $|dh|$ a.e. and for $|dh|$ a.e. $t \in (0,1)$ the family $(F \circ H_\lambda^{-1})'$ $(0 < \lambda < 1)$ is uniformly continuous at t. Here $H_\lambda = \lambda F + (1-\lambda)G$. In this case we shall say that (h, F) satisfies the condition (A′). Note that condition (A) is slightly weaker than (A′).

Now let $L_n^* = \sum_{i=1}^n a_n(i) Z_{i,n}$ denote a signed rank statistic with $a_n(i) = h[i/(n+1)]$, where $Z_{i,n} = 1$ if the ith smallest variable among $|X_1|, \dots, |X_n|$ comes from a positive X_j and $Z_{i,n} = 0$ otherwise. Following Pyke and Shorack (1968b) the sequences $\zeta_k = X_k 1_{\{X_k \ge 0\}}$, $\eta_k = -X_k 1_{\{X_k < 0\}}$, and $\xi_k = 1_{[0,\infty]}(X_k)$ are independent. Setting $\pi_0 = P(X_1 \ge 0)$ we define

$$\sum_n(u) = n^{-1/2}[un]^{1/2} T_{[un]}(h) + \frac{m_{[un]}}{\sqrt{n}} \int h(F_{[un]}(t)) \, dF(t)$$

$$- \frac{[un]}{\sqrt{n}} \pi_0 \int h(F_{\pi_0}(t)) \, dF(t)$$

$$= n^{-1/2} \sum_{i \le [un]} a_{[un]}(i) Z_{i,[un]} - \frac{[un]}{\sqrt{n}} \pi_0 \int h(F_{\pi_0}(t)) \, dF(t)$$

where $m_k = \operatorname{card}\{X_j \ge 0 | j \le k\}$ and $F_{\pi_0}(t) = \pi_0 F(t) + (1-\pi_0) G(t)$.

Note that $T_{[un]}(h)$ denotes the two-sample linear rank statistic based on the samples (ζ_k) and (η_k) with distributions

$$F(t) = \frac{P(0 \le X_1 \le t)}{1 - \pi_0} \quad \text{and} \quad G(t) = \frac{P(-t \le X_1 \le 0)}{\pi_0} \qquad (t \ge 0)$$

which have a random number of observations, more precisely

$$T_{[un]}(h) = [un]^{1/2}\left(\int_{-\infty}^{\infty} h\left(\hat{F}_{[un]}(t)\right) d\hat{H}_{[un]}(t) - \int_{-\infty}^{\infty} h\left(F_{[un]}(t)\right) dH_{[un]}(t)\right)$$

where

$$\hat{F}_{[un]}(t) = \frac{1}{[un]+1}\left\{\sum_{X_i \ge 0} 1_{\{X_i \le t\}} + \sum_{X_i < 0} 1_{\{-X_i \le t\}}\right\}$$

$$\hat{H}_{[un]}(t) = \frac{1}{[un]} \sum_{X_i \ge 0} 1_{\{X_i \le t\}}, \qquad H_{[un]} = \frac{1}{[un]} m_{[un]} F$$

and

$$F_{[un]}(t) = \frac{1}{[un]}\left(m_{[un]}F(t) + \left([un] - m_{[un]}\right)(1 - F(-t))\right)$$

In order to compare Theorem 5.2 with previous results, we state the following theorem from Denker and Puri (1985).

Theorem 5.3. Let (h, F) satisfy condition (A′). Then Σ_n converges weakly in $D([0,1])$ to the process

$$u \to Z(h)[u\pi_0, u(1-\pi_0)] - \sqrt{\pi_0(1-\pi_0)}\, W(u)\left[\int_{-0}^{1} F\left(F_{\pi_0}^{-1}(t)\right) dh(t) + \pi_0 \int_0^1 \left(F \circ F_{\pi_0}^{-1}\right)'(t)\left(G\left(F_{\pi_0}^{-1}(t)\right) - F\left(F_{\pi_0}^{-1}(t)\right)\right) dh(t)\right] \qquad (5.16)$$

where W denotes a standard Wiener process independent of $Z(h)$ and where the process $Z(h)$ has a representation

$$Z(h)[r,s] = \frac{s}{r+s}\int_{-\infty}^{\infty} \frac{dG}{dH_{r/(r+s)}}(t) K_1(r,t)\, dh\big(H_{r/(r+s)}(t)\big)$$

$$+ \frac{r}{r+s}\int_{-\infty}^{\infty} \frac{dG}{dH_{r/(r+s)}}(t) K_2(s,t)\, dh\big(H_{r/(r+s)}(t)\big)$$

$$(r, s \neq 0)$$

Here K_1 and K_2 are two independent Kiefer processes on $C([0,1]^2)$ with expectation 0 and covariance structure

$$EK_1(r,t)K_1(r',t') = \min(r,r')F(t)(1 - F(t'))$$

$$t \le t', \quad 0 \le r, \quad r' \le 1$$

$$EK_2(s,t)K_2(s',t') = \min(s,s')G(t)(1 - G(t'))$$

$$t \le t', \quad 0 \le s, \quad s' \le t$$

Our result in this section may be stated as follows:

Theorem 5.4. Let (h, F) satisfy condition (A). Then

$$u \to (2n)^{-1/2} L_{[un]} - (2n)^{-1/2}[un] \int g(H(t))\, dF(t) \qquad (5.17)$$

converges weakly in $D([0,1])$ to $(1/4\sqrt{2})Z(g)$, where $Z(g)$ is defined in (5.15) and g in (5.4).

Proof. Theorem 5.2 and Lemma 5.1. ■

If the scores $a_n(i)$ in L_n^* are given by the score function h as above, then we consider the score $\tilde{h}(t) = h(2t-1)$ $(t \ge \frac{1}{2})$ for L_n. The g corresponding to $\tilde{h}$ is then given by

$$g(t) = \begin{cases} h(2t-1) & t \ge \frac{1}{2} \\ -h(1-2t) & t < \frac{1}{2} \end{cases}$$

Since $L_n^* = \frac{1}{2}L_n + \frac{1}{2}\Sigma\, a_n(i)$, we find that Σ_n and $2^{-1/2}((2n)^{-1/2}L_{[un]} - (2n)^{-1/2}[un]\int g(H(t))\, dF(t)$ have the same limiting distribution. Consequently Σ_n converges weakly to $\frac{1}{8}Z(g)$ by (5.17), and the process $\frac{1}{8}Z(g)$ agrees with that one of (5.16).

ACKNOWLEDGMENT

We would like to express our sincere thanks to the referees for a number of useful comments and suggestions.

REFERENCES

Ahmad, I. and Lin, P. E. (1980). On the Chernoff-Savage theorem for dependent random sequences. *Ann. Inst. Statist. Math.* **32**, 211–222.

Antille, A. (1979). On the invariance principle for signed rank statistics. *Z. Wahrscheinlichkeitstheorie* **47**, 315–324.

Bowen, R. (1978). Bernoulli maps of the interval. *Israel J. Math.* **28**, 161–168.

Braun, H. I. (1976). Weak convergence of sequential linear rank statistics. *Ann. Statist.* **4**, 554–576.

Chernoff, H. and Savage, I. R. (1958). Asymptotic normality and efficiency of certain nonparametric test statistics. *Ann. Math. Statist.* **29**, 972–994.

Davidov, Y. A. (1970). The invariance principle for stationary processes. *Theory Prob. Appl.* **15**, 487–498.

Denker, M. and Keller, G. (1983). On *U*-statistics and v. Mises' statistics for weakly dependent processes. *Z. Wahrscheinlichkeitstheorie verw.Geb.* **64**, 505–522.

Denker, M. and Keller, G. (1985). Rigorous statistical procedures for data from dynamical systems. *J. Statist. Phys.* (to appear).

Denker, M. and Puri, M. L. (1985). A general invariance principle for Chernoff–Savage statistics. Technical Report, Indiana University, Mathematics Department.

Denker, M. and Rösler, U. (1985a). Some contributions to Chernoff–Savage theorems. *Statist. Decis.* **3**, 49–75.

Denker, M. and Rösler, U. (1985b). A moment estimate for rank statistics. *J. Statist. Plann. Infe* **12**, 269–284.

Dupač, V. and Hájek, J. (1969). Asymptotic normality of simple linear rank statistics under alternatives II. *Ann. Math. Statist.* **40**, 1992–2017.

Hájek, J. (1968). Asymptotic normality of simple linear rank statistics under alternatives. *Ann. Math. Statist.* **39**, 325–346.

Hofbauer, F. and Keller, G. (1982). Ergodic properties of invariant measures for piecewise monotonic transformations. *Math. Z.* **180**, 119–140.

Ibragimov, I. A. and Linnik, Y. V. (1971). *Independent and Stationary Sequences of Random Variables*, Gröningen: Wolters-Nordhoff.

Keller, G. (1982). Stochastic stability in some chaotic dynamical systems. *Mh. Math.* **94**, 313–333.

Lai, T. L. (1975). On Chernoff–Savage statistics and sequential rank tests. *Ann. Statist.* **3**, 825–845.

Pyke, R. and Shorack, G. R. (1968a). Weak convergence of a two-sample empirical process and a new approach to Chernoff–Savage theorems. *Ann. Math. Statist.* **39**, 755–771.

Pyke, R. and Shorack, G. R. (1968b). Weak convergence and a Chernoff–Savage theorem for random sample sizes. *Ann. Math. Statist.* **39**, 1675–1685.

Pyke, R. and Shorack, G. R. (1969). A note on Chernoff–Savage theorems. *Ann. Math. Statist.* **40**, 1116–1119.

Ralescu, S. and Puri, M. L. (1985). On the rate of convergence in the central limit theorem for signed rank statistics. *Adv. Appl. Math.* **6**, 23–51.

Schulze-Pillot, T. (1978). Über Invarianzprinzipien für lineare Rang–Statistiken. Dissertation, Universität Göttingen.

The Empiric Entropy, A New Approach to Nonparametric Entropy Estimation

Edward J. Dudewicz
Department of Mathematics
Syracuse University
Syracuse, New York

Edward C. van der Meulen
Department of Mathematics
Katholieke Universiteit Leuven
Leuven, Belgium

SUMMARY

Since the pioneering work of Shannon in the late 1940s [e.g., see Shannon (1948)], entropy has been of great theoretical and applied interest. Entropy principles play a key role in applications, for example, in communication theory (Gallager, 1968), statistical decision theory (Kullback, 1959), time-series analysis (Priestley, 1981), econometrics (Theil and Laitinen, 1980), ecology and genetics (Rao, 1982a, b), and modern spectroscopy (Ni and Scheraga, 1985). For references to other fields of applications, see Shore and Johnson (1980, 1981) and Rao and Nayak (1985).

Research supported by a Syracuse University Senate Research Grant, a Fellowship of the Onderzoeksfonds of the Katholieke Universiteit Leuven, the NATO Research Grants Programme (NATO Research Grant No. 1674), and partially by the Ministerie van Wetenschapsbeleid, Brussels, Belgium (Project GOA 83/88-53).

The entropy $H(f)$ of a probability density function (pdf) $f(x)$ usually needs to be estimated nonparametrically from a random sample $X_1, \ldots, X_n$ since commonly $f(\cdot)$ is unknown. In this chapter, we review previous work on estimation of $H(f)$ based on $X_1, \ldots, X_n$, and provide new entropy estimators called *empiric entropies* by an entropylike functional of the empiric pdf and generalizations thereof.

1. INTRODUCTION

The *entropy* $H(f)$ of a probability density function (pdf) $f(x)$ is defined by

$$H(f) = -\int_{-\infty}^{\infty} f(x) \log f(x) \, dx \tag{1.1}$$

In the literature various estimates for $H(f)$, based on a random sample $X_1, \ldots, X_n$ from the underlying distribution with pdf f, have been proposed and their properties studied. To briefly review this work we mention Tarasenko (1968), Dmitriev and Tarasenko (1973, 1974), Ahmad and Lin (1976), and Ivanov and Rozhkova (1981) on the one hand, and Vasicek (1976), Theil and Laitinen (1980), Theil (1980), and Dudewicz and van der Meulen (1981) on the other. Books and surveys, respectively, on density estimation which also deal with the topic of entropy estimation include those by Prakasa Rao (1983) and Wertz and Schneider (1979).

Below we review Vasicek's estimator (Section 2), analyze Theil's estimator (Section 3), introduce a first empiric entropy $H(\hat{f})$ and relate it to Vasicek's $m = 1$ estimator (Section 4), and introduce a second empiric entropy $H_2(n)$ and relate it to Vasicek's as a generalization/improvement (Section 5). The extension of our empiric entropies $H(\hat{f}) = H_1(n)$ and $H_2(n)$ to a *class* of empiric entropies, and their relation to Vasicek's estimator for general m is given in Section 6. Other methods for estimating entropy appearing in the literature are reviewed in Section 7, where also their relationship to our proposed entropy estimators is discussed. The analysis of the statistical properties (such as consistency) of the empiric entropies and their extension to the multivariate case will, because of limitations of space, be dealt with elsewhere.

2. VASICEK'S ESTIMATE

Given a random sample $X_1, \ldots, X_n$ $(n \geq 3)$ from a population with pdf f, let $Y_1 \leq Y_2 \leq \cdots \leq Y_n$ denote the order statistics, and define $Y_i =$

$Y_1 (i < 1)$ and $Y_i = Y_n (i > n)$. Vasicek (1976) introduced the entropy estimate

$$H_V(m, n) = \frac{1}{n} \sum_{i=1}^{n} \log\left\{ \frac{n}{2m} (Y_{i+m} - Y_{i-m}) \right\} \tag{2.1}$$

where m is a positive integer smaller than $n/2$, and proved that, if the variance of the underlying distribution is finite, $H_V(m, n) \to^P H(f)$ as $n \to \infty$, $m \to \infty$, $m/n \to 0$.

3. THEIL'S ESTIMATE

Theil (1980) evaluated the entropy of the pdf $\hat{f}(x)$, a fitted distribution. His $\hat{f}(x)$ is obtained from a random sample $X_1, \ldots, X_n$ of a population with pdf $f(x)$ by a *maximum entropy principle* [see Theil and Laitinen (1980)]. Since this work sheds light on the nature of our new estimators (to be introduced in Sections 4, 5, and 6), we now comment on it in some detail.

Let $X_1, \ldots, X_n$ $(n \geq 2)$ be a random sample from a population with pdf $f(x)$. Denote the order statistics by $Y_1 \leq Y_2 \leq \cdots \leq Y_n$ and their observed values by $y_1 \leq y_2 \leq \cdots \leq y_n$. Theil and Laitinen (1980) define intermediate points between successive observations by $\xi_i = \xi(y_i, y_{i+1})$, $i = 1, \ldots, n-1$, where $\xi(\cdot, \cdot)$ is a symmetric differentiable function of its two arguments with a value not outside their range. Each of the n intervals $I_1 = (-\infty, \xi_1)$, $I_i = [\xi_{i-1}, \xi_i)$ $(i = 2, \ldots, n-1)$, $I_n = [\xi_n, +\infty)$ contains exactly one of $y_1, \ldots, y_n$, hence a fraction $1/n$ of the mass of the observed distribution. Theil and Laitinen (1980) consider density estimates $\hat{f}(x)$ that preserve these observed fractions, that is, satisfy the mass-preserving constraint

$$\int_{I_i} \hat{f}(x)\, dx = \frac{1}{n} \qquad (i = 1, \ldots, n) \tag{3.1}$$

and are mean-preserving in that the mean over interval I_i must be a homogeneous linear function of y_{i-1}, y_i, y_{i+1} (for $i = 2, \ldots, n-1$), of y_1 and y_2 (for $i = 1$), and of y_{n-1} and y_n (for $i = n$). Of all such density estimates $\hat{f}(x)$, that which maximizes $H(\hat{f})$ is called the *maximum-entropy density estimate*, denoted $\hat{f}_{\mathrm{ME}}(x)$. Theil and Laitinen (1980) showed that

$$\hat{f}_{\mathrm{ME}}(x) = \begin{cases} \dfrac{1}{n(\xi_i - \xi_{i-1})} & \text{if } x \in I_i = [\xi_{i-1}, \xi_i) \quad (2 \le i \le n-1) \\ \dfrac{\lambda_1}{n} e^{\lambda_1(x-\xi_1)} & \text{if } x \in I_1 = (-\infty, \xi_1) \\ \dfrac{\lambda_n}{n} e^{-\lambda_n(x-\xi_{n-1})} & \text{if } x \in I_n = [\xi_{n-1}, +\infty) \end{cases} \tag{3.2}$$

with $\xi_i = (y_i + y_{i+1})/2$ $(1 \le i \le n-1)$, $\lambda_1 = 4/(y_2 - y_1)$, and $\lambda_n = 4/(y_n - y_{n-1})$.

Thus $\hat{f}_{\mathrm{ME}}(x)$ is uniform on each of the finite intervals $I_2, \ldots, I_{n-1}$, and exponential in the tail of intervals I_1 and I_n. The constrained ME choice of the endpoints ξ_i of the n intervals $I_1, \ldots, I_n$ amounts to choosing these intermediate points to be simply the averages of the two adjacent order statistics: $\xi_i = (y_i + y_{i+1})/2$. Theil and Laitinen (1980, p. 632, Fig. 1) plot the general form of $\hat{f}_{\mathrm{ME}}(x)$. However, it is more instructive to plot the corresponding cumulative distribution function (cdf) $\hat{F}_{\mathrm{ME}}(x)$, on the same axes as the empiric cdf $\hat{F}_0(x)$, as we do in Figure 1.

We see from Figure 1 that $\hat{F}_{\mathrm{ME}}(x)$ connects the midpoints of the horizontal segments of the graph of $\hat{F}_0(x)$ above the interval $[\xi_1, \xi_{n-1})$, while the tails come from exponential distributions over $(-\infty, \xi_1)$ and $[\xi_{n-1}, +\infty)$. We shall return to this idea (and its relation to the empiric pdf) in Sections 4 and 5.

Theil (1980) evaluated the entropy $H(\hat{f}_{\mathrm{ME}})$ of the fitted pdf $\hat{f}_{\mathrm{ME}}(x)$, and it is straightforward to show that

$$H(\hat{f}_{\mathrm{ME}}) = \frac{1}{n} \sum_{i=1}^{n} \log\left\{\frac{n}{2}(y_{i+1} - y_{i-1})\right\} + \frac{2 - 2\log 2}{n} \tag{3.3}$$

where $y_0 = y_1$ and $y_{n+1} = y_n$. While Theil (1980) did not so note, the entropy $H(\hat{f}_{\mathrm{ME}})$ of the fitted pdf $\hat{f}_{\mathrm{ME}}(x)$ can be regarded as an estimate of the entropy $H(f)$ of the underlying pdf f. As Theil (1980) does point out,

$$H(\hat{f}_{\mathrm{ME}}) = H_V(1, n) + \frac{2 - 2\log 2}{n} \tag{3.4}$$

relating this estimate of entropy to that of Vasicek, (2.1), with $m = 1$, and exceeding it by a positive amount which $\rightarrow 0$ as $n \rightarrow \infty$.

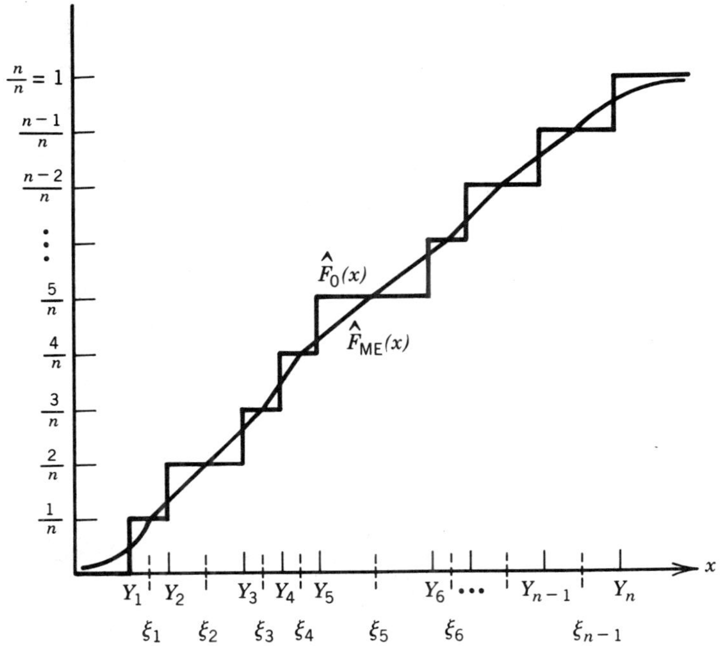

Figure 1. Maximum-entropy cdf $\hat{F}_{\mathrm{ME}}(x)$ and empiric cdf $\hat{F}_0(x)$, $n = 9$.

4. THE EMPIRIC ENTROPY OF ORDER 1

We propose to estimate the entropy $H(f)$ of an unknown pdf f as follows. First, estimate $f(x)$ by $\hat{f}_1(x)$, where $\hat{f}_1(x)$ is what we call the *empiric pdf*. Then estimate $H(f)$ by substituting $\hat{f}_1(x)$ into (1.1), obtaining the entropy estimate $H(\hat{f}_1)$, which we call the *empiric entropy of order one*.

Estimating $H(f)$ by $H(\hat{f})$ has great intuitive appeal, and while the empiric pdf $\hat{f}_1(x)$ is not so well known as the empiric cdf $\hat{F}_0(x)$, it is the "natural" density estimate to use in this process. Indeed, since recent work in bootstrapping [e.g., see Dudewicz (1985)] shows that for many purposes "smoothed" versions of $\hat{F}_0(x)$ do not (asymptotically) improve on $\hat{F}_0(x)$ itself, one might conjecture a strong role for $H(\hat{f}_1)$. [However, while $\hat{F}_0(x)$ may be "unique" as the natural empiric cdf, as we shall see in Section 5 $\hat{f}_1(x)$ is not unique as the natural empiric pdf. Hence we shall have additional empiric entropies $H(\hat{f})$ for other empiric-like pdf's $\hat{f}(x)$ that will merit attention.]

Let $X_1, \ldots, X_n$ $(n \geq 2)$ be a random sample from a population with pdf $f(x)$. Denote the order statistics by $Y_1 \leq Y_2 \leq \cdots \leq Y_n$, and define

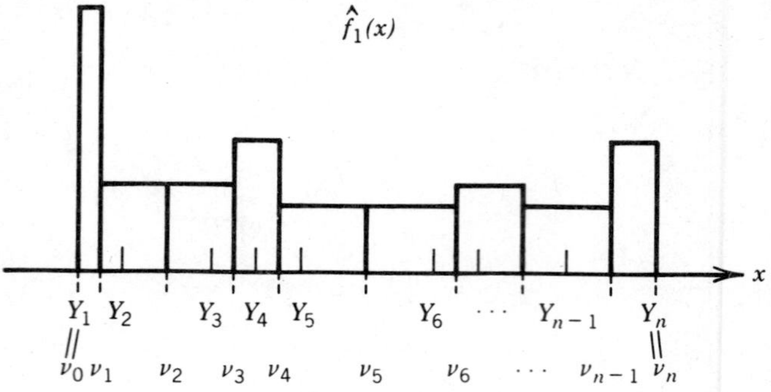

Figure 2. The empiric pdf $\hat{f}_1(x)$, $n = 9$.

$\nu_0 = Y_1$, $\nu_i = (Y_i + Y_{i+1})/2$ $(i = 1, 2, \ldots, n-1)$, and $\nu_n = Y_n$. Let I_i denote the interval $[\nu_{i-1}, \nu_i)$, and $\chi_{I_i}(x)$ its characteristic (indicator) function, for $i = 1, 2, \ldots, n$. Then the *empiric pdf of order one* is defined to be

$$\hat{f}_1(x) = \frac{1}{n}\sum_{i=1}^{n} \frac{1}{\nu_i - \nu_{i-1}} \chi_{I_i}(x)$$

$$= \frac{1}{n}\sum_{i=1}^{n} \frac{2}{Y_{i+1} - Y_{i-1}} \chi_{I_i}(x) \tag{4.1}$$

where $Y_0 = Y_1$ and $Y_{n+1} = Y_n$. The density estimate $\hat{f}_1(x)$ is sketched in Figure 2, while its corresponding cdf $\hat{F}_1(x)$ is sketched in Figure 3 together with the empiric cdf $\hat{F}_0(x)$.

We note that the graph of the cdf $\hat{F}_1(x)$ connects the midpoints of the horizontal lines in the graph of $\hat{F}_0(x)$ over the range (ν_1, ν_{n-1}), while on the left it connects the points $(Y_1, 0)$ and $(\nu_1, 1/n)$ by a straight line segment, and on the right it connects the points $(\nu_{n-1}, (n-1)/n)$ and $(Y_n, 1)$ by a straight line segment. Thus $\hat{f}_1(x)$ is uniform over each of the n intervals $[\nu_0, \nu_1)$, $[\nu_i, \nu_{i+1})$ $(i = 1, \ldots, n-1)$.

Relating $\hat{f}_1(x)$ to the $\hat{f}_{\mathrm{ME}}(x)$ of Section 3, note that $\hat{f}_{\mathrm{ME}}(x) = \hat{f}_1(x)$ for $\nu_1 \le x < \nu_{n-1}$, but, whereas $\hat{f}_1(x)$ restricts itself to uniform parts over the observed range of the data $X_1, \ldots, X_n$, on the other hand $\hat{f}_{\mathrm{ME}}(x)$ assumes a doubly infinite range of possible values with exponential tail weights.

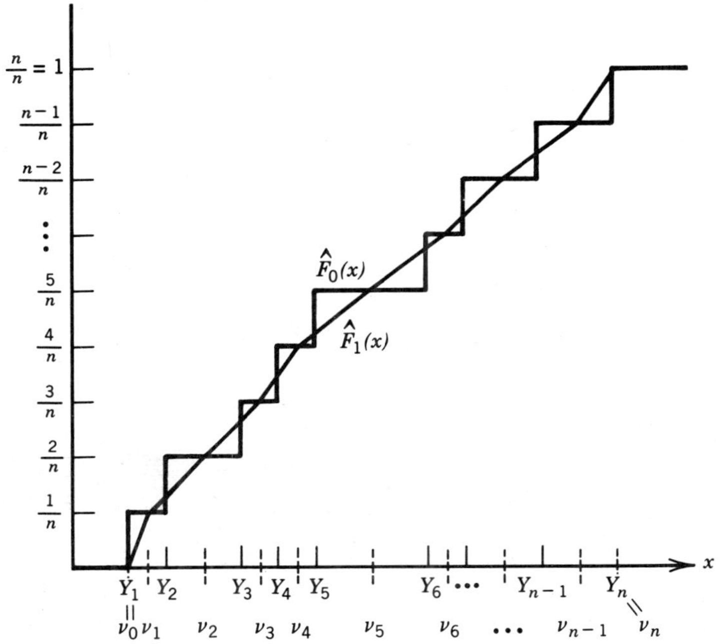

Figure 3. $\hat{F}_1(x)$ [cdf of the empiric pdf $\hat{f}_1(x)$] shown with the empiric cdf $\hat{F}_0(x)$.

The *empiric entropy of order 1*, and several equivalent formulas for it, are now obtained via use of (4.1) in (1.1), yielding

$$
\begin{aligned}
H_1(n) \equiv H(\hat{f}_1) &= -\int_{-\infty}^{\infty} \hat{f}_1(x) \log \hat{f}_1(x)\, dx \\
&= -\sum_{i=1}^{n} \int_{I_i} \frac{1}{n(\nu_i - \nu_{i-1})} \log \frac{1}{n(\nu_i - \nu_{i-1})}\, dx \\
&= \frac{1}{n} \sum_{i=1}^{n} \log\{n(\nu_i - \nu_{i-1})\} \\
&= \frac{1}{n} \sum_{i=1}^{n} \log\left\{\frac{n}{2}(Y_{i+1} - Y_{i-1})\right\} \qquad (4.2)
\end{aligned}
$$

Note that $H_V(1, n)$, the Vasicek $m = 1$ entropy estimator of equation (2.1), has the property that

$$H_V(1, n) = H_1(n) \tag{4.3}$$

thus showing Vasicek's estimator to be a natural one. Immediately, of course, the known results for Vasicek's estimator regarding asymptotic normality established by Dudewicz and van der Meulen (1981) carry over to $H_1(n)$.

5. THE EMPIRIC ENTROPY OF ORDER 2

We noted above that the empiric entropy (4.2) is not unique, since for every natural candidate $\hat{f}(x)$ for an empiric pdf there is another empiric entropy $H(\hat{f})$. We shall now develop a second empiric entropy $H_2(n)$ [recall that $H_1(n)$ was given in equation (4.2)], which will be generalized in Section 6 to a *family* of empiric entropies $H_m(n)$, $m = 1, 2, \ldots, [(n-1)/2]$, where $[t]$ denotes the largest integer $\leq t$.

Let $X_1, \ldots, X_n$ $(n \geq 5)$ be a random sample from a population with pdf $f(x)$. Denote the order statistics by $Y_1 \leq Y_2 \leq \cdots \leq Y_n$, define $Y_i = Y_1$ $(i < 1)$, $Y_i = Y_n$ $(i > n)$, and $\nu_i = (Y_i + Y_{i+2})/2$ for $i = -1, 0, 1, \ldots, n-1, n$. Also define intervals $I_i = [\nu_{i-2}, \nu_i)$ for $i = 1, \ldots, n$. Define functions

$$\begin{aligned} f_{21}(x) &= \frac{1}{n} \sum_{k=1}^{[(n+1)/2]} \frac{2}{\nu_{2k-1} - \nu_{2k-3}} \chi_{I_{2k-1}}(x) \\ &= \frac{1}{n} \sum_{k=1}^{[(n+1)/2]} \frac{4}{Y_{2k+1} - Y_{2k-3}} \chi_{I_{2k-1}}(x) \end{aligned} \tag{5.1}$$

and

$$\begin{aligned} f_{22}(x) &= \frac{1}{n} \sum_{k=1}^{[n/2]} \frac{2}{\nu_{2k} - \nu_{2k-2}} \chi_{I_{2k}}(x) \\ &= \frac{1}{n} \sum_{k=1}^{[n/2]} \frac{4}{Y_{2k+2} - Y_{2k-2}} \chi_{I_{2k}}(x) \end{aligned} \tag{5.2}$$

Note that $f_{21}(x)$ is constant on each of the intervals I_{2k-1} $(k = 1, 2, \ldots, [(n+1)/2])$, while $f_{22}(x)$ is constant on each of the intervals I_{2k} $(k = 1, 2, \ldots, [n/2])$.

For n *odd*, neither $f_{21}(x)$ nor $f_{22}(x)$ is a pdf since then

$$\int_{-\infty}^{\infty} f_{21}(x)\,dx = \frac{n+1}{n}, \qquad \int_{-\infty}^{\infty} f_{22}(x)\,dx = \frac{n-1}{n} \tag{5.3}$$

However, the average of f_{21} and f_{22},

$$\hat{f}_2(x) = \frac{f_{21}(x) + f_{22}(x)}{2} \tag{5.4}$$

is a pdf. For n *even*, all of f_{21}, f_{22}, and $\hat{f}_2$ are pdf's.

Formally, $\hat{f}_2(x)$ can be described as follows. With $\nu_i = (Y_i + Y_{i+2})/2$ $(i = -1, 0, 1, \ldots, n-1, n)$ as defined above, let J_i denote the interval $[\nu_{i-1}, \nu_i)$ for $i = 0, 1, \ldots, n$. Then

$$\hat{f}_2(x) = \begin{cases} \dfrac{1}{n}\left(\dfrac{1}{\nu_i - \nu_{i-2}} + \dfrac{1}{\nu_{i+1} - \nu_{i-1}}\right) & \text{if } x \in J_i = [\nu_{i-1}, \nu_i) \\ & (1 \le i \le n-1) \\ \dfrac{1}{n(\nu_1 - \nu_{-1})} & \text{if } x \in J_0 = [\nu_{-1}, \nu_0) \\ \dfrac{1}{n(\nu_n - \nu_{n-2})} & \text{if } x \in J_n = [\nu_{n-1}, \nu_n) \\ 0 & \text{if } x < \nu_{-1} \text{ or } x \ge \nu_n \end{cases} \tag{5.5}$$

An equivalent description of $\hat{f}_2(x)$ is given by

$$\hat{f}_2(x) = \begin{cases} \dfrac{2}{n}\left(\dfrac{1}{Y_{i+2} - Y_{i-2}} + \dfrac{1}{Y_{i+3} - Y_{i-1}}\right) & \text{if } x \in J_i = [\nu_{i-1}, \nu_i) \\ & (1 \le i \le n-1) \\ \dfrac{2}{n(Y_3 - Y_1)} & \text{if } x \in J_0 = [\nu_{-1}, \nu_0) \\ \dfrac{2}{n(Y_n - Y_{n-2})} & \text{if } x \in J_n = [\nu_{n-1}, \nu_n) \\ 0 & \text{if } x < \nu_{-1} \text{ or } x \ge \nu_n \end{cases} \tag{5.6}$$

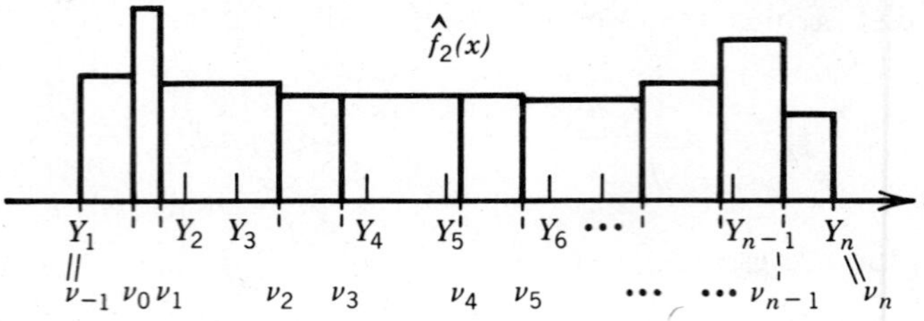

Figure 4. The empiric pdf $\hat{f}_2(x)$, $n = 9$.

Now, $\hat{f}_2(x)$, which is sketched in Figure 4, is also a reasonable candidate for *an empiric pdf*. Like $\hat{f}_1(x)$, it allows values only in the range of the data $X_1, \dots, X_n$ [since $\hat{f}_2(x) = 0$ outside the range of $X_1, \dots, X_n$], but does allow the full range [which $f_{21}(x)$ and $f_{22}(x)$ do not always do—if n is odd, f_{21} does, while if n is even, neither does]. Its main difference from $\hat{f}_1(x)$ is that, while $\hat{f}_1(x)$ arises from the empiric cdf $\hat{F}_0(x)$ by an "averaging" process over adjacent intervals, $\hat{f}_2(x)$ arises from an average of two "averaging" processes [those of $f_{21}(x)$ and $f_{22}(x)$] which attempt to better smooth out local roughness by averaging over intervals once removed from each other.

We remark at this point that the concept of *empirical density functions*, as opposed to that of *empirical distribution functions*, was also considered by Révész (1972), who put forward several possible definitions of an empirical density function. However, our empiric pdf's $\hat{f}_1$ and $\hat{f}_2$ *differ* from the empirical density functions considered by Révész (1972) in that the *endpoints of the intervals on which $\hat{f}_1$ and $\hat{f}_2$ are constant* are in principle *never* order statistics themselves [as is the case with the histogram-type empirical density functions defined by Révész (1972)], but are *always* functions (i.e., averages) of two order statistics, which may be adjacent (as in the case of $\hat{f}_1$) or one apart from each other (as in the case of $\hat{f}_2$).

The density estimates $\hat{f}_1$ and $\hat{f}_2$ also bear resemblance with the density estimates obtained by the histogram method, as proposed by Van Ryzin (1973). Such type of density estimates are referred to as estimators obtained by "the method of histograms with random partition" by Prakasa Rao (1983), and are called "variable histogram estimates" by Devroye and Györfi (1985). In a variable histogram estimate the endpoints of the intervals on which the density estimate is constant are allowed to depend on the data $X_1, \dots, X_n$, as is the case with $\hat{f}_1$ and $\hat{f}_2$. However, in the estimate studied by Van Ryzin (1973), and discussed in Prakasa Rao (1983, pp. 102–103), those endpoints are again taken to be actual order statistics, whereas, as pointed out above, in our estimates $\hat{f}_1$ and $\hat{f}_2$ the endpoints of the intervals of constancy are the ν_i's, which are averages of two order statistics. Moreover, in the context of a variable histogram estimate as considered by Devroye and Györfi (1985, p. 204) the constant value c_i of the density estimate on any particular interval J_i is usually taken to be $k_i/n\lambda(J_i)$, where k_i is the number of data points falling in J_i and λ is Lebesgue measure. An inspection of the definitions of $\hat{f}_1$ and $\hat{f}_2$ reveals that, whereas this latter property holds for $\hat{f}_1$, it does not hold for $\hat{f}_2$ and its subsequent generalizations $\hat{f}_m$. Therefore, although our $\hat{f}_2$ can be classified as a variable histogram estimate in its most general form, it does not seem that the density estimate $\hat{f}_2$ has been specifically considered before. Rather, $\hat{f}_2$ appears here motivated by our study of entropy estimation and as a result of our attempt to bring previous work on entropy estimation [such as that by Theil and Laitinen (1980) and Vasicek (1976)] into a general perspective.

Having introduced $\hat{f}_2$, we now define our (second, the first being in Section 4 above) empiric entropy

$$H_2(n) \equiv H(\hat{f}_2) = -\int_{-\infty}^{\infty} \hat{f}_2(x)\log \hat{f}_2(x)\,dx \tag{5.7}$$

To show connections with other entropy estimates, suppose that one estimated entropy based on $f_{21}(x)$, or $f_{22}(x)$ [which each may fail to be pdf's, as noted at equation (5.3)], or by the average of the "entropy" of these two. One would then be using

$$\begin{aligned} H_{21}(n) \equiv H(f_{21}) &= -\int_{-\infty}^{\infty} f_{21}(x)\log f_{21}(x)\,dx \\ &= \frac{2}{n}\sum_{k=1}^{[(n+1)/2]} \log\left\{\frac{n}{2}(\nu_{2k-1} - \nu_{2k-3})\right\} \\ &= \frac{2}{n}\sum_{k=1}^{[(n+1)/2]} \log\left\{\frac{n}{4}(Y_{2k+1} - Y_{2k-3})\right\} \end{aligned} \tag{5.8}$$

or

$$\begin{aligned} H_{22}(n) \equiv H(f_{22}) &= -\int_{-\infty}^{\infty} f_{22}(x)\log f_{22}(x)\,dx \\ &= \frac{2}{n}\sum_{k=1}^{[n/2]} \log\left\{\frac{n}{2}(\nu_{2k} - \nu_{2k-2})\right\} \\ &= \frac{2}{n}\sum_{k=1}^{[n/2]} \log\left\{\frac{n}{4}(Y_{2k+2} - Y_{2k-2})\right\} \end{aligned} \tag{5.9}$$

or

$$\frac{H_{21}(n) + H_{22}(n)}{2} = \frac{1}{n}\sum_{i=1}^{n} \log\left\{\frac{n}{4}(Y_{i+2} - Y_{i-2})\right\} \tag{5.10}$$

However, (5.10) is the Vasicek $m = 2$ entropy estimator of equation (2.1), $H_V(2, n)$.

Consider the case n even [so that $f_{21}(x)$ and $f_{22}(x)$ are both pdf's]. There is a well-known inequality in information theory [e.g., see Ash (1965)] which states that, for two arbitrary pdf's $f(x)$ and $g(x)$,

$$-\int_{-\infty}^{\infty} f(x)\log f(x)\,dx \le -\int_{-\infty}^{\infty} f(x)\log g(x)\,dx \tag{5.11}$$

and [if at least one side of (5.11) is finite] with equality iff $f \equiv g$. Thus

$$\begin{aligned}\frac{H_{21}(n) + H_{22}(n)}{2} &= -\frac{1}{2}\int_{-\infty}^{\infty} f_{21}(x)\log f_{21}(x)\,dx\\ &\quad -\frac{1}{2}\int_{-\infty}^{\infty} f_{22}(x)\log f_{22}(x)\,dx\\ &\le -\frac{1}{2}\int_{-\infty}^{\infty} f_{21}(x)\log \hat{f}_2(x)\,dx\\ &\quad -\frac{1}{2}\int_{-\infty}^{\infty} f_{22}(x)\log \hat{f}_2(x)\,dx\\ &= H(\hat{f}_2) = H_2(n)\end{aligned} \tag{5.12}$$

Thus (see the discussion following equation (5.10)) the Vasicek $m = 2$ entropy estimator $H_V(2, n)$ is always less than or equal to the empiric entropy $H_2(n)$, and may therefore be an unreasonable entropy estimator from the information-theoretic point of view.

When we evaluate $H_2(n)$ explicitly, we obtain with the use of (5.5) in (1.1) the following expression for the *empiric entropy of order* 2:

$$\begin{aligned}H_2(n) \equiv H(\hat{f}_2) &= -\int_{-\infty}^{\infty} \hat{f}_2(x)\log \hat{f}_2(x)\,dx\\ &= -\sum_{i=1}^{n-1}\int_{J_i}\frac{1}{n}\sum_{j=i}^{i+1}\frac{1}{\nu_j - \nu_{j-2}}\log\left(\frac{1}{n}\sum_{j=i}^{i+1}\frac{1}{\nu_j - \nu_{j-2}}\right)dx\\ &\quad -\int_{J_0}\frac{1}{n(\nu_1 - \nu_{-1})}\log\frac{1}{n(\nu_1 - \nu_{-1})}\,dx\\ &\quad -\int_{J_n}\frac{1}{n(\nu_n - \nu_{n-2})}\log\frac{1}{n(\nu_n - \nu_{n-2})}\,dx\end{aligned} \tag{5.13}$$

Now using (5.6), expression (5.13) can be written

$$H_2(n) = \frac{1}{n}\left(\frac{Y_2 - Y_1}{Y_3 - Y_1}\right)\log\left\{\frac{n}{2}(Y_3 - Y_1)\right\}$$

$$-\frac{1}{n}\sum_{i=1}^{n-1}\sum_{j=i}^{i+1}\left(\frac{Y_i + Y_{i+2} - Y_{i-1} - Y_{i+1}}{Y_{j+2} - Y_{j-2}}\right)$$

$$\times\log\left\{\frac{2}{n}\sum_{j=i}^{i+1}\left(\frac{1}{Y_{j+2} - Y_{j-2}}\right)\right\}$$

$$+\frac{1}{n}\left(\frac{Y_n - Y_{n-1}}{Y_n - Y_{n-2}}\right)\log\left\{\frac{n}{2}(Y_n - Y_{n-2})\right\} \tag{5.14}$$

Note that, whereas we have reason to believe that $H_2(n)$ is a better estimate of entropy than $H_V(2, n)$, a comparison of expressions (5.14) and (5.10) shows that $H_2(n)$ is substantially more complex than $H_V(2, n)$.

6. THE EMPIRIC ENTROPY OF ORDER M

We now give the generalization of the empiric entropies $H_1(n)$ and $H_2(n)$ to the family of empiric entropies $H_m(n)$, $m = 1, 2, \ldots, [(n-1)/2]$. We will call $H_m(n)$ the *empiric entropy of order m*.

Let $X_1, \ldots, X_n$ $(n \geq 2m + 1)$ be a random sample from a population with pdf $f(x)$. As before, denote the order statistics by $Y_1 \leq Y_2 \leq \cdots \leq Y_n$, and define $Y_i = Y_1$ if $i < 1$ and $Y_i = Y_n$ if $i > n$. Now let $\nu_i = (Y_i + Y_{i+m})/2$ for $i = -m+1, -m+2, \ldots, n-1, n$. Next define intervals $I_i = [\nu_{i-m}, \nu_i)$ for $i = 1, \ldots, n$. Then define m functions

$$f_{mj}(x) = \frac{1}{n}\sum_{k=1}^{[(n+m-j)/m]}\frac{m}{\nu_{mk+j-m} - \nu_{mk+j-2m}}\chi_{I_{km+j-m}}(x)$$

$$= \frac{1}{n}\sum_{k=1}^{[(n+m-j)/m]}\frac{2m}{Y_{mk+j} - Y_{mk+j-2m}}\chi_{I_{km+j-m}}(x) \tag{6.1}$$

$j = 1, \ldots, m$. In general $f_{mj}(x)$ is not a true pdf, but the average

$$\hat{f}_m(x) = \frac{1}{m}\sum_{j=1}^{m} f_{mj}(x) \tag{6.2}$$

is a pdf. Formally, $\hat{f}_m(x)$ can be described as follows. Let J_i denote the interval $[\nu_{i-1}, \nu_i)$ for $i = 2 - m, 2 - m + 1, \ldots, n - 1, n$. Then

$$\hat{f}_m(x) = \begin{cases} \dfrac{1}{n} \displaystyle\sum_{j=1}^{i+m-1} \dfrac{1}{\nu_j - \nu_{j-m}} & \text{if } x \in J_i, \quad 2 - m \le i \le 0 \\ \dfrac{1}{n} \displaystyle\sum_{j=i}^{i+m-1} \dfrac{1}{\nu_j - \nu_{j-m}} & \text{if } x \in J_i, \quad 1 \le i \le n + 1 - m \\ \dfrac{1}{n} \displaystyle\sum_{j=i}^{n} \dfrac{1}{\nu_j - \nu_{j-m}} & \text{if } x \in J_i, \quad n + 2 - m \le i \le n \\ 0 & \text{if } x < \nu_{1-m} \text{ or } x \ge \nu_n \end{cases} \tag{6.3}$$

Alternatively, $\hat{f}_m(x)$ can be written

$$\hat{f}_m(x) = \begin{cases} \dfrac{2}{n} \displaystyle\sum_{j=1}^{i+m-1} \dfrac{1}{Y_{j+m} - Y_{j-m}} & \text{if } x \in J_i, \quad 2 - m \le i \le 0 \\ \dfrac{2}{n} \displaystyle\sum_{j=i}^{i+m-1} \dfrac{1}{Y_{j+m} - Y_{j-m}} & \text{if } x \in J_i, \quad 1 \le i \le n + 1 - m \\ \dfrac{2}{n} \displaystyle\sum_{j=i}^{n} \dfrac{1}{Y_{j+m} - Y_{j-m}} & \text{if } x \in J_i, \quad n + 2 - m \le i \le n \\ 0 & \text{if } x < \nu_{1-m} \text{ or } x \ge \nu_n \end{cases} \tag{6.4}$$

Although we have omitted a sketch of $\hat{f}_m(x)$, it should be clear that for $m \ge 2$ $\hat{f}_m(x)$ will smooth out local roughness well, as it is the average of m averaging processes. The empiric entropy of order m is now defined as

$$H_m(n) \equiv H(\hat{f}_m) = -\int_{-\infty}^{\infty} \hat{f}_m(x) \log \hat{f}_m(x)\, dx \tag{6.5}$$

For general m, the connection between $H_m(n)$ and the Vasicek entropy estimator $H_V(m, n)$ is the following. If we define, in analogy with (5.8) and (5.9),

$$H_{mj}(n) \equiv H(f_{mj}) = -\int_{-\infty}^{\infty} f_{mj}(x) \log f_{mj}(x)\, dx \tag{6.6}$$

$j = 1, \ldots, m$, then, using (6.1), we obtain

$$H_{mj}(n) = \frac{m}{n} \sum_{k=1}^{[(n+m-j)/m]} \log\left\{\frac{n}{m}\left(v_{mk+j-m} - v_{mk+j-2m}\right)\right\}$$
$$= \frac{m}{n} \sum_{k=1}^{[(n+m-j)/m]} \log\left\{\frac{n}{2m}\left(Y_{mk+j} - Y_{mk+j-2m}\right)\right\} \quad (6.7)$$

It can be easily shown that the average of the m "entropies" $H_{mj}(n)$ exactly equals the Vasicek estimator $H_V(m, n)$, that is,

$$\frac{1}{m} \sum_{j=1}^{m} H_{mj}(n) = \frac{1}{n} \sum_{i=1}^{n} \log\left\{\frac{n}{2m}\left(Y_{i+m} - Y_{i-m}\right)\right\} \quad (6.8)$$

We thus have obtained a natural interpretation of the Vasicek estimator $H_V(m, n)$, as being the arithmetic average of m entropies, each individual one being the entropy of a pseudo-pdf.

Based on inequality (5.11) we also see that, at least for n an integer-multiple of m,

$$H_V(m, n) \leq H_m(n) \quad (6.9)$$

which makes us propose $H_m(n)$ as an estimator of entropy to be preferred over $H_V(m, n)$. To conclude this section, we give the analytical expression for $H_m(n)$. Using (6.4) in (6.5) we obtain the following expression for the empiric entropy of order m:

$$H_m(n) = -\frac{1}{n} \sum_{i=2-m}^{0} \sum_{j=1}^{i+m-1} \left(\frac{Y_{i+m} - Y_{i+m-1}}{Y_{j+m} - Y_{j-m}}\right)$$
$$\times \log\left\{\frac{2}{n} \sum_{j=1}^{i+m-1} \left(\frac{1}{Y_{j+m} - Y_{j-m}}\right)\right\}$$
$$-\frac{1}{n} \sum_{i=1}^{n+1-m} \sum_{j=i}^{i+m-1} \left(\frac{Y_i + Y_{i+m} - Y_{i-1} - Y_{i+m-1}}{Y_{j+m} - Y_{j-m}}\right)$$
$$\times \log\left\{\frac{2}{n} \sum_{j=i}^{i+m-1} \left(\frac{1}{Y_{j+m} - Y_{j-m}}\right)\right\}$$
$$-\frac{1}{n} \sum_{i=n+2-m}^{n} \sum_{j=i}^{n} \left(\frac{Y_i - Y_{i-1}}{Y_{j+m} - Y_{j-m}}\right) \log\left\{\frac{2}{n} \sum_{j=i}^{n} \left(\frac{1}{Y_{j+m} - Y_{j-m}}\right)\right\}$$
$$(6.10)$$

Clearly, when $m = 2$ (6.10) specializes to (5.14).

Previous studies have shown good power properties of statistical analyses based on $H_V(m, n)$, $1 \leq m < n/2$ [e.g., see Dudewicz and van der Meulen (1981) for some such results in uniformity testing]. The present results imply that even more powerful statistical procedures may result from the empiric entropies of general order m, which we are currently in the process of investigating.

7. REVIEW OF OTHER METHODS TO ESTIMATE ENTROPY AND THEIR STATISTICAL PROPERTIES

Having pointed out the connections between our proposed estimator of entropy $H_m(n)$ and the estimators of Vasicek (1976) and Theil and Laitinen (1980), we now briefly review the remaining literature on the topic of entropy estimation and make some additional remarks in relation to the new estimate we introduced.

Dmitriev and Tarasenko (1973) proposed to estimate the entropy functional $H(f)$, given by (1.1), by the estimator

$$I_n^* = -\int_{-b_n}^{b_n} g_n(x) \log g_n(x)\, dx \tag{7.1}$$

where $g_n(x)$ is a kernel-type density estimator [due to Rosenblatt (1956) and Parzen (1962)] given by

$$g_n(x) = \frac{1}{na_n} \sum_{i=1}^{n} K\left(\frac{x - X_i}{a_n}\right) \tag{7.2}$$

with K a suitable kernel satisfying certain well-known conditions, $\{a_n\}$ a sequence of positive numbers tending to 0, and $b_n \to \infty$ as $n \to \infty$. Dmitriev and Tarasenko (1973) showed that I_n^* converges to $H(f)$ with probability 1 if $a_n = n^{-1/4}$ and $b_n = G^{-1}(n^{1/10})$, where G is a known function such that

$$\sup_{|y| \leq x} 1/f(y) \leq G(x), \qquad -\infty < x < \infty \tag{7.3}$$

[For a proof of this fact, see Prakasa Rao (1983, p. 272).]

We note that, although Dmitriev and Tarasenko (1973) use essentially the same method as we do to estimate $H(f)$—namely, substituting a density estimate into (1.1)—our entropy estimate $H_m(n)$ differs from I_n^* because our density estimate $\hat{f}_m$ differs from g_n for the reasons explained in Section 5.

Dmitriev and Tarasenko (1974) proposed yet another possible approach to the nonparametric estimation of functionals, which is based on the use of quasi-U-statistics. In their article they advocate this method since the structure of quasi-U-statistics would be convenient for the practical computation of estimates. However, they do not specifically consider the estimation of the entropy functional this way.

Ahmad and Lin (1976) proposed estimating $H(f)$ by the quantity

$$\hat{I}_n = -\frac{1}{n}\sum_{i=1}^{n} \log g_n(X_i) \tag{7.4}$$

where $g_n(x)$ is again a Rosenblatt–Parzen type of kernel density estimate of the form (7.2). Ahmad and Lin (1976) showed, under certain regularity conditions on the sequence $\{a_n\}$, the unknown pdf $f(x)$, and the kernel function K, the consistency of $\hat{I}_n$ in the first and second means as an estimator of $H(f)$.

Continuing the investigations of Dmitriev and Tarasenko (1974) and Ahmad and Lin (1976), Ivanov and Rozhkova (1981) proposed yet another estimate for the entropy $H(f)$ that is analogous to (7.4). Their estimate is

$$I_n = -\frac{1}{n}\sum_{i=1}^{n} \log \left| g_n(X_i) - \frac{K(0)}{na_n} \right| \chi_{[-b_n, b_n]}(X_i) \tag{7.5}$$

where $g_n(x)$ is again a kernel-type density estimator of the form (7.2) with the special condition on the kernel function K that it be strictly positive for all $x \in \mathbb{R}$. Moreover, $\{a_n\}$ and $\{b_n\}$ are assumed to satisfy the usual conditions that $a_n \to 0$, $na_n \to \infty$, and $b_n \to \infty$ as $n \to \infty$. It is easily seen that I_n can be written in the following form:

$$I_n = -\frac{1}{n}\sum_{i=1}^{n} \log\left\{ \frac{1}{na_n} \sum_{j \neq i} K\left(\frac{X_i - X_j}{a_n} \right)\right\} \chi_{[-b_n, b_n]}(X_i) \tag{7.6}$$

Ivanov and Rozhkova (1981) showed that, under certain conditions on K, f, a_n, and b_n, I_n converges with probability 1 to $H(f)$ as $n \to \infty$. Moreover, they made an assertion regarding the rate of convergence of the moments $E|I_n - H(f)|^r$, $r \geq 1$, to 0. The latter result yields an upper bound for the bias of I_n (taking $r = 1$) and the standard deviation of I_n from $H(f)$ (taking $r = 2$).

From the discussion in Section 5 it follows that our proposed estimator of entropy $H_m(n)$ differs from those defined by (7.4) and (7.5) in various ways. First, the density estimate used in our case, $\hat{f}_m(x)$, differs from the density estimate used in (7.4) and (7.5) for reasons explained earlier. Moreover, the *methods* used to estimate the entropy functional differ in both cases. Our method, and also the one by Dmitriev and Tarasenko (1973), is based on substituting a suitable density estimate $f_n(x)$ into (1.1), yielding the estimator

$$\hat{\phi}_1 = -\int_{-\infty}^{\infty} f_n(x) \log f_n(x)\, dx \tag{7.7}$$

On the other hand, the method used by Ahmad and Lin (1976) and Ivanov and Rozhkova (1981) is based on estimators of the type

$$\hat{\phi}_2 = -\int_{-\infty}^{\infty} \log f_n(x)\, d\hat{F}_0(x) \tag{7.8}$$

where $\hat{F}_0(x)$ denotes the sample distribution function. Obviously, by using our density estimate $\hat{f}_m(x)$ instead of $f_n(x)$ in (7.8) we would generate yet another class of estimators of the entropy $H(f)$.

Hasminskii and Ibragimov (1978) noted that, from the point of view of asymptotic efficiency, estimators of smooth functionals $\phi(f(x))$ of the pdf of type (7.8) are to be preferred over those of type (7.7). However, in their discussion of smooth functionals no specific mention is made of the entropy functional as such. Hasminskii and Ibragimov (1978) also considered a modified estimator of type (7.8) and established asymptotic normality of this estimator under certain conditions on f and ϕ. Similar results were reported by Tarasenko (1976).

Another approach towards the nonparametric estimation of functionals of pdf's was taken by Wegman (1984). He developed a theory of functional optimization in Hilbert spaces, which is adapted from Tapia and Thompson (1978). Wegman (1984) proved an existence and uniqueness theorem for minimizing (or maximizing) a functional and remarked that the entropy functional fits the setting of that theorem.

Using yet another approach, Wertz (1985) investigated the problem of finding invariant estimators of functions that satisfy optimality conditions with respect to mean-integrated-squared error as risk criterion. Although the estimation of entropy under the restriction of invariance has not been specifically considered yet, this would certainly be an interesting direction of research to pursue.

To conclude this section, we finally mention an article by Vinod (1982), in which the maximum entropy technique proposed by Theil and Laitinen (1980), as discussed in Section 3, is slightly modified. In obtaining the maximum-entropy density estimate, Vinod (1982) added another constraint, requiring a fuzziness range $(-d, d)$ around each observation in which the true unknown random variable must lie with high probability. The resulting entropy estimate is then shown to be equal to

$$H(\hat{f}_{\text{ME2}}) = \log(nd) \tag{7.9}$$

which, according to Vinod (1982), appears to be smaller than the value $H(\hat{f}_{\text{ME}})$ given by formula (3.4).

8. CONCLUDING REMARKS

In this chapter we have put forward a new type of entropy estimator called *empiric entropy*, which is based on the principle of first estimating $f(x)$ by an empiric pdf $\hat{f}$ and then estimating $H(f)$ by $H(\hat{f})$. In Sections 4 and 5, two specific empiric pdf's $\hat{f}_1$ and $\hat{f}_2$ were introduced, and the corresponding empiric entropies $H(\hat{f}_1)$ and $H(\hat{f}_2)$ defined. Connections with other entropy estimators, such as the maximum entropy estimate $H(\hat{f}_{\text{ME}})$, suggested by the work of Theil and Laitinen (1980), and the Vasicek (1976) estimators $H_V(1, n)$ and $H_V(2, n)$ have been pointed out. It turns out that $\hat{f}_2$ is actually a new density estimate, belonging to the general class of variable histogram estimates. On information-theoretic grounds there is reason to believe that $H(\hat{f}_2)$ is a better estimator of entropy than $H_V(2, n)$, but further studies are needed in this regard, both theoretical and experimental.

In Section 6 the method of estimating $\hat{f}_2$ as described in Section 5 has been generalized to obtain an empiric pdf $\hat{f}_m$ for every integer $m \geq 3$ such that $m \leq [(n-1)/2]$. Corresponding to such $\hat{f}_m$, the empiric entropy $H_m(n)$ has been defined by $H(\hat{f}_m)$. The relationship between $H_m(n)$ and the Vasicek estimator $H_V(m, n)$ for a general m has also been pointed out in Section 6.

In Section 7 the relationship between $H_m(n)$ and other methods of estimating entropy which appear in the literature has been described.

Several properties of our estimators $\hat{f}_m$ and $H_m(n)$ are now being investigated. These investigations include the study of the possible consistency (and asymptotic normality) of $\hat{f}_m$ and $H_m(n)$, and also of the power properties of tests based on $H_m(n)$ [especially for testing uniformity on (0, 1)] in the vein of Dudewicz and van der Meulen (1981). The authors hope to report on their findings in this direction in a subsequent paper.

Another direction of research which the authors intend to explore is how the decision-theoretic criteria as put forward by Hasminskii and Ibragimov

(1978) and Wertz (1985) can be incorporated in our studies so as to yield entropy estimators satisfying certain desirable optimality criteria. It is hoped that the present investigations already shed some light on these related problems.

ACKNOWLEDGMENTS

The authors would like to thank one of the editors and both referees for helpful comments that led to an improvement of this paper and for bringing to their attention several of the references.

REFERENCES

Ahmad, I. A. and Lin, P. E. (1976). A nonparametric estimation of the entropy for absolutely continuous distributions. *IEEE Trans. Inform. Theory* **22**, 372–375.

Ash, R. B. (1965). *Information Theory*, New York: Interscience.

Devroye, L. and Györfi, L. (1985). *Nonparametric Density Estimation: The L1 View*, New York: Wiley.

Dmitriev, Yu. G. and Tarasenko, F. P. (1973). On the estimation of functionals of the probability density and its derivatives. *Theory Probab. Appl.* **18**, 628–633.

Dmitriev, Yu. G. and Tarasenko, F. P. (1974). On a class of nonparametric estimates of nonlinear functionals of a density. *Theory Probab. Appl.* **19**, 390–394.

Dudewicz, E. J. (Ed.) (1985). *The Frontiers of Modern Statistical Inference Procedures*, Vol. 10, Series in Mathematical and Management Sciences, Columbus, Ohio: American Sciences Press.

Dudewicz, E. J. and van der Meulen, E. C. (1981). Entropy-based tests of uniformity. *J. Amer. Statist. Assoc.* **76**, 967–974.

Gallager, R. G. (1968). *Information Theory and Reliable Communication*, New York: Wiley.

Hasminskii, R. Z. and Ibragimov, I. A. (1978). On the nonparametric estimation of functionals. In *Proceedings of the Second Prague Symposium on Asymptotic Statistics* (P. Mandl and M. Huskova, Eds.), Amsterdam: North-Holland, pp. 41–51.

Ivanov, A. V. and Rozhkova, M. N. (1981). Properties of the statistical estimate of the entropy of a random vector with a probability density (in Russian). *Probl. Peredachi Inform.* **17**, 33–43. Translated into English in *Problems Inform. Transmission* **17**, 171–178.

Kullback, S. (1959). *Information Theory and Statistics*, New York: Wiley.

Ni, F. and Scheraga, H. A. (1985). Resolution enhancement in spectroscopy by maximum entropy Fourier self-deconvolution, with applications to Raman spectra of peptides and proteins. *J. Raman Spectrosc.* **14** (to appear).

Parzen, E. (1962). On estimation of a probability density function and mode. *Ann. Math. Statist.* **33**, 1065–1076.

Prakasa Rao, B. L. S. (1983). *Nonparametric Functional Estimation*, New York: Academic Press.

Priestley, M. B. (1981). *Spectral Analysis and Time Series*, Vol. 1, New York: Academic Press.

Rao, C. R. (1982a). Diversity and dissimilarity coefficients: A unified approach. *J. Theor. Pop. Biol.* **21**, 24–43.

Rao, C. R. (1982b). Diversity: its measurement, decomposition, apportionment and analysis. *Sankhyā A* **44**, 1–22.

Rao, C. R. and Nayak, T. K. (1985). Cross entropy, dissimilarity measures, and characterizations of quadratic entropy. *IEEE Trans. Inform. Theory* **31**, 589–593.

Révész, P. (1972). On empirical density function. *Period. Math. Hungar.* **2**, 85–110.

Rosenblatt, M. (1956). Remarks on some nonparametric estimates of a density function. *Ann. Math. Statist.* **27**, 832–837.

Shannon, C. E. (1948). A mathematical theory of communication. *Bell Syst. Tech. J.* **27**, 379–423, 623–656. [Reprinted in *Key Papers in the Development of Information Theory* (D. Slepian, Ed.), New York: IEEE Press, pp. 5–29 (1974).]

Shore, J. E. and Johnson, R. W. (1980). Axiomatic derivation of the principle of maximum entropy and the principle of minimum cross-entropy. *IEEE Trans. Inform. Theory* **26**, 26–37.

Shore, J. E. and Johnson, R. W. (1981). Properties of cross-entropy minimization. *IEEE Trans. Inform. Theory* **27**, 472–482.

Tapia, R. A. and Thompson, J. R. (1978). *Nonparametric Probability Density Estimation*, Baltimore: The Johns Hopkins University Press.

Tarasenko, F. P. (1968). On the evaluation of an unknown probability density function, the direct estimation of the entropy from independent observations of a continuous random variable and the distribution-free entropy test of goodness of fit. *Proc. IEEE* **56**, 2052–2053.

Tarasenko, F. P. (1976). On interpretation of statistics as estimates of functionals. In *Proceedings of the 1975 IEEE–USSR Joint Workshop on Information Theory*, Piscataway, New Jersey: IEEE Service Center, pp. 212–215.

Theil, H. (1980). The entropy of the maximum entropy distribution. *Econom. Lett.* **5**, 145–148.

Theil, H. and Laitinen, K. (1980). Singular moment matrices in applied econometrics. *Multivariate Analysis—V* (P. R. Krishnaiah, Ed.), Amsterdam: North-Holland, pp. 629–649.

Vasicek, O. (1976). A test for normality based on sample entropy. *J. R. Statist. Soc. B* **38**, 54–59.

Van Ryzin, J. (1973). A histogram method of density estimation. *Comm. Statist.* **2**, 493–506.

Vinod, H. D. (1982). Maximum entropy measurement error estimates of singular covariance matrices in undersized samples. *J. Econometrics* **20**, 163–174.

Wegman, E. J. (1984). Optimal nonparametric function estimation. *J. Statist. Plann. Inf.* **9**, 375–387.

Wertz, W. (1985). On invariant curve estimators. *Probab. Math. Statist.* **5**, 31–44.

Wertz, W. and Schneider, B. (1979). Statistical density estimation: a bibliography. *Int. Statist. Rev.* **47**, 155–175.

Efficiency of a New Class of Linear Regression Estimates Obtained by Preliminary Nonparametric Estimation

P. Faraldo Roca and W. González Manteiga
Departamento de Estadistica Matematica
Universidad de Santiago de Compostela
Santiago de Compostela, Spain

SUMMARY

In this Chapter a new class of estimates for the parameters θ_1 and θ_2 of the linear regression model $Y = \alpha(X) + \eta = \theta_1 + \theta_2 X + \eta$ is introduced. The estimates are obtained by minimizing with respect to θ_1 and θ_2 a functional distance between the nonparametric estimation function and the line α. Consistency and asymptotic normality properties are obtained for such estimates and the calculation of the mean-square error for particular cases of the new estimates is used to prove that they can be more efficient than the classical least-squares estimates.

1. INTRODUCTION

Let a random sample $\{(X_1, Y_1), \ldots, (X_n, Y_n)\}$ of a random vector (X, Y) be given, following the model

$$Y = \theta_1 + \theta_2 X + \eta \tag{1.1}$$

where η is a random variable of mean 0, θ_1 and θ_2 are unknown parameters, and X is a real random variable. In this chapter we describe the estimation of θ_1 and θ_2 using an alternative methodology to the classical least-squares procedures.

The first step is to carry out a nonparametric estimation of the regression function $\alpha(x) = E\{Y|X = x\} = \theta_1 + \theta_2 x$ using one of the usual estimates [see, e.g., Collomb (1983)]. The estimates $\hat{\theta}_1$ and $\hat{\theta}_2$ are then defined as those values of θ_1 and θ_2 that minimize the functional

$$G(\theta_1, \theta_2) = \int (\hat{\alpha}_n(x) - \theta_1 - \theta_2 x)^2 \, d\Omega_n(x) \tag{1.2}$$

In this chapter we assume that the random error η in (1.1) has homogeneous variance σ^2.

The functional (1.2) is very general. The concrete form it adopts in particular cases depends on the nonparametric estimate $\hat{\alpha}_n$ that is employed and on the weighting function Ω_n. One such particular concrete form is determined by taking

$$\hat{\alpha}_n(x) = \sum_{i=1}^{n} Y_i 1_{\{X_i\}}(x) \quad \text{and} \quad \Omega_n(x) = \frac{1}{n} \sum_{i=1}^{n} 1_{(-\infty, x]}(X_i) \tag{1.3}$$

where 1_C is the indicator function of C and so Ω_n is the empirical distribution of $\{X_1, \ldots, X_n\}$ and $\{\hat{\theta}_1, \hat{\theta}_2\}$ coincides with the classical least-squares estimates. Assuming that the regression variable X has density function f, we can define

$$\hat{\alpha}_n(x) = \sum_{i=1}^{n} Y_i \delta_m(x, X_i) / \sum_{r=1}^{n} \delta_m(x, X_r) \tag{1.4}$$

where $0/0$ is defined as 0, and

$$\Omega_n(x) = \int_{-\infty}^{x} \hat{f}_n(t) \, d(t) = \frac{1}{n} \sum_{i=1}^{n} \int_{-\infty}^{x} \delta_m(t, X_i) \, dt$$

where $\hat{\alpha}_n$ and $\hat{f}_n$ are the nonparametric estimates of regression and density with $\{\delta_m\colon \mathbb{R} \times \mathbb{R} \to \mathbb{R}\}_{m=m(n)\to\infty}$ a sequence of measurable functions. Nearly all existing nonparametric estimates (histograms, orthogonal series, interpolations, kernel estimates, etc.) follow this general form, which can be found in Wertz (1978), Susarla and Walter (1981), González Manteiga (1982), and Collomb (1983).

In Section 2 we show the consistency and asymptotic normality of the new regression estimates $\{\hat{\theta}_1, \hat{\theta}_2\}$ obtained when $\hat{\alpha}_n$ and Ω_n are defined by

(1.4). In Section 3 we investigate their efficiency with respect to the least-squares estimates for some special cases, showing that the use of smoother nonparametric estimates than those of (1.3) may result in estimates for θ_1 and θ_2 that are more efficient than the least-squares estimates.

2. CONSISTENCY AND ASYMPTOTIC NORMALITY

The consistency properties of the new estimators $\{\hat{\theta}_1, \hat{\theta}_2\}$ for $\{\theta_1, \theta_2\}$ are the same as those of $\{\hat{\alpha}_n, \hat{f}_n\}$ for $\{\alpha, f\}$. For example, we can see that almost sure (a.s.) consistency of the estimators $\{\hat{\theta}_1, \hat{\theta}_2\}$ is ensured if Y is bounded with probability 1, in a way similar to that of Devroye (1981), who showed $\hat{\alpha}_n \to \alpha$ a.s. pointwise mod$[\mu]$, μ being the probability measured associated to X. Greblicki and Pawlak (1985) showed the former consistency under the less restrictive condition $E\{|Y|^{1+\gamma}\} < \infty$ for some $\gamma > 0$ using a recursive version of $\hat{\alpha}_n$.

In the following, S is a open set in the usual topology of $\mathbb{R}$ that contains the support of f, the fourth moment of X is assumed to exist, and the sequence (δ_m) fulfills the following conditions:

1. $\sup\limits_u \delta_m(x, u) = O(m) \quad$ for all $x \in \mathbb{R}$

2. $\delta_m(x, t) = \delta_m(t, x) \quad$ and $\quad \delta_m(x, t) \geq 0$ for all $(t, x) \in \mathbb{R}^2 \quad (2.1)$

3. $\int \delta_m(x, u)\, du = 1 \quad$ for all $x \in \mathbb{R}$ with $\delta_m(x, t) = 0$

$$\text{if } \|x - t\| > c\varepsilon_m$$

and

$$\varepsilon_m \to 0 \text{ if } m \to \infty,$$

$$c \in \mathbb{R}^+ \text{ being fixed}$$

These conditions are fulfilled for example by the histogram and kernel estimates with compact support that will be used in Section 3.

If we use $\hat{\alpha}_n$ and Ω_n defined by (1.4) the functional (1.2) is

$$G(\theta_1, \theta_2) = \int (\hat{\alpha}_n(x) - \theta_1 - \theta_2 x)^2 \hat{f}_n(x)\, dx \tag{2.2}$$

and considering that the support of f does not necessarily coincide with the support of $\hat{f}_n$, more generally than (2.2) we can use

$$G(\theta_1, \theta_2) = \int_S (\hat{\alpha}_n(x) - \theta_1 - \theta_2 x)^2 \hat{f}_n(x)\, dx \tag{2.3}$$

and then it is easily shown that the estimates $\{\hat{\theta}_1, \hat{\theta}_2\}$ defined by minimization of (2.3) are

$$\hat{\theta}_1 = (\overline{Z}\overline{V} - \overline{T}\overline{W})/(\overline{V}\overline{D} - \overline{W}^2) \quad \text{and} \quad \hat{\theta}_2 = (\overline{T}\overline{D} - \overline{Z}\overline{W})/(\overline{V}\overline{D} - \overline{W}^2) \tag{2.4}$$

where $\overline{Z}$, $\overline{V}$, $\overline{W}$, $\overline{T}$ and $\overline{D}$ are sample means of the random variables defined by

$$Z_i = \int_S Y_i \delta_m(x, X_i)\, dx, \quad V_i = \int_S x^2 \delta_m(x, X_i)\, dx, \quad W_i = \int_S x\delta_m(x, X_i)\, dx$$

$$T_i = \int_S Y_i x \delta_m(x, X_i)\, dx, \quad D_i = \int_S \delta_m(x, X_i)\, dx, \quad i = 1, \ldots, n \tag{2.5}$$

since the second moment of X exists, we may also define

$$Z_0 = \theta_1 + \theta_2 EX, \qquad V_0 = EX^2, \qquad W_0 = EX,$$

$$T_0 = \theta_1 EX + \theta_2 EX^2, \qquad D_0 = 1$$

and show the following result.

Lemma 2.1. If the fourth moment of X exists and the premises (2.1) are satisfied, then

1. If $EY^2 < \infty$, then $\overline{Z} \xrightarrow{P} Z_0$, $\overline{V} \xrightarrow{P} V_0$, $\overline{W} \xrightarrow{P} W_0$, $\overline{T} \xrightarrow{P} T_0$, and $\overline{D} \xrightarrow{P} D_0$.
2. If Y is bounded with probability 1 and $\overline{S}$ is a compact set then $\overline{Z} \to Z_0$ a.s.; $\overline{V} \to V_0$ a.s., $\overline{W} \to W_0$ a.s., $\overline{T} \to T_0$ a.s., and $\overline{D} \to D_0$ a.s.

3. If $\sqrt{n}(E\overline{Z} - Z_0, E\overline{V} - V_0, E\overline{W} - W_0, E\overline{T} - T_0)' \to 0$ and there exists $\gamma > 0$ such that $E|Y|^{2+\gamma} < \infty$, then

$$\sqrt{n}(\overline{Z} - Z_0, \overline{V} - V_0, \overline{W} - W_0, \overline{T} - T_0)' \xrightarrow{d} N_4(0, A)$$

where A is the symmetric matrix defined by

$$a_{11} = \sigma^2 + \theta_2 \text{Var } X, \qquad a_{12} = \theta_2(EX^3 - EXEX^2), \qquad a_{13} = \theta_2 \text{Var } X$$

$$a_{14} = \sigma^2 EX + \theta_1\theta_2 \text{Var } X + \theta_2^2(EX^3 - EXEX^2), \qquad a_{22} = \text{Var } X^2,$$

$$a_{23} = EX^3 - EXEX^2 \qquad a_{24} = \theta_1(EX^3 - EXEX^2) + \theta_2 \text{Var } X^2,$$

$$a_{33} = \text{Var } X, \qquad a_{34} = \theta_1 \text{Var } X + \theta_2(EX^3 - EXEX^2)$$

and

$$a_{44} = \sigma^2 EX^2 + \theta_1^2 \text{Var } X + \theta_2^2 \text{Var } X^2 + 2\theta_1\theta_2 \text{Cov}(X, X^2)$$

Proof. (1). We present only the proof for $T = \int_S Yx\delta_m(x, X)\,dx$, the arguments for the other variables in (2.5) are analogous. Consider the descomposition $\overline{T} - T_0 = \overline{T} - E\overline{T} + E\overline{T} - T_0 = \Delta_1 + \Delta_2$, where Δ_1 is the random part and Δ_2 the deterministic part. Clearly

$$\Delta_2 = E\left\{\int_S Yx\delta_m(x, X)\,dx\right\} - \{\theta_1 EX + \theta_2 EX^2\}$$

$$= E\left\{E\left\{\int_S Yx\delta_m(x, X)\,dx/X\right\}\right\} - \{\theta_1 EX + \theta_2 EX^2\}$$

$$= E\left\{\alpha(X)\int_S x\delta_m(x, X)\,dx\right\} - \{\theta_1 EX + \theta_2 EX^2\}$$

$$= \int \alpha(u)\int_S x\delta_m(x, u)\,dx f(u)\,du - \{\theta_1 EX + \theta_2 EX^2\}$$

$$= \int(\theta_1 + \theta_2 u)\int_S x\delta_m(u, x)\,dx f(u)\,du - \{\theta_1 EX + \theta_2 EX^2\}$$

using that $\int_S x\delta_m(u, x)\,dx \to u$ when $n \to \infty$, which is obvious by (2.1) and applying the dominated convergence theorem it is easy to see that $\Delta_2 \to 0$.

Writing $\Delta_1 = \overline{T} - E\overline{T}$,

$$E\Delta_1^2 = \frac{\operatorname{Var} T}{n} \sim \frac{ET^2}{n} = \frac{E\left\{\left(\int_S x\delta_m(x, X)\,dx\right)^2\right\}}{n} = O\left(\frac{1}{n}\right),$$

so that by Chebychev's inequality $\Delta_1 \to_P 0$ and this completes the proof of part 1).

(2) As in part (1) $\Delta_2 \to 0$. It remains to show that $\Delta_1 \to 0$ a.s., which follows from Bennett's inequality (Bennett, 1962) applied to the variables $R_i = T_i - ET_i$, $i = 1, \ldots, n$, which are bounded with probability 1 because Y is so bounded and $\overline{S}$ is a compact set, so that $P\{|\overline{T}| \geq \varepsilon\} \leq 2\exp\{-n\varepsilon^2 O(1)\}$ and the Borel–Cantelli lemma completes the proof.

(3) Considerer descomposition

$$\sqrt{n}\begin{bmatrix} \overline{Z} - Z_0 \\ \overline{V} - V_0 \\ \overline{W} - W_0 \\ \overline{T} - T_0 \end{bmatrix} = \sqrt{n}\begin{bmatrix} \overline{Z} - E\overline{Z} \\ \overline{V} - E\overline{V} \\ \overline{W} - E\overline{W} \\ \overline{T} - E\overline{T} \end{bmatrix} + \sqrt{n}\begin{bmatrix} E\overline{Z} - Z_0 \\ E\overline{V} - V_0 \\ E\overline{W} - W_0 \\ E\overline{T} - T_0 \end{bmatrix} = \Delta_1 + \Delta_2$$

By the hypothesis in part 3, it is only necessary to show that $\Delta_1 \to_d N_4(0, A)$.

Hence, we need only apply the generalized Liapunov criterion (Araujo and Gine, 1980, p. 41) to the vectors

$$H_{ni} = \sqrt{n}\begin{bmatrix} (Z_i - EZ_i)/n \\ (V_i - EV_i)/n \\ (W_i - EW_i)/n \\ (T_i - ET_i)/n \end{bmatrix}, \qquad i = 1, \ldots, n$$

It is necessary to prove

$$\sum_{i=1}^{n} E\{H_{ni}^r H_{ni}^j\} \to a_{rj}, \qquad r, j = 1, 2, 3, 4$$

and

$$\sum_{i=1}^{n} \int_{\{\|H_{ni}\| > \varepsilon\}} \|H_{ni}\|^2\, dP \to 0 \qquad \forall \varepsilon > 0$$

where H_{ni}^r represent the rth coordinate of H_{ni}, a_{rj} is the (r, j)th element of the matrix A, and P is the probabilistic measure induced by the vector H_{ni}.

To show the first condition, we develop for example the case of a_{44}. The other cases are similar.

$$\sum_{i=1}^{n} E\{H_{ni}^4 H_{ni}^4\} = \frac{1}{n}\sum_{i=1}^{n} E\{(T_i - ET_i)^2\} = E\{(T - ET)^2\} = \operatorname{Var} T$$

$$= E\left\{\left(Y\int_S x\delta_m(x, X)\,dx\right)^2\right\} - \left(E\left\{Y\int_S x\delta_m(x, X)\,dx\right\}\right)^2$$

$$= E\left\{E\left\{\left(Y\int_S x\delta_m(x, X)\,dx\right)^2 \Big/ X\right\}\right\}$$

$$- \left(E\left\{E\left\{Y\int_S x\delta_m(x, X)\,dx \Big/ X\right\}\right\}\right)^2$$

$$= E\left\{(\sigma^2 + \alpha^2(X))\left(\int_S x\delta_m(x, X)\,dx\right)^2\right\}$$

$$- \left(E\left\{\alpha(X)\int_S x\delta_m(x, X)\,dx\right\}\right)^2$$

Clearly the dominated convergence theorem can be applied because the fourth moment of X exists and then

$$\sum_{i=1}^{n} E\{H_{ni}^4 H_{ni}^4\} \to \sigma^2 EX^2 + \theta_1^2 \operatorname{Var} X + \theta_2^2 \operatorname{Var} X^2 + 2\theta_1\theta_2 \operatorname{Cov}(X, X^2)$$

The second condition is obtained by the following relations:

$$\sum_{i=1}^{n} \int_{\{\|H_{ni}\| > \varepsilon\}} \|H_{ni}\|^2\,dP = \sum_{i=1}^{n} \int_{\{\|H_{ni}/\varepsilon\| > 1\}} \varepsilon^2 \left\|\frac{H_{ni}}{\varepsilon}\right\|^2 dP$$

$$\le \sum_{i=1}^{n} \int_{\{\|H_{ni}/\varepsilon\| > 1\}} \varepsilon^2 \left\|\frac{H_{ni}}{\varepsilon}\right\|^{2+\gamma} dP$$

$$\le \frac{1}{\varepsilon^\gamma}\sum_{i=1}^{n} E\{\|H_{ni}\|^{2+\gamma}\}$$

$$= \frac{1}{\varepsilon^\gamma}\sum_{i=1}^{n} E\left\{\left[\left(\frac{Z_i - EZ_i}{\sqrt{n}}\right)^2 + \left(\frac{V_i - EV_i}{\sqrt{n}}\right)^2\right.\right.$$

$$\left.\left. + \left(\frac{W_i - EW_i}{\sqrt{n}}\right)^2 + \left(\frac{T_i - ET_i}{\sqrt{n}}\right)^2\right]^{(2+\gamma)/2}\right\}$$

$$\sim n^{-\gamma/2} \to 0$$

and the proof is then completed. ■

The expressions (2.4) allow the $\hat{\theta}_i$, $i = 1, 2$, to be expressed in the form

$$\hat{\theta}_1 = g_1(\overline{Z}, \overline{V}, \overline{W}, \overline{T}, \overline{D}) \quad \text{and} \quad \hat{\theta}_2 = g_2(\overline{Z}, \overline{V}, \overline{W}, \overline{T}, \overline{D}) \tag{2.6}$$

where

$$g_1(z, v, w, t, d) = (zv - tw)/(vd - w^2)$$

and

$$g_2(z, v, w, t, d) = (td - zw)/(vd - w^2)$$

are continuous in the region $\{(z, v, w, t, d)^t / vd - w^2 > 0\}$ to which $(Z_0, V_0, W_0, T_0, D_0)^t$ belongs (since Var $X > 0$).

Using the expressions (2.6) and Lemma 2.1 we can show the following theorem.

Theorem 2.1. With the same conditions as in Lemma 2.1.

1. If $EY^2 < \infty$, then $\hat{\theta}_i \to_P \theta_i$ $(i = 1, 2)$.
2. If Y is bounded with probability 1 and $\overline{S}$ is a compact set, then $\hat{\theta}_i \to \theta_i$ a.s. $(i = 1, 2)$.
3. If $E\{|Y|^{2+\gamma}\} < \infty$ for some $\gamma > 0$, then

$$\sqrt{n}\left[\begin{bmatrix} \hat{\theta}_1 \\ \hat{\theta}_2 \end{bmatrix} - \begin{bmatrix} \theta_1 \\ \theta_2 \end{bmatrix}\right] \xrightarrow{d} N\left[\begin{bmatrix} 0 \\ 0 \end{bmatrix}, \sigma^2\left(E\left\{\begin{pmatrix} 1 \\ X \end{pmatrix}(1, X)\right\}\right)^{-1}\right]$$

Proof. Parts 1 and 2 are proved using the results of Lemma 2.1 and the expressions (2.6) for $\hat{\theta}_1$ and $\hat{\theta}_2$.

To prove part 3, note that since $\sqrt{n}(\overline{D} - E\overline{D}) = o_p(1)$, the solutions $\hat{\theta}_1$ and $\hat{\theta}_2$ of the minimization problem associated to (2.3) have the same asymptotic distribution as those solutions of the system

$$\begin{bmatrix} \int_S \hat{\alpha}_n(x) \hat{f}_n(x)\, dx \\ \int_S x \hat{\alpha}_n(x) \hat{f}_n(x)\, dx \end{bmatrix} = \begin{bmatrix} 1 & \int_S x \hat{f}_n(x)\, dx \\ \int_S x \hat{f}_n(x)\, dx & \int_S x^2 \hat{f}_n(x)\, dx \end{bmatrix} \begin{bmatrix} \theta_1 \\ \theta_2 \end{bmatrix}$$

so that

$$\begin{bmatrix} \hat{\theta}_1 \\ \hat{\theta}_2 \end{bmatrix} = \begin{bmatrix} 1 & \overline{W} \\ \overline{W} & \overline{V} \end{bmatrix}^{-1} \begin{bmatrix} \overline{Z} \\ \overline{T} \end{bmatrix} = \begin{bmatrix} h_1(\overline{Z}, \overline{V}, \overline{W}, \overline{T}) \\ h_2(\overline{Z}, \overline{V}, \overline{W}, \overline{T}) \end{bmatrix} = \begin{bmatrix} g_1(\overline{Z}, \overline{V}, \overline{W}, \overline{T}, 1) \\ g_2(\overline{Z}, \overline{V}, \overline{W}, \overline{T}, 1) \end{bmatrix}$$

Hence the application of the generalized δ method (Bishop et al., 1975, p. 493) to the vector $(\overline{Z}-Z_0, \overline{V}-V_0, \overline{W}-W_0, \overline{T}-T_0)^t$ shows that

$$\sqrt{n}\left[\begin{bmatrix}\hat{\theta}_1\\ \hat{\theta}_2\end{bmatrix}-\begin{bmatrix}\theta_1\\ \theta_2\end{bmatrix}\right]\xrightarrow{d} N_2\left[\begin{bmatrix}0\\ 0\end{bmatrix}, B^tAB\right],$$

where B is the matrix

$$\begin{bmatrix}\dfrac{\delta h_1}{\delta z} & \dfrac{\delta h_2}{\delta z}\\ \dfrac{\delta h_1}{\delta v} & \dfrac{\delta h_2}{\delta v}\\ \dfrac{\delta h_1}{\delta w} & \dfrac{\delta h_2}{\delta w}\\ \dfrac{\delta h_1}{\delta t} & \dfrac{\delta h_2}{\delta t}\end{bmatrix}$$

at the point $(Z_0, V_0, W_0, T_0)^t$. The result follows on calculating the product B^tAB. ■

3. ONE STUDY OF EFFICIENCY FOR PARTICULAR CASES

Part 3 of Theorem 2.1 ensures that the new estimates have the same asymptotic covariance matrix as the classical least-squares estimates (which we shall henceforth denote by $\hat{\theta}_1^*$ and $\hat{\theta}_2^*$). In the present section we compare the new estimates with least-squares estimates for finite samples of size n by examining the mean-square error nonconditionals M.S.E. $\{\hat{\theta}_i\}$ and M.S.E. $\{\hat{\theta}_i^*\}$ or the conditional M.S.E. $\{\hat{\theta}_i|X_1,\ldots,X_n\}$ and M.S.E. $\{\hat{\theta}_i^*|X_1,\ldots,X_n\}$. Hence we shall see in some examples with finite samples how the efficiency in terms of the M.S.E. depends on the nature of the nonparametric estimators employed and on the open set S.

Example 3.1 (Efficiency Using Kernel Estimates). If we take $S=(-\infty,\infty)$ and $\delta_m(x,u)=(1/\varepsilon_m)K((x-u)/\varepsilon_m)$, where K is a kernel function belonging to the family $H=\{K|K$ symmetric, positive, bounded and for some c, $\int_{-c}^{c} K(z)\,dz = 1\}$, then

$$E\{\hat{\theta}_1 | X_1, \ldots, X_n\} = \theta_1 + \frac{\theta_2 \overline{X}}{S^2} \varepsilon_m^2 \int_{-c}^{c} z^2 K(z)\,dz + o(\varepsilon_m^2)$$

$$E\{\hat{\theta}_2 | X_1, \ldots, X_n\} = \theta_2 - \frac{\theta_2}{S^2} \varepsilon_m^2 \int_{-c}^{c} z^2 K(z)\,dz + o(\varepsilon_m^2)$$

$$\operatorname{Var}\{\hat{\theta}_1 | X_1, \ldots, X_n\} = \frac{\sigma^2}{n} \left\{ \frac{(\sum_{i=1}^{n} X_i^2/n) S^2 - 2\overline{X}^2 \varepsilon_m^2 \int_{-c}^{c} z^2 K(z)\,dz}{S^4} \right\} + \frac{o(\varepsilon_m^2)}{n}$$

$$\operatorname{Var}\{\hat{\theta}_2 | X_1, \ldots, X_n\} = \frac{\sigma^2}{n} \left\{ \frac{S^2 - 2\varepsilon_m^2 \int_{-c}^{c} z^2 K(z)\,dz}{S^4} \right\} + \frac{o(\varepsilon_m^2)}{n}$$

Using the definition of M.S.E.,

$$\text{M.S.E. } \{\hat{\theta}_i | X_1, \ldots, X_n\} \sim (p_{i0} \varepsilon_m^2)^2 + \frac{p_{i1} - p_{i2} \varepsilon_m^2}{n}, \qquad i = 1, 2$$

for suitably defined p_{i0}, p_{i1}, and p_{i2} (e.g., $p_{10} = (\theta_2 \overline{X}/S^2) \int_{-c}^{c} z^2 K(z)\,dz$). Here

$$S^2 = \frac{1}{n} \sum_{i=1}^{n} (X_i - \overline{X})^2 \quad \text{and} \quad \overline{X} = \frac{1}{n} \sum_{i=1}^{n} X_i$$

Hence judicious choice of the window parameter ε_n produces estimates that are more efficient than the unbiased least-squares estimates $\hat{\theta}_1^*$ and $\hat{\theta}_2^*$, for which

$$\text{M.S.E. } \{\hat{\theta}_i^* | X_1, \ldots, X_n\} = \operatorname{Var}\{\hat{\theta}_i^* | X_1, \ldots, X_n\}$$

$$= \begin{cases} \dfrac{\sigma^2}{n} \left\{ \dfrac{\sum_{i=1}^{n} X_i^2/n}{S^2} \right\} = \dfrac{p_{11}}{n}, & i = 1 \\[2ex] \dfrac{\sigma^2}{n} \left\{ \dfrac{1}{S^2} \right\} = \dfrac{p_{21}}{n}, & i = 2 \end{cases}$$

In particular, for the minimum M.S.E., M.S.E_{op}, which corresponds to the value $\hat{\varepsilon}_m(\text{i}) = (p_{i2}/2p_{i0}^2 n)^{1/2}$ of the window parameter,

$$\text{M.S.E}_{\text{op}}\{\hat{\theta}_i | X_1, \ldots, X_n\} \sim \frac{p_{i1}}{n} - \frac{p_{i2}^2}{4p_{i0}^2 n}$$

$$< \text{Var}\{\hat{\theta}_i^* | X_1, \ldots, X_n\}, \qquad i = 1, 2$$

In the nonconditional case

$$\text{M.S.E}_{\text{op}}\{\hat{\theta}_i\} = (\text{Bias}\,\hat{\theta}_i)^2 + \text{Var}\,\hat{\theta}_i = \frac{q_{i1}}{n} - \frac{q_{i2}^2}{4q_{i0}^2 n^2} + o\left(\frac{1}{n^2}\right), \qquad i = 1, 2$$

where q_{ij} are the expressions p_{ij} substituting S^2, $\overline{X}$ and $\Sigma_{r=1}^n X_r^2/n$ by $\text{Var}\,X$, EX, and EX^2.

The new regression estimates are in this case more efficient than the least-squares estimates, at least using windows close to optimal.

Example 3.2 (Efficiency Using Histogram Estimates). If we take $S = (0, 1)$ the support of f, f uniform in $(0, 1)$ and $\delta_m(x, u) = m\Sigma_{j=1}^m 1_{I_j}(x) 1_{I_j}(u)$ with

$$I_j = \left[\frac{j-1}{m}, \frac{j}{m}\right), \qquad j = 2, \ldots, m-1,$$

$$I_1 = \left(0, \frac{1}{m}\right), \quad \text{and} \quad I_m = \left[\frac{m-1}{m}, 1\right).$$

If for convenience we consider here the model $Y = \theta_2 X + \eta$ and $\hat{f}_n(x) = 1_{(0,1)}(x)$ in (2.3), then using the optimal $\hat{m}$

$$\text{M.S.E}_{\text{op}}\{\hat{\theta}_2\} = \frac{3\sigma^2}{n} - \frac{9}{4}\frac{\sigma^2}{n^2} + o\left(\frac{1}{n^2}\right)$$

which again shows one M.S.E. greater than the M.S.E. of the least-squares estimator, for which $\text{Var}\{\hat{\theta}_2^*\} \simeq 3\sigma^2/n$.

Example 3.3 (A Case of Inefficiency). If $S = (a, b)$ is the support of f and $\delta_m(x, u) = (1/\varepsilon_m)K((x-u)/\varepsilon_m)$, with K as in the first example, then the frontier effect of the tails $(a, a + c\varepsilon_m]$ and $[b - c\varepsilon_m, b)$ comes into play. As a result, although the new estimates defined in this situation are asymptotically as efficient as the least-squares estimates (as we have already seen), they are less efficient for finite samples, since

$$\operatorname{Var}\{\hat{\theta}_1\} = \frac{\sigma^2}{n}\left\{\frac{EX^2}{\operatorname{Var} X} + \frac{(EX^2 - aEX)^2 f(a) + (EX^2 - bEX)^2 f(b)}{\operatorname{Var} X} \times \int_0^c (1 - \bar{K}(z))^2 \, dz \varepsilon_m\right\} + \frac{o(\varepsilon_m)}{n}$$

$$\operatorname{Var}\{\hat{\theta}_2\} = \frac{\sigma^2}{n}\left\{\frac{1}{\operatorname{Var} X} + \frac{(a - EX)^2 f(a) + (b - EX)^2 f(b)}{\operatorname{Var} X} \times \int_0^c (1 - \bar{K}(z))^2 \, dz \varepsilon_m\right\} + \frac{o(\varepsilon_m)}{n}$$

with

$$\bar{K}(z) = \int_{-c}^{z} K(t) \, dt, \qquad z \in [-c, c].$$

The calculus necessary for the Examples 3.1, 3.2, and 3.3 are realized using classical methods of expansion for expectations and variances [see, e.g., Cramér (1958)].

Remark 3.1. In this paper we developed the case of a straight line in the linear general model, the generalization to higher dimensional linear parameters could be realized considering the functional

$$G(\theta) = \int \left(\hat{\alpha}_n(x) - A^t(x)\theta\right)^2 \hat{f}_n(x) \, dx \quad \text{with} \quad \theta \in \mathbb{R}^p$$

where $A(x) \in \mathbb{R}^p$ and $\{\hat{\alpha}_n, \hat{f}_n\}$ are q-dimensional nonparametric estimators. Examples of this generalization are the polynomial regression or the regression multiple with associated functionals

$$G(\theta_1, \ldots, \theta_p) = \int \left(\hat{\alpha}_n(x) - \left(\theta_1 + \theta_2 x + \cdots + \theta_p x^{p-1}\right)\right)^2 \hat{f}_n(x) \, dx$$

(clearly $q = 1$)

$$G(\theta_1, \ldots, \theta_p) = \int \left(\hat{\alpha}_n(x_1, \ldots, x_{p-1}) - \left(\theta_1 + \theta_2 x_1 + \cdots + \theta_p x_{p-1}\right)\right)^2 \times \hat{f}_n(x_1, \ldots, x_{p-1}) \, dx_1 \cdots dx_{p-1}$$

(clearly $q = p - 1$).

REMARK 3.2. Similar results as in the examples could be deduced comparing by the multidimensional M.S.E.

$$\text{M.S.E.}\left\{\begin{bmatrix}\hat{\theta}_1\\ \hat{\theta}_2\end{bmatrix}\right\} = E\left\{\begin{bmatrix}\hat{\theta}_1-\theta_1\\ \hat{\theta}_2-\theta_2\end{bmatrix}[\hat{\theta}_1-\theta_1 \quad \hat{\theta}_2-\theta_2]\right\} + D\begin{bmatrix}\hat{\theta}_1\\ \hat{\theta}_2\end{bmatrix}$$

where here

$$D\begin{bmatrix}\hat{\theta}_1\\ \hat{\theta}_2\end{bmatrix}$$

is the covariance matrix of $(\hat{\theta}_1, \hat{\theta}_2)^t$. However, the optimal election of the window would be function of every element in the matrix

$$\text{M.S.E.}\left\{\begin{bmatrix}\hat{\theta}_1\\ \hat{\theta}_2\end{bmatrix}\right\}$$

ACKNOWLEDGMENT

We thank the referees for their good suggestions, which have led to considerable improvements in the exposition of our results, and the editor, Professor Dr. Wertz, for his interest.

REFERENCES

Araujo, A. and Gine, E. (1980). *The Central Limit Theorem*. New York: Wiley.

Bennett, G. (1962). Probability inequalities for the sum of independent random variables. *J. Amer. Statist. Assoc.* **57**, 33–45.

Bishop, Y. M. N., Fienberg, S. E. and Holland, P. W. (1975). *Discrete Multivariate Analysis: Theory and Practice*, Cambridge, Massachusetts: MIT Press.

Collomb, G. (1983). Nonparametric regression on up-to-date bibliography. Sixth International Summer on problems of model choice and parameter estimation in regression analysis, Sellin/Rügen.

Cramér, H. (1958). *Mathematical Methods of Statistics*. Princeton, New Jersey: Princeton University Press.

Devroye, L. (1981). On the almost everywhere convergence of nonparametric regression functions estimates. *Ann. Statist.* **9**, 1310–1319.

González Manteiga, W. (1982). Construcción axiomática y distribuciones asintóticas para estimadores no paramétricos de las funciones de densidad y de regresión. Tesis Doctoral. Santiago de Compostela.

Greblicki, W. and Pawlak, M. (1985). Necessary and sufficient consistency conditions for a recursive kernel regression estimate. *J. Multivari. Anal.* (to appear).

Susarla, V. and Walter, G. (1981). Estimation of a multivariate density function using delta sequences. *Ann. Statist.* **9**, 347–355.

Wertz, W. (1978). *Statistical Density Estimation. A survey*. Göttingen: Vandenhoeck & Ruprecht.

Nonparametric Confidence Intervals in Regression: The Bootstrap and Randomization Methods

J. Robinson
Department of Mathematical Statistics
University of Sydney
Sydney, New South Wales, Australia

SUMMARY

The bootstrap method is used to approximate the distribution of estimates of parameters. Its major use has been in the estimation of bias and standard deviation and in obtaining approximate confidence intervals. The bootstrap percentile method is described and compared with the method of inversion of permutation tests, which yields exact conditional confidence intervals. This suggests an alternative bootstrap confidence interval, which is obtained by analogy to the permutation method. These methods are compared in a numerical example. A percentile method based on the bootstrap of a standardized estimate is also considered. An asymptotic comparison of the competing bootstrap methods and the exact conditional confidence intervals indicates that bootstrapping the standardized estimate gives the best approximation to the exact methods.

1. INTRODUCTION

The bootstrap method (Efron, 1979, 1982) is a technique for approximation of the distribution of estimates obtained from samples. In the simplest case we have a sample $X_1, \ldots, X_n$ from a population with distribution function F and an estimate $\theta(F_n)$ of some parameter $\theta(F)$, where F_n is the empirical distribution function of $X_1, \ldots, X_n$. We consider samples $X_1^*, \ldots, X_n^*$ taken from the population with distribution function F_n and we find the distribution of $\theta(F_n^*)$, the estimate of $\theta(F_n)$ from this bootstrap sample, where F_n^* is the empirical distribution function of $X_1^*, \ldots, X_n^*$, either by theoretical means or by Monte Carlo approximation using a large number of repetitions. This bootstrap distribution can be used to find the bootstrap expectation and standard deviation, which are used to estimate the bias and standard deviation of the original estimate $\theta(F_n)$. Efron (1982) also suggested that $\alpha/2$ and $1 - \alpha/2$ quantiles of the bootstrap distribution could be used to give an approximate confidence interval.

In Section 2 we consider estimation of the slope in simple linear regression. The bootstrap could be applied either to pairs or residuals, but we concentrate on bootstrapping residuals. We also consider a variant of the bootstrap where, at the Monte Carlo stage, sampling without replacement is used.

We then consider the exact confidence interval obtained by inverting a permutation test and by analogy obtain an alternative bootstrap confidence interval. These methods are compared numerically on the law school data of Efron (1982).

It has been suggested in a series of papers of K. Singh et al. that bootstrapping a standardized estimate would give a better approximation to the confidence interval for a mean (Singh, 1981; Babu and Singh, 1983, 1985; Abramovitch and Singh, 1985). Such a method is described for the regression case, and finally, asymptotic approximations are used to compare the methods discussed above. These show that the bootstrap of the standardized estimate gives confidence intervals that are closest to the exact interval with coverage errors correct to order less than n^{-1}.

2. BOOTSTRAP CONFIDENCE INTERVALS FOR REGRESSION

Consider the simple linear regression model

$$y_i = \alpha + \beta x_i + \varepsilon_i, \quad i = 1, \ldots, n$$

where $\varepsilon_1, \ldots, \varepsilon_n$ are an independent sample from the distribution function F. We consider only β and its least-squares estimate

$$\hat{\beta} = \sum_{i=1}^{n} x_i y_i \Big/ \sum_{i=1}^{n} x_i^2$$

where we assume for simplicity that $\sum_{i=1}^{n} x_i = 0$. We shall consider bootstrap replicates β^* based on F_n, the empirical distribution function of $\hat{\varepsilon}_i = y_i - \bar{y} - \hat{\beta}x_i$, $i = 1, \ldots, n$.

In the sequel only the case in which $x_1, \ldots, x_n$ are regarded as fixed will be considered. Freedman (1981) looks at this case separately from the case of random pairs (y, x) and suggests that a bootstrap on pairs is appropriate only in the latter case, whereas the bootstrap on the residuals is appropriate in the case of fixed $x_1, \ldots, x_n$. Another approach is to use the bootstrap on residuals if the linear model with homogeneous error variance is strongly supported either in the random or fixed cases. The bootstrap on pairs might be expected to give good estimates of the distribution of $\hat{\beta}$ under less restrictive conditions, as is implicit in Freedman (1981).

Let $\varepsilon_1^*, \ldots, \varepsilon_n^*$ be an independent random sample drawn from F_n, the empirical distribution function of $\hat{\varepsilon}_1, \ldots, \hat{\varepsilon}_n$, and let $y_i^* = x_i\hat{\beta} + \varepsilon_i^*$, $i = 1, \ldots, n$. Then

$$\beta^* = \sum_{i=1}^{n} x_i y_i^* \Big/ \sum_{i=1}^{n} x_i^2 = \hat{\beta} + \sum_{i=1}^{n} x_i \varepsilon_i^* \Big/ \sum_{i=1}^{n} x_i^2$$

If E_* and V_* represent expectation and variance over the bootstrap distribution function of $\varepsilon_1^*, \ldots, \varepsilon_n^*$, then

$$E_*\beta^* = \hat{\beta}, \qquad V_*\beta^* = \sigma^{*2} \sum_{i=1}^{n} x_i^2$$

where $\sigma^{*2} = n^{-1}\sum_{i=1}^{n}\hat{\varepsilon}_i^2$.

Let $\beta_1^* \leq \cdots \leq \beta_B^*$ be the ordered values of a sample of B values of β^* obtained by Monte Carlo simulation. (Note that it would be possible if $B = n^n$ to obtain the entire distribution of β^* if n is small, but generally the calculation is too formidable.) Let $\kappa = [B\alpha/2]$, where $[\cdot]$ denotes the integer part, then the percentile methods gives an approximation to an α-level confidence interval as $(\beta_\kappa^*, \beta_{B-\kappa+1}^*)$.

An alternative approach to the bootstrap in this case would be to use sampling without replacement from $\hat{\varepsilon}_1, \ldots, \hat{\varepsilon}_n$ instead of the sampling with replacement used to obtain the bootstrap sample. Let $\varepsilon_1^\#, \ldots, \varepsilon_n^\#$ be a sample of $\hat{\varepsilon}_1, \ldots, \hat{\varepsilon}_n$ drawn without replacement. Then these are simply a random permutation of $\hat{\varepsilon}_1, \ldots, \hat{\varepsilon}_n$. Let $y_i^\# = x_i\hat{\beta} + \varepsilon_i^\#$, $i = 1, \ldots, n$, and

$$\beta^\# = \sum_{i=1}^{n} x_i y_i^\# \Big/ \sum_{i=1}^{n} x_i^2 = \hat{\beta} + \sum_{i=1}^{n} x_i \varepsilon_i^\# \Big/ \sum_{i=1}^{n} x_i^2$$

Here, in the obvious notation

$$E_\#\beta^\# = \hat{\beta} \quad \text{and} \quad V_\#\beta^\# = (n-1)^{-1} \sum_{i=1}^{n} \hat{\varepsilon}_i^2 \Big/ \sum_{i=1}^{n} x_i^2$$

Again, we can obtain a Monte Carlo approximation to the α-level confidence interval as $(\beta_{\kappa}^{\#}, \beta_{B-\kappa+1}^{\#})$, where the notation is as above. This procedure is not advocated here, but it is convenient to consider it in order to clarify the connection between the bootstrap percentile approximation and the confidence interval of Section 3.

3. CONFIDENCE INTERVALS FROM PERMUTATION TESTS

First, we shall describe briefly the method of obtaining confidence intervals from permutation tests. A detailed discussion is given in Maritz (1981) and Gabriel and Hall (1983). A permutation test of the null hypothesis H_0: $\beta = \beta_0$ may be based on the test statistic

$$T(\beta_0) = \sum_{i=1}^{n} x_i \varepsilon_i^0 \Big/ \sum_{i=1}^{n} x_i^2$$

where $\varepsilon_i^0 = y_i - \bar{y} - x_i \beta_0$, $i = 1, \dots, n$. Under H_0, the conditional distribution of $\varepsilon_1^0, \dots, \varepsilon_n^0$ given their order statistics $\varepsilon^0(1), \dots, \varepsilon^0(n)$ is that of $\varepsilon^0(R_1), \dots, \varepsilon^0(R_n)$, where $R_1, \dots, R_n$ is a uniform random permutation of $1, \dots, n$. So we can obtain the exact conditional distribution of the test statistic given these order statistics. Put

$$\beta^{\#}(\varepsilon_i^0) = \sum_{i=1}^{n} x_i \varepsilon^0(R_i) \Big/ \sum_{i=1}^{n} x_i^2$$

Then, given $\varepsilon^0(1), \dots, \varepsilon^0(n)$, this takes $n!$ values with equal probability. One of these is

$$\hat{\beta} - \beta_0 = \sum_{i=1}^{n} x_i \varepsilon_i^0 \Big/ \sum_{i=1}^{n} x_i^2$$

the observed value. The observed significance level or p value is $2\min(P_0^+, P_0^-, \frac{1}{2})$, where P_0^+ and P_0^- are proportions of $\beta^{\#}(\varepsilon_i^0)$ greater than and less than $\hat{\beta} - \beta_0$, respectively. Thus an α-level confidence interval for β is the set

$$\left\{ \beta_0 : \tfrac{1}{2}\alpha < P_{\#}\left(\beta^{\#}(\varepsilon_i^0) \geq \hat{\beta} - \beta_0\right) < 1 - \tfrac{1}{2}\alpha \right\}$$

where $P_{\#}$ denotes probability from the distribution of $R_1, \dots, R_n$.

Now if $\beta^{\#}, \hat{\varepsilon}_1, \ldots, \hat{\varepsilon}_n$ are as given in Section 2,

$$\hat{\varepsilon}_i = \varepsilon_i^0 - (\hat{\beta} - \beta_0)x_i$$

and, since $\beta^{\#}(\varepsilon_i^0) = \beta^{\#} - \hat{\beta}$,

$$\beta^{\#}(\varepsilon_i^0) = \beta^{\#} - \hat{\beta} + (\hat{\beta} - \beta_0)\beta^{\#}(x_i)$$

Now

$$\{\beta_0 : \beta^{\#}(\varepsilon_i^0) \geq \hat{\beta} - \beta_0\} = \{\beta_0 : \beta^{\#} - \hat{\beta} \geq (\hat{\beta} - \beta_0)(1 - \beta^{\#}(x_i))\}$$
$$= \{\beta_0 : (\beta^{\#} - \hat{\beta})/(1 - \beta^{\#}(x_i)) \geq \hat{\beta} - \beta_0\}$$

$(\beta^{\#} - \hat{\beta})/(1 - \beta^{\#}(x_i))$ is a "pivotal" quantity that does not depend on β_0. Thus the confidence region is the interval between the κth and $(n! - \kappa + 1)$th order statistics of $\hat{\beta} - (\beta^{\#} - \hat{\beta})/(1 - \beta^{\#}(x_i))$, where $\kappa = [n!\alpha/2]$. It is worthwhile noting that the standard (Hodges–Lehmann) nonparametric estimate in this case is not $\hat{\beta}$, the solution of $T(\hat{\beta}) = 0$, but

$$\hat{\beta} - \text{median}[(\beta^{\#} - \hat{\beta})/(1 - \beta^{\#}(x_i))]$$

4. AN ALTERNATIVE BOOTSTRAP CONFIDENCE INTERVAL

The confidence interval of Section 3 suggests an adjustment to the bootstrap approximation given by the percentile method by noting that the bootstrap is obtained by sampling with replacement in place of sampling without replacement. Then the adjusted confidence interval is obtained by adding $\hat{\beta}$ to the $\frac{1}{2}\alpha$ quantiles of

$$-(\beta^* - \hat{\beta})/(1 - \beta^*(x_i))$$

which can be estimated from a Monte Carlo simulation as before.

5. A NUMERICAL EXAMPLE

The data used in Table 1 below are the law school data given in Efron (1982). For these data, $\hat{\beta} = 0.452$. We give the 10% confidence limits L, U, and the Hodges–Lehmann estimate in the first two cases or the median of the sampled distribution in the other cases, all based on resampling 1000

Table 1. Estimates and Confidence Intervals

Method	Estimate	Lower C.I.	Upper C.I.
Permutation Test Inverse	0.46	0.28	0.62
Adjusted bootstrap	0.46	0.28	0.62
Percentile method without replacement	0.45	0.29	0.59
Bootstrap percentile method	0.45	0.29	0.61
Normal approximation	0.45	0.27	0.63
Bootstrap on pairs	0.46	0.31	0.57

times. The results indicate that for these data all methods give quite similar results. Perhaps the most surprising result is that the narrowest confidence interval is given by the bootstrap over pairs. A possible explanation of this is that for this data the largest residuals occur in the middle of the line, so the variance of the bootstrap distribution estimated from a result of Freedman (1981) in his Theorem 3.1 as

$$\sum_{i=1}^{n} x_i^2 \hat{\varepsilon}_i^2 \Big/ \left(\sum_{i=1}^{n} x_i^2 \right)^2$$

is smaller than $n^{-1}\Sigma_{i=1}^{n}\hat{\varepsilon}_i^2 / \Sigma_{i=1}^{n} x_i^2$ since $\hat{\varepsilon}_i^2$ is small for large x_i^2.

6. THE BOOTSTRAP OF THE STANDARDIZED STATISTIC

In studies of the bootstrap for sample means it has been suggested by Singh (1981), Babu and Singh (1983, 1985), and Abramovitch and Singh (1985) that improvements in the accuracy of the bootstrap percentile method could be achieved by bootstrapping a standardized pivotal quantity. The same method could be applied to the regression situation.

A confidence interval for β is $(\hat{\beta} - u_L(F)s/(\Sigma_i x_i^2)^{1/2},\ \hat{\beta} - u_U(F)s/(\Sigma_i x_i^2)^{1/2}$, where $s^2 = (n-2)^{-1}[\Sigma_i y_i^2 - n\bar{y}^2 - \hat{\beta}^2 \Sigma_i x_i^2] = (n-2)^{-1}\Sigma_i \hat{\varepsilon}_i^2$, $u_U(F)$ is the solution of

$$P\left(\frac{\hat{\beta} - \beta}{\left[(n-2)^{-1}\Sigma_i \hat{\varepsilon}_i^2 / \Sigma_i x_i^2\right]^{1/2}} \le u_U(F) \right) = \frac{1}{2}\alpha$$

with $u_L(F)$ being defined similarly, and where

$$\hat{\varepsilon}_i = \varepsilon_i - \sum_i \varepsilon_i \Big/ n - (\hat{\beta} - \beta)x_i, \qquad i = 1, \ldots, n$$

so that $\hat{\beta} - \beta = \Sigma_i \varepsilon_i x_i / \Sigma_i x_i^2$, it being assumed that $\varepsilon_1, \ldots, \varepsilon_n$ are independent with distribution function F. The bootstrap approximation for $u_U(F)$ is given by $u_U^* = u_U^*(F_n)$, the solution of

$$P_*\left(\frac{\beta^* - \hat{\beta}}{s^*/\left(\Sigma_i x_i^2\right)^{1/2}} \leq u_U^*\right) = \frac{1}{2}\alpha$$

where

$$\beta^* - \hat{\beta} = \sum_i \hat{\varepsilon}_{I_i} x_i \Big/ \sum_i x_i^2$$

and

$$s^{*2} = (n-2)^{-1}\left[\sum_i \hat{\varepsilon}_{I_i}^2 - \left(\sum_i \hat{\varepsilon}_{I_i}\right)^2 \Big/ n - (\beta^* - \hat{\beta})^2 \sum_i x_i^2\right]$$

and where P_* denotes a probability taken only with respect to $I_1, \ldots, I_n$, which are independent and uniform on $(1, 2, \ldots, n)$. Note that in this probability $\hat{\varepsilon}_1, \ldots, \hat{\varepsilon}_n$ are regarded as fixed, whereas in the previous probability $\varepsilon_1, \ldots, \varepsilon_n$ are regarded as random. We can obtain u_U^* from the exact bootstrap distribution by obtaining all n^n possible values of $(I_1, \ldots, I_n)$, which occur with equal probability, or we can approximate it by taking a large sample of size B, say. When u_L^* is similarly obtained we can approximate the confidence interval for β given above by $(\hat{\beta} - u_L^* s/(\Sigma_i x_i^2)^{1/2}, \hat{\beta} + u_U^* s/(\Sigma_i x_i^2)^{1/2})$.

We now have three competing bootstrap approximate confidence intervals. To compare them with the exact confidence interval based on u_L and u_U above, and with the exact conditional interval obtained from the permutation test, we need to use some asymptotic approximations based on Edgeworth expansions, which we will consider in Section 7.

7. ASYMPTOTIC APPROXIMATIONS TO THE CONFIDENCE INTERVALS

Edgeworth approximations for confidence interval for the sample mean based on the bootstrap were considered by Singh (1981), Babu and Singh

(1983, 1985), and Abramovitch and Singh (1985). They were also considered for confidence intervals from permutation tests for paired comparisons and two sample comparisons by John and Robinson (1983).

Let $a_1, \ldots, a_n$ and $c_1, \ldots, c_n$ be sequences with

$$\sum_i a_i = \sum_i c_i = 0, \qquad \sum_i a_i^2 = \sum_i c_i^2 = 1$$

and let

$$T_n = (n-1)^{1/2} \sum_i a_{R_i} c_i$$

where $(R_1, \ldots, R_n)$ is a uniform random permutation of $(1, \ldots, n)$. Consider the formal Edgeworth expansion for $P(T_n \le t)$,

$$\Pi_T(t) = \Phi(t) - \phi(t)\left[\frac{K_3(T)}{6}(t^2 - 1) + \frac{K_4(T)}{24}(t^3 - 3t) + \frac{K_3(T)^2}{72}(t^5 - 10t^3 + 15t)\right]$$

where

$$K_3(T) = n^{1/2} \sum_i a_i^3 \sum_i c_i^3$$

and

$$K_4(T) = n\left(\sum_i a_i^4 - 3n^{-1}\right)\left(\sum_i c_i^4 - 3n^{-1}\right) - 6n^{-1}$$

In the sequel we shall assume that this is a valid asymptotic approximation in the cases considered, in the sense that the error is $o(n^{-1})$. Unfortunately there does not appear to be a theorem of sufficient generality to cover all the cases of interest to us. So we can either restrict attention to the case of two samples, where either $a_1, \ldots, a_n$ or $c_1, \ldots, c_n$ take only two values, proved by Bickel and Van Zwet (1978) and Robinson (1978), and the case where $a_1, \ldots, a_n$ or $c_1, \ldots, c_n$ takes values $J(1/(n+1))$, $\ldots, J(n/(n+1))$, considered by Does (1983), or we can accept for the moment the following conjecture, which includes the results above and seems to be of sufficient generality for our purpose.

Conjecture. For the sequences $a_1, \ldots, a_n$ and $c_1, \ldots, c_n$ considered above, if

$$n^{3/2} \sum_i |a_i|^5 \sum_i |c_i|^5 = o(n^{-1})$$

and, if, for λ denoting Lebesque measure, there is a $\delta > 0$ such that for $(d_1, \ldots, d_n)$ either $(a_1, \ldots, a_n)$ or $(c_1, \ldots, c_n)$,

$$\lambda\{x : \exists_j |x - d_j| < \zeta\} \geq \delta n \zeta \quad \text{for some} \quad \zeta \geq n^{-3/2} \log n$$

then

$$P(T_n \leq t) = \Pi_T(t) + o(n^{-1})$$

It should be noted that the results obtained below in the bootstrap could be obtained by appealing to the results of Bhattacharya and Ghosh (1978), since in this case the formal Edgeworth expansions could be obtained by regarding the statistics as functions of sums of independent random variables. This method will not give the conditional expansion needed for the permutation test approach, however, so in the sequel all results will be based on the above conjecture, which is known to be true under the restricted conditions discussed before its statement.

From the definition of u_U^* given above, we can write

$$P_*\left(\sum_i \hat{\varepsilon}_{I_i} x_i \left[(n-1)^{-1} \sum_i \left(\hat{\varepsilon}_{I_i} - \frac{\sum_i \hat{\varepsilon}_{I_i}}{n}\right)^2 \sum_i x_i^2\right]^{-1/2} \leq t^*\right) = \frac{1}{2}\alpha$$

where $t^* = [(n-1)/(n-2)]^{1/2} u_U^* (1 + u_U^{*2}/(n-2))^{1/2}$. The value of t^* can be approximated by finding an Edgeworth expansion for the left-hand side and inverting the relation to obtain a Cornish–Fisher expansion for t^*. The Edgeworth expansion is most easily found by first finding an expansion for the conditional distribution given the order statistics of $\hat{\varepsilon}_{I_1}, \ldots, \hat{\varepsilon}_{I_n}$, and then taking the conditional expectation given $\hat{\varepsilon}_1, \ldots, \hat{\varepsilon}_n$. As stated above the same expression is obtained by using the methods of Bhattacharya and Ghosh (1978). If we assume that F has an absolutely continuous component with density that is positive on an interval, and moments of order 6, then this gives

$$\Pi_T(t^*) + o_P(n^{-1}) = \tfrac{1}{2}\alpha$$

where Π_T is given above with $a_i = \hat{\varepsilon}_i (\sum_i \hat{\varepsilon}_i^2)^{-1/2}$ and $c_i = x_i (\sum_i x_i^2)^{-1/2}$, and where we use $o_P(n^{-1})$ to indicate that the asymptotic expansion is valid to $o(n^{-1})$ on a set of probability $1 - o(n^{-1})$. Solving this equation for t^* gives, for $z = \Phi^{-1}(\frac{1}{2}\alpha)$,

$$t^* = z + \tfrac{1}{6}(z^2 - 1)K_3(T) + \tfrac{1}{24}(z^3 - 3z)K_4(T)$$
$$- \tfrac{1}{36}(2z^3 - 5z)K_3(T)^2 + o_P(n^{-1})$$

So the expression for u_U^* is

$$u_U^* = t^* + \tfrac{1}{2}(z^3 - z)n^{-1} + o_P(n^{-1})$$

The same procedure gives an expression for $u_U(F)$ as

$$u_U(F) = z + \tfrac{1}{6}(z^2 - 1)K_3 + \tfrac{1}{2}(z^3 - z)n^{-1}$$
$$+ \tfrac{1}{24}(z^3 - 3z)K_4 - \tfrac{1}{36}(2z^3 - 5z)K_3^2 + o(n^{-1})$$

where

$$K_3 = \frac{\mu_3 \Sigma_i x_i^3}{\sigma^3 \left(\Sigma_i x_i^2\right)^{3/2}} \quad \text{and} \quad K_4 = \left(\frac{\mu_4}{\sigma^4} - 3\right)\left(\frac{\Sigma_i x_i^4}{\left(\Sigma_i x_i^2\right)^2} - \frac{3}{n}\right) - \frac{6}{n},$$

for $\sigma^3 = E\varepsilon_i^2$, $\mu_3 = E\varepsilon_1^3$, $\mu_4 = E\varepsilon_1^4$. So

$$u_U^* - u_U(F) = \frac{1}{6}(z^2 - 1)\left[\frac{\Sigma_i x_i^3}{\left(\Sigma_i x_i^2\right)^{3/2}}\right]\frac{\Sigma_i\left(\hat{\varepsilon}_i^3 - \mu_3\right)}{n\sigma^3} + o_P(n^{-1})$$

This is a random variable of $O_P(n^{-1})$ with expectation of $o(n^{-1})$. So the coverage probability of the bootstrap estimate using this method is correct to $o(n^{-1})$.

For the exact confidence interval from the permutation test we find a random variable $v = v(\hat{\varepsilon})$ such that

$$P_{\#}\left(\frac{\Sigma_i \hat{\varepsilon}_{R_i} x_i}{\Sigma_i x_i^2} \le v\left(1 - \frac{\Sigma_i x_{R_i} x_i}{\Sigma_i x_i^2}\right)\right) = \frac{1}{2}\alpha$$

where P denotes a probability taken only with respect to $(R_1, \ldots, R_n)$, a uniform random permutation of $(1, \ldots, n)$. If $u_U^{\#} = v[\Sigma_i \hat{\varepsilon}_i^2 / (n-2)\Sigma_i x_i^2]^{-1/2}$, then we want to compare $u_U^{\#}$ and u_U^*. This probability is $P_{\#}(W \le w) = \frac{1}{2}\alpha$, where $w = [(n-1)/(n-2)]^{1/2} u_U^{\#}(1 + u_U^{\#2}/$

$(n-2))^{-1/2}$ and

$$W = \frac{\Sigma_i(\hat{\varepsilon}_{R_i} + vx_{R_i})x_i}{\left[(n-1)^{-1}\Sigma_i(\hat{\varepsilon}_i + vx_i)^2\Sigma_i x_i^2\right]^{1/2}}$$

If we assume that F has an absolutely continuous component with density positive on an interval and moments of order 6, then

$$\Pi_W(w) + o_P(n^{-1}) = \tfrac{1}{2}\alpha$$

where Π_W is defined above with $a_i = (\hat{\varepsilon}_i + vx_i)[\Sigma_i(\hat{\varepsilon}_i + vx_i)^2]^{-1/2}$ and $c_i = x_i(\Sigma_i x_i^2)^{-1/2}$. This gives in the same way as before

$$u_U^{\#} = z + \tfrac{1}{6}(z^2-1)K_3(W) + \tfrac{1}{24}(z^3-3z)K_4(W)$$
$$-\tfrac{1}{36}(2z^3-5z)K_3(W)^2 + \tfrac{1}{2}(z^3-z)n^{-1} + o_P(n^{-1})$$

where

$$K_3(W) = \frac{\Sigma_i(\hat{\varepsilon}_i + vx_i)^3}{\left[\Sigma_i\hat{\varepsilon}_i + v^2\Sigma_i x_i^2\right]^{3/2}} = K_3(T) + \gamma z + o_P(n^{-1})$$

and

$$K_4(W) = K_4(T) + o_P(n^{-1})$$

where

$$\gamma = \frac{1}{2}\frac{\Sigma_i\hat{\varepsilon}_i^2 x_i \Sigma_i x_i^3}{\left[\Sigma_i\hat{\varepsilon}_i^2(\Sigma_i x_i^2)^2\right]}$$

Thus

$$u_U^{\#} - u_U^{*} = \tfrac{1}{6}(z^3-z)\gamma + o_P(n^{-1})$$

and γ is a random variable of $O_P(n^{-1})$ with expectation 0.

Finally, it is of interest to examine the alternative bootstrap estimate suggested earlier. We want to find v^+ such that

$$P_*\left(\frac{\Sigma_i\hat{\varepsilon}_{I_i}x_i}{\Sigma_i x_i^2} \le v^+\left(1 - \frac{\Sigma_i x_{I_i}x_i}{\Sigma_i x_i^2}\right)\right) = \frac{1}{2}\alpha$$

If $u_U^+ = v^+[\Sigma_i\varepsilon_i^2/(n-2)\Sigma_i x_i^2]^{-1/2}$, then we need to compare u_U^+ with

$u_U(F)$ and $u_U^{\#}$. This probability is

$$P_*\left(\frac{\Sigma_i(\hat{\varepsilon}_{I_i} + v^+ x_{I_i})x_i}{\left[(n-1)^{-1}\Sigma_i(\hat{\varepsilon}_i + v^+ x_i)^2 \Sigma_i x_i^2\right]^{1/2}} \leq y\right) = \frac{1}{2}\alpha$$

for $y = [(n-1)/(n-2)]^{1/2} u_U^+(1 + u_U^{+2}/(n-2))^{-1/2}$. Using the Edgeworth expansion for sums of independent random variables and inverting as above we obtain

$$u_U^+ = z + \tfrac{1}{6}(z^2 - 1)\beta_3 + \tfrac{1}{24}(z^3 - 3z)\beta_4$$
$$- \tfrac{1}{36}(2z^3 - 5z)\beta_3^2 + \tfrac{1}{2}(z^3 - z)n^{-1} + o_P(n^{-1})$$

where

$$\beta_3 = K_3(T) + \gamma z + o_P(n^{-1})$$

but

$$\beta_4 = n\left[\frac{\Sigma_i \hat{\varepsilon}_i^4}{\left(\Sigma_i \hat{\varepsilon}_i^2\right)^2}\right]\left[\frac{\Sigma_i x_i^4}{\left(\Sigma_i x_i^2\right)^2} - \frac{3}{n}\right] + o_P(n^{-1})$$

So the difference between $u_U^{\#}$ and u_U^+ is

$$\tfrac{1}{8}(z^3 - 3z)\left[\frac{\Sigma_i x_i^4}{\left(\Sigma_i x_i^2\right)^2} - \frac{1}{n}\right] + o_P(n^{-1})$$

a quantity of $O_P(n^{-1})$ but which also has expectation of $O(n^{-1})$. Thus the coverage probability of the alternative bootstrap confidence interval differs from α by a quantity of $O(n^{-1})$. Note that the alternative bootstrap confidence interval adjusts for the random difference between the bootstrap and the exact conditional level, but there is no apparent way to obtain a correct standardization that will equate the terms of $O(n^{-1})$. The sample percentile approximation would contain both these sources of error.

Evidently, confidence intervals based on the standardized bootstrap or on the permutation procedure are best and, while they differ by a quantity of $O_P(n^{-1})$, their coverage probabilities are the same to $o(n^{-1})$, and agree, to that order, with the coverage probability based on the distribution of $(\hat{\beta} - \beta)(\Sigma_i x_i^2)^{1/2}/s$.

ACKNOWLEDGMENT

The work for Sections 6 and 7 was prompted by comments made at Bilbao for which I am most grateful.

REFERENCES

Abramovitch, L. and Singh, K. (1985). Edgeworth corrected pivotal statistics, and the bootstrap. *Ann. Statist.* **13**, 116–132.

Babu, G. J. and Singh, K. (1983). Inference on means using the bootstrap. *Ann. Statist.* **11**, 999–1003.

Babu, G. J. and Singh, K. (1985). On one term Edgeworth correction by Efron's bootstrap. *Sankhyā A* **46**, 219–232.

Bhattacharya, R. N. and Ghosh, J. K. (1978). On the validity of the formal Edgeworth expansion. *Ann. Statist.* **6**, 434–451.

Bickel, P. J. and Van Zwet, W. R. (1978). Asymptotic expansions for the power of distribution free tests in the two sample problem. *Ann. Statist.* **6**, 937–1004.

Does, R. J. M. M. (1983). An Edgeworth expansion for simple linear rank statistics under the null hypothesis. *Ann. Statist.* **11**, 607–624.

Efron, B. (1979). Bootstrap methods: another look at the jackknife. *Ann. Statist.* **7**, 1–26.

Efron, B. (1982). *The Jackknife, the Bootstrap and Other Resampling Plans*. SIAM Regional Conference Series in Applied Mathematics, No. **38**.

Freedman, D. A. (1981). Bootstrapping regression models. *Ann. Statist.* **9**, 1218–1228.

Gabriel, K. R. and Hall, W. J. (1983). Rerandomization inference on regression and shift effects: computationally feasible methods. *J. Amer. Statist. Assoc.* **78**, 827–836.

John, R. D. and Robinson, J. (1983). Significance levels and confidence intervals for permutation tests. *J. Statist. Comput. Simul.* **16**, 161–173.

Maritz, J. S. (1981). *Distribution Free Statistical Methods*. London: Chapman and Hall.

Robinson, J. (1978). An asymptotic expansion for samples from a finite population. *Ann. Statist.* **6**, 1005–1011.

Singh, K. (1981). On the asymptotic accuracy of Efron's bootstrap. *Ann. Statist.* **9**, 1187–1195.

PART 4

Density Estimation

Density Estimation: Robustness versus Consistency

Antonio Cuevas
Departamento de Estadística e Investigación Operativa
Facultad de Ciencias Matemáticas
Universidad Complutense de Madrid
Madrid, Spain

SUMMARY

We prove the incompatibility between robustness (in Hampel's sense) and consistency in density estimation. This feature means a significant difference with respect to parametric estimation. We specially study the robustness of kernel estimates.

1. INTRODUCTION

Let $(\mathscr{F}, d_P)$ be the metric space of probability distributions on $\mathscr{B}(\mathbb{R}^k)$ (Borel σ-algebra on $\mathbb{R}^k$) endowed with the Prohorov metric d_P (that induces the weak topology on $\mathscr{F}$). For each absolutely continuous distribution $F \in \mathscr{F}$, let f denote the corresponding density function (d.f.) and let $(\mathscr{D}, L_1)$ be the space of d.f.'s on $\mathbb{R}^k$ endowed with the L_1 metric

$$L_1(f_1, f_2) = \int_{\mathbb{R}^k} |f_1(x) - f_2(x)|\, dx$$

We identify, as usual, f_1 and f_2 if $f_1 = f_2$ almost everywhere with respect to Lebesgue measure.

Weak convergence will be denoted by $F_n \to^{w}_{n\to\infty} F$.

Definition 1.1. A sequence of density estimates $\{T_n\}$ (that is, a sequence of measurable functions T_n: $(\mathbb{R}^k)^n \to \mathscr{D}$) is called *qualitatively robust* at $F \in \mathscr{F}$ iff

$$\forall \varepsilon > 0\ \exists \delta > 0 | \forall G, \forall n\colon \{d_P(F,G) < \delta \Rightarrow d_P^*(\mathscr{L}_F(T_n), \mathscr{L}_G(T_n)) < \varepsilon\},$$

where $\mathscr{L}_F(T_n)$ denotes the probability distribution induced by T_n on $\mathscr{B}(\mathscr{D})$ (Borel σ-algebra on $\mathscr{D}$) under F, and d_P^* is the Prohorov metric in the space of probability measures on $\mathscr{B}(\mathscr{D})$.

This definition has been proposed by Hampel (1968, 1971) to formalize the notion of robustness for $\mathbb{R}^P$-valued estimates. Here we simply study the applicability of Hampel's idea in the context of density estimation.

2. ROBUSTNESS VERSUS CONSISTENCY

As an auxiliary tool, we need the following lemma, which is essentially due to Hampel (1971, Lemma 3 and Theorem 2); we present it here with a slightly different statement appropriate for density estimation.

Lemma 2.1. (i) Let $\mathscr{E} \subset \mathscr{F}$ be endowed with the relative topology. Suppose that $\{T_n\}$ is qualitatively robust at $F \in \mathscr{E}$ and consistent (in probability) for all $G \in \mathscr{E}_1 = U(F) \cap \mathscr{E}$, where $U(F)$ is a neighborhood of F in $\mathscr{F}$. Then the functional T_∞: $\mathscr{E}_1 \to \mathscr{D}$ defined by $T_\infty(G) = \lim_{n\to\infty} T_n$ (in probability, under G) is continuous at F.

(ii) If there exists a functional T: $\mathscr{F} \to \mathscr{D}$ such that T_n is the restriction of T to the set $\mathscr{F}_n$ of empirical distributions associated with samples of size n (briefly, $T_n = T|\mathscr{F}_n$), then the continuity of T on $\mathscr{F}$ is a sufficient condition for qualitative robustness of $\{T_n\}$ on $\mathscr{F}$.

Proof. (i) If we suppose that $T_\infty|\mathscr{E}_1$ is not continuous at F, then there exists $0 < \varepsilon < \frac{1}{2}$ and $\{G_r\} \subset \mathscr{E}_1$ such that $G_r \to^{w}_{r\to\infty} F$ and $L_1(T_\infty(G_r), T_\infty(F)) \geq 2\varepsilon$. This implies $d_P^*(T_\infty(G_r), T_\infty(F)) \geq 2\varepsilon$, where $T_\infty(G_r)$, $T_\infty(F)$ denote here the corresponding degenerate distributions. From the robustness, given $\varepsilon > 0$, there exists $\delta > 0$ such that $d_P(F, G_r) < \delta$ implies $d_P^*(\mathscr{L}_F(T_n), \mathscr{L}_G(T_n)) < \varepsilon$. Since $G_r \to^{w}_{r\to\infty} F$, there exists $r_0 \in \mathbb{N}$ such that $d_P(F, G_r) < \delta$ for all $r \geq r_0$ and, therefore, $d_P^*(\mathscr{L}_{G_r}(T_n), \mathscr{L}_F(T_n)) < \varepsilon$ for all $r \geq r_0$ and $n \in \mathbb{N}$, which leads to a contradiction because $\mathscr{L}_F(T_n) \to^{w}_{n\to\infty} T_\infty(F)$ and $\mathscr{L}_{G_r}(T_n) \to^{w}_{n\to\infty} T_\infty(G_r)$.

(ii) This result has been also proved by Hampel (1971, Theorem 2) for $\mathbb{R}^P$-valued estimates. His proof is valid here with trivial changes: it suffices to replace the metric in $\mathbb{R}^P$ by the metric L_1 considered in $\mathscr{D}$. ∎

Now we can establish the incompatibility between robustness and consistency for univariate density estimates (that is, $k = 1$):

Theorem 2.1. Let $\mathscr{E}$ denote the set of absolutely continuous distributions in $\mathscr{F}$. If $\{T_n\}$ is a sequence of density estimates such that, for all $F \in \mathscr{E}$, $T_n \to_{n\to\infty} f$ (in probability, under F), then, for all $F \in \mathscr{E}$, $\{T_n\}$ is not qualitatively robust at F.

Proof. From the above lemma, it suffices to prove that the functional T_∞: $\mathscr{E} \to \mathscr{D}$ defined by $T_\infty(F) = f$, is discontinuous for all $F \in \mathscr{E}$:

Let f be an arbitrary d.f.; we shall construct a sequence of d.f.'s such that the corresponding sequence of distribution functions $\{F_r\}$ fulfils $F_r \to^{w}_{r\to\infty} F$ and, however,

$$\int_{\mathbb{R}} |f_r(x) - f(x)|\, dx \underset{r\to\infty}{\not\to} 0$$

Since, for all $r \in \mathbb{N}$,

$$\int_{\mathbb{R}} f(x)\, dx = \int_{\mathbb{R}} f(x) \cos^2 rx\, dx + \int_{\mathbb{R}} f(x) \sin^2 rx\, dx = 1$$

the sequences of integrals in the second member of this expression cannot converge both simultaneously to zero as $r \to \infty$. We may suppose, without loss of generality, that

$$\int_{\mathbb{R}} f(x) \cos^2 rx\, dx \underset{r\to\infty}{\not\to} 0$$

Let us define

$$k_r = \left[\int_{\mathbb{R}} f(x)(1 + \cos rx)\, dx\right]^{-1} \quad \text{and} \quad f_r(x) = k_r f(x)(1 + \cos rx)$$

As a direct consequence of the well-known Riemann–Lebesgue lemma [see, for example, Pryce (1973, p. 132)], we have

$$\int_B f(x) \cos rx\, dx \underset{r\to\infty}{\to} 0$$

for all Borel sets $B \subset \mathbb{R}$; therefore

$$\int_B f_r(x)\, dx \underset{r\to\infty}{\to} \int_B f(x)\, dx$$

and the weak convergence $F_r \xrightarrow[r\to\infty]{w} F$ is proved. On the other hand,

$$\int_{\mathbb{R}} |f_r(x) - f(x)|\, dx = \int_{\mathbb{R}} |k_r f(x)\cos rx - (1 - k_r)f(x)|\, dx$$

$$\geq \left| \int_{\mathbb{R}} |k_r f(x)\cos rx|\, dx - \int_{\mathbb{R}} |(1 - k_r)f(x)|\, dx \right| \quad (2.1)$$

Now, since $k_r \to_{r\to\infty} 1$ and

$$\int_{\mathbb{R}} f(x)|\cos rx|\, dx \geq \int_{\mathbb{R}} f(x)\cos^2 rx\, dx \underset{r\to\infty}{\nrightarrow} 0$$

the last member in (2.1) does not converge to 0 and, therefore,

$$\int_{\mathbb{R}} |f_r(x) - f(x)|\, dx \underset{r\to\infty}{\nrightarrow} 0 \quad \blacksquare$$

3. ROBUSTNESS AND CONSISTENCY OF KERNEL ESTIMATES

The kernel estimate associated to $K: \mathbb{R} \to \mathbb{R}$ (kernel function) is given by

$$T_n(X_1, \ldots, X_n; x) = \hat{f}_n(x) = \frac{1}{n\varepsilon_n} \sum_{i=1}^{n} K\left(\frac{x - X_i}{\varepsilon_n}\right) \quad (3.1)$$

We shall study the robustness of this important class of estimates. Roughly speaking, the situation can be summarized as follows: under extremely general conditions, the usual kernel estimates with $\varepsilon_n \downarrow 0$ show a radical nonrobustness; however, when the "bandwidth" ε_n is constant ($\varepsilon_n = \varepsilon$ for all n), the sequence $\{T_n\}$ is robust under mild conditions. This is formalized in the following result.

Theorem 3.1. Let $\{T_n\}$ be the sequence of estimates defined in (3.1), and let $\mathscr{E}$ denote the set of absolutely continuous distributions in $\mathscr{F}$.

(i) If $\lim_{n\to\infty}\varepsilon_n = 0$, $\lim_{n\to\infty} n\varepsilon_n = \infty$, and K is a non-negative measurable function on $\mathbb{R}$ with $\int_{\mathbb{R}} K(x)\, dx = 1$, then, for all $F \in \mathscr{E}$, the sequence $\{T_n\}$ is not robust at F.

(ii) If $\varepsilon_n = \varepsilon > 0$ for all n, and K is a continuous and bounded density function, then $\{T_n\}$ is robust for all $F \in \mathscr{F}$.

Proof. (i) Under these hypotheses, Devroye (1983) has proved an interesting result concerning the consistency of kernel estimates; this result includes the L_1 consistency (almost surely). Therefore, the nonrobustness follows directly from Theorem 2.1.

(ii) We have $T_n = T|\mathscr{F}_n$, where $T: \mathscr{F} \to \mathscr{D}$ is defined by

$$T(F) = \varepsilon^{-1} \int_{\mathbf{R}} K\left(\frac{\cdot - u}{\varepsilon}\right) dF(u)$$

Let $F \in \mathscr{F}$ and $\{F_r\} \subset \mathscr{F}$ be such that $F_r \to^{w}_{r\to\infty} F$. Then since K is continuous and bounded,

$$T(F_r)(x) = \varepsilon^{-1} \int_{\mathbf{R}} K\left(\frac{x-u}{\varepsilon}\right) dF_r(u) \underset{r\to\infty}{\to} \varepsilon^{-1} \int_{\mathbf{R}} K\left(\frac{x-u}{\varepsilon}\right) dF(u)$$

Now, since the almost sure convergence of d.f.'s (to a d.f.) implies their L_1 convergence (Scheffé, 1947), we obtain $T(F_r) \to_{r\to\infty} T(F)$. Therefore, T is continuous on $\mathscr{F}$ and, from part (ii) of Lemma 2.1, we conclude the robustness of $\{T_n\}$. ∎

4. DISCUSSION AND COMMENTS

The concept of sampling distribution for a density estimate is a fundamental component in our study; this concept does not play a role as important as the analogous one in parametric inference and it presents a considerably more difficult mathematical handling. In this respect, let us observe that the robustness can be established (as well as in the parametric case) through sufficient conditions which do not require knowledge of sampling distributions (Theorem 3.1 provides an example).

On the other hand, the practical meaning of qualitative robustness in density estimation may seem, perhaps, of less importance than in parametric theory because the hypotheses concerning underlying distribution are more relevant in this latter case (the assumption of normality is a typical example); however, it may be also interesting to consider the infinitesimal notion of robustness in density estimation since it is a structural property providing a qualitative information about the stability of the estimates and it allows us to answer such questions as, What is the behavior of a density estimate if the underlying distribution is not absolutely continuous? Besides, from a theoretical viewpoint, the comparative study of an idea (Hampel's concept of robustness, in our case) in different contexts may reveal some interesting features; thus, the incompatibility between robustness and consistency that we have proved for density estimates does not occur necessarily in parametric inference: for instance, the *sample median* $T_n = T|\mathscr{F}_n$ is robust and consistent under F to the *populational median* $T(F)$ whenever $F^{-1}(\frac{1}{2})$ contains no more than one point. As we have seen, this peculiarity of density estimation is basically due to the noncontinuity of the *natural parametrization* $F \mapsto T(F) = f$.

REFERENCES

Devroye, L. (1983). The equivalence of weak, strong and complete convergence in L_1 for kernel density estimates. *Ann. Statist.* **11**, 896–904.

Hampel, F. R. (1968). Contributions to the theory of robust estimation. Ph.D. thesis, University of California, Berkeley.

Hampel, F. R. (1971). A general qualitative definition of robustness. *Ann. Math. Statist.* **42**, 1887–1896.

Pryce, J. D. (1973). *Basic Methods of Linear Functional Analysis*. London: Hutchinson University Library.

Scheffé, H. (1947). A useful convergence theorem for probability distributions. *Ann. Math. Statist.* **18**, 434–438.

The Histogram with Random Partition

Jean-Pierre Lecoutre
Laboratoire de Statistique Théorique et Appliquée
Université de Paris VI
Paris, France

SUMMARY

We consider a density estimator based on k statistically equivalent blocks. Under appropriate assumptions, we prove the L^2 consistency of this estimator and we describe the asymptotic limiting behavior of the integrated mean-square error. A functional form for the optimum k expressed in terms of the sample size and the underlying probability distribution is also obtained. Finally, the estimator is shown to be asymptotically normal.

1. INTRODUCTION

Let $X, X_1, \ldots, X_n$ be independent random variables (r.v.), identically distributed with absolutely continuous distribution function F and density function f. A class of nonparametric estimators of f based on statistically equivalent blocks (Tukey, 1947; Anderson, 1966) has been proposed by Gessaman (1970), Van Ryzin (1970) and Révész (1972). Results on their consistency are found in Van Ryzin (1973) and Kim and Van Ryzin (1975) and asymptotic normality is proved in Bauer (1975). The asymptotic distribution of the maximum absolute deviation is described in Tusnady (1974), Bera (1977), and Reiss (1978). Lecoutre (1975) in the case $X \in [0, 1]$

and Abou-Jaoudé (1977) in the case $X \in \mathbb{R}^d$ have given a necessary and sufficient condition for strong uniform consistency.

Let $X_{(1)}, \dots, X_{(n)}$ be the order statistic of $(X_1, \dots, X_n)$. Let us suppose $n \geq 2$. Choose arbitrarily an integer $k \in [2, n]$ and let $\nu = [n/k]$. The interval $[X_{(1)}, X_{(n)}]$ is divided into k disjoint intervals

$$[X_{(1)}, X_{(R(1))}[, \dots, [X_{(R(q-1))}, X_{(R(q))}[, \dots, [X_{(R(k-1))}, X_{(n)}]$$

so that each of these blocks contains ν or $\nu + 1$ points of the sample. The histogram with random partition f_n is defined by

$$f_n(x) = \begin{cases} \dfrac{\nu_{nq}}{n\lambda_{nq}} & \text{if } x \in \Pi_{nq}, 1 \leq q \leq k \\ 0 & \text{otherwise} \end{cases}$$

where $\Pi_{nq} = (X_{(R(q-1))}, X_{(R(q))})$, $\lambda_{nq} = X_{(R(q))} - X_{(R(q-1))}$,

$$\nu_{nq} = \sum_{i=1}^{n} 1_{\Pi_{nq}}(X_i)$$

(1_A denotes the indicator function of the set A). For histogram estimates with a fixed cell width h_n, the important problem of finding the optimal h_n has been solved by Scott (1979) and Freedman and Diaconis (1981) in the one-dimensional case and by Lecoutre (1985) in the multidimensional case. One object of this Chapter is to give similar results for the histogram with random cells. We need some preliminary definitions.

Definition 1.1. For a fixed point x, the *mean-square error* (MSE) of f_n is defined by

$$\Delta_n(x) = E[f_n(x) - f(x)]^2$$

Definition 1.2. If the MSE tends to zero as $n \to \infty$, f_n is *consistent in quadratic mean*:

$$f_n(x) \overset{\text{q.m.}}{\to} f(x).$$

Definition 1.3. The *integrated mean-square error* (IMSE) of an estimate f_n of $f \in L^2$ is defined by

$$J_n(f) = E\int [f_n(x) - f(x)]^2 \, dx$$

Definition 1.4. An estimate is said to be *consistent with respect to IMSE* if $J_n(f) \to 0$ as $n \to \infty$.

Definition 1.5. A sequence of r.v. Z_n is said to *converge almost completely* (a.c.) to a r.v. Z,

$$Z_n \overset{\text{a.c.}}{\to} Z,$$

if

$$\sum P\{|Z_n - Z| > \varepsilon\} < \infty \quad \text{for each } \varepsilon > 0$$

Definition 1.6 (Bertrand-Retali, 1978). Let g be a real function defined on $\mathbb{R}$. If g_n^+ and g_n^- are two sequences of step functions on a sequence $\prod_n$ of partitions of $\mathbb{R}$, Riemann-integrable on $\mathbb{R}$ and such that

$$g_n^- \leq g_{n+1}^- \leq g \leq g_{n+1}^+ \leq g_n^+,$$

and that

$$\lim_n \int g_n^+(x)\,dx = \lim_n \int g_n^-(x)\,dx = I \quad \text{as } n \to \infty$$

we say that I is the *generalized Riemann integral* (G.R.I.) of g. In this case, g is Lebesgue-integrable on $\mathbb{R}$ with

$$\int g(x)\,dx = I$$

and

$$\forall \varepsilon > 0,\ g_\varepsilon(x) = \sup\{g(t): |t - x| \leq \varepsilon\}$$

is Lebesgue-integrable on $\mathbb{R}$ and $\int g_\varepsilon(x)\,dx$ remains bounded as $\varepsilon \to 0$. We say that $g \in L_2'$ if the G.R.I. of g^2 exists.

We assume that the support of the probability distribution of X is $\mathbb{R}$. Let us note that in this case

$$X_{(1)} \overset{\text{a.c.}}{\to} -\infty \quad \text{and} \quad X_{(n)} \overset{\text{a.c.}}{\to} +\infty \quad \text{as} \quad n \to \infty \tag{1.1}$$

2. RESULTS

Theorem 2.1. Let x be a point with $f(x) > 0$ and suppose f has a bounded derivative in a neighborhood of x. Then, if $k \to \infty$ and $k/n \to 0$

as $n \to \infty$

$$f_n(x) \xrightarrow{\text{q.m}} f(x)$$

Theorem 2.2. Suppose f is bounded, positive, and differentiable on $\mathbb{R}$ with a bounded derivative. Suppose $f \in L^2$. Then, if $k \to \infty$ and $k/n \to 0$ as $n \to \infty$, f_n is consistent with respect to IMSE.

The next theorem provides the optimal rate of convergence for the IMSE.

Theorem 2.3. Suppose that f is bounded, positive, and twice differentiable on $\mathbb{R}$ with bounded derivatives; $f \in L_2'$ and $f'/f \in L_2'$. Then, if $k \to \infty$ and $k/n \to 0$ as $n \to \infty$

$$J_n(f) = \frac{k}{n}\int f^2(x)\,dx + \frac{1}{12k^2}\int \frac{f'^2(x)}{f^2(x)}\,dx + o((k/n) + k^{-2}) \quad (2.1)$$

and the value of k that minimizes the asymptotic IMSE

$$J_n'(f) = \frac{k}{n}\int f^2(x)\,dx + \frac{1}{12k^2}\int \frac{f'^2(x)}{f^2(x)}\,dx$$

for a fixed value of n is given by

$$k^* = \left[\left(\left(n\int\left(\frac{f'^2(x)}{f^2(x)}\right)dx\right)\bigg/\left(6\int f^2(x)\,dx\right)\right)^{1/3}\right] \quad (2.2)$$

leading to

$$J_n^*(f) = \frac{3}{2}\left(\left(\int\left(\frac{f'^2(x)}{f^2(x)}\right)dx\right)\bigg/\left(6\int f^2(x)\,dx\right)\right)^{1/3} n^{-2/3}(1 + o(1))$$

REMARK 2.1. Note that when $f(x)$ is small, the bias of this estimate appears to be larger than its variance because of the factor $f(x)$ in the denominator. We shall see that in the case of densities having a simple exponential form, $f(x) = ce^{-ax^b}$, with $c, a, b > 0$, the formula (2.1) is not valid because f'^2/f^2 is not integrable if $b \geq \frac{1}{2}$.

Theorem 2.4 is concerned with the asymptotic normality of $f_n(x)$.

Theorem 2.4. Suppose f is positive, differentiable, and has a bounded derivative in a neighborhood of x. Then, if $k^3/n \to \infty$ and $k/n \to 0$ as $n \to \infty$

$$\sqrt{\frac{n}{k}}\,\frac{f_n(x) - f(x)}{f(x)} \to N(0,1) \quad \text{in distribution}$$

3. PROOFS

Proof of Theorem 2.1. Conditional on $E_n = \{X_{(1)} \le x \le X_{(n)}\}$, let Π_{nq} be the cell containing x and $t \in \Pi_{nq}$; by Taylor's expansion

$$f(t) = f(x) + (t - x)f'(x + \theta(t - x)), \qquad \theta \in]0,1[;$$

integration over Π_{nq} implies that

$$p_{nq} = \int_{\Pi_{nq}} f(t)\,dt = \lambda_{nq} f(x) + O(\lambda_{nq}^2)$$

Therefore

$$f_n(x) = \frac{\nu_{nq} f(x)}{np_{nq}}(1 + O(p_{nq})) \tag{3.1}$$

The r.v. p_{nq} is known to have a β distribution with parameters ν_{nq} and $n - \nu_{nq} + 1$ (cf. Wilks, 1962, p. 238). It follows that

$$Ep_{nq} = \frac{\nu_{nq}}{n+1}, \qquad E\frac{1}{p_{nq}} = \frac{n}{\nu_{nq} - 1}$$

We have $Ef_n(x) = E(f_n(x)|E_n)P(E_n)$ and we can deduce from (1.1) and (3.1) that

$$Ef_n(x) = f(x)(1 + o(1)) \tag{3.2}$$

and

$$f_n(x) - Ef_n(x) = \frac{\nu_{nq} f(x)}{n}\left(\frac{1}{p_{nq}} - E\left(\frac{1}{p_{nq}}\right)\right) + \frac{\nu}{n}O(p_{nq} - Ep_{nq})$$

Using the moments of the β distribution, we obtain

$$Vf_n(x) = \frac{f^2(x)}{\nu}(1 + o(1)) \tag{3.3}$$

Theorem 2.1 follows from (3.2) and (3.3). ■

Proof of Theorem 2.2. We can write

$$\int (f_n(x) - f(x))^2\,dx = \int_{-\infty}^{X_{(1)}} f^2(x)\,dx + \sum_{q=1}^{k} I_{nq}(f) + \int_{X_{(n)}}^{+\infty} f^2(x)\,dx$$

with

$$I_{nq}(f) = \int_{\Pi_{nq}} (f_n(x) - f(x))^2\,dx$$

Then taking expectations, one finds that

$$J_n(f) = E\int_{-\infty}^{X_{(1)}} f^2(x)\,dx + \sum_{q=1}^{k} J_{nq}(f) + E\int_{X_{(n)}}^{+\infty} f^2(x)\,dx$$

with

$$J_{nq}(f) = EI_{nq}(f).$$

If M is an upper bound for f,

$$E\int_{-\infty}^{X_{(1)}} f^2(x)\,dx \le MEF(X_{(1)}) = \frac{M}{n+1}$$

likewise

$$E\int_{X_{(n)}}^{+\infty} f^2(x)\,dx = O\left(\frac{1}{n}\right)$$

Now, we deduce from (3.1), for $x \in \Pi_{nq}$,

$$(f_n(x) - f(x))^2 = \left(\frac{\nu_{nq}}{np_{nq}} - 1\right)^2 f^2(x)$$

$$+ \left(2\frac{\nu_{nq}}{n}\left(\frac{\nu_{nq}}{np_{nq}} - 1\right) + \frac{\nu_{nq}^2}{n^2}\right) f^2(x)O(1).$$

Thus

$$I_{nq}(f) \le \left(\frac{\nu_{nq}}{np_{nq}} - 1\right)^2 Mp_{nq} + \left(2\frac{\nu_{nq}}{n}\left(\frac{\nu_{nq}}{np_{nq}} - 1\right) + \frac{\nu_{nq}^2}{n^2}\right) Mp_{nq}O(1)$$

It follows from the distribution of p_{nq} that

$$J_{nq}(f) = O\left(\frac{1}{n}\right) + O\left(\frac{\nu^3}{n^3}\right);$$

therefore

$$\sum_{q=1}^{k} J_{nq}(f) = O\left(\frac{k}{n}\right) + O\left(\frac{1}{k^2}\right) \qquad ■$$

Proof of Theorem 2.3. We have seen that

$$J_n(f) = \sum_{q=1}^{k} J_{nq}(f) + O\left(\frac{1}{n}\right) \tag{3.4}$$

We shall study the term $I_{nq}(f)$.

Expanding f about the center $\bar{x}$ of the cell Π_{nq}, we have

$$f(x) = f(\bar{x}) + (x - \bar{x})f'(\bar{x}) + \tfrac{1}{2}(x - \bar{x})^2 f''(y), \qquad y \in \Pi_{nq}$$

Integration over Π_{nq} implies that

$$p_{nq} = \lambda_{nq} f(\bar{x}) + \lambda_{nq}^3 M_q$$

where

$$|M_q| \le \tfrac{1}{24} \sup_{x \in \Pi_{nq}} |f''(x)|$$

It follows that

$$\lambda_{nq} = \frac{p_{nq}}{f(\bar{x})} - M_q \frac{p_{nq}^3}{f^4(\bar{x})} + o\left(p_{nq}^3\right) \tag{3.5}$$

Then

$$f_n(x) = \frac{\nu_{nq} f(\bar{x})}{np_{nq}} + \frac{\nu_{nq} M_q p_{nq}(1 + o(1))}{nf^2(\bar{x})}$$

and

$$f_n(x) - f(x) = \left(\frac{\nu_{nq}}{np_{nq}} - 1\right) f(\bar{x}) + \frac{\nu_{nq} M_q}{nf^2(\bar{x})}(1 + o(1)) p_{nq}$$

$$-(x - \bar{x}) f'(\bar{x}) - \tfrac{1}{2}(x - \bar{x})^2 f''(y)$$

Using (3.5), we may write

$$I_{nq}(f) = \frac{f^2(\bar{x})\lambda_{nq}}{\nu_{nq}} + \frac{f'^2(\bar{x})}{12 f^2(\bar{x})}\left(\frac{\nu_{nq}}{n}\right)^2 \lambda_{nq} + Y_{nq}$$

where

$$Y_{nq} = \left(\left(\frac{\nu_{nq}}{np_{nq}} - 1\right)^2 - \frac{1}{\nu_{nq}}\right) f(\bar{x}) p_{nq}\left(1 + O\left(p_{nq}^2\right)\right) + \frac{f'^2(\bar{x})}{12 f^2(\bar{x})}\lambda_{nq}$$

$$\times \left(p_{nq}^2 - \frac{\nu_{nq}^2}{n^2}\right) + \left[\frac{p_{nq}^2}{f^2(\bar{x})} + 2\frac{\nu_{nq} M_q p_{nq}}{nf^2(\bar{x})}\right]\left(\frac{\nu_{nq}}{n} - p_{nq}\right)$$

$$+ \frac{\nu_{nq}^2}{n^2} O\left(p_{nq}^3\right) + O\left(p_{nq}^4\right)$$

Taking expectations, one finds that

$$J_{nq}(f) = EZ'_{nq} + EZ''_{nq} + O(1/n\nu) + O(\nu/n^2)$$

with

$$Z'_{nq} = f^2(\bar{x})\lambda_{nq}/\nu_{nq}, \qquad Z''_{nq} = (\nu_{nq}/n)^2\left(f'^2(\bar{x})/12 f^2(\bar{x})\right)\lambda_{nq}$$

It follows from (3.4)

$$J_n(f) = EZ'_n + EZ''_n + O\left(\frac{1}{n}\right) \tag{3.6}$$

with

$$Z'_n = \sum_{q=1}^{k} Z'_{nq}, \qquad Z''_n = \sum_{q=1}^{k} Z''_{nq}$$

We now give a lemma which proof is immediate.

Lemma 3.1. Let Z_n be a sequence of i.i.d. r.v. such that $0 < Z_n < A$ and

$$\frac{Z_n}{a_n} \overset{\text{a.c.}}{\to} 1 \quad \text{as} \quad n \to \infty$$

where a_n is a sequence of positive numbers. Then, for each $\varepsilon > 0$,

$$a_n(1-\varepsilon)(1-\varepsilon'_n) < EZ_n < a_n(1+\varepsilon)(1-\varepsilon_n) + A\varepsilon_n$$

where

$$\varepsilon_n = P\{Z_n > a_n(1+\varepsilon)\} \quad \text{and} \quad \varepsilon'_n = P\{Z_n < a_n(1-\varepsilon)\}.$$

The existence of the G.R.I. of f^2 and f'^2/f^2 implies that

$$\frac{n}{k} Z'_n \overset{\text{a.c.}}{\to} \int f^2(x)\, dx \quad \text{and} \quad k^2 Z''_n \overset{\text{a.c.}}{\to} \frac{1}{12} \int \frac{f'^2(x)}{f^2(x)}\, dx. \tag{3.7}$$

To apply the lemma, it remains to prove that Z'_n and Z''_n are bounded. Using (3.5)

$$Z'_{nq} = f(\bar{x}) \frac{p_{nq}}{\nu_{nq}} \left(1 + O\left(p_{nq}^2\right)\right) \le \frac{2Mp_{nq}}{\nu_{nq}}$$

Thus

$$0 \le Z'_n \le 2M/\nu$$

Likewise, f'/f being overbounded, there exists a positive number M' such that

$$Z''_{nq} \le \left(\frac{\nu_{nq}}{n}\right)^2 M' p_{nq}$$

Thus

$$0 \le Z''_n \le 2M'\left(\frac{\nu}{n}\right)^2$$

Applying Lemma 3.1, it follows from (3.6) and (3.7) that

$$J_n(f) = \frac{k}{n}\int f^2(x)\,dx + \frac{1}{12k^2}\int \frac{f'^2(x)}{f^2(x)}\,dx + o\left(\frac{k}{n} + k^{-2}\right)$$

It is easy to see that the asymptotic IMSE is minimized by taking $k = k^*$ given by (2.2).

Proof of Theorem 2.4. Because of (1.1), $\sqrt{n/k}\,P\{(x < X_{(1)}) \cup (X_{(n)} < x)\} \to 0$ as $n \to \infty$, and then it suffices to consider the case when x belongs to $\Pi_{nq}, 1 \le q \le k$.

We have seen in the proof of Theorem 2.1 that

$$p_{nq} = \lambda_{nq} f(x) + M'_q \lambda^2_{nq} \quad \text{with } |M'_q| \le \tfrac{1}{2} \sup_{x \in \Pi_{nq}} |f'(x)|$$

Therefore

$$f_n(x) - f(x) = \left(\frac{\nu_{nq}}{np_{nq}} - 1\right) f(x) + \frac{\nu_{nq}}{np_{nq}} M'_q \lambda_{nq} \tag{3.8}$$

By Chebychev's inequality, we have

$$\frac{\nu_{nq}}{np_{nq}} \to 1 \text{ in probability} \quad \textit{as } \mathrm{n} \to \infty \tag{3.9}$$

and from Theorem 2.1

$$f_n(x) \overset{\text{q.m.}}{\to} f(x) \quad \text{as } n \to \infty.$$

Thus

$$\sqrt{\frac{n}{k}}\frac{\nu_{nq}}{np_{nq}}M_q'\lambda_{nq} = \frac{\nu_{nq}}{np_{nq}}M_q'\sqrt{\frac{n}{k^3}}\,(1 + o(1))\frac{1}{f_n(x)} \to 0 \text{ in probability}$$

as $n \to \infty$

It follows from (3.8) and (3.9) that $\sqrt{n/k}\,(f_n(x) - f(x))$ has the same limit distribution as

$$Z_{nq} = \frac{n}{\sqrt{\nu_{nq}}}\left(\frac{\nu_{nq}}{n} - p_{nq}\right)$$

We have

$$P\{Z_{nq} < t\} = 1 - P\left\{p_{nq} < \frac{\nu_{nq}}{n} - \frac{t\sqrt{\nu_{nq}}}{n}\right\}$$

The density of Z_{nq} is then

$$h(t) = \frac{\sqrt{\nu_{nq}}/n}{\beta(\nu_{nq}, n - \nu_{nq} + 1)}\left(\frac{\nu_{nq}}{n} - \frac{t\sqrt{\nu_{nq}}}{n}\right)^{\nu_{nq}-1}$$

$$\times\left(1 - \frac{\nu_{nq}}{n} + \frac{t\sqrt{\nu_{nq}}}{n}\right)^{n-\nu_{nq}}$$

Using Stirling's approximation yields

$$\log h(t) = -\frac{t^2}{2} - \log\sqrt{2\pi} + O\left(\frac{\nu}{n}\right) + o(1)$$

This concludes the proof.

REFERENCES

Abou-Jaoudé, S. (1977). Sur la convergence L^1 et L^∞ de certains estimateurs d'une densité de probabilité. Thèse de l'Univ. P. & M. Curie.

Anderson, T. W. (1966). Some nonparametric multivariate procedures based on statistically equivalent blocks. In *Multivariate Analysis*, Orlando, Florida: Academic Press, pp. 5–27.

Bauer, W. (1975). On pointwise nonparametric estimation of a density function. *Periodica Math. Hung.* **6**, 59–67.

Bera, M. (1977). Lois limites des écarts extrêmes associés aux histogrammes et à diverses statistiques d'ordre dans l'estimation d'une densité de probabilité. Thèse 3ème cycle de l'Univ. P & M. Curie.

Bertrand-Retali, M. (1978). Convergence uniforme d'un estimateur de la densité par la méthode du noyau. *Rev. Roum. Math. Pures Appl.*, **23**, 361–385.

Freedman, D. and Diaconis, P. (1981). On the histogram as a density estimator: L^2 theory. *Z. Wahrsch. verw. Gebiete* **57**, 453–476.

Gessaman, M. (1970). A consistent nonparametric multivariate density estimator based on statistically equivalent blocks. *Ann. Math. Statist.* **41**, 1344–1346.

Kim, B. K. and Van Ryzin, J. (1975). Uniform consistency of a histogram density estimator and modal estimation. *Comm. Statist.* **4**, 303–315.

Lecoutre, J. P. (1975). Convergence et optimisation de certains estimateurs de la densité. Thèse 3ème cycle de l'Univ. P & M. Curie.

Lecoutre, J. P. (1985). The L^2-optimal cell width for the histogram. *Statist. Prob. Lett.* **3**, 303–306.

Reiss, R. D. (1978). Approximate distribution of the maximum deviation of histograms. *Metrika* **25**, 9–26.

Révész, P. (1972). On empirical density function. *Period. Math. Hung.* **2**, 85–110.

Scott, D. W. (1979). On optimal and data-based histograms. *Biometrika* **66**, 605–610.

Tukey, J. (1947). Nonparametric estimation II. Statistically equivalent blocks and tolerance regions. The continuous case. *Ann. Math. Statist.* **18**, 529–539.

Tusnady, G. (1974). On testing density functions. *Period. Math. Hung.* **5**, 161–169.

Van Ryzin, J. (1970). On a histogram method of density estimation. Technical Report No. 226, Dept. of Statistics, University of Wisconsin.

Van Ryzin, J. (1973). A histogram method of density estimation. *Comm. Statist.* **2**, 493–506.

Wilks, S. S. (1962). *Mathematical Statistics*. New York: Wiley.

The Consistency of Variable-Knot Density Estimates

Armin Schilling and Winfried Stute
Mathematisches Institut
Universität Giessen
Giessen, Federal Republic of Germany

SUMMARY

In this paper we show that variable-knot kernel density estimates are almost surely uniformly convergent at a certain rate. The result is established by proving that they are close to fixed-radii estimators, with a bandwidth function depending on the unknown density f. This expresses the fact that variable-knot estimators are self-adjusting to various level regions of f.

1. INTRODUCTION AND MAIN RESULTS

Kernel estimates have played a dominant role in density estimation since the now classical papers by Rosenblatt (1956) and Parzen (1962). See Wertz (1978) for a survey. For the sake of reference, such estimates are of the form

$$f_n(x) = \int a_n^{-1} K\left(\frac{x-y}{a_n}\right) F_n\, dy$$

where K is a probability kernel, $(a_n)_n$ is a band sequence tending to 0, and

$$F_n(y) = n^{-1} \sum_{i=1}^{n} 1_{(-\infty, y]} \circ \xi_i, \qquad y \in \mathbb{R}$$

is the empirical distribution function pertaining to an independent sample $\xi_1, \ldots, \xi_n$ from an unknown density f to be estimated. The nearest neighbor (NN) kernel estimate is obtained when replacing a_n by $R(x) \equiv R_{k,n}(x)$, the distance of x to its kth nearest neighbor within the sample. Usually $k = k(n) \to \infty$ in such a way that $k/n \to 0$. Moore and Yackel (1977) showed that NN estimates are related to fixed-width estimates, where a_n corresponds to k/n, at least as far as consistency is concerned. As to limit distributions, there is a significant difference, however, in that NN estimates are more efficient than fixed-width estimates in low-density regions.

Nearest-neighbor estimates have the drawback of not being a density. In fact $\int f_n(x)\, dx = \infty$. Victor (1976) and Breiman et al. (1977) independently proposed a so-called variable-knot estimator, which results from replacing the overall bandwidth $R(x)$ by the bandwidth function $y \to R(y)$:

$$f_n(x) = \int R^{-1}(y) K\left(\frac{x-y}{R(y)}\right) F_n\, dy \tag{1.1}$$

It is easily checked that f_n is a true density. Moreover, it is hoped that variable-knot kernel estimates share the property of NN estimates of appropriately responding to variations in the magnitude of f. Though several simulations are reported [see Breiman et al. (1977) and Raatgever and Duin (1978)] we are not aware of any result showing that f_n is consistent.

In this chapter our main result states that $f_n \to f$ uniformly on bounded intervals at a rate that is related to the exact rate of convergence in fixed-radii estimation. This stems from the fact that $f_n(x)$ is seen to be close to

$$\hat{f}_n(x) = \int b_n^{-1}(x) K\left(\frac{x-y}{b_n(x)}\right) F_n\, dy \tag{1.2}$$

where $b_n(x) = k/2nf(x)$. Hence for each x, $\hat{f}_n(x)$ is a fixed-width estimator, but with an a_n unknown to the statistician. This again reflects the property of R-based estimates of being self-adjusting to the level of f.

Throughout the paper we assume that

K is a continuously differentiable probability kernel of bounded support. (K)

$k(n) \in \mathbb{N}$ is such that

$$\lim_{n \to \infty} kn^{-a} = c \quad \text{for some } c > 0 \text{ and } 0 < a < \tfrac{4}{5} \tag{k}$$

Let J be some finite closed interval on which f is to be estimated. We assume that f is twice continuously differentiable on a neighborhood of J, say $U \supset J$, with $0 < m \leq f \leq M < \infty$ there (J)

Theorem 1.1. Under (K), (k), and (J) we have with probability 1

$$\sup_{x \in J} |f_n(x) - \hat{f}_n(x)| = \mathcal{O}\left[(n^{-a} \ln n)^{1/2} + n^{a-1}\right] \tag{1.3}$$

where f_n and $\hat{f}_n$ are defined in (1.1) and (1.2).

Application of results known for fixed-radii estimates yields the following

Corollary 1.1. Under (K), (k), and (J)

$$\sup_{x \in J} |f_n(x) - f(x)| = \mathcal{O}\left[(n^{-a} \ln n)^{1/2} + n^{a-1}\right]$$

almost surely.

2. LEMMAS AND PROOFS

We need the following result on the convergence of the naive NN estimate

$$g_n(x) = \frac{k}{2nR_{k,n}(x)}$$

Lemma 2.1 (Mack, 1983). Under the assumptions of Theorem 1.1 we have

$$\sup_{x \in J} |g_n(x) - f(x)| = \mathcal{O}\left[(k^{-1} \ln n)^{1/2}\right] \quad \text{a.s.}$$

This suggests that $R_{k,n}(x) \to 0$ as $b_n(x) = k/2nf(x)$.

In the following we assume that $K \equiv 0$ outside $[-1, 1]$. Write $\|K\|$ and $\|K'\|$ for the sup-norm of K and K', respectively.

For the proof of Theorem 1.1, we have to consider the sets

$$A_{k,n}(x) = \{y\colon |x - y| \leq R(y)\} \quad \text{where } R(y) \equiv R_{k,n}(y)$$

$$B_n(x) = \{y \in J\colon |x - y| \leq b_n(y)\}$$

$$C_n(x) = \{y\colon |x - y| \leq b_n(x)\}$$

Denote by $Y_{l,k}(x)$ $[Y_{r,k}(x)]$ the kth nearest neighbor of x within those sample points lying left [right] of x. For the sake of definiteness, put $Y_{l,k}(x) = -\infty[Y_{r,k}(x) = +\infty]$ if no such k points exist. Clearly,

$$A_{k,n}(x) \subset \{y\colon Y_{l,k}(x) \le y \le Y_{r,k}(x)\} \tag{2.1}$$

so that each $A_{k,n}(x)$ contains at most $2k$ sample points. Recall $U \supset J$, a neighborhood of J. Let I be any closed interval such that $J \subset I^0 \subset I \subset U$. We need the following lemma.

Lemma 2.2. For almost all $\omega \in \Omega$, there exists $N(\omega) \ge 1$ such that

$$A_{k,n}(x) \subset I \quad \text{for all } x \in J \text{ and } n \ge N \tag{2.2}$$

Proof. By (2.1), it suffices to show $Y_{l,k}(x_l)$, $Y_{r,k}(x_r) \in I$ eventually, where $[x_1, x_r] \equiv J$. As to $Y_{l,k}(x_l)$, we prove that for an arbitrarily small $0 < \delta$

$$x_l - Y_{l,k}(x_l) \le \delta \quad \text{for } n \ge N_1 \text{ (say)} \tag{2.3}$$

with probability 1. Assume on the contrary that for some $\omega \in \Omega$ and some sequence $(n_j)_j$ tending to infinity

$$x_l - Y_{l,k_{n_j}}(x_l) > \delta \quad \text{for all } j$$

We then have (with μ_n and μ being the measure for F_n and f)

$$\mu_{n_j}[x_l - \delta, x_l] \le \mu_{n_j}\Big[Y_{l,k(n_j)}(x_l), x_l\Big] \le k(n_j)/n_j \to 0$$

while on the other hand

$$\mu[x_l - \delta, x_l] \ge m \cdot \delta > 0$$

But $\mu_{n_j} \to \mu$ (uniformly) on the class of intervals with probability 1, a contradiction. In summary this shows (2.3). A similar argument applies to $Y_{r,k}$, so that (2.2) holds. ∎

Let us now consider the sets $B_n(x)$. For trivial reasons,

$$B_n(x) \subset \{y \in J\colon |x - y| \le k/2nm\}, \qquad x \in J$$

Since $k_n/n \to 0$, this gives

$$B_n(x) \subset I \quad \text{for all } x \in J,\ n \ge N'$$

Similarly

$$C_n(x) \subset I \quad \text{for all } x \in J,\ n \geq N''$$

Take some $n \geq \max(N', N'')$. Since $K \equiv 0$ outside $[-1, 1]$, we may write

$$f_n(x) = \int_{A_{k,n}(x)} R^{-1}(y) K\left(\frac{x-y}{R(y)}\right) F_n\, dy$$

and

$$\hat{f}_n(x) = \int_{C_n(x)} b_n^{-1}(x) K\left(\frac{x-y}{b_n(x)}\right) F_n\, dy$$

Check that

$$\begin{aligned}
f_n(x) - \hat{f}_n(x) &= \int_{A_{k,n}(x)} R^{-1}(y)\left[K\left(\frac{x-y}{R(y)}\right) - K\left(\frac{x-y}{b_n(y)}\right)\right] F_n\, dy \\
&\quad + \int_{A_{k,n}(x)} [R^{-1}(y) - b_n^{-1}(y)] K\left(\frac{x-y}{b_n(y)}\right) F_n\, dy \\
&\quad + \int_{A_{k,n}(x)} b_n^{-1}(y) K\left(\frac{x-y}{b_n(y)}\right) F_n\, dy - \int_{B_n(x)} b_n^{-1}(y) K\left(\frac{x-y}{R(y)}\right) F_n\, dy \\
&\quad + \int_{B_n(x)} b_n^{-1}(y)\left[K\left(\frac{x-y}{R(y)}\right) - K\left(\frac{x-y}{b_n(y)}\right)\right] F_n\, dy \\
&\quad + \int_{B_n(x)} b_n^{-1}(y)\left[K\left(\frac{x-y}{b_n(y)}\right) - K\left(\frac{x-y}{b_n(x)}\right)\right] F_n\, dy \\
&\quad + \int_{B_n(x)} [b_n^{-1}(y) - b_n^{-1}(x)] K\left(\frac{x-y}{b_n(x)}\right) F_n\, dy \\
&\quad + \int_{B_n(x)} b_n^{-1}(x) K\left(\frac{x-y}{b_n(x)}\right) F_n\, dy - \int_{C_n(x)} b_n^{-1}(x) K\left(\frac{x-y}{b_n(y)}\right) F_n\, dy \\
&\quad + \int_{C_n(x)} b_n^{-1}(x)\left[K\left(\frac{x-y}{b_n(y)}\right) - K\left(\frac{x-y}{b_n(x)}\right)\right] F_n\, dy \\
&\equiv \mathrm{I} + \mathrm{II} + \cdots + \mathrm{VIII}
\end{aligned}$$

The theorem will be proved by showing that, uniformly in $x \in J$, each of the eight summands admits the indicated bound.

Lemma 2.3. Uniformly in $x \in J$,

$$|\mathrm{I}| = \mathcal{O}\big((n^{-a}\ln n)^{1/2}\big) \quad \text{with probability 1}$$

Proof. Fix some x and apply the mean value theorem to get

$$|\mathrm{I}| \le \|K'\| \int_{A_{k,n}(x)} \left|\frac{x-y}{R(y)}\right| \cdot \left|R^{-1}(y) - b_n^{-1}(y)\right| F_n\, dy$$

But

$$\left|\frac{x-y}{R(y)}\right| \le 1 \quad \text{on } A_{k,n}(x),$$

so that by Lemma 2.1 (applied to I) and Lemma 2.2

$$|\mathrm{I}| \le 4\|K'\| \sup_{y \in I} |g_n(y) - f(y)| = \mathcal{O}\big((n^{-a}\ln n)^{1/2}\big). \quad \blacksquare$$

Lemma 2.4. Uniformly in $x \in J$,

$$|\mathrm{II}| = \mathcal{O}\big((n^{-a}\ln n)^{1/2}\big) \quad \text{with probability 1}$$

Proof. Check that

$$|\mathrm{II}| \le \|K\| \sup_{y \in I} |R^{-1}(y) - b_n^{-1}(y)| 2k/n$$

and use Lemma 2.1 again. ■

Lemma 2.5. Uniformly in $x \in J$,

$$|\mathrm{III}| = \mathcal{O}\big((n^{-a}\ln n)^{1/2}\big) \quad \text{with probability 1}$$

Proof. Putting

$$D_{1,n}(x) = A_{k,n}(x) \setminus B_n(x)$$

$$D_n(x) = A_{k,n}(x) \cap B_n(x)$$

$$D_{2,n}(x) = B_n(x) \setminus A_{k,n}(x)$$

we get

$$|\mathrm{III}| = \left| \int_{D_{1,n}(x)} b_n^{-1}(y) K\left(\frac{x-y}{b_n(y)}\right) F_n\,dy \right.$$

$$+ \int_{D_n(x)} b_n^{-1}(y) \left[K\left(\frac{x-y}{b_n(y)}\right) - K\left(\frac{x-y}{R(y)}\right) \right] F_n\,dy$$

$$\left. + \int_{D_{2,n}(x)} b_n^{-1}(y) K\left(\frac{x-y}{R(y)}\right) F_n\,dy \right|$$

The first and third integrals vanish, since $K \equiv 0$ outside $[-1,1]$. As to the middle summand, apply the mean value theorem and the fact that

$$\left| \frac{x-y}{b_n(y)} \right| \leq 1 \quad \text{on } D_n(x)$$

to get

$$|\mathrm{III}| \leq \|K'\| \int_{A_{k,n}(x)} |R^{-1}(y) - b_n^{-1}(y)| F_n\,dy$$

Proceed as in Lemma 2.3. ■

Lemma 2.6. Uniformly in $x \in J$,

$$|\mathrm{IV}| = \mathcal{O}\left((n^{-a}\ln n)^{1/2}\right) \quad \text{with probability 1}$$

Proof. Observe that

$$|\mathrm{IV}| \leq \|K'\| \int_{B_n(x)} |R^{-1}(y) - b_n^{-1}(y)| F_n\,dy$$

which for all large n is less than or equal to

$$\|K'\| \frac{2n}{k} \sup_{y \in I} |g_n(y) - f(y)| \sup_{x \in J} \int_{C_n'(x)} F_n\,dy$$

with

$$C_n'(x) = \{ y \in J \colon |x-y| \leq k/2nm \}$$

But

$$\int_{C_n'(x)} F_n(dy) = \int_{C_n'(x)} F\,dy + \int_{C_n'(x)} (F_n - F)\,dy$$

For all large n, the first integral is less than or equal to $Mk/nm = \mathcal{O}(k/n)$. The second integral may be bounded by using Theorem 2.15 of Stute (1982a). It follows that

$$\sup_{x \in J} \left| \int_{C_n'(x)} (F_n - F)\,dy \right| = \mathcal{O}\left(\sqrt{n^{a-2}\ln n}\right)$$

with probability 1. In summary, by Lemma 2.1,

$$\begin{aligned} |\mathrm{IV}| &= \mathcal{O}\left(\frac{n}{k}\left(\frac{\ln n}{n^a}\right)^{1/2}\left[\frac{k}{n} + \sqrt{n^{a-2}\ln n}\right]\right) \\ &= \mathcal{O}\left((n^{-a}\ln n)^{1/2}\right). \quad \blacksquare \end{aligned}$$

Lemma 2.7. Uniformly in $x \in J$,

$$|\mathrm{V}| = \mathcal{O}(n^{a-1}) \quad \text{with probability 1}$$

Proof. We have

$$\begin{aligned} |\mathrm{V}| &\leq \frac{2n}{k}\|K'\| \int_{B_n(x)} |f(y) - f(x)| F_n\,dy \\ &\leq \frac{2n}{k}\|K'\| \int_{C_n'(x)} |f(y) - f(x)| F_n\,dy \\ &\leq \sup_{z \in J} |f'(z)| \frac{\|K'\|}{m} \int_{C_n'(x)} F_n\,dy \end{aligned}$$

Argue as in the proof of Lemma 2.6 to get the desired bound. ■

Lemma 2.8. Uniformly in $x \in J$,

$$|\mathrm{VI}| = \mathcal{O}(n^{a-1}) \quad \text{with probability 1}$$

Proof. As for Lemma 2.7. ■

Lemma 2.9. Uniformly in $x \in J$,

$$|\mathrm{VII}| = \mathcal{O}(n^{a-1}) \quad \text{with probability 1}$$

Proof. Similar to the proof of Lemma 2.5, verify that with $E_n(x) = B_n(x) \cap C_n(x)$

$$|\mathrm{VII}| \le \|K'\| \int_{E_n(x)} |b_n^{-1}(x) - b_n^{-1}(y)| F_n \, dy$$

$$\le \frac{2n}{K} \|K'\| \int_{B_n(x)} |f(y) - f(x)| F_n \, dy$$

Now proceed as for V. ■

Lemma 2.10. Uniformly in $x \in J$,

$$|\mathrm{VIII}| = \mathcal{O}(n^{a-1}) \quad \text{with probability 1}$$

Proof. Similar to Lemma 2.7. ■

Collecting error terms gives the proof of the theorem. We are now in a position to give the

Proof of the Corollary. Write

$$|f_n(x) - f(x)| \le |f_n(x) - \hat{f}_n(x)| + |\hat{f}_n(x) - \mathbb{E}\hat{f}_n(x)| + |\mathbb{E}\hat{f}_n(x) - f(x)|$$

Use (1.3) to bound the first summand. The second term may be dealt with as in Stute (1982b), giving

$$\sup_{x \in J} |\hat{f}_n(x) - \mathbb{E}\hat{f}_n(x)| = \mathcal{O}\left(\left(\frac{\ln n}{n^a}\right)^{1/2}\right) \quad \text{with probability 1}$$

Finally, integration by parts and Taylor expansion immediately yield

$$\sup_{x \in J} |\mathbb{E}\hat{f}_n(x) - f(x)| = \mathcal{O}(n^{a-1})$$

completing the proof of the corollary. ■

REFERENCES

Breiman, L., Meisel, W. and Purcell, E. (1977). Variable kernel estimates of multivariate densities. *Technometrics* **19**, 135–144.

Mack, Y. P. (1983). Rate of strong uniform convergence of k-NN density estimates. *J. Statist. Plan. Inf.* **8**, 185–192.

Moore, D. S. and Yackel, J. W. (1977). Consistency properties of nearest neighbor density function estimators. *Ann. Statist.* **5**, 143–154.

Parzen, E. (1962). On estimation of a probability density function and mode. *Ann. Math. Statist.* **33**, 1065–1076.

Raatgever, J. W. and Duin, R. P. W. (1978). On the variable kernel model for multivariate nonparametric density estimation. *COMPSTAT*, 524–594.

Rosenblatt, M. (1956). Remarks on some nonparametric estimates of a density function. *Ann. Math. Statist.* **27**, 832–837.

Stute, W. (1982a). The oscillation behavior of empirical processes. *Ann. Probab.* **10**, 86–107.

Stute, W. (1982b). A law of the logarithm for kernel density estimators. *Ann. Probab.* **10**, 414–422.

Victor, N. (1976). Nonparametric allocation rules. In *Decision Making and Medical Care* (F. T. De Dombal and F. Grémy, Eds.), Amsterdam: North-Holland, pp. 515–527.

Wertz, W. (1978). *Statistical density estimation. A survey*. Göttingen: Vandenhoeck und Ruprecht.

Invariant Density Estimators in Homogeneous Spaces

Wolfgang Wertz
Institut für Statistik und Wahrscheinlichkeitstheorie
Technische Universität
Vienna, Austria

SUMMARY

In this chapter we treat the problem of estimating an unknown probability density that is defined on a homogeneous space, based on n independent observations. It is proved that, under certain conditions, for every density estimator $\hat{f}$ an invariant density estimator exists whose maximal risk does not exceed the maximal risk of $\hat{f}$. The proof of this fact uses the fixed-point property of amenable groups.

1. NOTATION AND INTRODUCTION

Let X be a locally compact Hausdorff space, $\mathfrak{X}$ the σ-algebra of its Borel sets, G a σ-compact locally compact group with Borel sets $\mathfrak{G}$. We shall assume G to have the Lindelöf property. Let G act transitively on X, that is, there is a mapping $(g, x) \mapsto gx$ from $G \times X$ to X, continuous separately in each variable, with $g_1(g_2x) = (g_1g_2)x$ and $ex = x$ (e denotes the unit element of G). Furthermore,

$$\forall x_1, x_2 \in \mathfrak{X} \quad \exists g \in G: \quad x_2 = gx_1$$

It is known that these properties imply the continuity of the mapping

$(g, x) \mapsto gx$ (Ellis, 1957) and furthermore, $x \mapsto gx$ is a homeomorphism for every $g \in G$. Let $gA := \{gx : x \in A\}$ for $g \in G$ and $A \in \mathfrak{X}$.

Under these conditions X is a homogeneous space and can be represented as G/H with a closed subgroup H of G. We assume the existence of a relatively invariant measure μ on $(X, \mathfrak{X})$, that is,

$$\mu(gA) = \chi(g)\mu(A) \quad \forall g \in G, \forall A \in \mathfrak{X}$$

with a fixed positive function χ that is necessarily a continuous extension to G of the quotient δ/Δ of the modular functions δ and Δ of H and G, respectively (Gaal, 1973, Theorem V.3.15). Let $\mu(K) < \infty$ for every compact set K; hence μ is σ-finite.

For any measurable function φ we denote the left and right translate by ${}_g\varphi(x) := \varphi(gx)$ and $\varphi_g(x) := \varphi(xg)$, respectively. If $(Y, \mathfrak{Y}, \eta)$ is an arbitrary measure space, we shall write usually $L_p(\eta)$ instead of $L_p(Y, \mathfrak{Y}, \eta)$ for the L_p-space. A distinction between equivalence classes and individual functions will be made only in case of doubt. $\|\cdot\|_p = \|\cdot\|_{p,\eta}$ denotes the norm of $L_p(\eta)$.

Let $p > 1$ be a fixed number, $f \in L_p(\mu)$ a probability density with respect to μ, ${}^gf := \chi(g^{-1})_{g^{-1}}f$, which means $x \mapsto {}^gf(x) = \chi(g^{-1}) \cdot f(g^{-1}x)$ for every $g \in G$, P_g the corresponding probability measures, and $P := P_e$. We consider the problem of estimating ${}^gf$ based on n independent observations, say $x^n := (x_1, \ldots, x_n)$ of a random variable distributed with density ${}^gf$. Of course, G acts on X^n by $gx^n := (gx_1, \ldots, gx_n)$. Let

$$f_n(x^n) := \prod_{i=1}^{n} f(x_i)$$

and hence

$$({}^gf)_n(x^n) = \chi^n(g^{-1}) \prod_{i=1}^{n} f(g^{-1}x_i)$$

We consider now only strongly $\mathfrak{X}^n$-measurable functions $\tilde{f}: X^n \mapsto B := L_p(\mu)$. Write $\Phi(\psi) := \int |\psi|^p \, d\mu \; \forall \psi \in L_p(\mu)$ and

$$L_{p,g}(B) := \left\{\tilde{f}: \int \Phi \circ \tilde{f} dP_g^n < \infty\right\}$$

Every $\tilde{f} \in \mathfrak{L}_p := \bigcap_{g \in G} L_{p,g}(B)$ has a representation $\hat{f}(x^n, x) = [\tilde{f}(x^n)](x)$ by an $\mathfrak{X}^n \otimes \mathfrak{X}$-measurable function $\hat{f}: X^n \times X \mapsto X$ such that $\int_{X^n \times X} |\hat{f}|^p \, dP_g^n \otimes \mu < \infty$. For every $g \in G$, this means that $\hat{f}$ belongs to

$\bigcap_{g \in G} L_p(P_g^n \otimes \mu)$. In the following the symbol $\mathfrak{L}_p$ will be used for the corresponding space of functions $\hat{f}$ as well.

The *σ-topology* is the weakest topology in $\mathfrak{L}_p$ such that all linear functionals of the form

$$l(\hat{h}, P_g)(\hat{f}) = \int_{X^n}\left[\int_X \hat{f}(x^n, x) \cdot \hat{h}(x^n, x)\, d\mu(x)\right] dP_g^n(x^n)$$

with $\hat{h} \in L_q(P_g^n \otimes \mu)$ and $g \in G$ are continuous. The level sets

$$\left\{\hat{f} \in \mathfrak{L}_p \colon \iint |\hat{f}(x^n, x) - \alpha(x; g)|^p\, d\mu(x)\, dP_g^n(x^n) \le \beta(g)\ \forall g \in G\right\}$$

are σ-compact for any function $\beta : G \mapsto [0, \infty)$ and function $\alpha : X \times G \mapsto \mathrm{R}$ with $\alpha(\cdot\,; g) \in L_p(\mu)\ \forall g \in G$. Furthermore, the mappings

$$\hat{f} \mapsto \iint |\hat{f}(x^n, x) - \alpha(x; g)|^p\, d\mu(x)\, dP_g^n(x^n)$$

are σ-lower semicontinuous.

The results quoted above are special cases of a general theory developed in Kozek (1976, 1980) and Kozek and Wertz (1981, 1982). We emphasize that, for sake of brevity, the discussion has been informal here.

Any $\tilde{f} \in \mathfrak{L}_p$ (and hence its representation $\hat{f}$ as a function of $n + 1$ variables) is called a *density estimator*. A density estimator $\hat{f}$ is said to be *χ-invariant* if

$$\hat{f}(x^n, x) = \chi(g) \cdot \hat{f}(gx^n, gx) \quad \forall g \in G\ \forall x^n \in X^n\ \forall x \in X$$

holds. The *risk* of $\hat{f}$ at P_g will be defined by

$$R(\hat{f}, g) := \chi^{p-1}(g) \cdot \int_{X^n \times X} |\hat{f} - {}^g f|^p\, dP_g^n \otimes \mu$$

$$= \chi^{p-1}(g) \int_{X^n}\int_X |\hat{f}(x^n, x) - \chi(g^{-1}) f(g^{-1}x)|^p\, d\mu(x) \chi^n(g^{-1})$$

$$\times \prod_{i=1}^{n} f(g^{-1}x_i)\, d\mu^n(x^n)$$

Since $\int f^p\, d\mu < \infty$ by assumption, $R(\hat{f}, g) < \infty$ is equivalent to $\hat{f} \in L_p(P_g^n \otimes \mu)$.

The following theorem can be generalized to more general curve-estimation problems; however, for the sake of simplicity we restrict ourselves to the case of density estimation. A similar result has been proved in Wertz (1976) under different assumptions by application of the Hunt–Stein method. The formulation given here seems to be more natural from the point of view of statistics.

2. EXISTENCE OF χ-INVARIANT ESTIMATORS

First a few facts from the theory of amenable groups are needed; for details, refer to Pier (1984) or Bondar and Milnes (1981).

Let G be any locally compact group and S a compact, convex subset of a locally convex Hausdorff topological vector space E. Let G act on S [here the action is denoted by $(g, s) \mapsto T_g(s) = gs$] and let the mapping $s \mapsto T_g(s)$ be affine, that is, $T_g(\alpha s_1 + (1-\alpha)s_2) = \alpha T_g(s_1) + (1-\alpha)T_g(s_2)$ for every $\alpha \in [0,1]$ and $s_1, s_2 \in S$. Then the pair (G, S) is called an *affine flow*.

The group G has the *fixed-point property* if every affine flow (G, S) has a fixed point, that is, $\exists s_0 \in S$: $T_g(s_0) = s_0 \ \forall g \in G$.

Lemma 2.1 (Pier, 1984, Theorem 5.4). Every amenable group has the fixed-point property.

Theorem 2.1. Let the assumption specified in Lemma 2.1 be satisfied and

$$F_p := \left\{ \hat{f} \in \mathfrak{L}_p \colon \sup_{g \in G} R(\hat{f}, g) < \infty \right\}.$$

Furthermore, let G be σ compact and amenable and $g \mapsto {}_g f$ L_1-continuous. That means

$$\lim_{g \mapsto e} \int_X |{}_g f(x) - f(x)| \, d\mu(x) = 0 \tag{2.1}$$

Then for every $\hat{f} \in F_p$, there exists a χ-invariant $\hat{f}_0 \in F_p$ with

$$\sup_{g \in G} R(\hat{f}_0, g) \le \sup_{g \in G} R(\hat{f}, g)$$

Proof. Let $\hat{f}_1 \in F_p$ be fixed. We define $[T_g(\hat{f})](x^n, x) := \chi(g) \cdot \hat{f}(gx^n, gx)$. It is clear that the mapping T_g is affine for every $g \in G$. We further set

$$S := \left\{ \hat{f} \in \bigcap_{g \in G} L_p\left(P_g^n \otimes \mu\right) : R(\hat{f}, h) \le \sup_{g \in G} R(\hat{f}_1, g) \ \forall h \in G \right\}$$

First, we show that $\hat{f} \in S$ implies $T_g \hat{f} \in S$ for every $g \in G$. Now, for every $h \in G$ we have

$$\chi^{p-1}(h) \cdot \int_{X^n \times X} |T_g(\hat{f}) - {}^h f|^p \, dP_h^n \otimes \mu$$

$$= \chi^{p-1}(h) \int_{X^n} \int_X |\chi(g) \hat{f}(gx^n, gx) - \chi(h^{-1}) f(h^{-1}x)|^p \cdot \chi^n(h^{-1})$$

$$\cdot \prod_{i=1}^n f(h^{-1}x_i) \, d\mu^n(x^n) \, d\mu(x)$$

$$= \chi^{p-1}(gh) \int_{X^n} \int_X |\hat{f}(x^n, x) - \chi(h^{-1}g^{-1}) f(h^{-1}g^{-1}x)|^p \cdot \chi^n(h^{-1}g^{-1})$$

$$\cdot \prod_{i=1}^n f(h^{-1}g^{-1}x_i) \, d\mu^n(x^n) \, d\mu(x)$$

$$= \chi^{p-1}(gh) \int_{X^n \times X} |\hat{f} - {}^{gh} f|^p \, dP_{gh}^n \otimes \mu \le \sup_{g \in G} R(\hat{f}_1, g)$$

Furthermore, S is compact in the σ-topology. The convexity of S is trivial.

Second, the σ-continuity of $\hat{f} \mapsto T_g \hat{f}$ for every $g \in G$ follows easily from the definition of the σ-topology: let $g \in G$, $\hat{h} \in L_q(P_h)$, and $\hat{f}_1, \hat{f}_2 \in F_p \subseteq \mathfrak{L}_p$; then $l(\hat{h}, P_h)(T_g \hat{f}_1 - T_g \hat{f}_2) = \int_{g^{-1}} \hat{h}(\hat{f}_1 - \hat{f}_2) \, dP_{gh}^n \otimes \mu$. Now ${}_{g^{-1}}\hat{h} \in L_q(P_{gh}^n \otimes \mu)$ and $|l({}_{g^{-1}}\hat{h}, P_{gh})(\hat{f}_1 - \hat{f}_2)| < \varepsilon$ implies $|l(\hat{h}, P_h)(T_g \hat{f}_1 - T_g \hat{f}_2)| < \varepsilon$.

Third, one has to prove the continuity of the mapping $g \mapsto T_g(\hat{f})$ for every $\hat{f} \in F_p$.

By definition of the σ-topology, it is sufficient to verify the continuity of $g \mapsto l(\hat{h}, P_h)[T_g(\hat{f})]$ for every $\hat{h} \in L_q(P_h^n \otimes \mu)$. Obviously, it suffices to establish this for $h = e$. For fixed $\varepsilon > 0$, there is an $\hat{h}_0 \in L_q(P^n \otimes \mu)$ such that

$$\hat{h}_0 \text{ has compact support } K, \qquad \|\hat{h}_0 - \hat{h}\|_q := \|\hat{h}_0 - \hat{h}\|_{q, P^n \otimes \mu} < \varepsilon \quad (2.2)$$

Of course, $\hat{h}_0$ is bounded.

We put $\psi(g) := \chi(g) \cdot \int_{X^n \times X} {}_g\hat{f} \cdot \hat{h}\, dP^n \otimes \mu$ and consider

$$
\begin{aligned}
|\psi(g) - \psi(e)| & \\
&= \left| \int_{X^n \times X} [\chi(g)\hat{f}(gx^n, gx) - \hat{f}(x^n, x)] \right. \\
&\qquad \left. \cdot \hat{h}(x^n, x) \cdot f_n(x^n)\, d\mu^n \otimes \mu(x^n, x) \right| \\
&= \left| \int_{X^n \times X} \hat{f} \cdot \left[{}_{g^{-1}}\hat{h}({}^g f)_n - \hat{h} f_n \right] d\mu^n \otimes \mu \right| \\
&\le \int_{X^n \times X} |\hat{f}| \cdot |{}_{g^{-1}}\hat{h}({}^g f)_n - {}_{g^{-1}}\hat{h}_0({}^g f)_n|\, d\mu^n \otimes \mu \\
&\quad + \int_{X^n \times X} |\hat{f}| \cdot |{}_{g^{-1}}\hat{h}_0({}^g f)_n - \hat{h}_0 f_n|\, d\mu^n \otimes \mu \\
&\quad + \int_{X^n \times X} |\hat{f}| \cdot |\hat{h}_0 f_n - \hat{h} f_n|\, d\mu^n \otimes \mu \\
&=: I_1 + I_2 + I_3
\end{aligned}
$$

We estimate the three integrals I_1, I_2, and I_3 separately.

$$
I_1 = \int |\hat{f}| \cdot |{}_{g^{-1}}\hat{h} - {}_{g^{-1}}\hat{h}_0|\, dP_g^n \otimes \mu \le \|f\|_{p,\, P_g^n \otimes \mu} \cdot \|{}_{g^{-1}}\hat{h} - {}_{g^{-1}}\hat{h}_0\|_{q,\, P_g^n \otimes \mu}.
$$

Since

$$
\|{}_{g^{-1}}\hat{h} - {}_{g^{-1}}\hat{h}_0\|_{q,\, P_g^n \otimes \mu}^q = \chi(g) \cdot \|\hat{h} - \hat{h}_0\|_q^q,
$$

since χ is continuous and by (2.2) there is a neighborhood U of $e \in G$ such that $I_1 < C_1 \varepsilon$ for every $g \in U$. By the same consideration $I_3 < C_2 \varepsilon$ with a suitable constant C_2.

$$
\begin{aligned}
I_2 &= \int_{X^n \times X} |\hat{f}| \cdot |{}_{g^{-1}}\hat{h}_0({}^g f)_n - \hat{h}_0 f_n|\, d\mu^n \otimes \mu \\
&\le \int_{X^n \times X} |\hat{f}|\, |{}_{g^{-1}}\hat{h}_0({}^g f)_n - \hat{h}_0({}^g f)_n|\, d\mu^n \otimes \mu \\
&\quad + \int_{X^n \times X} |\hat{f}|\, |\hat{h}_0({}^g f)_n - \hat{h}_0 f_n|\, d\mu^n \otimes \mu
\end{aligned} \tag{2.3}
$$

The first term of (2.3) equals

$$\chi^{1/q}(g^{-1}) \cdot \int_{X^n \times X} \chi^{1/q}(g) |\hat{f}| \cdot |_{g^{-1}}\hat{h}_0 - \hat{h}_0| \cdot ({}^g f)_n \, d\mu^n \otimes \mu$$

$$\leq \left[\int_{X^n \times X} \chi^{p-1}(g) |\hat{f}|^p \, dP_g^n \otimes \mu \right]^{1/p}$$

$$\cdot \left[\int_{X^n \times X} \chi(g^{-1}) |_{g^{-1}}\hat{h}_0 - \hat{h}_0|^q \cdot ({}^g f)_n \, d\mu^n \otimes \mu \right]^{1/q}$$

$$\leq \left[\sup_{g \in G} \chi^{p-1}(g) \cdot \|\hat{f}\|^p_{p, P_g^n \otimes \mu} \right]^{1/p} \cdot \left[\int |\hat{h}_0 - {}_g\hat{h}_0|^q f_n \, d\mu^n \otimes \mu \right]^{1/q}$$

The first factor is finite. Since $\hat{h}_0$ has compact support K, for every compact neighborhood V of $e \in G$, ${}_g\hat{h}_0$ vanishes outside the compact set VK for every $g \in V$. (Without restricting the generality, we can assume U to be compact.) The second factor is bounded by

$$\left[\int |\varepsilon_1 I_{VK}|^q f_n \, d\mu^n \otimes \mu \right]^{1/q} \leq \left[\int \varepsilon_1^q I_{\pi(VK)} \, d\mu \cdot \int f_n \, d\mu^n \right]^{1/q} < \varepsilon$$

provided $\varepsilon_1 > 0$ is chosen properly and V such that $|\hat{h}_0 - {}_g\hat{h}_0| < \varepsilon_1$ for every $g \in V$. This is possible because of the uniform continuity of $\hat{h}_0$. Here and henceforth I denotes indicator functions and $\pi: X^n \times X \to X$ the projection mapping.

The second term of (2.3) can be estimated as follows:

$$\left[\int_{X^n \times X} |\hat{f}|^p |({}^g f)_n - f_n| \, d\mu^n \otimes \mu \right]^{1/p} \cdot \left[\int_{X^n \times X} |\hat{h}_0|^q |({}^g f)_n - f_n| \, d\mu^n \otimes \mu \right]^{1/q}$$

$$\leq \left[\|\hat{f}\|^p_{p, P_g^n \otimes \mu} + \|\hat{f}\|^p_P \right]^{1/p}$$

$$\cdot \left[\|\hat{h}_0\|^q_\infty \cdot \mu[\pi(VK)] \cdot \int |({}^g f)_n - f_n| \, d\mu^n \right]^{1/q}$$

By the assumptions of $\hat{f}$, the continuity of χ and condition (2.1) this can be made smaller than ε by choosing g in an appropriately small neighborhood of e.

Hence $|\psi(g) - \psi(e)|$ can be made arbitrary small if g is chosen in a sufficiently small compact neighborhood of e. Therefore ψ is continuous.

Since all conditions of the lemma are satisfied, there is an $\hat{f} \in F_p$ with $T_g \hat{f} = \hat{f}$ for all $g \in G$. However, one has to be careful with the interpretation of this result, because it refers to equivalence classes of functions. In fact it reads as follows:

$$\forall g \in G \; \exists N_g \in \mathfrak{X}^n \otimes \mathfrak{X}:$$

$$P_h^n \otimes \mu(N_g) = 0 \quad \forall h \in G, \chi(g)\hat{f}(gx^n, gx) = \hat{f}(x^n, x)$$

$$\forall (x^n, x) \in X^n \times X - N_g$$

The following argument, yielding in fact a χ-invariant estimator, is a generalization of the method presented in Wertz (1976). Refer to this paper for details.

Choose a probability measure λ that is equivalent to left Haar measure on G. Then $\lambda(A) = 0$ implies $\lambda(gA) = \lambda(Ag) = 0$ for every set $A \in \mathfrak{G}$. Let us denote $z := (x^n, x)$, $Z := X^n \times X$, and $gz := (gx^n, gx)$. Furthermore,

$$I := \left\{ z \in Z : \int_G \chi(h)\hat{f}(hz)\, d\lambda(h) = \chi(g)\hat{f}(gz) \text{ for } \lambda\text{-almost all } g \in G \right\}$$

$$= \left\{ z \in Z : \int_G \left| \int_G \chi(h)\hat{f}(hz)\, d\lambda(h) - \chi(g)\hat{f}(gz) \right| d\lambda(g) = 0 \right\}.$$

The function $z \mapsto \int_G \chi(h)\hat{f}(hz)\, d\lambda(h)$ is defined and finite for $P_g^n \otimes \mu$-almost all $z \in Z$ and for every $g \in G$ and the set I is measurable. Hence the function $\hat{f}_0$ defined by

$$\hat{f}_0(z) := \begin{cases} \int_G \chi(g)\hat{f}(gz)\, d\lambda(g) & \text{if } z \in I \\ 0 & \text{if } z \notin I \end{cases}$$

is $\mathfrak{X}^n \otimes \mathfrak{X}$ measurable.

Let $D := \{(z, g) \in Z \times G : \chi(g)\hat{f}(gz) \neq \hat{f}(z)\}$. Then the sections of D are $D_z = \{g \in G : \chi(g)\hat{f}(gz) \neq \hat{f}(z)\}$ and $D_g = \{z \in Z : \chi(g)\hat{f}(gz) \neq \hat{f}(z)\}$. Since clearly $D_g \subset N_g$, we have:

$$\int_Z \lambda(D_z)\, dP_g^n \otimes \mu(z) = P_g^n \otimes \mu \otimes \lambda(D) = \int_G P_g^n \otimes \mu(D_h)\, d\lambda(h) = 0$$

Hence $N := \{z \in Z : \lambda(D_z) \neq 0\}$ has $P_g^n \otimes \mu$-measure zero for every $g \in G$.

Since it is easily proved that $\mathbf{C}N \subset I$, for every $z \in \mathbf{C}N$,

$$\hat{f}_0(z) = \int_G \chi(h)\hat{f}(hz)\,d\lambda(h) = \int_{\mathbf{C}D_z} \chi(h)\hat{f}(hz)\,d\lambda(h)$$

$$= \int_{\mathbf{C}D_z} \hat{f}(z)\,d\lambda(h) = \hat{f}(z)$$

follows and $\hat{f}_0$ and $\hat{f}$ coincide almost everywhere.

By definition of I:

$$\forall z \in I\ \exists a(z) \in R \wedge \exists N_z \in \mathfrak{G}:$$

$$\lambda(N_z) = 0, \quad \chi(g)\hat{f}(gz) = a(z)\ \forall g \in \mathbf{C}N_z$$

hence

$$\forall z \in I\ \forall g \in G\ \forall h \in (\mathbf{C}N_z)g^{-1}: \quad \chi(hg)\hat{f}(hgz) = a(z)$$

If $z \in I$ and $g \in G$, then for every $h \in (\mathbf{C}N_z)g^{-1}$ we have

$$\chi(h)\hat{f}(hgz) = \chi(g^{-1})a(z) = \chi(g^{-1}) \cdot \int_{(\mathbf{C}N_z)g^{-1}} a(z)\,d\lambda(h)$$

$$= \chi(g^{-1}) \cdot \int_{(\mathbf{C}N_z)g^{-1}} \chi(hg)\hat{f}(hgz)\,d\lambda(z)$$

$$= \int_G \chi(h)\hat{f}[h(gz)]\,d\lambda(h)$$

Since $\lambda(N_z g^{-1}) = 0$ this implies that gz belongs to I: hence I is invariant with respect to G in this sense.

Finally, we establish the χ invariance of $\hat{f}_0$: If $z \notin I$, then $gz \notin I$ and $\chi(g)\hat{f}_0(gz) = 0 = \hat{f}_0(z)$. On the other hand, $z \in I$ implies $gz \in I$ and

$$\chi(g)\hat{f}_0(gz) = \chi(g)\int_G \chi(h)\hat{f}(hgz)\,d\lambda(h) = \int_G \chi(h)\hat{f}(hz)\,d\lambda(h) = \hat{f}_0(z)$$

The risks of $\hat{f}_0$ and of $\hat{f}$ are equal because $\hat{f}_0 = \hat{f}[P_g^n \otimes \mu]$ for every $g \in G$. ■

REMARK 2.1. Condition (2.1) is fulfilled if X is a topological group and μ Haar measure. It is easy to see that (2.1) is equivalent to the L_1-continuity of $g \mapsto {}^g f$ in the case X is homogeneous.

REMARK 2.2. The choice of the (weighted) risk $R(\hat{f}, g)$ is natural in this connection. Otherwise, if $\chi \not\equiv 1$, the maximum risk of χ-invariant estimator turns out to be infinite in general (Wertz, 1979).

REMARK 2.3. Under our conditions the existence of a minimax density estimator follows (Kozek and Wertz, 1981, 1982). Hence the theorem implies the existence of a χ-invariant minimax estimator. On the other hand, it is possible to construct density estimators minimizing the risk $\hat{f} \mapsto R(\hat{f}, g)$ for every $g \in G$ (Wertz, 1979, 1982, 1985). Under the above assumptions these estimators are minimax estimators. Their Bayes properties are discussed in Wertz (1982, 1985).

REFERENCES

Bondar, J., and Milnes, P. (1981). Amenability: a survey for statistical applications of Hunt-Stein and related conditions on groups. *Z. Wahrsch. Verw. Gebiete* **57**, 103–128.

Ellis, R. (1957). Locally compact transformation groups. *Duke Math. J.* **24**, 119–125.

Gaal, S. (1973). *Linear Analysis and Representation Theory*. Grundlehren Math. Wiss. Vol. 108, Berlin, Heidelberg, and New York: Springer-Verlag.

Kozek, A. (1976). Orlicz spaces of functions with values in Banach spaces. *Comment. Math. Prace Mat.* **19**, 259–288.

Kozek, A. (1980). Convex integral functionals on Orlicz spaces. *Comment. Math. Prace Mat.* **21**, 109–135.

Kozek, A., and Wertz, W. (1981). On the existence of minimal complete classes of estimators. *The First Pannonian Symposium on Math. Statistics* (*Bad Tatzmannsdorf*, 1979), Lecture Notes in Statist., **8**, Berlin, Heidelberg and New York: Springer-Verlag, pp. 140–155.

Kozek, A. and Wertz, W. (1982). Estimators with values in Banach spaces under convex loss and applications to curve estimation problems. *Nonparametric statistical inference*, Vol. I, II, Colloq. Math. Soc. János Bolyai, **32**, Amsterdam: North-Holland, pp. 567–586.

Pier, J.-P. (1984). *Amenable Locally Compact Groups*. New York: Wiley.

Wertz, W. (1976). Invariant density estimation. *Monatsh. Math.* **81**, 315–324.

Wertz, W. (1979). Über ein nichtparametrisches Schätzproblem. *Metrika* **26**, 157–167.

Wertz, W. (1982). Invariantly optimal curve estimators with respect to integrated mean error risk. *Probability and Statistical Inference*, Dordrecht and Boston, Massachusetts: Reidel, pp. 373–384.

Wertz, W. (1985). On invariant curve estimators. *Probab. Math. Statist.* **5**, 31–44.

PART 5

Testing and Estimation

A Test of Mixture of Two Known Populations

M. Salomé E. Cabral Neto
Departamento de Estatística
Centro de Estatística e Aplicacoès
Faculdade de Ciências
Universidade de Lisboa
Lisbon, Portugal

SUMMARY

In some experiments the analysis of a sample (frequency polygon) seems to suggest the existence of a mixture of two populations. In these cases arises the problem of deciding about the hypothesis of mixture.

Based on the estimator T_n, developed for the mixture parameter θ of two populations when we have a scale mixture with known parameters δ_1 and δ_2, or a location mixture with known parameters λ_1 and λ_2, a test is devised for that purpose; the asymptotic power function is given and some examples are analyzed.

1. INTRODUCTION

In Cabral Neto (1984) an estimator T_n was developed for the parameter θ of a mixture of two populations in the cases where the distribution function of the mixture is given by

$$F(x; \theta, \delta_1, \delta_2) = \theta F(x/\delta_1) + (1 - \theta) F(x/\delta_2) \qquad 0 \le \theta \le 1 \quad \text{(scale mixture)} \quad (1.1a)$$

$$F(x; \theta, \lambda_1, \lambda_2) = \theta F(x - \lambda_1) + (1 - \theta) F(x - \lambda_2) \quad \text{(location mixture)} \quad (1.1b)$$

where δ_1 and λ_1 are known; $\delta_1 > \delta_2$; $\lambda_1 < \lambda_2$; δ_2, λ_2 are also known and with the conditions

$$\text{(i)} \qquad \lambda_1 = \lambda_2 = \lambda \text{ known or } \lambda = 0 \text{ for simplicity} \tag{1.2a}$$

$$\text{(ii)} \qquad \delta_1 = \delta_2 = \delta \text{ known or } \delta = 1 \text{ for simplicity} \tag{1.2b}$$

Based on this estimator we can test if

$$\theta = 0 \quad \text{or} \quad \theta = 1 \quad \text{or} \quad 0 < \theta < 1 \tag{1.3}$$

that is, to decide if our sample is coming from one population only ($\theta = 0$ or $\theta = 1$) or from a mixture of two populations ($0 < \theta < 1$). The test is used very easily; we may say that is *computable*, and its domain of application is large. This work is divided in three sections. In Section 2 we give the construction of the test and the decision rule as well as the asymptotic power function in both cases (1.1a) and (1.1b). Some examples are analyzed in Section 3. We may note yet that in regard to test problems as well as to estimate problems for mixture, a very important concept is the one of *identifiability*. Before the construction of a test procedure is tried, it must be verified that the alternative hypothesis is identifiable. For scale and location mixtures, Behboodian (1975) proves that finite location mixtures are identifiable. He also proves that finite scale mixtures are identifiable if the distribution $F(z)$ has rth moment for some $r > 0$.

We have also been influenced by Tiago de Oliveira (1963) and Everitte and Hand (1981).

2. TEST OF MIXTURE

2.1. Test of Mixture for Scale Mixture

Let $(X_1, \ldots, X_n)$ be independent and identically distributed (i.i.d.) random variables with distribution function given by (1.1a), where δ_1 and δ_2 are known and it is assumed, without loss of generality,

1. $\delta_1 > \delta_2$
2. $\lambda_1 = \lambda_2 = 0$

Let us now consider the transformation

$$Y = X^{a_n} \tag{2.1}$$

where a_n will be a convenient function of n, $a_n > 0$ and $a_n \to +\infty$ and the statistics

$$T_n = \frac{\bar{Y}_n}{A_n} \tag{2.2}$$

where

$$\bar{Y}_n = \frac{1}{n}\sum_{i=1}^{n} X_i^{a_n}, \qquad A_n = \delta_1^{a_n}\mu(a_n),$$

and

$$\mu(t) = \int_{-\infty}^{+\infty} z^t\, dF(z), \qquad z > 0 \tag{2.3}$$

As we have proved in Cabral Neto (1984),

Theorem 2.1. The statistics T_n is a consistent estimator of θ if

$$\lim_n \frac{\mu(2a_n)}{n\mu^2(a_n)} = 0 \tag{2.4}$$

Theorem 2.2. The estimator T_n is asymptotically normal with mean $\mu(\theta) = [\theta + (1-\theta)Q_n]$ and variance $\sigma_n^2(\theta)/(nA_n^2)$ if

$$\lim_n \frac{\mu'_{3,n} - 3\mu'_{1,n}\sigma_n^2(\theta) - (\mu'_{1,n})^3}{\sqrt{n}\,\sigma_n^2(\theta)} = 0 \tag{2.5}$$

Theorem 2.3. The estimator T_n is asymptotically normal with mean θ $(\theta \neq 0)$ and variance $\sigma_n^2(\theta)/(nA_n^2)$ if:

1. $\lim_n(\mu'_{3,n} - 3\mu'_{1,n}\sigma_n^2(\theta - (\mu'_{1,n})^3)/(\sqrt{n}\,\sigma_n^3(\theta)) = 0.$
2. $\lim_n \mu^2(a_n)/\mu(2a_n) = c, \quad 0 \le c < 1.$
3. $\lim_n \sqrt{n}\, Q_n = 0$

where

$$\mu'_{k,n} = \left[\theta + (1-\theta)Q_n^k\right]\delta_1^{a_n}\mu(ka_n), \qquad \sigma_n^2(\theta) = \mu'_{2,n} - \mu'^2_{1,n},$$

$$Q_n = \left(\frac{\delta_2}{\delta_1}\right)^{a_n}$$

All these conditions, of course, are restrictions on $\{a_n\}$ that will be determinated in each example, as we shall see in Section 3.

On the basis of this estimator we can test the trilemma

$$\theta = 0 \quad \text{or} \quad \theta = 1 \quad \text{or} \quad 0 < \theta < 1 \tag{2.6}$$

to what we suggest the decision rule:

If

$$T_n \le t'_n \tag{2.7a}$$

we accept the hypothesis $\theta = 0$;

if

$$t'_n < T_n \le t_n \tag{2.7b}$$

we accept the hypothesis of existence of mixture, $0 < \theta < 1$;

if

$$T_n > t_n \tag{2.7c}$$

we accept the hypothesis $\theta = 1$ with $0 < t'_n < t_n < 1$.

The problem now is to determine the values t'_n and t_n. Making use of Theorem 2.2 we know that

$$\frac{\sqrt{n}\,(T_n - \mu(\theta))}{\sigma_n(\theta)/A_n} \approx N(0,1) \tag{2.8}$$

is asymptotically $N(0,1)$, where $N(0,1)$ means the normal distribution with mean 0 and variance 1 if (2.5) is verified. If 1, 2, and 3 of Theorem 2.3 hold, we have ($\theta \neq 0$)

$$\frac{\sqrt{n}\,(T_n - \theta)}{\sigma_n(\theta)/A_n} \approx N(0,1) \tag{2.9}$$

As our objective is to give a test whose domain of application is as large as possible and the size n of the sample a "reasonable" value, we decide to make use of (2.8), obviously.

Let us fix then an asymptotic level of significance α, and take λ_α, the solution of the equation

$$N(y) = \int_{-\infty}^{y} \frac{1}{\sqrt{2\pi}} \exp(-x^2/2)\, dx = 1 - \alpha \tag{2.10}$$

$\lambda_\alpha + \lambda_{1-\alpha} = 0$ and $\lambda_\alpha > 0$ if $\alpha > \frac{1}{2}$.

Let

$$P_R(\theta) = \text{Probability}\{\text{decided mixture}|\theta\}$$

We must have

$$1 - P_R(0) = 1 - \alpha \quad \text{and} \quad 1 - P_R(1) = 1 - \alpha \tag{2.11}$$

The asymptotic solution of (2.11) is given by

$$t_n = \mu(1) - \frac{\lambda_\alpha \sigma_n(1)}{\sqrt{n}\, A_n} \quad \text{and} \quad t'_n = \mu(0) + \frac{\lambda_\alpha \sigma_n(0)}{\sqrt{n}\, A_n}$$

with $t_n \to 1$ and $t'_n \to 0$.

The asymptotic power function is given by

$$\tilde{P}_R(\theta) = N\left(\frac{\sqrt{n}\,(t_n - \mu(\theta))}{\sigma_n(\theta)/A_n}\right) - N\left(\frac{\sqrt{n}\,(t'_n - \mu(\theta))}{\sigma_n(\theta)/A_n}\right)$$

As $(\sigma_n(\theta)/\sqrt{n}\,(A_n)) \to_n 0$ and $\mu(\theta)$ is an increasing function of θ, we obtain

$$\lim_n \tilde{P}_R(\theta) = \begin{cases} 1 & \text{if } \theta \neq 0, 1 \\ \alpha & \text{if } \theta = 0 \text{ or } \theta = 1 \end{cases}$$

The test leading to the solution of the trilemma is a consistent one.

2.2. Test of Mixture for Location Mixture

Let now $(X_1, \ldots, X_n)$ be i.i.d. random variables with distribution function given by (1.1b), where λ_1 and λ_2 are known, $F(x - \lambda_1)$ and $F(x - \lambda_2)$ have the same support and it is assumed, without loss of generality:

1. $\lambda_1 < \lambda_2$.
2. $\delta_1 = \delta_2 = 1$.

Consider the transformation

$$Y = \exp(-b_n X) \tag{2.12}$$

where b_n is a function of n, $b_n > 0$, and $b_n \to +\infty$.

The estimator of θ is, in this case, the statistics

$$T_n = \bar{Y}_n / A_n \tag{2.13}$$

with

$$\bar{Y}_n = \frac{1}{n} \sum_{i=1}^{n} \exp(-b_n X_i), \qquad A_n = \exp(-b_n \lambda_1) \Phi(b_n)$$

where

$$\Phi(t) = \int_{-\infty}^{+\infty} \exp(-tz)\, dF(z) \tag{2.14}$$

defined, at least, in some neighborhood of the origin.

As in the last paragraph, Theorems 2.1, 2.2, and 2.3 hold (Cabral Neto, 1984) replacing a_n by b_n, $\mu(ka_n)$ by $\Phi(kb_n)$, $(\delta_2/\delta_1)^{a_n}$ by $\exp(-b_n(\lambda_2 - \lambda_1))$, and $\delta_1^{a_n}$ by $\exp(-b_n\lambda_1)$.

The decision rule to the trilemma (2.6) is, once more, given by (2.7), and the asymptotic solution of (2.11) as well as the asymptotic power function have the same expression of the previous paragraph, with the correct replacements.

Before giving the applications, we may note that T_n is not an unbiased estimator of θ and, if we center it, we fall in the method of fractional moments and $a_n(b_n)$ is no more a function of n but, as we shall see, the variation of the moment with n is useful in the text.

3. APPLICATIONS

To illustrate this test in both cases, scale and location mixtures, we use a mixture of two exponential distributions, with dispersions parameters and a mixture of two Gumbel distributions with locations parameters, respectively.

In each case we give the conditions to a_n and b_n as well as the a_n and b_n considered in the test.

3.1. Scale Mixture

Let $(X_1, \ldots, X_n)$ be i.i.d. random variables whose density function is a mixture of two exponential densities with dispersion parameters δ_1 and δ_2 $(\delta_1 > \delta_2)$.

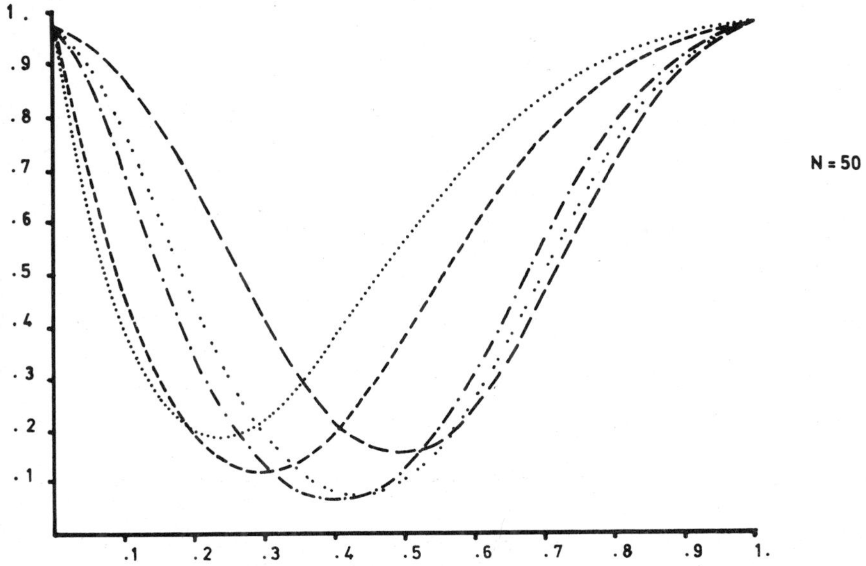

Figure 1. Graph of $1 - \tilde{P}_R(\theta)$ for several a_n when $\delta_1 = 3.5$ and $\delta_2 = 1$. The a_n considered in the graph are (from left to right of each curve) 1.55, 1.25, 0.60, 0.40, 0.05.

In this case, conditions (2.4) and (2.5) lead us to the asymptotic solution $a_n \approx k \log(n)$, $k < c_1/(6 \log(3/2))$, $c_1 \leq 1$, which is a possible solution to numerical work, and simulation results show that $\tilde{P}_R(\theta)$ provides a good approximation of $P_R(\theta)$ when $c_1 < 0.975$ and $n \geq 30$.

To choose the most satisfactory a_n we made a numerical study of the function $1 - \tilde{P}_R(\theta)$, using increments of 0.05, and we conclude that, for n and $d = \delta_1/\delta_2$ fixed, the function behaves as illustrated in Figures 1–4.

This study shows us that there is not an optimal value of a_n, but it leads us to investigate for which value of a_n the function $1 - \tilde{P}_R(\theta)$ has its lowest minimum. When d and n take small values, we conclude that this minimum is attained in a neighborhood of $\theta = 0.4$ (Figures 1 and 2). Besides, when n increases we observe that the value of a_n that gives the lowest minimum for $1 - \tilde{P}_R(\theta)$, a_n^*, increases with n. For example, when $\delta_1 = 3.5$ and $\delta_2 = 1$ we have $n = 30$, $a_n^* = 0.55$; $n = 50$, $a_n^* = 0.60$; $n = 70$, $a_n^* = 0.70$; $n = 100$, $a_n^* = 0.75$. We may yet note that the value of a_n^* considered is very close to the value a_n' that gives the maximum efficiency to T_n at the point $\theta = 0.9$ ($a_n' - a_n^* \leq 0.1$)

When d or n becomes larger, $1 - \tilde{P}_R(\theta)$ attains, very quickly, the value 0 in more than one point θ, and for several a_n (see Figures 3 and 4). In this

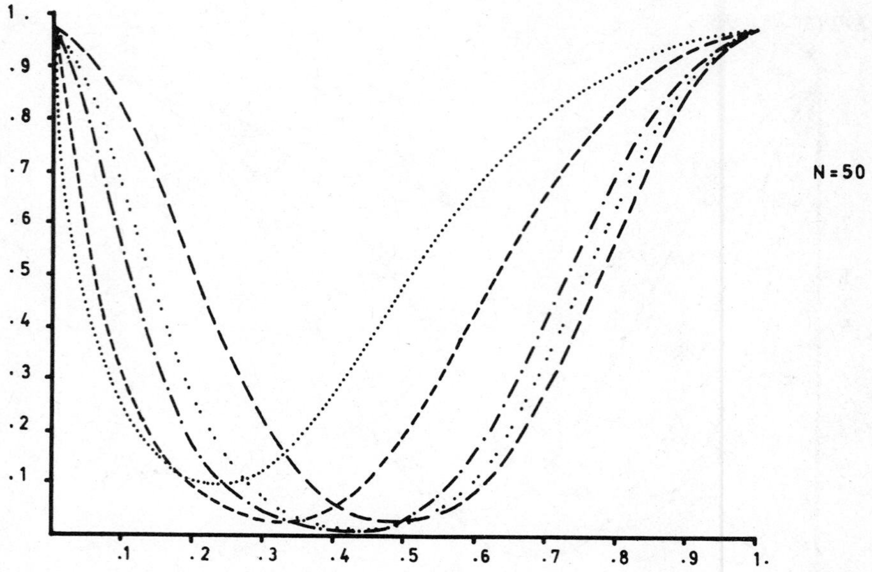

Figure 2. Graph of $1 - \tilde{P}_R(\theta)$ for several a_n when $\delta_1 = 5$ and $\delta_2 = 1$. The a_n considered in the graph are (from left to right of each curve) 1.55, 1.10, 0.55, 0.35, 0.05.

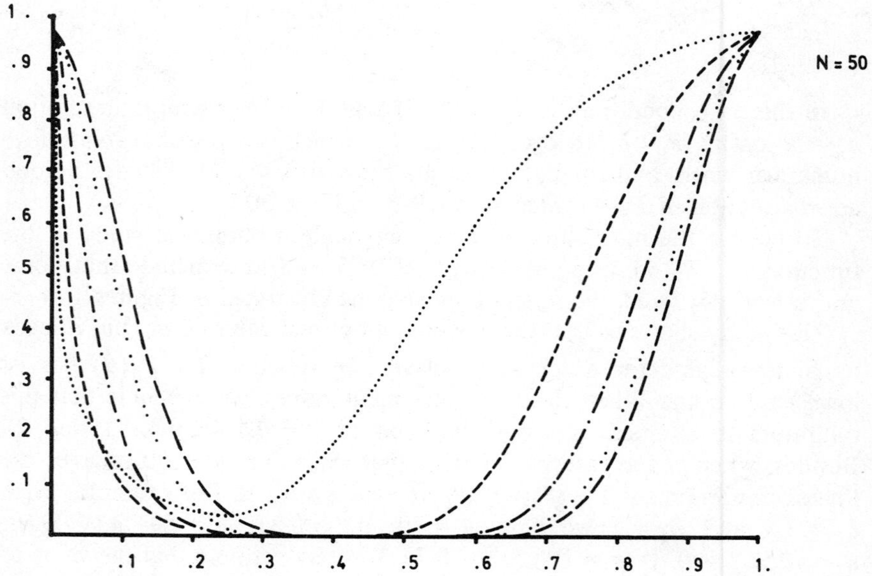

Figure 3. Graph of $1 - \tilde{P}_R(\theta)$ for several a_n when $\delta_1 = 20$ and $\delta_2 = 1$. The a_n considered in the graph are (from left to right of each curve) 1.55, 0.80, 0.40, 0.15, 0.05.

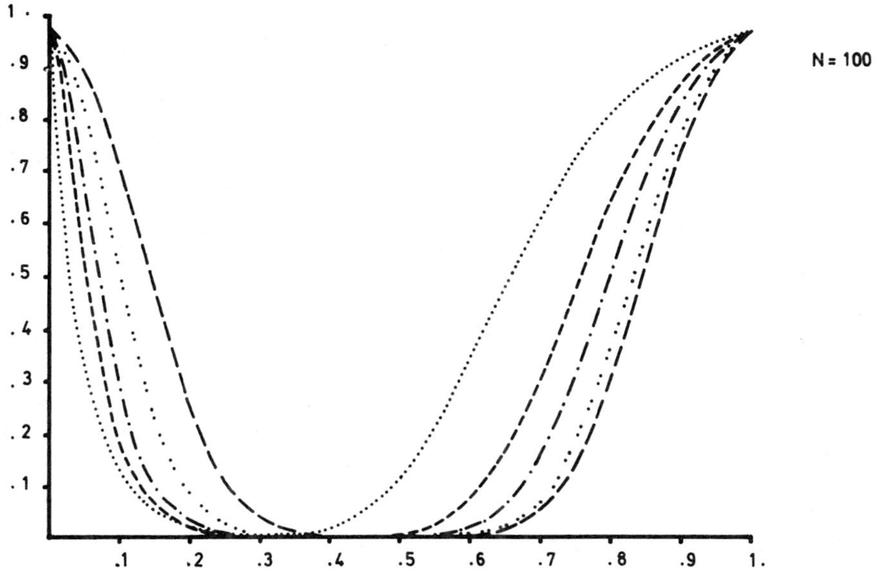

Figure 4. Graph of $1 - \tilde{P}_R(\theta)$ for several a_n when $\delta_1 = 5$ and $\delta_2 = 1$. The a_n considered in the graph are (from left to right of each curve) 1.55; 1.0, 0.70, 0.35, 0.05.

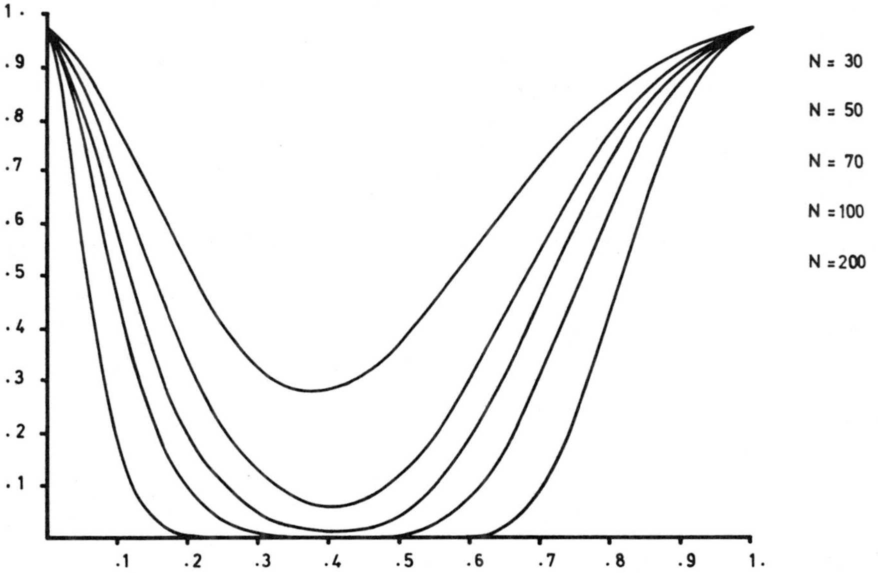

Figure 5. Graph of $1 - \tilde{P}_R(\theta)$ for several n's and $\delta_2 = 3.5$ and $\delta_2 = 1$. The a_n^* considered in each curve (n) are (from top down) 0.55, 0.60; 0.70, 0.75, 0.90. Level of significance $\alpha = 0.025$.

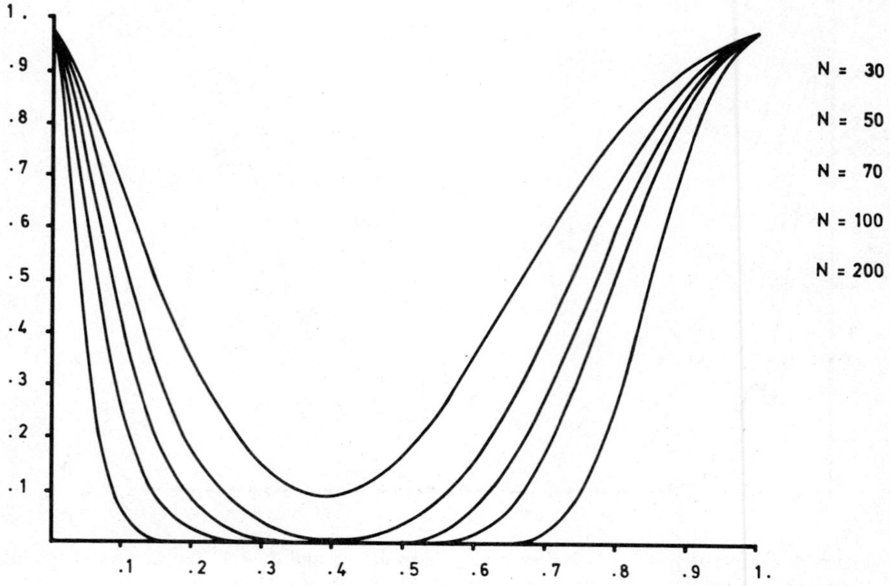

Figure 6. Graph of $1 - \tilde{P}_R(\theta)$ for several n's and $\delta_1 = 5$ and $\delta_1 = 1$. The a_n^* considered in each curve (n) are (from top down) 0.50, 0.55, 0.60, 0.70, 0.80. Level of significance $\alpha = 0.025$.

case we cannot speak about a lowest minimum, so the selection of a_n^* must be different.

In an attempt to make the criterion uniform, we choose the value a_n that gives the maximum efficiency to T_n at the point $\theta = 0.9$.

In a first approximation of the test we may use the same value a'' for several n's, but in this case we do not attain the lowest minimum for every n. For example, $\delta_1 = 3.5$, $\delta_2 = 1$ and we use the moment $a'' = 0.70$ for $n = 30, 50, 70, 100$; we must have for $n = 30$, $1 - \tilde{P}_R(0.4) = 0.2953$, while if we use $a_n^* = 0.55$ we have for $n = 30$, $1 - \tilde{P}_R(0.4) = 0.2822$.

In Figures 5 and 6 we have the graph of $1 - \tilde{P}_R(\theta)$ for some examples of mixture of two exponential distributions and we can see the behavior of the test with n and d.

3.2. Location Mixture

Let $(X_1, \ldots, X_n)$ be i.i.d. random variables whose density function is a mixture of two Gumbel densities with location parameters λ_1 and λ_2 ($\lambda_1 < \lambda_2$). The conditions of consistency and asymptotical normality of T_n lead us to the previous asymptotic solution (in Section 3.1) replacing a_n by

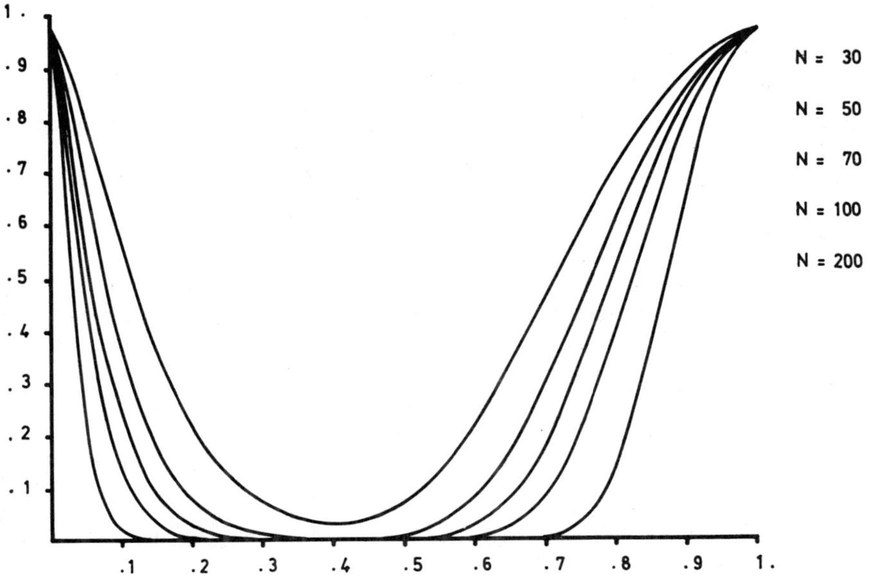

Figure 7. Graph of $1 - \tilde{P}_R(\theta)$ for several n's and $\lambda_1 = 0$ and $\lambda_2 = 2$. The a_n^* considered in each curve (n) are (from top down) 0.45, 0.55, 0.60, 0.65, 0.70. Level of significance $\alpha = 0.025$.

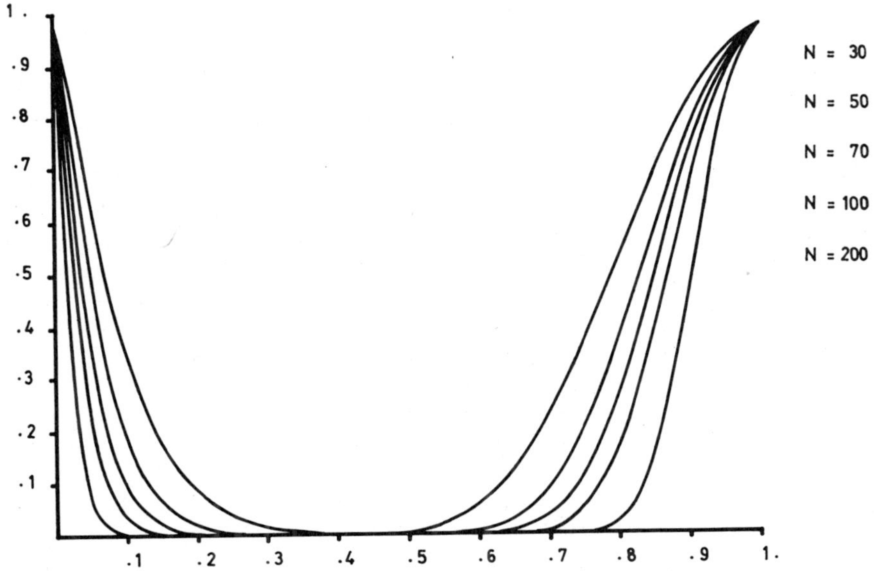

Figure 8. Graph of $1 - \tilde{P}_R(\theta)$ for several n's and $\lambda_1 = 0$ and $\lambda_2 = 3$. The a_n^* considered in each curve (n) are (from top down) 0.35, 0.40, 0.45, 0.50, 0.55. Level of significance $\alpha = 0.025$.

b_n. $\tilde{P}_R(\theta)$ and $1 - \tilde{P}_R(\theta)$ behave with b_n, n, and $d'(= \lambda_2 - \lambda_1)$ in the same way as they did in Section 3.1 with respect to a_n, n, and d, so the criterion adopted here, to choose b_n is the same we use in 3.1 to select a_n. Figures 7 and 8 give some examples.

ACKNOWLEDGMENTS

I am grateful to Professor Dr. J. Tiago de Oliveira for guidance and for several suggestions which very much improved the presentation of this paper.

REFERENCES

Behboodian, J. (1975). Structural Properties and Statistics of Finite Mixtures. *Statistical Distributions in Scientific Work*, (G. P. Patil et al., Eds), Reidel, pp. 103–112.

Cabral Neto, M. S. (1984). An Estimation Technique for the Mixture Parameter. *III Colóquio de Estatística e Investigação Operacional*, 72–87.

Everitt, B. S. and Hand, D. J. 91981). *Finite Mixtures Distributions*. London: Chapman and Hall.

Tiago De Oliveira, J. (1963). Some Elementary Tests for Mixtures of Discrete Distributions. *Proc. Int. Symp. on Classical and Contagious Discrete Distributions*, 379–384.

Efficiencies of Some Disjoint Spacings Tests Relative to a χ^2 Test

S. Rao Jammalamadaka
Department of Mathematics
University of California at Santa Barbara
Santa Barbara, California

Ram C. Tiwari
Department of Mathematics
Indian Institute of Technology
Bombay, India

SUMMARY

As is well known, the goodness-of-fit problem can be reduced, through a probability integral transformation, to testing randomness or uniformity on the interval [0, 1]. Among tests based symmetrically on the m-step spacings ($m \geq 1$), a generalized Greenwood statistic based on the sum of squares of the spacings is known to be locally most powerful. On the other hand, one may consider a χ^2 test statistic with an expected frequency of m in each cell, which is again locally most powerful among tests based symmetrically on the observed frequencies. This comparison is justified since, while the Greenwood test compares the observed and expected cell lengths, holding the observed number in each cell to m, the χ^2 test compares the observed and expected frequencies, holding the expected number in each cell to m. By considering a suitable sequence of alternatives, we compare the asymptotic relative efficiencies of these two tests as well as a third entropy-type test based on the m-step spacings. These results generalize some earlier work of the authors (Jammalamadaka and Tiwari, 1985).

1. INTRODUCTION

Let $X_1, X_2, \ldots, X_{n-1}$ be independent random variables (r.v.) with a common continuous distribution function F. The goodness-of-fit problem of testing whether a specified distribution generated the observations can be reduced to testing if the observations have a uniform distribution, through the probability integral transformation. Thus we may (and shall) assume, without any loss of generality, that the support of F is $[0,1]$ and that the null hypothesis of interest is

$$H_0\colon f(x) = 1, \qquad x \in [0,1] \tag{1.1}$$

where f denotes the probability density function. One classical approach is to use the χ^2 procedure. Suppose we have N classes

$$\left(\frac{i-1}{N}, \frac{i}{N}\right], \qquad i = 1, \ldots, N$$

and O_i denotes the observed frequency in the ith class. The χ^2 statistic is given by

$$\begin{aligned} T_{n,N} &= \frac{N}{n} \sum_{i=1}^{N} \left(O_i - \frac{n}{N}\right)^2 \\ &= \frac{N}{n}\left(\sum_{i=1}^{N} O_i^2\right) - n \end{aligned} \tag{1.2}$$

For reasons that will become clear soon, we shall be interested in the case where $n, N \to \infty$ such that the expected frequency in each class $n/N \to m$, $0 < m < \infty$.

An alternative approach to testing (1.1) is to construct tests based on spacings. Let $\{X'_k\}$ denote the order statistics, with the notation $X'_0 = 0$ and $X'_k = 1 + X'_{k-n}$ for $k \geq n$, circularly for convenience. Then the non-overlapping (or disjoint) m-step spacings ($1 \leq m < n$) are defined by

$$D_{k\cdot m}^{(m)} = X'_{k\cdot m} - X'_{(k-1)\cdot m}, \qquad k = 1, \ldots, [n/m] \tag{1.3}$$

where $[x]$ denotes the integer part of x. Since we are concerned with asymptotic results as $n \to \infty$ with m fixed, there is no loss of generality in assuming $[n/m]$ is an integer N. When $m = 1$, $\{D_k^{(1)}\}$ are simply $\{D_k\}$ are called one-step or simple spacings. The order statistics as well as the spacings should have an extra subscript n, which we suppress throughout this chapter for notational simplicity. Tests based on simple spacings, for the goodness-of-fit problem, have been considered in the literature. See, for instance, Pyke (1965), Rao and Sethuraman (1975), and references contained there. Del Pino (1979), following the approach taken by Rao and Sethuraman (1975), studies tests based on disjoint m spacings defined in (1.3). We consider two such tests, namely,

$$V_n^{(m)} = \frac{1}{N} \sum_{i=1}^{N} (nD_{i \cdot m})^2 \tag{1.4}$$

and an entropy-type statistic

$$E_n^{(m)} = \frac{1}{N} \sum_{i=1}^{N} \left(nD_{i \cdot m}^{(m)}\right) \log\left(nD_{i \cdot m}^{(m)}\right) \tag{1.5}$$

The statistic in (1.4) is a generalization of the so-called Greenwood statistic [see, e.g., Rao and Kuo (1984)], and the one in (1.5) for the special case $m = 1$ has been considered earlier by Gebert and Kale (1969) and more recently by Jammalamadaka and Tiwari (1985). The statistic $V_n^{(m)}$ may be seen as being equivalent to

$$\frac{1}{m} \sum_{i=1}^{N} \left[m - nD_{i \cdot m}^{(m)}\right]^2$$

which in this form may be thought of as the dual of the χ^2 statistic (1.2) with the observed frequencies in each cell (of length $D_{i \cdot m}^{(m)}$) being fixed at m.

To compare the asymptotic relative efficiencies of these three tests, we need to consider their distribution theory under the following sequence of alternatives:

$$A_n\colon f_n(x) = 1 + \frac{l(x)}{n^{1/4}}, \qquad 0 \le x \le 1 \tag{1.6}$$

where $l(\cdot)$ is square-integrable and continuously differentiable on $[0, 1]$. This sequence of alternatives has been considered before. See, for instance, Rao and Sethuraman (1975) and Del Pino (1979).

Under the sequence of alternatives (1.6), the asymptotic normality of the three statistics is established in Section 2 and comparison of asymptotic efficiencies made in Section 3.

2. ASYMPTOTIC NORMALITY OF $V_n^{(m)}$, $E_n^{(m)}$, AND $T_{n,N}$

To establish the asymptotic normality of $V_n^{(m)}$ and $E_n^{(m)}$, we use the following result of Del Pino (1979). See also Rao and Kuo (1984).

Theorem 2.1 (Del Pino, 1979). Under the alternatives (1.6), if $h(\cdot)$ is a function satisfying some regularity conditions (Del Pino, 1979) and S is a Gamma$(m, 1)$ random variable with density $s^{m-1}e^{-s}/\Gamma m$ for $s \geq 0$, then

$$N^{-1/2} \sum_{i=1}^{N} \left[h\left(nD_{i \cdot m}^{(m)}\right) - Eh(S)\right] \xrightarrow{d} N(\mu, \sigma^2)$$

where

$$\mu = \left(\int_0^1 l^2(t)\, dt\right) \text{Cov}\left(h(S), (S - m - 1)^2\right) / 2\sqrt{m} \tag{2.1}$$

and

$$\sigma^2 = \text{Var}(h(S)) - \text{Cov}^2(h(S), S)/m \tag{2.2}$$

The statistic $V_n^{(m)}$ as a special case of this with $h(x) = x^2$ has already been studied by Del Pino (1979) and Rao and Kuo (1984), and we state the needed result.

Theorem 2.2. Under the sequence of alternatives (1.6), the random variable

$$\sqrt{N}\left(V_n^{(m)} - m(m+1)\right) \xrightarrow{d} N(\mu_1, \sigma_1^2)$$

where

$$\mu_1 = \sqrt{m} \cdot (m+1)\left(\int_0^1 l^2(t)\, dt\right) \quad \text{and} \quad \sigma_1^2 = 2m(m+1)$$

Now we consider $E_n^{(m)}$ in (1.8) and establish the following

Theorem 2.3. Under the sequence of alternatives (1.6), the random variable

$$\sqrt{N}\left[E_n^{(m)} - m\left(1 + \frac{1}{2} + \cdots + \frac{1}{m} - \gamma\right)\right] \xrightarrow{d} N(\mu_2, \sigma_2^2)$$

where

$$\mu_2 = \frac{\sqrt{m}}{2}\left(\int_0^1 l^2(t)\, dt\right)$$

and

$$\sigma_2^2 = m(m+1)\left\{\frac{\pi^2}{6} - \sum_{j=1}^{m}\frac{1}{j^2}\right\} - m$$

(Here γ is the Eüler constant, 0.5772... .)

Proof. It can be checked that the function $h(x) = x \log x$ satisfies the simple set of sufficient conditions [see Eq. (3.3), p. 1061, of Del Pino (1979)]. Then to apply Theorem 2.1, we need to evaluate the centering constant $E(S \log S)$ and the asymptotic mean and variance in (2.1) and (2.2) for this case. Evaluation of the required integrals turns out to be quite complex but manageable. For instance

$$\begin{aligned} E(S \log S) &= \frac{1}{\Gamma m}\int_0^{\infty} e^{-x}\cdot x^m \cdot \log x \cdot dx \\ &= \frac{1}{\Gamma m}\left\{m\int_0^{\infty} e^{-x}\cdot x^{m-1}\cdot \log x \cdot dx + \Gamma m\right\} \\ &= m\left\{1 + \frac{1}{2} + \cdots + \frac{1}{m} - \gamma\right\} \end{aligned}$$

For getting μ, we need

$$\begin{aligned} &\operatorname{Cov}\left(S \log S, (S-m-1)^2\right) \\ &\quad = E\left\{S^3 \log S + (m+1)^2 S \log S - 2(m+1)S^2 \log S\right\} \\ &\qquad - E(S\log S)E\left(S^2 + (m+1)^2 - 2(m+1)S\right) \\ &\quad = \left\{(m+2)(m+1)m\left(1 + \frac{1}{2} + \cdots + \frac{1}{m+2} - \gamma\right)\right. \\ &\qquad + (m+1)^2 m\left(1 + \frac{1}{2} + \cdots + \frac{1}{m} - \gamma\right) \\ &\qquad \left. - 2(m+1)^2 m\left(1 + \frac{1}{2} + \cdots + \frac{1}{m+1} - \gamma\right)\right\} \\ &\qquad - m\left\{1 + \frac{1}{2} + \cdots + \frac{1}{m} - \gamma\right\}\left\{(m+1)m + (m+1)^2 - 2(m+1)m\right\} \\ &\quad = m \end{aligned}$$

and hence μ_2 follows from (2.1). Similarly, one can show that the expression (2.2) for σ^2 reduces to the one given in Theorem 2.3. ■

From these two theorems, 2.2 and 2.3, one obtains the asymptotic null distributions under (1.1) by setting $l(x) \equiv 0$ as well as the special case for simple spacings by taking $m = 1$.

Finally, we consider the asymptotic distribution of the χ^2 statistic $T_{n,N}$ in (1.2) under the alternatives (1.6). For this, we use Theorem 2.1 of Holst and Rao (1980, p. 25) on the asymptotic distribution of statistics based on multinomial frequencies. The proof of the following result is essentially similar to that of Theorem 2.7 of Jammalamadaka and Tiwari (1985) and is omitted.

Theorem 2.4. Under the alternatives (1.6), the random variable

$$N^{-1/2}\left\{T_{n,N} - N\left(1 + \frac{\sqrt{n}}{N}\int_0^1 l^2(t)\,dt\right)\right\}$$

has an asymptotic normal distribution with mean 0 and variance 2 as $n, N \to \infty$ such that $n/N \to m$, finite.

3. ASYMPTOTIC RELATIVE EFFICIENCIES OF $V_n^{(m)}$, $E_n^{(m)}$, AND $T_{n,N}$

The Pitman asymptotic relative efficiency (ARE) of a test relative to another is defined to be the limit of the inverse ratio of sample sizes required to obtain the same limiting power at a sequence of alternatives converging to the null. Under certain regularity conditions [see, for example, Fraser (1957)], which include a condition about the type of alternatives, asymptotic normality of the test statistic under a sequence of alternatives, etc., the "efficacy" of a test statistic is given by (μ_Δ^4/σ^4), where μ_Δ and σ are the mean and variance of the limiting normal distribution under the sequence of alternatives, when the test statistic has been normalized to have a limiting standard normal distribution under the hypothesis. In such a situation, the ARE of one test with respect to another is simply the ratio of their efficacies.

From Theorem 2.2, the efficacy of the test statistic $V_n^{(m)}$ is

$$\text{Eff}\left(V_n^{(m)}\right) = \frac{m^2(m+1)^4\left(\int_0^1 l^2(t)\,dt\right)^4}{4m^2(m+1)^2}$$

$$= \frac{(m+1)^2}{4}\cdot\left(\int_0^1 l^2(t)\,dt\right)^4 \qquad (3.1)$$

Similarly from Theorem 2.3, the efficacy of $E_n^{(m)}$ is

$$\operatorname{Eff}\left(E_n^{(m)}\right) = \frac{m\left(\int_0^1 l^2(t)\,dt\right)^4}{16\left\{m(m+1)\left[\dfrac{\pi^2}{6} - \Sigma_{j=1}^{m}\dfrac{1}{j^2}\right] - m\right\}^2}$$

$$= \frac{\left(\int_0^1 l^2(t)\,dt\right)^4}{16\left\{(m+1)\left[\dfrac{\pi^2}{6} - \Sigma_{j=1}^{m}\dfrac{1}{j^2}\right] - 1\right\}^2}. \tag{3.2}$$

And finally from Theorem 2.4, the efficacy of the χ^2 statistic $T_{n,N}$ is

$$\operatorname{Eff}(T_{n,N}) = \frac{m^2}{4}\left(\int_0^1 l^2(t)\,dt\right)^4 \tag{3.3}$$

Since all these efficacies depend on the alternatives only through the multiplying constant $(\int_0^1 l^2(t)\,dt)^4$, it makes the comparison easy. It is clear that $V_n^{(m)}$ based on the sum of squares of disjoint m spacings is superior to the χ^2 statistic or the entropy-type statistic $E_n^{(m)}$. Also, the test based on $E_n^{(m)}$ is asymptotically more efficient than the χ^2 statistic with expected frequency of m in each cell. Thus spacings tests seem preferable to comparable χ^2 procedures. One can make a table of the relative efficiencies of these three tests using (3.1), (3.2), and (3.3) for various values of m. These conclusions agree with the relative ordering obtained earlier by the authors (Jammalamadaka and Tiwari, 1985) for the case $m = 1$.

REFERENCES

Del Pino, G. E. (1979). On the asymptotic distribution of k-spacings with applications to goodness of fit tests, *Ann. Statist.* **7**, 1058–1065.

Fraser, D. A. S. (1957). *Nonparametric Methods in Statistics*, New York: Wiley.

Gebert, J. B. and Kale, B. K. (1969). Goodness of fit tests based on discriminatory information. *Statist. Hefte* **10**, 192–200.

Holst, L. and Rao, J. S. (1980). Asymptotic theory for some families of two-sample nonparametric statistics. *Sankhyā A* **42**, 19–52.

Jammalamadaka, S. Rao and Tiwari, R. C. (1985). Asymptotic comparison of three tests for goodness of fit. *J. Statist. Plan. Inf.* **12**, 295–304.

Pyke, R. (1965). Spacings, *J.R. Statist. Soc. B* **27**, 395–449.

Rao, J. S. and Kuo, M. (1984). Asymptotic results on the Greenwood statistic and some of its generalizations. *J. R. Statist. Soc. B* **46**, 228–237.

Rao, J. S. and Sethuraman, J. (1975). Weak convergence of empirical distribution functions of random variables subject to perturbations and scale factors. *Ann. Statist.* **3**, 299–313.

Pseudorandom Numbers Generated by the Doubly Multiplicative Congruential Method

Haruo Miyazaki
Department of Computer Science
Gunma University
Gunma, Japan

SUMMARY

A new method of generating random numbers using doubly multiplicative congruential method is described. It is pointed out that the advantages of pseudorandom generator presented here are like all simple multiplicative congruential methods, subjected to stringent tests for randomness and achieving very long period. This generator allows well for the applications of recent large-scale and high-precision Monte Carlo simulations.

1. INTRODUCTION

Many methods for the generation of pseudorandom numbers have been suggested, accepted as a means of arithmetically producing long-period sequences that pass most randomness tests, and widely used in the fields of Monte Carlo simulation. The most commonly used present methods for generating pseudorandom numbers are the procedures that produce a nonrandom sequence of numbers according to some arithmetic formula, such as multiplicative congruential methods and mid-square methods. It is said that the random numbers generated by any method are good if they are uniformly distributed, statistically independent, reproducible, necessarily fast, and require minimum memory capacity. However, all these properties are rarely satisfied, and some compromise must be necessary. Some attentions must be paid to the use and application of these pseudorandom numbers, especially for

1. J. von Neumann's mid-square method likewise proved slow and awkward for statistical analysis. Additionally the sequences tend to cyclicity, and once a zero is encountered the sequence terminates.
2. The multiplicative congruential methods are designed to satisfy as many of these requirements, but it has been said that the sequences generated by these congruential methods suffer from defects the assumption of independence of successive random numbers being crucial and the periods of every congruential method being not so long.

We present a method to generate random numbers that are more uniformly and randomly distributed with fairly long period. This approach can be used to improve the reduction of autocorrelation of the random number series and may be useful for the Monte Carlo simulation to require more independency in the series of random numbers, which is called here as doubly multiplicative congruential method.

2. DOUBLY MULTIPLICATIVE CONGRUENTIAL METHOD

2.1. The Generator

The main idea of this paper is presented in the following formula

$$X_{n+1} = aX_n \pmod{P}, \qquad 0 \le X_{n+1} < P \tag{2.1}$$

$$X_{n+2} = bX_{n+1} \pmod{Q}, \qquad 0 \le X_{n+2} < Q, \qquad n = 2m \tag{2.2}$$

taking a set of positive integers (a, b, P, Q, X_0) and generating a sequence of integers $\{X_{n+2}\}$ as a pseudorandom number series by relations (2.1) and (2.2).

It may be good for the generation of a sequence of random numbers with a long period to select the integers a, b, P, Q being relatively prime in pairs, where X_0 is an appropriately given initial value (Hull and Dobell,

Table 1. The Number of Rejected Times in 100 Repeated Tests (significant level $\alpha = 0.05$)

	Test					
	Frequency	Run	D^2	Combination	Gap	Random Walk
$X_{n+1} = 2456949 X_n \pmod{2^{24}}$ $(X_0 = 1234567)$	4	9	2	5	7	6
$X_{n+1} = 3513381 X_n \pmod{2^{25}}$ $(X_0 = 1234567)$	6	4	4	5	5	4
$X_{n+1} = 11257 X_n \pmod{10^8 + 7}$ $X_{n+2} = 12553 X_{n+1} \pmod{10^8 + 37}$ $(X_0 = 12345678)$	3	1	2	2	2	1

1962). But the conditions on these integers (a, b, P, Q, X_0) are not so easy to realize and to decide the optimal combination from the side of mathematical theory. Considering past experience on the generation of the usual multiplicative methods, (Rubinstein, 1981; Bratley et al., 1983), we select some sets of integers (a, b, P, Q, X_0) and obtain the class of generators that might be expected to exhibit some satisfactory statistical behaviors.

2.2. Some Selected Generators and Statistical Tests

We describe some statistical tests for checking uniformity and independency of the sequence of pseudorandom numbers produced by the doubly multiplicative congruential method. As mentioned before, a sequence of pseudorandom numbers is completely deterministic, and the random numbers on the unit interval (0, 1) can be obtained by

$$U_m = X_{n+2}/Q \tag{2.3}$$

But insofar as it passes the set of statistical tests, it may be treated as one of random numbers as a sample from $U(0, 1)$.

Our object here is to provide some results of statistical tests and to compare the doubly multiplicative congruential method to the usual multiplicative congruential method recommended by Borosh and Niederreiter (1983).

Goodness of Fit

Goodness of fit of the generated numbers to the uniform distribution $U(0, 1)$ was tested as shown in Table 1. These tests are tried on run test, D^2 test, combination test, gap test, and random walk test. Every value of test statistics was revealed to have not so many unduly large discrepancies from theoretical values as compared to the usual multiplicative congruential method.

Table 2. Serial Correlations of Lag 1 to Lag 5 Every Interval in the Random Number Series (from the first to 1000000th numbers in every series)

	Lag				
	1	2	3	4	5
$X_{n+1} = 2456949X_n \pmod{2^{24}}$	0.0591	0.0442	0.0554	0.0516	0.0533
$X_{n+1} = 3513381X_n \pmod{2^{25}}$	−0.0223	0.0321	−0.0116	−0.0337	−0.0083
$X_{n+1} = 11257X_n \pmod{10^8 + 7}$ $X_{n+2} = 12553X_{n+1} \pmod{10^8 + 37}$	−0.0044	−0.0016	−0.0033	−0.0018	−0.0030

Test of Randomness

Serial correlations of lag 1, lag 2, lag 3, lag 4, and lag 5 are computed and displayed in Table 2. None of the serial correlations for every generator was significant at the 0.5 level, but the values for the new method are relatively fairly smaller than for the other methods as shown in Table 2.

3. PERIOD OF THE DOUBLY MULTIPLICATIVE CONGRUENTIAL METHOD

It is very difficult to obtain the full or maximum period by number-theoretical approaches, providing a set of integers (a, b, P, Q, X_0). Then we calculate the period of the doubly multiplicative congruential generator

```
C     LENGTH OF CYCLE
      DOUBLE PRECISION A,B,C,D,F
      DOUBLE INTEGER I,J,K
      DATA F/(initial value)/
      A=F
      B=F
    1 CALL R(A)
      CALL R(B)
      CALL R(B)
      IF (A.NE.B) GO TO 1
      I=0
    2 I=I+1
      CALL R(B)
      IF (A.NE.B) GO TO 2
      A=F
      B=F
      DO 20 J=1,I
      CALL R(B)
   20 CONTINUE
      K=0
    3 K=K+1
      CALL R(A)
      CALL R(B)
      IF (A.NE.B) GO TO 3
      WRITE(6,600) I,K
  600 FORMAT(////1H0,'LENGTH OF CYCLE'/1H0,
     *'C=',I10,5X,'L=',I10)
      STOP
      END

      SUBROUTINE R(N)
```

Figure 1. Algorithm to obtain the length of tail and cycle of the doubly multiplicative congruential method.

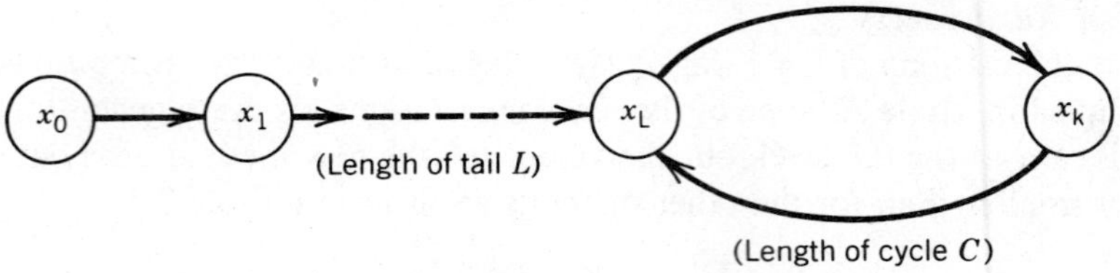

Figure 2. Period of the generator.

Table 3. Periods of the Generators

	L	C	$L+C$
$X_{n+1} = 2456949X_n \pmod{2^{24}}$	0	4194304	4194304
$X_{n+1} = 3513383X_n \pmod{2^{25}}$	0	8388608	8388608
$X_{n+1} = 11257X_n \pmod{10^8 + 7}$ $X_{n+2} = 12553X_{n+1} \pmod{10^8 + 37}$	2476661	16626756	19103417

using an algorithm which is suggested by Knuth (1971). The algorithm is shown in Figure 1. The period of the new generator is defined as $L + C$, which consists of the length of tail L and the length of cycle C, as shown in Figure 2 to be fairly long. The values of L and C for every generator are presented in Table 3, and every generator is thought to be good for practical use in simulation.

4. CONCLUDING REMARK

The pseudorandom uniform number generator based on the doubly multiplicative congruential method, as described above, appears to possess satisfactory statistical properties. It maintains the usual advantages of multiplicative congruential method as the long cycle and ease of implementation on a computer, and does not make any substantial additional demand on the computer memory. The possible existence of autocorrelation in every lag of the sequence of numbers may have small value.

ACKNOWLEDGMENTS

The author wishes to thank professor M. Soshiroda, Waseda University, for his good suggestions in this research.

REFERENCES

Borosh, I. and Niederreiter, H. (1983). Optimal multipliers for pseudorandom number generation by the linear congruential method. *BIT* **23**, 65–74.

Bratley Fox, B. L., and Schrage, L. E. (1983). *A Guide to Simulation*, New York: Springer-Verlag.

Hull, T. E., and Dobell, A. R. (1962). Random number generators, *SIAM Rev*., **4**.

Knuth, D. E. (1971). *The Art of Computer Programming*; *Seminumerical Algorithms*, Reading, Massachusetts: Addison-Wesley.

Rubinstein, R. Y. (1981). *Simulation and the Monte Carlo Method*, New York: Wiley.

Further Results on Testing a Poisson Hypothesis Against Compound Poisson Alternatives

Prem S. Puri
Department of Statistics
Purdue University
West Lafayette, Indiana

SUMMARY

In this chapter we extend the author's previous results for testing a Poisson hypothesis against compound Poisson alternatives based on a sample $X_1, X_2, \ldots, X_n$ of i.i.d. random variables to the case where the X_i's, although mutually independent, are not necessarily identically distributed. In particular, we address ourselves to those situations where for each compound Poisson point process $\{X_i(t) : t \geq 0\}$ all we are able to observe is the total number of events that have occurred by the end of a time period of length t_i, and are otherwise unable to monitor both the points and the sizes of various jumps of the process over time. The approach adopted here is that of Bartoo and Puri (1967); this is an extension of Neyman's (1959) theory of $C(\alpha)$-tests.

1. INTRODUCTION AND STATEMENT OF THE PROBLEM

This chapter continues the author's work in Puri (1985). The problem considered there was to develop a suitable procedure for testing a Poisson hypothesis against compound Poisson alternatives based on a sample

---

These investigations were supported in part by the U.S. National Science Foundation Grant No. DMS-8504319.

$X_1, X_2, \dots, X_n$ of independent and identically distributed (i.i.d.) non-negative integer-valued random variables (r.v.'s). The common distribution of these r.v.'s was taken to be compound Poisson with probability generating function (p.g.f.) given by

$$G(s) \equiv E(s^X) = \exp\{-\lambda(1 - h(s|\xi))\}, \qquad |s| \le 1 \tag{1.1}$$

where λ is a positive constant and for each $\xi \ge 0$, $h(s|\xi)$ is a p.g.f. given by

$$h(s|\xi) = \sum_{k=1}^{\infty} R(k|\xi)s^k, \qquad |s| \le 1 \tag{1.2}$$

with

$$h(0|\xi) = 0, \qquad R(k|\xi) \ge 0, \qquad \sum_{k=1}^{\infty} R(k|\xi) = 1$$

Here ξ is a non-negative parameter and is such that, for $k \ge 1$,

$$\lim_{\xi \to 0} R(k|\xi) = \delta_{1k}, \qquad \lim_{\xi \to 0} h(s|\xi) \equiv h(s|\xi = 0) \equiv s \tag{1.3}$$

where δ_{1k} is the Kronecker delta. With this, the above-mentioned hypothesis testing problem corresponds to testing H_0: $\xi = 0$ against the alternative H_1: $\xi > 0$ in the presence of λ as the nuisance parameter. The reader may refer to Puri (1985) for a detailed motivation to the above formulation along with some applications. In Puri (1985), an optimal $C(\alpha)$ test was developed for the Poisson hypothesis H_0 against H_1 based on Neyman's $C(\alpha)$-test theory (see Neyman, 1959).

Again in many practical situations the X_i's represent the numbers of events taking place over possibly different lengths t_i of time, based on independently conducted experiments. In particular here [as well as in Puri (1985)] we are addressing ourselves to those situations where we are unable to monitor both the jump points of a compound Poisson point process over time and the sizes of the various jumps. All we are able to observe is the total number of events that have occurred by the end of a time period of length t. The reader may refer to the author's joint work with Neyman in the area of radiation biology for precisely such a situation arising in the case of UV radiation [see Neyman and Puri (1976, 1981) and, in particular, Puri (1982)].

Since, typically, different X_i's correspond to observations taken over different lengths of time, we need to reconsider the original problem of testing our hypothesis H_0 for the case where the X_i's are independent, but not necessarily identically distributed. For this, we need to use the Bartoo–Puri generalization of Neyman's $C(\alpha)$-test theory, which handles precisely such situations [see Bartoo and Puri (1967)]. A brief summary of this generalization is outlined in Section 2. In the light of the above remarks, we provide a reformulation of our problem where we allow not only the period of observation but also the p.g.f. h of (1.2) to vary from observation to observation. Furthermore, it is assumed that the p.g.f.'s h_i otherwise depend not only on the same parameter ξ under test but also on the same nuisance parameter λ. Thus we have $X_1, X_2, \ldots, X_n$, mutually independent non-negative integer-valued r.v.'s with the p.g.f. G_i of X_i given by

$$G_i(s) \equiv E(s^{X_i}) = \exp\{-\lambda t_i(1 - h_i(s|\xi, \lambda))\}, \qquad |s| \leq 1 \tag{1.4}$$

$i = 1, 2, \ldots, n$, with $\xi \geq 0$, $\lambda > 0$,

$$h_i(s|\xi, \lambda) = \sum_{k=1}^{\infty} R_i(k|\xi, \lambda) s^k, \qquad |s| \leq 1 \tag{1.5}$$

$$R_i(k|\xi, \lambda) \geq 0, \qquad \sum_{k=1}^{\infty} R_i(k|\xi, \lambda) \equiv 1 \tag{1.6}$$

and for every $\lambda > 0$, $k \geq 1$ and $i = 1, 2, \ldots, n$,

$$\lim_{\xi \to 0} R_i(k|\xi, \lambda) = \delta_{1k}, \qquad \lim_{\xi \to 0} h_i(s|\xi, \lambda) \equiv h_i(s|\xi = 0, \lambda) \equiv s \tag{1.7}$$

where the t_i are known lengths of the corresponding time intervals. Our object is to develop an optimal $C(\alpha)$ test based on $X_1, X_2, \ldots, X_n$, for testing H_0: $\xi = 0$ against H_1: $\xi > 0$. For this, the needed generalization of Neyman's $C(\alpha)$-test theory is summarized in Section 2.

2. THE BARTOO–PURI GENERALIZATION OF NEYMAN'S $C(\alpha)$-TEST THEORY

Neyman (1959) developed his $C(\alpha)$-test theory keeping in mind those situations where the distributions of the observable random variables turn out to be highly nonstandard and often involve many nuisance parameters, $\theta_1, \ldots, \theta_r$, besides the parameter ξ under test. Also, the estimators available for the nuisance parameters may not be too good and in particular may be biased. Under these circumstances, for testing the hypothesis, say, H_0: $\xi = \xi_0$, against the alternative hypothesis, say, H_1: $\xi > \xi_0$ (in the presence of nuisance parameters) Neyman developed tests that are locally asymptotically most powerful in a class of so-called $C(\alpha)$ tests and are based on a sample of i.i.d. r.v.'s, $X_1, X_2, \ldots, X_n$. Below we present a summary of the needed generalization of Neyman's theory where the observable random variables $\{X_{nk}: k = 1, 2, \ldots, n\}$ are independently but not necessarily identically distributed. The reader may find further details of these and other generalizations of Neyman's theory in Bartoo and Puri (1967) and Bühler and Puri (1966).

Let X_{nk}, $k = 1, 2, \ldots, n$; $n = 1, 2, \ldots$, be a double sequence of independent r.v.'s with each X_{nk} having a probability density $p_{n,k}(x; \xi, \boldsymbol{\theta})$ with respect to some σ-finite measure γ_{nk}, which is independent of the parameters ξ and $\boldsymbol{\theta}$, where $\xi \in [0, a)$ for some $a > 0$ and $\boldsymbol{\theta} = (\theta_1, \theta_2, \ldots, \theta_r) \in \Theta$, Θ being an open set in R^r. Also we assume that the support of the distribution of each X_{nk} is independent of ξ and $\boldsymbol{\theta}$. For convenience we let the null hypothesis be H_0: $\xi = 0$, which is to be tested against H_1: $\xi > 0$, in the presence of the nuisance parameters $\boldsymbol{\theta} = (\theta_1, \theta_2, \ldots, \theta_r) \in \Theta$. We impose conditions C_1–C_3 given below on the sequence $p_{n,k}(x; \xi, \boldsymbol{\theta})$ of density functions.

(C_1) *The derivatives*

$$\varphi_{nk(j)}(x; \boldsymbol{\theta}) = \left. \frac{\partial \ln p_{n,k}(x; \xi, \boldsymbol{\theta})}{\partial \theta_j} \right|_{\xi=0}, \qquad j = 1, 2, \ldots, r \tag{2.1}$$

and

$$\varphi_{nk(\xi)}(x; \boldsymbol{\theta}) = \left. \frac{\partial \ln p_{n,k}(x; \xi, \boldsymbol{\theta})}{\partial \xi} \right|_{\xi=0} \tag{2.2}$$

exist for arbitrary $\boldsymbol{\theta} \in \Theta$, *the derivative in* (2.2) *being the right-hand derivative. If* $p_{nk} = 0$, *we define* $\varphi_{nk(\xi)} = \varphi_{nk(j)} = 0$, $j = 1, 2, \ldots, r$. *We assume that*

$$E_{0,\boldsymbol{\theta}}\left[\varphi^2_{nk(\xi)}(X_{nk}; \boldsymbol{\theta})\right], \qquad E_{0,\boldsymbol{\theta}}\left[\varphi^2_{nk(j)}(X_{nk}; \boldsymbol{\theta})\right], \qquad j = 1, 2, \ldots, r \tag{2.3}$$

are finite for every n and k and for all $\boldsymbol{\theta} \in \Theta$, *where the subscripts* $0, \boldsymbol{\theta}$ *under the expectation sign indicate that the expectations are obtained under* H_0.

(C_2) *Under* H_0, *whatever be* $\boldsymbol{\theta} \in \Theta$ *and n and k, the quantities* $\varphi_{nk(\xi)}(X_{nk}; \boldsymbol{\theta})$ *and* $\varphi_{nk(j)}(X_{nk}; \boldsymbol{\theta})$, $j = 1, 2, \ldots, r$, *are linearly independent with positive probability.*

Let $\mathbf{a}_n^0(\boldsymbol{\theta}) = (a_{n1}^0(\boldsymbol{\theta}),\dots, a_{nr}^0(\boldsymbol{\theta}))$ be a vector that minimizes the variance of

$$\sum_{k=1}^{n} \varphi_{nk(\xi)}(X_{nk}; \boldsymbol{\theta}) - \sum_{j=1}^{r} a_{nj}(\boldsymbol{\theta}) \sum_{k=1}^{n} \varphi_{nk(j)}(X_{nk}; \boldsymbol{\theta}) \tag{2.4}$$

under H_0 for each n and fixed $\boldsymbol{\theta}$. The symbol $S_n^{*2}(\boldsymbol{\theta})$ will denote this minimum variance. Note that in view of conditions C_1–C_2 the values $a_{nj}^0(\boldsymbol{\theta})$ are always determined and $S_n^{*2}(\boldsymbol{\theta})$ is always positive. We now add condition C_3.

(C_3) *We assume that* $\mathbf{a}^0(\boldsymbol{\theta}) = (a_1^0, a_2^0,\dots, a_r^0) = \lim_{n\to\infty} \mathbf{a}_n^0(\boldsymbol{\theta})$ *exists and that the sequence* $\{p_{nk}(x; \xi, \boldsymbol{\theta})\}$ *is regular enough for the sequence*

$$f_{nk}^*(x, \boldsymbol{\theta}) = \varphi_{nk(\xi)}(x; \boldsymbol{\theta}) - \sum_{j=1}^{r} a_j^0(\boldsymbol{\theta})\varphi_{nk(j)}(x; \boldsymbol{\theta}), \qquad k = 1,2,\dots, n \tag{2.5}$$

to form an expectation centered Cramér sequence. [*See Neyman* (1959) *and, in particular, Bartoo and Puri* (1967) *for their definition.*]

It may be remarked here that the "regularity conditions" assumed for the Cramér sequences are similar to the ones imposed by Cramér (1946) in his treatment of consistency of maximum-likelihood estimates. Consequently following Neyman (1959), Bartoo and Puri (1967) referred to these as Cramér sequences of functions. While avoiding the statement of these conditions in detail here in defining these sequences, we refer the reader to Bartoo and Puri (1967) for their definition.

Let $B(\alpha)$ be an arbitrary measurable set on the real line whose indicator function is continuous almost everywhere and is such that

$$(2\pi)^{-1/2}\int_{B(\alpha)} \exp(-x^2/2)\,dx = \alpha \tag{2.6}$$

Let $\{f_{nk}(x, \boldsymbol{\theta})\}$ be an expectation-centered Cramér sequence such that for $j = 1,2,\dots, r$

$$\lim_{n\to\infty} n^{-1} \sum_{k=1}^{n} \operatorname{Cov}_{0,\boldsymbol{\theta}}\{f_{nk}(X_{nk}, \boldsymbol{\theta}), \varphi_{nk(j)}(X_{nk}; \boldsymbol{\theta})\} = 0. \tag{2.7}$$

Here again the subscripts $0, \boldsymbol{\theta}$ of Cov and of Var indicate that the covariances and the variances, respectively, are obtained under H_0. A typical member of Neyman's class of $C(\alpha)$ tests is now defined for each pair $\{f_{nk}(x, \boldsymbol{\theta})\}$, an expectation-centered Cramér sequence and a set $B(\alpha)$, by rejecting H_0 whenever

$$Z_n(\hat{\boldsymbol{\theta}}_n) \in B(\alpha) \tag{2.8}$$

where

$$Z_n(\hat{\boldsymbol{\theta}}_n) = [S_n(\hat{\boldsymbol{\theta}}_n)]^{-1} \sum_{k=1}^{n} f_{nk}(X_{nk}, \hat{\boldsymbol{\theta}}_n) \tag{2.9}$$

$$S_n^2(\boldsymbol{\theta}) = \sum_{k=1}^{n} \operatorname{Var}_{0,\boldsymbol{\theta}}\{f_{nk}(X_{nk}, \boldsymbol{\theta})\} \tag{2.10}$$

and $\hat{\boldsymbol{\theta}}_n = (\hat{\theta}_{n1}, \ldots, \hat{\theta}_{nr})$ is a locally $\sqrt{n}$ consistent estimator for $\boldsymbol{\theta}$ defined by Neyman (1959) to be such that for every $j = 1, 2, \ldots, r$, and for some constants A_j, the quantities

$$|\hat{\theta}_{nj} - \theta_j - A_j \xi|\sqrt{n} \tag{2.11}$$

remain bounded in probability, as $n \to \infty$, for all $(\xi, \boldsymbol{\theta})$.

Using Neyman's (1959) local asymptotic optimality criterion based on a class Γ of local alternatives $\{\xi_n\}$ such that $\sqrt{n}\,\xi_n$ remains bounded as $n \to \infty$, a test that is optimal in the above class $C(\alpha)$ of tests corresponds to rejecting H_0: $\xi = 0$ in favor of the alternative hypothesis H_1: $\xi > 0$, whenever

$$Z_n^*(\hat{\boldsymbol{\theta}}_n) = [S_n^*(\hat{\boldsymbol{\theta}}_n)]^{-1} \sum_{k=1}^{n} f_{nk}^*(X_{nk}, \hat{\boldsymbol{\theta}}_n) > z_{1-\alpha} \tag{2.12}$$

Furthermore, for the local alternatives Γ, its asymptotic power is given by

$$1 - \Phi(z_{1-\alpha} - \xi_n S_n^*(\theta)) \tag{2.13}$$

where $z_{1-\alpha}$ satisfies $\Phi(z_{1-\alpha}) = 1 - \alpha$, and Φ is the cumulative distribution function of the standard normal distribution. It may be remarked here that an asymptotically equivalent optimal $C(\alpha)$ test results if the $a_j^0(\boldsymbol{\theta})$'s are replaced by $a_{nj}^0(\boldsymbol{\theta})$, $j = 1, 2, \ldots, r$. The test function so obtained has the same asymptotic properties as those of $Z_n^*(\hat{\boldsymbol{\theta}}_n)$.

3. AN OPTIMAL $C(\alpha)$-TEST FOR THE POISSON HYPOTHESIS

We return to our problem introduced in Section 1 and apply the $C(\alpha)$-test theory of Section 2 to arrive at a solution. For our random variables X_i of

(1.4) with $i = 1, 2, \ldots, n$, let

$$p_i(m|\xi, \lambda) = P(X_i = m|\xi, \lambda), \qquad m = 0, 1, 2, \ldots \tag{3.1}$$

Then it can be easily seen [see Katti (1967)] that for every i, these probabilities are recursively related to each other and to the probabilities $R_i(k|\xi, \lambda)$ of (1.6) as follows:

$$\begin{cases} p_i(0|\xi, \lambda) = \exp(-\lambda t_i) \\ p_i(m+1|\xi, \lambda) = \dfrac{\lambda t_i}{m+1} \sum_{k=0}^{m} (m-k+1) p_i(k|\xi, \lambda) \\ R_i(m-k+1|\xi, \lambda), \qquad m \geq 0 \end{cases} \tag{3.2}$$

Also in view of (1.7), we have for $m \geq 0$,

$$\lim_{\xi \to 0} p_i(m|\xi, \lambda) = p_i(m|0, \lambda) = \frac{(\lambda t_i)^m}{m!} \exp(-\lambda t_i) \tag{3.3}$$

We add further conditions A_1–A_3 on the probabilities $R_i(k|\xi, \lambda)$ and the p_i's given below.

(A_1) *For $i = 1, 2, \ldots, n$, and $k \geq 1$, the probabilities $R_i(k|\xi, \lambda)$ are twice differentiable with respect to ξ for $\xi \geq 0$ and also with respect to λ for $\lambda > 0$ (the derivatives with respect to ξ at $\xi = 0$ are to be considered as right-hand derivatives), and these differentiations are valid under the summation sign of*

$$\sum_{k=1}^{\infty} R_i(k|\xi, \lambda) \equiv 1 \tag{3.4}$$

We denote these derivatives by $R_{i\xi}(k|\xi, \lambda)$, $R_{i\lambda}(k|\xi, \lambda), \ldots$. It is assumed that for $i = 1, 2, \ldots, n$ and for all $\lambda > 0$, $\sum_{j=1}^{\infty} jR_{i\xi}(j|0, \lambda)$ is finite, with $R_{i\xi}(1|0, \lambda) < 0$. Furthermore,

$$R_{i\lambda}(j|0, \lambda) \equiv 0, \qquad j = 1, 2, \ldots, \quad \forall \lambda > 0 \tag{3.5}$$

where

$$R_{i\xi}(j|0, \lambda) = \left.\frac{\partial R_i(j|\xi, \lambda)}{\partial \xi}\right|_{\xi=0}, \qquad R_{i\lambda}(j|0, \lambda) = \left.\frac{\partial R_i(j|\xi, \lambda)}{\partial \lambda}\right|_{\xi=0} \tag{3.6}$$

REMARK 3.1. Note that in view of (1.6) and (1.7), it is easily seen that $R_{i\xi}(1|0, \lambda) \le 0$ and $R_{i\xi}(j|0, \lambda) \ge 0$ for $j \ge 2$. Moreover, condition A_1 implies that

$$-R_{i\xi}(1|0, \lambda) = \sum_{j=2}^{\infty} R_{i\xi}(j|0, \lambda), \qquad i = 1, 2, \ldots, n \tag{3.7}$$

holds with $R_{i\xi}(1|0, \lambda) < 0$. Again condition (3.5) holds for most of the standard cases. For instance, in the case of the negative binomial distribution viewed as a compound Poisson distribution, the corresponding probabilities $R_i(k|\xi, \lambda)$ depend both on ξ as well as λ and in particular they satisfy (3.5) [see concluding remarks in Puri (1985)]. Basically all that condition (3.5) requires is that for all $j \ge 1$ and $\lambda > 0$, in

$$\lim_{\xi \to 0} \lim_{\delta \to 0} \frac{R_i(j|\xi, \lambda + \delta) - R_i(j|\xi, \lambda)}{\delta} \tag{3.8}$$

an interchange of the two limits be allowed. Finally, it follows from (3.2) and condition A_1 that for every $i = 1, 2, \ldots, n$ and $m \ge 0$, the probabilities $p_i(m|\xi, \lambda)$ are twice differentiable with respect to $\xi \ge 0$ and also for $\lambda > 0$.

(A_2) *It is assumed that the above differentiations of $p_i(m|\xi, \lambda)$ are permitted under the summation sign of*

$$\sum_{m=0}^{\infty} p_i(m|\xi, \lambda) = 1 \tag{3.9}$$

We define for $i = 1, 2, \ldots, n$,

$$\varphi_{i\xi}(m|\lambda) = \left.\frac{\partial \ln p_i(m|\xi, \lambda)}{\partial \xi}\right|_{\xi=0}, \qquad \varphi_{i\lambda}(m|\lambda) = \left.\frac{\partial \ln p_i(m|\xi, \lambda)}{\partial \lambda}\right|_{\xi=0} \tag{3.10}$$

(A_3) *We assume that the t_i are bounded both from above as well as away from 0. In particular, $n^{-1}\Sigma_{i=1}^{n} t_i$ and*

$$\frac{1}{n} \sum_{i=1}^{n} t_i \left[\sum_{j=1}^{\infty} j R_{i\xi}(j|0, \lambda) \right] \tag{3.11}$$

both converge as $n \to \infty$. Finally, it is assumed that both $\{\varphi_{i\xi}(m|\lambda) : i = 1, 2 \ldots\}$ and $\{\varphi_{i\lambda}(m|\lambda) : i = 1, 2, \ldots\}$ form Cramér sequences [see Bartoo and Puri (1967) for their definition], which implies that the limit

$$\lim_{n \to \infty} \frac{1}{n} \sum_{i=1}^{n} E_{0,\lambda}\left[\varphi_{i\xi}(X_i|\lambda)\right]^2 \tag{3.12}$$

exists and is strictly positive for all $\lambda > 0$.

As in Puri (1985) standard calculations using (3.2) and (3.5) lead us to the following lemma.

Lemma 3.1. For $i = 1, 2, \ldots$

$$\begin{cases} \varphi_{i\xi}(m|\lambda) = \sum_{j=0}^{m} \frac{m!}{(m-j)!}(\lambda t_i)^{-(j-1)} R_{i\xi}(j|0,\lambda), & m \geq 1 \\ \varphi_{i\xi}(0|\lambda) = 0 \end{cases} \tag{3.13}$$

$$\varphi_{i\lambda}(m|\lambda) = \left(\frac{m}{\lambda} - t_i\right), \qquad m \geq 0 \tag{3.14}$$

where we have adopted the convention that

$$R_{i\xi}(0|0,\lambda) = R_{i\lambda}(0|0,\lambda) = 0$$

REMARK 3.2. As was shown in Puri (1985), subject to (1.7) and from condition A_1, it follows that for every i,

$$E_{0,\lambda}\left[\varphi_{i\xi}(X_i|\lambda)\right]^2 > 0 \tag{3.15}$$

and that under the hypothesis H_0: $\xi = 0$, the random quantities $\varphi_{i\xi}(X_i|\lambda)$ and $\varphi_{i\lambda}(X_i|\lambda)$ are linearly independent with a positive probability for every $\lambda > 0$, a condition needed for C_2 as part of the existence of an optimal $C(\alpha)$ test for the present case.

The next lemma gives further expressions that are needed for the construction of the optimal $C(\alpha)$ test. The proof is omitted as it involves lengthy but routine calculations using (3.13) and (3.14).

Lemma 3.2. For $i = 1, 2, \ldots,$ we have

$$\operatorname{Var}_{0,\lambda}(\varphi_{i\lambda}(X_i|\lambda)) = t_i/\lambda \tag{3.16}$$

$$\operatorname{Cov}_{0,\lambda}(\varphi_{i\xi}(X_i|\lambda), \varphi_{i\lambda}(X_i|\lambda)) = t_i \sum_{j=0}^{\infty} jR_{i\xi}(j|0,\lambda) \tag{3.17}$$

$$a_n^0(\lambda) = \lambda\left(\sum_{i=1}^{n} t_i\right)^{-1} \sum_{i=1}^{n} t_i\left[\sum_{j=0}^{\infty} jR_{i\xi}(j|0,\lambda)\right] \tag{3.18}$$

$$\operatorname{Var}_{0,\lambda}(\varphi_{i\xi}(X_i|\lambda)) = V_i(\lambda t_i) \tag{3.19}$$

where

$$V_i(\lambda t_i) = (\lambda t_i)^2 \sum_{k=1}^{\infty} R_{i\xi}(k|0,\lambda) \sum_{l=1}^{\infty} R_{i\xi}(l|0,\lambda) B_{l,k}(\lambda t_i) \tag{3.20}$$

and

$$B_{l,k}(\lambda t_i) = \sum_{r=0}^{\min(l,k)} \binom{k}{r}\binom{l}{r}(r!)(\lambda t_i)^{-r} \tag{3.21}$$

On comparing the expression (3.20) with the corresponding one of Puri (1985, Eq. (31)) we note that a factor of λ^2 was inadvertently left out there. Finally, applying the $C(\alpha)$-test theory of Section 2 [in particular (2.12) and (2.13) and Lemmas 3.1 and 3.2], the following theorem gives the desired optimal $C(\alpha)$ test for our problem.

Theorem 3.1. Subject to assumptions A_1–A_3, an optimal $C(\alpha)$ test for testing H_0: $\xi = 0$ against H_1: $\xi > 0$ is to reject H_0 whenever

$$\tilde{Z}_n(\hat{\lambda}) > z_{1-\alpha} \tag{3.22}$$

where

$$\tilde{Z}_n(\hat{\lambda}) = [\tilde{S}_n(\hat{\lambda})]^{-1} \sum_{i=1}^{n} g_{ni}(X_i, \hat{\lambda}) \tag{3.23}$$

$$g_{ni}(X_i, \lambda) = \varphi_{i\xi}(X_i|\lambda) - a_n^0(\lambda)\varphi_{i\lambda}(X_i|\lambda) \tag{3.24}$$

$$\tilde{S}_n^2(\lambda) = \sum_{i=1}^{n} V_i(\lambda t_i) - \lambda\left(\sum_{i=1}^{n} t_i\right)^{-1}\left[\sum_{i=1}^{n} t_i \sum_{j=0}^{\infty} jR_{i\xi}(j|0, \lambda)\right]^2 \tag{3.25}$$

and $\hat{\lambda}$ stands for a locally $\sqrt{n}$ consistent estimator of the nuisance parameter λ.

Of particular importance is the special case when $R_{i\xi}(j|0, \lambda) = 0$ for $j \geq 3$ and for all $i = 1, 2, \ldots, n$. This condition is satisfied for several well-known compound Poisson alternatives arising in practice, such as negative binomial distributions and Neyman type A distributions, among others. Furthermore, the test statistic (3.23) simplifies somewhat for this case. The next theorem deals with this special case.

Theorem 3.2. In addition to the conditions of theorem 3.1, if $\forall i = 1, 2, \ldots,$ $R_{i\xi}(j|0, \lambda) = 0$ for $j \geq 3$, also holds, the test statistic (3.23) reduces to

$$\tilde{Z}_n(\hat{\lambda}) = [\tilde{S}_n(\hat{\lambda})]^{-1}\left\{\sum_{i=1}^{n}\left[\frac{X_i(X_i-1)}{\hat{\lambda}t_i} - X_i\right]R_{i\xi}(2|0, \hat{\lambda}) - a_n^0(\hat{\lambda})\sum_{i=1}^{n}\left(\frac{X_i}{\hat{\lambda}} - t_i\right)\right\} \tag{3.26}$$

with the asymptotic power of the optimal $C(\alpha)$ test, for the local alternatives $\{\xi_n\}$, given by

$$1 - \Phi\left(z_{1-\alpha} - \xi_n \tilde{S}_n(\lambda)\right) \tag{3.27}$$

where

$$a_n^0(\lambda) = \lambda\left[\sum_{i=1}^{n} t_i\right]^{-1}\left[\sum_{i=1}^{n} t_i R_{i\xi}(2|0, \lambda)\right]$$

$$(3.28)$$

$$\tilde{S}_n^2(\lambda) = 2\sum_{i=1}^{n}\left[R_{i\xi}(2|0, \lambda)\right]^2 + \lambda\sum_{i=1}^{n} t_i\left(R_{i\xi}(2|0, \lambda)\right)^2$$

$$-\lambda\left[\sum_{i=1}^{n} t_i\right]^{-1}\left[\sum_{i=1}^{n} t_i R_{i\xi}(2|0, \lambda)\right]^2 \tag{3.29}$$

Finally, the next theorem further specializes the test to the case where the $R_{i\xi}(2|0, \lambda)$ are all equal for $i \geq 1$.

Theorem 3.3. If, subject to the conditions of theorem 3.2, the $R_{i\xi}(2|0, \lambda)$ are all equal for $i \geq 1$, the test statistic (3.26) further reduces to

$$\tilde{Z}_n(\hat{\lambda}) = (2n)^{-1/2}\sum_{i=1}^{n}(\hat{\lambda}t_i)^{-1}\left\{(X_i - \hat{\lambda}t_i)^2 - X_i\right\} \tag{3.30}$$

Furthermore, for the local alternatives $\{\xi_n\}$, the asymptotic power of the optimal $C(\alpha)$ test is given by (3.27) with

$$\tilde{S}_n^2(\lambda) = 2n\left[R_\xi(2|0, \lambda)\right]^2 \tag{3.31}$$

4. CONCLUDING REMARKS

For the case considered in Theorem 3.3, note that the asymptotic power for the local alternatives [as is exhibited through (3.31)] does not depend on the lengths t_i of the various observation periods. This implies, under the assumption of Theorem 3.3, the following interesting observation regarding the design of our experiments. For instance, if the total length of the periods $\Sigma_{i=1}^{n} t_i$ is fixed, it would be advisable to take a larger n with smaller t_i's, for example, by doubling the number n of observations and at the same time cutting down the average period length per observation to half that of the original one. Intuitively this sounds reasonable, since for small t_i's most of the time the positive X_i's would constitute the results of single jumps of the process, thereby enabling us to distinguish more effectively between the two situations, namely, the one where the jump sizes are 1 (Poisson process case) and the other where the jump sizes are greater than 1 with a positive probability (compound Poisson process case).

Again for the same case of Theorem 3.3, it is interesting to note that the optimal test based on the test statistic (3.30) is independent of the common p.g.f. $h(s|\xi, \lambda)$. This property has been referred to as the *robustness of optimality* property by Neyman [see Bartoo and Puri (1967) for more comments on this point].

In the above tests one needs to have a locally $\sqrt{n}$ consistent estimator of the nuisance parameter λ. One possible way of constructing such an estimate for the most general case of Theorem 3.1 is by using the fact that $P(X_i = 0) = \exp(-\lambda t_i)$, $i = 1, 2, \ldots, n$. Let $\eta_i = 1$ if $X_i = 0$ and $\eta_i = 0$ otherwise, $i = 1, 2, \ldots, n$. The likelihood function based on $\eta_1, \eta_2, \ldots, \eta_n$ is then given by

$$L(\lambda|\eta_1, \eta_2, \ldots, \eta_n) = \exp\left(-\lambda \sum_{i=1}^{n} t_i\eta_i\right) \prod_{i=1}^{n} \{1 - \exp(-\lambda t_i)\}^{1-\eta_i} \qquad (4.1)$$

The desired estimate of λ is now simply the maximum-likelihood estimate based on (4.1) and is given by the unique solution for λ of the equation

$$\sum_{i=1}^{n} \eta_i t_i = \sum_{i=1}^{n} (1 - \eta_i) t_i [\exp(\lambda t_i) - 1]^{-1} \qquad (4.2)$$

We need hardly mention that other, more involved, estimates of λ that use more information about the X_i are also possible.

Finally, the approach adopted here for the testing of the Poisson hypothesis falls within the classical α-level hypothesis-testing framework. The question of testing the Poisson hypothesis is currently being studied in the present context from a decision-theoretic as well as a Bayesian point of view. The results pertaining to these investigations will be reported elsewhere.

REFERENCES

Bartoo, J. B. and Puri, P. S. (1967). On optimal asymptotic tests of composite statistical hypotheses. *Ann. Math. Statist.* **38**, 1845–1852.

Bühler, W. J. and Puri, P. S. (1966). On optimal asymptotic tests of composite hypotheses with several constraints. *Z. Wahrscheinlichkeitstheorie* **5**, 71–88.

Cramér, H. (1946). *Mathematical Methods of Statistics*, Princeton, New Jersey: Princeton University Press.

Katti, S. K. (1967). Infinite divisibility of integer valued random variables. *Ann. Math. Statist.* **38**, 1306–1308.

Neyman, J. (1959). Optimal asymptotic tests of composite statistical hypotheses. *The Herald Cramér Volume*. Stockholm: Almquist and Wiksell, and New York: Wiley, pp. 213–234.

Neyman, J. and Puri, P. S. (1976). A structural model of radiation effects in living cells. *Proc. Nat. Acad. Sci. U.S.A.* **73**, 3360–3363.

Neyman, J. and Puri, P. S. (1981). A hypothetical stochastic mechanism of radiation effects in single cells. *Proc. R. Soc. B* **213**, 139–160.

Puri, P. S. (1982). A hypothetical stochastic mechanism of radiation effects in single cells: Some further thoughts and results. In *Probability Models and Cancer*, *Proceedings of the Inter-Disciplinary Cancer Study Conference* (LeCam, L. and Neyman, J., Eds.) New York: North Holland, pp. 171–187.

Puri, P. S. (1985). On an optimal $C(\alpha)$-test of Poisson hypothesis against compound Poisson alternatives. In *Statistical Theory and Data Analysis* (Matusita, K., Ed.) Amsterdam: North-Holland, pp. 565–582.

Matrix Derivatives of Patterned Matrices and Statistical Applications

D. S. Tracy and K. G. Jinadasa
Department of Mathematics and Statistics
University of Windsor
Windsor, Ontario, Canada

SUMMARY

Matrix derivatives are often used for estimation problems in multivariate analysis. Patterned matrices, where relationships exist between the elements of the matrix, pose special problems in matrix differentiation. A general method is suggested to take care of these, using representations of differentials. The method is illustrated on the least-squares estimation of the three-mode factor-analysis model.

1. INTRODUCTION

In estimation problems in multivariate analysis, econometrics and psychometrics, maximum-likelihood and generalized least-squares methods call for maximization or minimization of functions of matrix arguments. The use of matrix derivatives results in mathematically efficient methods. By their very nature, covariance matrices are symmetric, and in covariance structures, other types of patterns appear between the elements of a matrix.

Such relationships between the elements create problems in matrix differentiation. Nel (1980) has proposed a method to handle derivatives of patterned matrices. This method is not applicable when nonzero constants appear in the matrix. In certain models, like confirmatory factor analysis, however, identifiability of parameters forces a situation with nonzero constants.

To overcome the problems, we have proposed a procedure (Tracy and Jinadasa, 1985) in which we first obtain the matrix derivative ignoring the pattern in the argument matrix, and then postmultiply it by a matrix related to the pattern. For this purpose, we study matrix derivatives as representations of differentials of matrix-valued functions of matrix arguments with respect to formal bases. Our method is applicable in very general situations, and is simpler to handle even when Nel's method is applicable.

Dwyer (1967) gave formulas for derivatives of matrix functions of symmetric matrices (i.e., for the symmetric patterned matrices) and McCulloch (1982) has proposed an extension of these formulas using a result given by his Theorem 1 (p. 680). We find that there is a possibility that his Theorem 1 could lead to a contradictory argument.

Finally, we consider the least-squares estimation of the three-mode factor-analysis model (Bentler and Lee, 1979) and calculate the necessary matrix derivatives required to obtain the gradient of the objective function and for the iterative procedure of obtaining the estimators.

2. DEFINITIONS

Definition 2.1. A matrix is said to be *patterned* if one or more of the following is true:

1. Some elements are constant.
2. Some elements are repeated, possibly with a change of sign.
3. Some elements are functions of the other elements.

Otherwise, it is said to be *independent and variable*.

Definition 2.2. vec X is the column vector obtained by stacking all the columns of X, one above the other, starting with the first column. This is the same as X_c (Tracy and Dwyer, 1969) and packing operator $P(X)$ (MacRae, 1974).

Definition 2.3. vech X is the column vector obtained from vec X after eliminating all the constant elements, repeated elements, and elements that are functions of (dependent on) some other elements, thus retaining only the distinct and independent elements (Henderson and Searle, 1979). This is the same as vecp X (Nel, 1980).

Definition 2.4. Suppose we rearrange the rows of the identity matrix of order mn by taking every mth row, starting with the first, then every mth row, starting with the second, and so on. This matrix, first introduced by Tracy and Dwyer (1969), is denoted $P_{n,m}$ and called the *permuted identity matrix* (MacRae, 1974; Izenman, 1975) or *commutation matrix* (Magnus and Neudecker, 1979).

Definition 2.5. A real-valued function f defined on an open interval I is said to be *differentiable at a point c in I* if

$$\lim_{h \to 0} \frac{f(c+h) - f(c)}{h}$$

exists. The limit is denoted f_c'.

Definition 2.6. Let f be a function defined on an open subset A of a normed linear space M, taking values in a normed linear space N. Let c be any point in A. If there exists a linear function f_c' on M into N satisfying

$$\lim_{\|h\| \to 0} \frac{f(c+h) - f(c) - f_c'(h)}{\|h\|} = 0 \in N \qquad (2.1)$$

then f is said to be *differentiable at the point c*, and f_c' is said to be the *total differential of f at c*.

Let u be a unit vector. Then $f_c'(u)$ is the directional derivative of f at c in the direction of u. Partial derivatives are the directional derivatives along the basis vectors of the space. Thus the partial derivatives depend on the bases of the underlying spaces (range and domain) of the function.

Sometimes, for convenience, the domain of the function may be changed, keeping the range fixed, using composition of functions. If we ignore this change of domain after obtaining the derivative, however, it may lead to confusion.

Take, for example, $f(x, 2x) = 4x^2$. If we consider the domain to be a subspace of R^2, $8x$ is the directional derivative in the direction of $(1, 2)$. If we consider the domain to be R, however, considering f as a composition of two functions, then $8x$ is actually the derivative of this composite function. If we wanted to calculate the total derivative of $f(x, 2x) = 4x^2$, we get $(8/\sqrt{5})x$. This shows that the total derivative of the composite function equals the directional derivative of the original function but not the total derivative of the original function. Thus when we change the domain for convenience, the derivative obtained after that must be identified properly.

3. MATRIX DERIVATIVES

3.1. Independent and Variable Matrices

Let X be an $m \times n$ independent and variable matrix with real entries, and let $Y = F(X)$ be a $p \times q$ matrix, also with real entries. Let M and N be the collections of all $m \times n$ and $p \times q$ matrices, respectively, with real entries. Then it is well known that M and N are normed linear spaces (we may take the trace norm). When X is independent and variable, the domain of the function Y is M itself. So the differential can be defined at any point c of M, if it exists, as f_c', satisfying condition (2.1).

Since f_c' is linear, we can obtain a matrix representation if we fix bases in M and N. This representation is not unique and depends on the bases chosen. Yet it is needed in practice since the existence of an abstract function itself is not sufficient, and we need representations to work with. Thus, in optimization problems, when closed-form solutions do not exist, in order to use iterative procedures, we need to use the gradients of the functions. (Sometimes the second-order derivative is also needed.) There precisely we have to use the representations.

We shall identify some of the matrix derivatives that appear in the literature as different representations of the above differential.

Let $\{\varepsilon_{ij} : i = 1,\dots,m,\ \ j = 1,\dots,n\}$ and $\{\delta_{kl} : k = 1,\dots,p,\ \ l = 1,\dots,q\}$ be the usual bases in M and N, respectively. (ε_{ij} is an $m \times n$ matrix with only the ijth element 1 and the rest 0. Similarly δ_{kl}.) Usual bases are chosen for convenience.

Now

$$f_c'(\varepsilon_{ij}) = \lim_{t\to 0} \frac{f(c + t\varepsilon_{ij}) - f(c)}{t}$$

Denoting by f_{kl} the coefficient of δ_{kl} in the linear expansion of a matrix in terms of the basis $\{\delta_{kl}\}$, we have

$$f(c + t\varepsilon_{ij}) = \sum_{kl} f_{kl}(c + t\varepsilon_{ij})\delta_{kl}, \qquad f(c) = \sum_{kl} f_{kl}(c)\delta_{kl}$$

Thus,

$$\begin{aligned} f_c'(\varepsilon_{ij}) &= \lim_{t\to 0}\left\{\sum_{kl}\left\{\frac{f_{kl}(c + t\varepsilon_{ij}) - f_{kl}(c)}{t}\right\}\delta_{kl}\right\} \\ &= \sum_{kl}\left\{\lim_{t\to 0}\frac{f_{kl}(c + t\varepsilon_{ij}) - f_{kl}(c)}{t}\right\}\delta_{kl} \\ &= \sum_{kl}\frac{\partial f_{kl}(c)}{\partial x_{ij}} \cdot \delta_{kl} \end{aligned}$$

Thus $\partial f_{kl}(c)/\partial x_{ij}$ is the element corresponding to ε_{ij} and δ_{kl}. We arrange $\{\varepsilon_{ij}\}$ in the order $\{\varepsilon_{11}, \varepsilon_{21}, \ldots, \varepsilon_{m1}, \ldots, \varepsilon_{1n}, \ldots, \varepsilon_{mn}\}$ and similarly $\{\delta_{kl}\}$. Then the representation obtained with respect to these two bases is $\partial \operatorname{vec} Y/\partial \operatorname{vec} X$ (Nel, 1980). This is the same as $\partial Y_c/\partial X_c$ (Tracy and Dwyer, 1969).

If, instead, we obtain all the elements of the representation corresponding to each δ_{kl} and place them in the order of the indices (ij) to form an $m \times n$ matrix and then construct an $mp \times nq$ matrix by letting its klth block be the matrix corresponding to δ_{kl}, we obtain

$$\frac{\partial Y}{\partial X} = Y \otimes \frac{\partial}{\partial X}$$

(MacRae, 1974). This is another representation of the above differential. If we interchange the roles of the two bases $\{\varepsilon_{ij}\}$ and $\{\delta_{kl}\}$ in the above, we obtain $\varepsilon Y/\varepsilon X$ (Rogers, 1980).

Thus the various derivatives that appear in the literature are different arrangements of the representation of the general differential of the matrix function when the argument matrix is variable and independent.

3.2. Patterned Matrices

Matrix derivatives are well established in the literature for the case of variable and independent matrices. But when the argument matrix is patterned, such as symmetric, different methods are proposed in the literature, and Nel (1980) points out that sometimes these methods lead to confusion. We have proposed (Tracy and Jinadasa, 1985) a simple general procedure that is applicable to any kind of pattern. For completeness, we outline the procedure here.

Let $Y = f(X)$ be a function of the patterned matrix A. First obtain $\partial \operatorname{vec} Y/\partial \operatorname{vec} X$ ignoring the pattern of X. Then obtain $J = \partial \operatorname{vec} X/\partial \operatorname{vech} X$ imposing the pattern of X. The required matrix derivative is

$$\frac{\partial \operatorname{vec} Y}{\partial \operatorname{vech} X} = \frac{\partial \operatorname{vec} Y}{\partial \operatorname{vec} X} \cdot J$$

Whatever the pattern of X may be, it is easy to obtain the matrix J. In this case it is to be noted that the domain of the function is changed.

In Section 4 we show that, if we do not identify the domain correctly, certain results in the literature could cause confusion.

4. A MODIFICATION PROPOSED BY McCULLOCH

McCulloch (1982) uses the notation of MacRae (1974) for the matrix derivative

$$\frac{dY}{dX} = Y \otimes \frac{\partial}{\partial X}$$

He regards the $p \times p$ symmetric matrix Σ as a function of vech V, where

$$\sigma_{ij}(\text{vech}\, V) = \begin{cases} V_{ij}, & i \leq j \\ V_{ji}, & i > j \end{cases}$$

Thus vech Σ = vech V, and according to MacRae (1974, (2.1))

$$\frac{d\, \text{vech}\, V}{d\, \text{vech}\, \Sigma} = \text{vec}\, I_{p(p+1)/2}$$

Using $*$ for the star product (MacRae, 1974, p. 339),

$$\frac{df(\Sigma)}{d\, \text{vech}\, \Sigma} = \frac{df(\Sigma)}{d\, \text{vech}\, V} * \frac{d\, \text{vech}\, V}{d\, \text{vech}\, \Sigma} = \frac{df(\Sigma)}{d\, \text{vech}\, V} * \text{vec}\, I_{p(p+1)/2}$$

$$= \frac{df(\Sigma)}{d\, \text{vech}\, V}$$

This shows that

$$\frac{df(\Sigma)}{d\, \text{vech}\, \Sigma} = \frac{df(\Sigma)}{d\, \text{vech}\, V}$$

Theorem 1 of McCulloch (1982) gives

$$\frac{df(\Sigma)}{d\, \text{vech}\, V} = G'G\, \text{vech} \frac{df(\Sigma)}{d\Sigma}$$

But since $f(\Sigma)$ is scalar-valued,

$$\text{vech} \frac{df(\Sigma)}{d\Sigma} = \frac{df(\Sigma)}{d\, \text{vech}\, \Sigma}$$

Thus

$$G'G \frac{df(\Sigma)}{d\, \text{vech}\, \Sigma} = \frac{df(\Sigma)}{d\, \text{vech}\, \Sigma}$$

Since $f(\Sigma)$ is an arbitrary function of Σ, the above expression implies that $G'G = I$. Apparently this is not true, according to the definition of G (McCulloch 1982, (2.1)).

As an example, consider

$$X = \begin{bmatrix} x_{11} & x_{12} \\ x_{12} & x_{22} \end{bmatrix}$$

Then

$$\operatorname{vec} X = \begin{bmatrix} x_{11} \\ x_{12} \\ x_{12} \\ x_{22} \end{bmatrix}, \qquad \operatorname{vech} X = \begin{bmatrix} x_{11} \\ x_{12} \\ x_{22} \end{bmatrix}$$

and vec $X = G$ vech X with

$$G = \begin{bmatrix} 1 & 0 & 0 \\ 0 & 1 & 0 \\ 0 & 1 & 0 \\ 0 & 0 & 1 \end{bmatrix}$$

Then

$$G'G = \begin{bmatrix} 1 & 0 & 0 \\ 0 & 2 & 0 \\ 0 & 0 & 1 \end{bmatrix} \neq I_3$$

The reason this theorem leads to this type of result is that the domain of the function is not properly identified. Mathematically there is no difference between vech V and vech Σ. In obtaining $df(\Sigma)/d$ vech Σ, certainly we find the derivative of the composite function. Introduction of vech V leads to the above contradictory result.

5. A RESULT ABOUT VEC($A \otimes B$)

We now prove a useful result related to the vector of Kronecker product of two matrices. This is helpful later in obtaining the derivatives of Kronecker products. Let $A_{m \times n}$ and $B_{p \times q}$ be two arbitrary matrices. Then according to Neudecker and Wansbeek (1983), with commutation matrix P,

$$\begin{aligned} \operatorname{vec}(A \otimes B) &= (I_n \otimes P_{m,q} \otimes I_p)(\operatorname{vec} A \otimes \operatorname{vec} B) && (5.1) \\ &= (I_n \otimes P_{m,q} \otimes I_p) \operatorname{vec}[\operatorname{vec} B \cdot (\operatorname{vec} A)'] \\ &= (I_n \otimes P_{m,q} \otimes I_p)(\operatorname{vec} A \otimes I_{pq}) \operatorname{vec} B && (5.2) \end{aligned}$$

$$\begin{aligned} \operatorname{vec}(A \otimes B) &= (I_n \otimes P_{m,q} \otimes I_p) P_{pq,mn} \operatorname{vec}[\operatorname{vec} A (\operatorname{vec} B)'] \\ &= (I_n \otimes P_{m,q} \otimes I_p) P_{pq,mn} (\operatorname{vec} B \otimes I_{mn}) \operatorname{vec} \mathrm{A} && (5.3) \end{aligned}$$

6. THREE-MODE FACTOR-ANALYSIS MODEL

6.1. The Model

In this section we consider the three-mode factor-analysis model considered by Bentler and Lee (1979). The model is

$$x = \mu + (A \otimes B)G\gamma + \zeta$$

where x is a pq vector of observations; μ is a vector of population means; A, B, and G are parameter matrices of order $p \times a$, $q \times b$, and $ab \times c$, respectively; γ and ζ are random vectors of dimension c and pq, respectively. The covariance matrix of the model is

$$\Sigma = (A \otimes B)G\phi G'(A' \times B') + Z,$$

where $E(\gamma\gamma') = \phi$, $E(\zeta\zeta') = Z$ and $E(\gamma\zeta') = 0$, ϕ and Σ are symmetric.

For this model to be identifiable, Bentler and Lee (1979) proposed the following specification of parameter matrices: $\phi = I$, G is lower triangular, that is, $G_{ij} = 0$ if $i < j$, $A' = (A_1', A_2')_3$ and $B' = (B_1', B_2')$, where A_1 and B_1 are identity matrices. The restrictions make the corresponding matrices patterned. Corresponding to each pattern we can obtain the related J matrix very easily. In this case each J matrix will be the unit matrix with certain rows repeated and/or zero rows inserted.

6.2. Generalized Least-Squares Estimation

Let S be the sample variance covariance matrix. The objective function to be minimized in the generalized least-squares estimation is

$$\begin{aligned} Q &= \tfrac{1}{2}\operatorname{tr}\left[(S - \Sigma)S^{-1}\right]^2 \\ &= \tfrac{1}{2}\left[\operatorname{vec}(S - \Sigma)\right]' S^{-1} \otimes S^{-1} \operatorname{vec}(S - \Sigma) \end{aligned}$$

Since S^{-1} is positive definite, so is $S^{-1} \otimes S^{-1}$. Taking its symmetric square root, we can write $Q = \frac{1}{2}[h(\Sigma)]'h(\Sigma)$, where $h(\Sigma) = U^{1/2}\operatorname{vec}(S - \Sigma)$ and $S^{-1} \otimes S^{-1} = U^{1/2}U^{1/2}$. In order to implement the Gauss–Newton algorithm we only have to calculate the Jacobian matrix $\partial h(\Sigma)/\partial\boldsymbol{\theta}$, where $\boldsymbol{\theta}$ is the vector of all the parameters involved in Σ, that is,

$$\boldsymbol{\theta} = [(\operatorname{vech} A)', (\operatorname{vech} B)', (\operatorname{vech} G)', (\operatorname{vech} \phi)', (\operatorname{vech} Z)']'$$

But

$$\frac{\partial h(\Sigma)}{\partial \boldsymbol{\theta}} = -U^{1/2}\frac{\partial \text{ vec } \Sigma}{\partial \boldsymbol{\theta}}$$

Thus

$$\frac{\partial \text{ vec } \Sigma}{\partial \boldsymbol{\theta}} = \left(\frac{\partial \text{ vec } \Sigma}{\partial \text{ vech } A}, \frac{\partial \text{ vec } \Sigma}{\partial \text{ vech } B}, \frac{\partial \text{ vec } \Sigma}{\partial \text{ vech} G}, \frac{\partial \text{ vec } \Sigma}{\partial \text{ vech} \phi}, \frac{\partial \text{ vec } \Sigma}{\partial \text{ vech } Z} \right)$$

But for any patterned matrix H,

$$\frac{\partial \text{ vec } \Sigma}{\partial \text{ vech } H} = \frac{\partial \text{ vec } \Sigma}{\partial \text{ vec } H} \cdot (J_H)$$

where (J_H) is the matrix corresponding to the pattern of H, and $\partial \text{ vec } \Sigma / \partial \text{ vec } H$ is calculated ignoring the pattern of H.

Now we shall calculate the required main derivatives of

$$\Sigma = (A \otimes B)G\phi G'(A' \otimes B') + Z.$$

By using (5.1) and (5.3), we obtain

$$\begin{aligned} &\text{vec}\left[(A \otimes B)G\phi G'(A \otimes B)\right] \\ &\quad = \left\{\left[\text{vec}(G\phi G')\right] \otimes I_{p^2q^2a^2b^2}\right\}\left(I_{ab} \otimes P_{pq,\,ab} \otimes I_{pq}\right) \\ &\qquad \times\left(I_a \otimes P_{p,\,b} \otimes I_{pa} \otimes P_{p,\,b} \otimes I_p\right)\left(\text{vec}\, A \otimes I_{qb} \otimes \text{vec}\, A \otimes I_{qb}\right) \\ &\qquad \times\left(\text{vec}\, B \otimes \text{vec}\, B\right) \end{aligned} \tag{6.1}$$

Similarly, we get using (5.1) and (5.3),

$$\begin{aligned} &\text{vec}\left[(A \otimes B)G\phi G'(A' \otimes B')\right] \\ &\quad = \left\{\left[\text{vec}(G\phi G')\right]' \otimes I_{p^2q^2a^2b^2}\right\}\left(I_{ab} \otimes P_{pq,\,ab} \otimes I_{pq}\right) \\ &\qquad \times\left(I_a \otimes P_{p,\,b} \otimes I_{qa} \otimes P_{p,\,b} \otimes I_q\right)\left(P_{qb,\,pa} \otimes P_{qb,\,pa}\right) \\ &\qquad \times\left[\text{vec}\, B \otimes I_{pq} \otimes \text{vec}\, B \otimes I_{pq}\right]\left(\text{vec}\, A \otimes \text{vec}\, A\right) \end{aligned} \tag{6.2}$$

It is well known that, for any A,

$$\frac{\partial(\operatorname{vec} A \otimes \operatorname{vec} A)}{\partial \operatorname{vec} A} = \left(I_{p^2q^2} + P_{pa,pa}\right)\left(\operatorname{vec} A \otimes I_{pq}\right) \tag{6.3}$$

Using (6.1) and (6.3), we obtain

$$\frac{\partial \operatorname{vec} \Sigma}{\partial \operatorname{vec} B} = \left\{\left[\operatorname{vec}(G\phi G')\right]' \otimes I_{p^2q^2a^2b^2}\right\}\left(I_{ab} \otimes P_{pq,ab} \otimes I_{pq}\right)$$

$$\times\left(I_a \otimes P_{p,b} \otimes I_{pa} \otimes P_{p,b} \otimes I_p\right)\left(\operatorname{vec} A \otimes I_{qb} \otimes \operatorname{vec} A \otimes I_{qb}\right)$$

$$\times\left(I_{q^2b^2} + P_{qb,qb}\right)\left(\operatorname{vec} B \otimes I_{qb}\right)$$

Similarly (6.2) and (6.3) give $\partial \operatorname{vec} \Sigma/\partial \operatorname{vec} A$.

Since

$$\frac{\partial \operatorname{vec}(G\phi G')}{\partial \operatorname{vec} G} = \left(I_{a^2b^2} + P_{ab,ab}\right)\left(G\phi \otimes I_{ab}\right) \quad \text{and} \quad \frac{\partial \operatorname{vec}(G\phi G')}{\partial \operatorname{vec} \phi} = G\phi G$$

we have

$$\frac{\partial \operatorname{vec} \Sigma}{\partial \operatorname{vec} G} = (A \otimes B \otimes A \otimes B)\left(I_{a^2b^2} + P_{ab,ab}\right)\left(G\phi \otimes I_{ab}\right)$$

and

$$\frac{\partial \operatorname{vec} \Sigma}{\partial \operatorname{vec} \phi} = (A \otimes B \otimes A \otimes B)(G \otimes G)$$

Finally $\partial \operatorname{vec} \Sigma/\partial \operatorname{vec} Z = I$.

Now according to the identification conditions we can define the patterns of all the matrices involved, and hence the J matrices also could be found. Then each of the above derivatives must be multiplied by the corresponding J matrix. In order to impose the patterns, Bentler and Lee (1979) had to introduce very cumbersome notation. Our method of handling the derivatives of patterned matrices makes things easier and gives very compact formulas for the derivatives.

REFERENCES

Bentler, P. M. and Lee, S. Y. (1979). A statistical development of three-mode factor analysis. *Br. J. Math. Statist. Psych.* **32**, 87–104.

Dwyer, P. S. (1967). Some applications of matrix derivatives in multivariate analysis. *J. Amer. Statist. Assoc.* **62**, 607–625.

Henderson, H. V. and Searle, S. R. (1979). Vec and vech operators for matrices, with some uses in jacobians and multivariate statistics. *Can. J. Statist.* **7**, 65–81.

Izenman, A. J. (1975). Reduced-rank regression for the multivariate linear model. *J. Mult. Anal.* **5**, 248–264.

MacRae, E. C. (1974). Matrix derivatives with an application to an adaptive linear decision problem. *Ann. Statist.* **2**, 337–346.

Magnus, J. R. and Neudecker, H. (1979). The commutation matrix: some properties and applications. *Ann. Statist.* **7**, 381–394.

McCulloch, C. E. (1982). Symmetric matrix derivatives with applications. *J. Amer. Statist. Assoc.* **77**, 679–682.

Nel, D. G. (1980). On matrix differentiation in statistics. *S. Afr. Statist. J.* **14**, 137–193.

Neudecker, H. and Wansbeek, T. (1983). Some results on commutation matrices with statistical applications. *Can. J. Statist.* **11**, 221–231.

Rogers, G. S. (1980). *Matrix Derivatives*. New York: Marcel Dekker.

Tracy, D. S. and Dwyer, P. S. (1969). Multivariate maxima and minima with matrix derivatives. *J. Amer. Statist. Assoc.* **64**, 1576–1594.

Tracy, D. S. and Jinadasa, K. G. (1985). On some misinterpretations of matrix derivatives. Unpublished manuscript.

The Hurwitz Distribution

José Pérez Vilaplana
Seminar of Statistics
Department of Applied Mathematics
Faculty of Science
University of the Basque Country
Bilbao, Vizcaya, Spain

SUMMARY

We introduce a new statistical distribution, the *Hurwitz distribution*, which is defined in terms of the expansion of the Hurwitz zeta function $\zeta(s + 1, a + 1)$. In this distribution, the probability that the random variable takes the value $k + a$ is proportional to the quantity $(k + a)^{-(s+1)}$, with $s > 0$, where k is a positive integer and a is a fixed real number, $0 < a \leq 1$. We study this distribution with its basic properties (e.g., distribution function, probability generating function, characteristic function, moments). Applying the integral representation of the Hurwitz zeta function, we demonstrate that this distribution is asymptotically convergent to the geometric distribution. We also prove that the maximum-likelihood estimator of the parameter s is an acceptable estimation of s. Finally, a table for individual probabilities for the values of $s = 1(0.5)5$, and $a = 0.1(0.1)1$, and an auxiliary table for the estimation of s are given.

1. INTRODUCTION

Estoup (1916) and Zipf (1949) defined independently the zeta distribution or Zipf–Estoup law (Johnson and Kotz, 1969, pp. 240ff.). Afterwards, it has been used with relative success in statistical mechanics and in actuarial applications (Seal, 1947, 1952). In this chapter we introduce a new statistical distribution, the *Hurwitz distribution*, which can be defined in terms of the expansion of the Hurwitz zeta function $\zeta(s, a)$, where $s > 1$ and a is a real number such that $0 < a \leq 1$ (Hurwitz, 1882). This distribution, similar to the Zipf–Estoup law, is useful when the discrete variate is defined in **R**. In certain cases, it can replace the discrete Pareto law satisfactorily. Our results mainly concern the study of its properties and the estimation of its parameters. Section 2 contains the basic definitions and some elementary properties. The results in Section 3 refer to the calculation of the probability distribution function and the characteristic function in terms of the integral representation of the Hurwitz zeta function. It is proved that the distribution is asymptotically convergent to the geometric distribution. In Section 4, the exact expressions of the rth factorial and about the mean moments are given. Also obtained are the estimation of the parameters and characteristics of the computed tables of this distribution.

2. THE HURWITZ DISTRIBUTION

A random variable X is said to have a *Hurwitz distribution* if its probability mass function is given by

$$P(X = k) = \begin{cases} c(k+a)^{-(s+1)}, & k = 1, 2, 3, \ldots \\ 0, & \text{elsewhere} \end{cases}$$

where $0 < a \leq 1$ and $s > 0$. Since the sum of all the above probabilities must be equal to 1,

$$\frac{1}{c} = \sum_{j=1}^{\infty} (j+a)^{-(s+1)} = \zeta(s+1, a+1)$$

from the functional equation

$$\zeta(s+1, a+1) + a^{-(s+1)} = \zeta(s+1, a)$$

In this distribution, we call s and a the *shape* and *scale parameters*, respectively. If $a = 1$, the relationship between the Zipf–Estoup law, $Z(s+1)$, and the Hurwitz distribution, $Z(s+1, 2)$, verify

$$\zeta(s+1, 2) \cdot Z(s+1, 2) = \zeta(s+1) \cdot Z(s+1) - 1$$

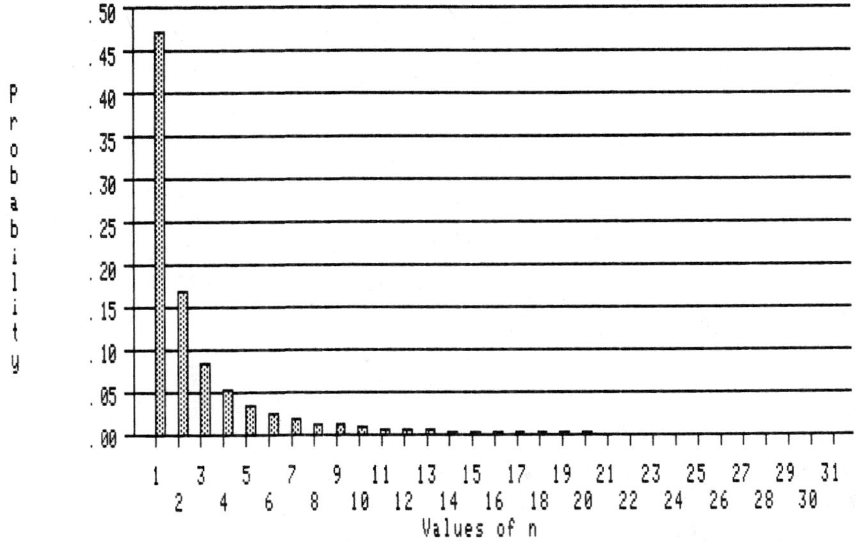

Figure 1

A plot of this function is given in Figure 1 for $s = 1$ and $a = 0.5$. For values of s greater than 5, the probabilities for $k > 3$ are practically 0. (See Table 1.)

From the relation

$$P(X = k)/P(X = k - 1) = [(k + a - 1)/(k + a)]^{s+1}$$
$$= [1 - 1/(k + a)]^{(s+1)} < 1$$

we can see that $P(X = k - 1) > P(X = k)$. As k tends to infinity, this ratio approaches 1. Therefore, the *mode* is always $X = 1$, and the *skewness* of the distribution is always positive.

3. INTEGRAL REPRESENTATION

From the definition of the gamma function results

$$(a + 1 + k)^{-(s+1)}\Gamma(s + 1)$$
$$= \int_0^\infty x^s \exp[-(a + 1 + k)x]\, dx \tag{3.1}$$

Table 1. Individual Terms, Hurwitz's Distribution

s = 1

n	a=0.1	a=0.2	a=0.3	a=0.4	a=0.5	a=0.6	a=0.7	a=0.8	a=0.9	a=1.0
1	0.577	0.548	0.522	0.498	0.475	0.455	0.436	0.419	0.403	0.388
2	0.158	0.163	0.167	0.169	0.171	0.172	0.173	0.173	0.173	0.172
3	0.073	0.077	0.081	0.084	0.087	0.090	0.092	0.094	0.096	0.097
4	0.042	0.045	0.048	0.050	0.053	0.055	0.057	0.059	0.061	0.062
5	0.027	0.029	0.031	0.033	0.035	0.037	0.039	0.040	0.042	0.043
6	0.019	0.021	0.022	0.024	0.025	0.027	0.028	0.029	0.031	0.032
7	0.014	0.015	0.017	0.018	0.019	0.020	0.021	0.022	0.023	0.024
8	0.011	0.012	0.013	0.014	0.015	0.016	0.017	0.018	0.018	0.019
9	0.008	0.009	0.010	0.011	0.012	0.013	0.013	0.014	0.015	0.016
10	0.007	0.008	0.008	0.009	0.010	0.010	0.011	0.012	0.012	0.013
11	0.006	0.006	0.007	0.008	0.008	0.009	0.009	0.010	0.010	0.011
12	0.005	0.005	0.006	0.006	0.007	0.007	0.008	0.008	0.009	0.009
13	0.004	0.005	0.005	0.005	0.006	0.006	0.007	0.007	0.008	0.008
14	0.004	0.004	0.004	0.005	0.005	0.005	0.006	0.006	0.007	0.007
15	0.003	0.003	0.004	0.004	0.004	0.005	0.005	0.005	0.006	0.006
16	0.003	0.003	0.003	0.004	0.004	0.004	0.005	0.005	0.005	0.005
17	0.002	0.003	0.003	0.003	0.003	0.004	0.004	0.004	0.005	0.005
18	0.002	0.002	0.003	0.003	0.003	0.003	0.004	0.004	0.004	0.004
19	0.002	0.002	0.002	0.003	0.003	0.003	0.003	0.003	0.004	0.004
20	0.002	0.002	0.002	0.002	0.003	0.003	0.003	0.003	0.003	0.004
21	0.002	0.002	0.002	0.002	0.002	0.002	0.003	0.003	0.003	0.003
22	0.001	0.002	0.002	0.002	0.002	0.002	0.002	0.003	0.003	0.003
23	0.001	0.001	0.002	0.002	0.002	0.002	0.002	0.002	0.003	0.003
24	0.001	0.001	0.001	0.002	0.002	0.002	0.002	0.002	0.002	0.002
25	0.001	0.001	0.001	0.002	0.002	0.002	0.002	0.002	0.002	0.002
26	0.001	0.001	0.001	0.001	0.002	0.002	0.002	0.002	0.002	0.002
27	0.001	0.001	0.001	0.001	0.001	0.002	0.002	0.002	0.002	0.002
28	0.001	0.001	0.001	0.001	0.001	0.001	0.002	0.002	0.002	0.002
29	0.001	0.001	0.001	0.001	0.001	0.001	0.001	0.002	0.002	0.002
30	0.001	0.001	0.001	0.001	0.001	0.001	0.001	0.001	0.002	0.002
31	0.001	0.001	0.001	0.001	0.001	0.001	0.001	0.001	0.001	0.002
32	0.001	0.001	0.001	0.001	0.001	0.001	0.001	0.001	0.001	0.001
33	0.001	0.001	0.001	0.001	0.001	0.001	0.001	0.001	0.001	0.001
34	0.001	0.001	0.001	0.001	0.001	0.001	0.001	0.001	0.001	0.001
35	0.001	0.001	0.001	0.001	0.001	0.001	0.001	0.001	0.001	0.001
36	0.001	0.001	0.001	0.001	0.001	0.001	0.001	0.001	0.001	0.001
37	0.001	0.001	0.001	0.001	0.001	0.001	0.001	0.001	0.001	0.001
38	0.000	0.001	0.001	0.001	0.001	0.001	0.001	0.001	0.001	0.001
39	0.000	0.001	0.001	0.001	0.001	0.001	0.001	0.001	0.001	0.001
40	0.000	0.000	0.001	0.001	0.001	0.001	0.001	0.001	0.001	0.001
41	0.000	0.000	0.001	0.001	0.001	0.001	0.001	0.001	0.001	0.001
42	0.000	0.000	0.000	0.001	0.001	0.001	0.001	0.001	0.001	0.001
43	0.000	0.000	0.000	0.001	0.001	0.001	0.001	0.001	0.001	0.001
44	0.000	0.000	0.000	0.000	0.001	0.001	0.001	0.001	0.001	0.001
45	0.000	0.000	0.000	0.000	0.001	0.001	0.001	0.001	0.001	0.001
46	0.000	0.000	0.000	0.000	0.000	0.001	0.001	0.001	0.001	0.001
47	0.000	0.000	0.000	0.000	0.000	0.001	0.001	0.001	0.001	0.001
48	0.000	0.000	0.000	0.000	0.000	0.000	0.001	0.001	0.001	0.001
49	0.000	0.000	0.000	0.000	0.000	0.000	0.001	0.001	0.001	0.001
50	0.000	0.000	0.000	0.000	0.000	0.000	0.000	0.001	0.001	0.001

Table 1 (*continued*)

s = 1

n	a=0.1	a=0.2	a=0.3	a=0.4	a=0.5	a=0.6	a=0.7	a=0.8	a=0.9	a=1.0
51	0.000	0.000	0.000	0.000	0.000	0.000	0.000	0.001	0.001	0.001
52	0.000	0.000	0.000	0.000	0.000	0.000	0.000	0.000	0.001	0.001
53	0.000	0.000	0.000	0.000	0.000	0.000	0.000	0.000	0.001	0.001
54	0.000	0.000	0.000	0.000	0.000	0.000	0.000	0.000	0.000	0.001
55	0.000	0.000	0.000	0.000	0.000	0.000	0.000	0.000	0.000	0.000

s = 1.5

n	a=0.1	a=0.2	a=0.3	a=0.4	a=0.5	a=0.6	a=0.7	a=0.8	a=0.9	a=1.0
1	0.716	0.689	0.663	0.638	0.615	0.593	0.572	0.553	0.535	0.518
2	0.142	0.151	0.159	0.166	0.171	0.176	0.180	0.183	0.186	0.188
3	0.054	0.059	0.065	0.069	0.074	0.078	0.082	0.085	0.089	0.092
4	0.027	0.030	0.033	0.036	0.039	0.042	0.045	0.048	0.050	0.052
5	0.015	0.018	0.020	0.022	0.024	0.026	0.028	0.030	0.031	0.033
6	0.010	0.011	0.013	0.014	0.016	0.017	0.019	0.020	0.021	0.023
7	0.007	0.008	0.009	0.010	0.011	0.012	0.013	0.014	0.015	0.016
8	0.005	0.006	0.006	0.007	0.008	0.009	0.010	0.010	0.011	0.012
9	0.004	0.004	0.005	0.005	0.006	0.007	0.007	0.008	0.009	0.009
10	0.003	0.003	0.004	0.004	0.005	0.005	0.006	0.006	0.007	0.007
11	0.002	0.003	0.003	0.003	0.004	0.004	0.005	0.005	0.005	0.006
12	0.002	0.002	0.002	0.003	0.003	0.003	0.004	0.004	0.004	0.005
13	0.001	0.002	0.002	0.002	0.003	0.003	0.003	0.003	0.004	0.004
14	0.001	0.001	0.002	0.002	0.002	0.002	0.003	0.003	0.003	0.003
15	0.001	0.001	0.001	0.002	0.002	0.002	0.002	0.002	0.003	0.003
16	0.001	0.001	0.001	0.001	0.002	0.002	0.002	0.002	0.002	0.002
17	0.001	0.001	0.001	0.001	0.001	0.001	0.002	0.002	0.002	0.002
18	0.001	0.001	0.001	0.001	0.001	0.001	0.001	0.002	0.002	0.002
19	0.001	0.001	0.001	0.001	0.001	0.001	0.001	0.001	0.002	0.002
20	0.001	0.001	0.001	0.001	0.001	0.001	0.001	0.001	0.001	0.001
21	0.000	0.001	0.001	0.001	0.001	0.001	0.001	0.001	0.001	0.001
22	0.000	0.000	0.001	0.001	0.001	0.001	0.001	0.001	0.001	0.001
23	0.000	0.000	0.000	0.001	0.001	0.001	0.001	0.001	0.001	0.001
24	0.000	0.000	0.000	0.001	0.001	0.001	0.001	0.001	0.001	0.001
25	0.000	0.000	0.000	0.000	0.001	0.001	0.001	0.001	0.001	0.001
26	0.000	0.000	0.000	0.000	0.000	0.001	0.001	0.001	0.001	0.001
27	0.000	0.000	0.000	0.000	0.000	0.000	0.001	0.001	0.001	0.001
28	0.000	0.000	0.000	0.000	0.000	0.000	0.000	0.001	0.001	0.001
29	0.000	0.000	0.000	0.000	0.000	0.000	0.000	0.000	0.001	0.001
30	0.000	0.000	0.000	0.000	0.000	0.000	0.000	0.000	0.001	0.001
31	0.000	0.000	0.000	0.000	0.000	0.000	0.000	0.000	0.000	0.001
32	0.000	0.000	0.000	0.000	0.000	0.000	0.000	0.000	0.000	0.000

Table 1 (*continued*)

s = 2

n	a=0.1	a=0.2	a=0.3	a=0.4	a=0.5	a=0.6	a=0.7	a=0.8	a=0.9	a=1.0
1	0.807	0.783	0.760	0.737	0.715	0.694	0.674	0.655	0.636	0.619
2	0.116	0.127	0.137	0.146	0.154	0.162	0.168	0.174	0.179	0.183
3	0.036	0.041	0.046	0.051	0.056	0.061	0.065	0.070	0.074	0.077
4	0.016	0.018	0.021	0.024	0.026	0.029	0.032	0.035	0.037	0.040
5	0.008	0.010	0.011	0.013	0.015	0.016	0.018	0.020	0.021	0.023
6	0.005	0.006	0.007	0.008	0.009	0.010	0.011	0.012	0.013	0.014
7	0.003	0.004	0.004	0.005	0.006	0.006	0.007	0.008	0.009	0.010
8	0.002	0.002	0.003	0.003	0.004	0.004	0.005	0.006	0.006	0.007
9	0.001	0.002	0.002	0.002	0.003	0.003	0.004	0.004	0.004	0.005
10	0.001	0.001	0.002	0.002	0.002	0.002	0.003	0.003	0.003	0.004
11	0.001	0.001	0.001	0.001	0.002	0.002	0.002	0.002	0.003	0.003
12	0.001	0.001	0.001	0.001	0.001	0.001	0.002	0.002	0.002	0.002
13	0.000	0.001	0.001	0.001	0.001	0.001	0.001	0.001	0.002	0.002
14	0.000	0.000	0.001	0.001	0.001	0.001	0.001	0.001	0.001	0.001
15	0.000	0.000	0.000	0.001	0.001	0.001	0.001	0.001	0.001	0.001
16	0.000	0.000	0.000	0.000	0.001	0.001	0.001	0.001	0.001	0.001
17	0.000	0.000	0.000	0.000	0.000	0.001	0.001	0.001	0.001	0.001
18	0.000	0.000	0.000	0.000	0.000	0.000	0.001	0.001	0.001	0.001
19	0.000	0.000	0.000	0.000	0.000	0.000	0.000	0.000	0.001	0.001
20	0.000	0.000	0.000	0.000	0.000	0.000	0.000	0.000	0.000	0.001
21	0.000	0.000	0.000	0.000	0.000	0.000	0.000	0.000	0.000	0.000

s = 2.5

n	a=0.1	a=0.2	a=0.3	a=0.4	a=0.5	a=0.6	a=0.7	a=0.8	a=0.9	a=1.0
1	0.868	0.847	0.827	0.807	0.788	0.769	0.750	0.732	0.714	0.697
2	0.090	0.102	0.112	0.122	0.132	0.141	0.149	0.156	0.163	0.169
3	0.023	0.027	0.032	0.036	0.041	0.045	0.049	0.054	0.058	0.062
4	0.009	0.011	0.013	0.015	0.017	0.019	0.021	0.024	0.026	0.028
5	0.004	0.005	0.006	0.007	0.008	0.010	0.011	0.012	0.014	0.015
6	0.002	0.003	0.003	0.004	0.005	0.005	0.006	0.007	0.008	0.009
7	0.001	0.002	0.002	0.002	0.003	0.003	0.004	0.004	0.005	0.005
8	0.001	0.001	0.001	0.002	0.002	0.002	0.002	0.003	0.003	0.004
9	0.001	0.001	0.001	0.001	0.001	0.001	0.002	0.002	0.002	0.002
10	0.000	0.000	0.001	0.001	0.001	0.001	0.001	0.001	0.002	0.002
11	0.000	0.000	0.000	0.001	0.001	0.001	0.001	0.001	0.001	0.001
12	0.000	0.000	0.000	0.000	0.000	0.001	0.001	0.001	0.001	0.001
13	0.000	0.000	0.000	0.000	0.000	0.000	0.001	0.001	0.001	0.001
14	0.000	0.000	0.000	0.000	0.000	0.000	0.000	0.000	0.001	0.001
15	0.000	0.000	0.000	0.000	0.000	0.000	0.000	0.000	0.000	0.000

Table 1 (*continued*)

s = 3

n	a=0.1	a=0.2	a=0.3	a=0.4	a=0.5	a=0.6	a=0.7	a=0.8	a=0.9	a=1.0
1	0.908	0.892	0.875	0.858	0.841	0.824	0.808	0.791	0.775	0.759
2	0.068	0.079	0.089	0.099	0.109	0.118	0.127	0.135	0.143	0.150
3	0.014	0.018	0.021	0.025	0.028	0.032	0.036	0.040	0.044	0.047
4	0.005	0.006	0.007	0.009	0.010	0.012	0.014	0.016	0.018	0.019
5	0.002	0.003	0.003	0.004	0.005	0.005	0.006	0.007	0.008	0.009
6	0.001	0.001	0.002	0.002	0.002	0.003	0.003	0.004	0.004	0.005
7	0.001	0.001	0.001	0.001	0.001	0.002	0.002	0.002	0.003	0.003
8	0.000	0.000	0.001	0.001	0.001	0.001	0.001	0.001	0.002	0.002
9	0.000	0.000	0.000	0.000	0.001	0.001	0.001	0.001	0.001	0.001
10	0.000	0.000	0.000	0.000	0.000	0.000	0.001	0.001	0.001	0.001
11	0.000	0.000	0.000	0.000	0.000	0.000	0.000	0.000	0.001	0.001
12	0.000	0.000	0.000	0.000	0.000	0.000	0.000	0.000	0.000	0.000

s = 3.5

n	a=0.1	a=0.2	a=0.3	a=0.4	a=0.5	a=0.6	a=0.7	a=0.8	a=0.9	a=1.0
1	0.936	0.923	0.909	0.895	0.880	0.866	0.851	0.837	0.822	0.808
2	0.051	0.060	0.070	0.079	0.088	0.097	0.106	0.115	0.123	0.130
3	0.009	0.011	0.014	0.017	0.019	0.023	0.026	0.029	0.032	0.036
4	0.003	0.003	0.004	0.005	0.006	0.007	0.009	0.010	0.012	0.013
5	0.001	0.001	0.002	0.002	0.003	0.003	0.004	0.004	0.005	0.006
6	0.000	0.001	0.001	0.001	0.001	0.001	0.002	0.002	0.002	0.003
7	0.000	0.000	0.000	0.000	0.001	0.001	0.001	0.001	0.001	0.002
8	0.000	0.000	0.000	0.000	0.000	0.000	0.001	0.001	0.001	0.001
9	0.000	0.000	0.000	0.000	0.000	0.000	0.000	0.000	0.000	0.001
10	0.000	0.000	0.000	0.000	0.000	0.000	0.000	0.000	0.000	0.000

s = 4

n	a=0.1	a=0.2	a=0.3	a=0.4	a=0.5	a=0.6	a=0.7	a=0.8	a=0.9	a=1.0
1	0.955	0.944	0.933	0.922	0.910	0.897	0.885	0.872	0.859	0.846
2	0.038	0.046	0.054	0.062	0.071	0.079	0.088	0.096	0.104	0.111
3	0.005	0.007	0.009	0.011	0.013	0.016	0.018	0.021	0.024	0.026
4	0.001	0.002	0.002	0.003	0.004	0.005	0.005	0.006	0.008	0.009
5	0.000	0.001	0.001	0.001	0.001	0.002	0.002	0.003	0.003	0.003
6	0.000	0.000	0.000	0.000	0.001	0.001	0.001	0.001	0.001	0.002
7	0.000	0.000	0.000	0.000	0.000	0.000	0.000	0.001	0.001	0.001
8	0.000	0.000	0.000	0.000	0.000	0.000	0.000	0.000	0.000	0.000

Table 1 (*continued*)

s = 4.5

n	a=0.1	a=0.2	a=0.3	a=0.4	a=0.5	a=0.6	a=0.7	a=0.8	a=0.9	a=1.0
1	0.968	0.960	0.951	0.942	0.932	0.921	0.910	0.899	0.888	0.877
2	0.028	0.034	0.041	0.049	0.056	0.064	0.071	0.079	0.087	0.094
3	0.003	0.004	0.006	0.007	0.009	0.011	0.013	0.015	0.017	0.019
4	0.001	0.001	0.001	0.002	0.002	0.003	0.003	0.004	0.005	0.006
5	0.000	0.000	0.000	0.001	0.001	0.001	0.001	0.001	0.002	0.002
6	0.000	0.000	0.000	0.000	0.000	0.000	0.000	0.001	0.001	0.001
7	0.000	0.000	0.000	0.000	0.000	0.000	0.000	0.000	0.000	0.000

s = 5

n	a=0.1	a=0.2	a=0.3	a=0.4	a=0.5	a=0.6	a=0.7	a=0.8	a=0.9	a=1.0
1	0.977	0.971	0.964	0.956	0.948	0.939	0.930	0.921	0.911	0.901
2	0.020	0.026	0.031	0.038	0.044	0.051	0.058	0.065	0.072	0.079
3	0.002	0.003	0.004	0.005	0.006	0.007	0.009	0.010	0.012	0.014
4	0.000	0.001	0.001	0.001	0.001	0.002	0.002	0.003	0.003	0.004
5	0.000	0.000	0.000	0.000	0.000	0.001	0.001	0.001	0.001	0.001
6	0.000	0.000	0.000	0.000	0.000	0.000	0.000	0.000	0.000	0.000

Therefore, we can obtain an integral representation for the *probability distribution function* as follows,

$$P(X < k) = c \sum_{j=1}^{k} (a + j)^{-(s+1)} = c \sum_{j=0}^{k-1} (a + 1 + j)^{-(s+1)} \tag{3.2}$$

where $1/c = \zeta(s + 1, a + 1)$. Applying equality (3.1) to (3.2) we obtain

$$P(X < k) = c[\Gamma(s + 1)]^{-1} \int_0^\infty x^s \sum_{j=0}^{k-1} \exp[-(a + 1 + j)x] \, dx$$

$$= c[\Gamma(s + 1)]^{-1} \int_0^\infty x^s e^{-(a+1)x}(1 - e^{kx})/(1 - e^{-x}) \, dx$$

where $1/c = \zeta(s + 1, a + 1)$.

Multiplying both sides of (3.1) by $\exp[it(k+1)]$,

$$\exp[it(k+1)](a+1+k)^{-(s+1)}\Gamma(s+1)$$
$$= \int_0^\infty x^s\{\exp[-(a+1+k)x]$$
$$\times \exp[it(k+1)]\}\,dx$$

Hence,

$$c\sum_{j=1}^{\infty}(a+1+j)^{-(s+1)}\exp[it(j+1)]$$
$$= c[\Gamma(s+1)]^{-1}\lim_{N\to\infty}\sum_{j=0}^{N}\int_0^\infty x^s\{\exp[-(a+1+j)x]$$
$$\times \exp[it(j+1)]\}\,dx$$

where $i^2 = -1$ and $1/c = \zeta(s+1, a+1)$.

Therefore,

$$\varphi(t) = c[\Gamma(s+1)]^{-1}$$
$$\times \lim_{N\to\infty}\int_0^\infty x^s\left\{\sum_{j=0}^{N}\exp[-(a+j+1)x + it(j+1)]\right\}dx$$
$$= c[\Gamma(s+1)]^{-1}\lim_{N\to\infty}\int_0^\infty\left\{\frac{x^s\exp[it-(1+a)x]}{1-\exp(it-x)}\right\}dx$$
$$-\int_0^\infty x^s\left\{\frac{\exp[it-(1+a)x+(it-x)(N+1)]}{1-\exp(it-x)}\right\}dx$$

The second integral on the right-hand side of the above expression is less than or equal to

$$\int_0^\infty x^{s-1}\exp[-(a+1+N)x]\,dx = \frac{\Gamma(s)}{(a+1+N)^s} \tag{3.3}$$

Since (3.3) tends to 0 as $N \to \infty$, the other integral must also approach 0.

Then the expression of the *characteristic function* of this distribution is

$$\varphi(t) = c[\Gamma(s+1)]^{-1} \int_0^\infty x^s \left\{ \frac{\exp[-(a+1)x + it]}{1 - \exp(it - x)} \right\} dx$$

$$= ce^{it}[\Gamma(s+1)]^{-1} \int_0^\infty x^s \left\{ \frac{\exp[-(a+1)x]}{1 - \exp(it - x)} \right\} dx \qquad (3.4)$$

This function can be expressed with Lerch's notation (Whittaker and Watson, 1927) as

$$\varphi(t) = ce^{it}\phi\left(\frac{t}{2\pi}, a+1, s+1\right)$$

where $1/c = \zeta(s+1, a+1)$.

After separating the real and imaginary parts of the integrand of (3.4) and applying the mean-value theorem, we arrive at the conclusion that this distribution is asymptotically convergent to the geometric distribution with

$$p = \frac{\zeta(s+1, a+1)}{\zeta(s, a+1) - a\zeta(s+1, a+1)}$$

4. PROBABILITY GENERATING FUNCTION

The *probability-generating function* of the Hurwitz distribution is

$$g_x(z) = \sum_{j=0}^{\infty} f_x(j)z^j = \sum_{j=1}^{\infty} \frac{z^j(j+a)^{-(s+1)}}{\zeta(s+1, a+1)}$$

$$= \frac{z\zeta(s+1, a+1, z)}{\zeta(s+1, a+1)} \qquad (4.1)$$

where $\zeta(s, a, z)$ is the generalized Hurwitz zeta function or Lerch zeta function as considered by Lerch (1887).

From (4.1) the rth *factorial moment* is

$$D_z^r g_x(z)|_{z=1} = [\zeta(s+1, a+1)]^{-1} \sum_{j=1}^{\infty} j^{(r)}(j+a)^{-(s+1)}$$

and the rth *moment about* 0 is

$$\mu'_r = c\left[\sum_{j=1}^{r}(-1)^r\binom{r}{j}a^r\zeta(s-r+1+j, a+1)\right]$$

for $r < s$ and $1/c = \zeta(s+1, a+1)$. If $r \geq s$ the moment is infinite.

In particular,

$$EX = \frac{\zeta(s, a+1) - a\zeta(s+1, a+1)}{\zeta(s+1, a+1)}$$

and

$$VX = \frac{\zeta(s-1, a+1) - a\zeta(s, a+1)}{\zeta(s+1, a+1)} - EX[EX + a]$$

Therefore, for $r > 2$, the rth *moment about the mean* is

$$\begin{aligned}\mu_r = c\{&\zeta(s-r+1, a+1) - a\zeta(s-r+2, a+1)\\ &-ra[\zeta(s-r+2, a+1) - a\zeta(s-r+3, a+1)]\\ &+r(r-1)a^2[\zeta(s-r+3, a+1) - a\zeta(s-r+4, a+1)]\\ &-\cdots\} + (-1)^r a^{r-2}[(r-1)(VX + E^2X) + (r-2)aEX]\end{aligned}$$

Hence,

$$\mu_3 = c[\zeta(s-2, a+1) - a\zeta(s-1, a+1)] - a[2(VX + E^2X) + aEX]$$

and

$$\begin{aligned}\mu_4 = c\{&[\zeta(s-3, a+1) - a\zeta(s-2, a+1)]\\ &-3a[\zeta(s-2, a+1) - a\zeta(s-1, a+1)]\}\\ &+a^2[3(VX + E^2X) + 2aEX]\end{aligned}$$

with $1/c = \zeta(s+1, a+1)$.

If a is known and $x_1, x_2, \ldots, x_n$ are n observed values (supposed independent), the simplest way of estimating the parameter s is to equate the sample mean to the corresponding population value. Thus, we calculate the solution of s from the equation

$$\frac{\zeta(s, a+1)}{\zeta(s+1, a+1)} = \overline{X} + a \tag{4.2}$$

where $\overline{X} = \sum x_i/n$. This method of estimation allows us to obtain an acceptably accurate solution for the estimator of the parameter s for all $s > 1$.

However, the maximum-likelihood estimator $\hat{s}$ of s which satisfies the equation

$$D = -\frac{\zeta'(s+1, a+1)}{\zeta(s+1, a+1)} = \sum_{i=1}^{n} \frac{\log(x_i + a)}{n} \tag{4.3}$$

gives us a more accurate solution of the value of $\hat{s}$ making use of the values in Table 2.

Since

$$D_s^2 \log L\left(\frac{x}{s}\right)$$

$$= \frac{n^{-1}[\zeta(s+1, a+1)]^2}{\{\zeta''(s+1, a+1)\zeta(s+1, a+1) - [\zeta'(s+1, a+1)]^2\}} \tag{4.3}$$

is true, if we apply the integral representation of $\zeta(s+1, a+1)$ to (4.3),

Table 2. Hurwitz's Distribution

$$D = \frac{-\zeta'(s+1, a+1)}{\zeta(s+1, a+1)}$$

s \ a	0.1	0.2	0.3	0.4	0.5	0.6	0.7	0.8	0.9	1.0
1.0	0.695	0.809	0.913	1.008	1.096	1.177	1.253	1.324	1.391	1.454
1.5	0.406	0.514	0.612	0.703	0.787	0.866	0.939	1.008	1.073	1.134
2.0	0.277	0.380	0.474	0.562	0.643	0.719	0.790	0.857	0.921	0.981
2.5	0.208	0.308	0.399	0.484	0.563	0.637	0.707	0.772	0.834	0.893
3.0	0.169	0.265	0.354	0.437	0.514	0.586	0.654	0.719	0.078	0.837
3.5	0.144	0.238	0.325	0.406	0.482	0.553	0.620	0.683	0.743	0.800
4.0	0.128	0.221	0.307	0.386	0.461	0.531	0.597	0.659	0.718	0.774
4.5	0.118	0.209	0.294	0.372	0.446	0.515	0.580	0.641	0.700	0.755
5.0	0.111	0.202	0.285	0.363	0.435	0.504	0.568	0.629	0.686	0.741

we obtain

$$V\hat{s} \to \frac{VX}{n} + o(n)$$

as n increases.

If a is unknown, the estimation of s and a can be found only by trial and error using the method of moments.

Table 1 gives probabilities for the following values of $s = 1(0.5)5$ and $a = 0.1(0.1)1$ with three significant figures. The values in Table 2 are the quotient of the function $\zeta'(s + 1, a + 1)$ and the function $\zeta(s + 1, a + 1)$. The function $\zeta(s + 1, a + 1)$ was computed from the asymptotic formula given by Whittaker and Watson (1927), and the function $\zeta'(s + 1, a + 1)$ by numerical differentiation. All the calculated values will be in error by less than 10^{-4}. The values of all the individual probabilities, $\zeta(s + 1, a + 1)$ and $\zeta'(s + 1, a + 1)$ were computed up to 16 decimals by a UNIVAC 1100 computer with an error less than 10^{-9}. We have deposited a copy of the tables for the individual and accumulative probabilities with five significant figures in the UMT repository of the American Mathematical Society. (Department of Computer Sciences, Purdue University, West Lafayette, Indiana 47907, U.S.A.).

ACKNOWLEDGMENT

The author acknowledges the Computer Center of the University of the Basque Country at Lejona for providing him with computer time for this research. This work was supported in part by UPV/EHU Research Grant JPV-84/85. Reproduction in whole or in part is permitted for any purpose of the Spanish Government.

REFERENCES

Apostol, T. M. (1976) *Introduction to Analytic Number Theory*, Berlin: Springer-Verlag.

Atoji, M. and Clark, F. L. (1965). Tables of the generalized Riemann zeta-functions. *J. Assoc. Comput. Mach.*, **19**, 692.

Campbell, R. (1966). *Les Integrales Euleriennes et Leurs Applications*, Paris: Dunod.

Estoup, J. B. (1916). *Les Gammes Stenographiques*. Paris: Institut Stenographique.

Good, I. J. (1957). Distribution of word frequencies. *Nature* **179**, 595.

Hansen, E. R. and Patrick, M. L. (1962). Some relations and values for the generalized Riemann zeta-function. *J. Assoc. Comput. Mach.* **16**, 265–274.

Hurwitz, A. (1882). Einige Eigenschaften der Dirichletschen Functionen $F(s) = \Sigma(D/n)(1/n^s)$, die bei der Bestimmung der Classenzahlen binarer quadratischer Formen auftreten. *Z. Math. Phys.* **27**, 86–101.

Johnson, N. L. and Kotz, S. (1969). *Discrete Distributions*, Boston: Houghton-Mifflin.

Kerimov, M. K. (1980). Methods of computing the Riemann zeta-function and some generalizations of it. *Zh. vӯchisl. Mat. mat. Fiz.* **20**, 1580–1597 (in Russian).

Lerch, E. (1887). Note sur la fonction $R(w, z, s) = \Sigma_s^\infty[\exp(2k\pi iz)]/(w+k)^s$. *Acta Math.* **11**, 19–24.

Seal, H. L. (1947). A probability distribution of deaths at age x when policies are counted instead of lives. *Skandivisk Aktuarietidskrift* **30**, 18–43.

Seal, H. L. (1952). The maximum likelihood fitting of the discrete Pareto law. *J. Inst. Actuaries* **78**, 115–121.

Shreider, Yu. A. (1967). On the possibility of a theoretical model of statistical laws for texts (towards a basis for Zipf's law). *Problemy Peredači Informacii* **3**, 57–63 (in Russian).

Vilaplana, J. P. and Llena, J. (1986). *On the Convergence of the Hurwitz's Distribution to the Geometric Distribution*. In 1st *World Congress of the Bernoulli Society for Mathematical Statistics and Probability* to be held at Tashkent, September 8–12, 1986.

Whittaker, E. T. and Watson, G. N. (1927). *Course of Modern Analysis*, 4th ed. Cambridge: University Press.

Zipf, G. K. (1949). *Human Behaviour and the Principle of Least Effort*, Reading, Massachusetts: Addison-Wesley.

C. R. Rao's IPM Method: A Geometric Approach

Hans Joachim Werner
Institute for Econometrics and Operations Research
Econometrics Unit
University of Bonn
Bonn, Federal Republic of Germany

SUMMARY

The inverse partitioned matrix (IPM) method, introduced by C. R. Rao (1971, 1972a), is one of the methods that have hitherto been proposed for estimating parameters in the general Gauss–Markov model. Following Hall and Meyer (1975) and Baksalary and Kala (1980), our first aim is to have a fresh look at the computational aspects of this method. Afterwards, we demonstrate how the concept of *weak bicomplementarity* [see Werner (1986)] can be used to reveal new and simpler estimation formulas.

1. INTRODUCTION

In this chapter, we consider the general Gauss–Markov (GGM) model

$$(y, X\beta, \sigma^2 V) \tag{1.1}$$

where y is an $n \times 1$ vector of random variables with expectation $\mathscr{E}(y) = X\beta$ and dispersion matrix $\mathscr{D}(y) = \sigma^2 V$. The $n \times m$ matrix X and the non-negative definite symmetric (nnds) matrix V are known. No assumptions on

Research sponsored by the Deutsche Forschungsgemeinschaft (SFB 303, SFB 72) at the University of Bonn.

the rank of X or the rank of V are made. The unknown parameters are the coefficients β and the positive scalar σ^2.

The inverse partitioned matrix (IPM) method has been proposed by Rao (1971, 1972a) as a method for calculating the best linear unbiased estimator (BLUE) $\hat{\gamma}$ of a (linearly unbiasedly) estimable function $\gamma = L\beta$, its dispersion matrix $\mathscr{D}(\hat{\gamma})$, and the minimum norm quadratic unbiased estimator (MINQUE) $\hat{\sigma}^2$ of σ^2. The basis of this method is a g-inverse,

$$G = \left(\begin{array}{c|c} G_1 & G_2 \\ \hline G_3 & G_4 \end{array}\right)$$

of the partitioned matrix

$$B = \left(\begin{array}{c|c} V & X \\ \hline X' & O \end{array}\right) \tag{1.2}$$

It is convenient to denote by Γ_1 the set of all those matrices G_1 appearing in the northwest corner of some g-inverse of B. The sets Γ_2, Γ_3 and Γ_4 are defined accordingly. The statement of the IPM method is summarized in the next theorem.

Theorem 1.1 (Rao, 1971). For any $G_i \in \Gamma_i$, we have

1. If $\gamma = L\beta$ is estimable, then a BLUE of γ is given by $\hat{\gamma} = L\hat{\beta}$, where $\hat{\beta} = G_3 y$ or $\hat{\beta} = G_2' y$. Moreover, $\mathscr{D}(\hat{\gamma}) = -\sigma^2 L G_4 L'$.
2. The MINQUE $\hat{\sigma}^2$ of σ^2 is given by $\hat{\sigma}^2 = y'G_1 y/\mathrm{tr}(VG_1)$.

Evidently the crucial point in performing estimation with the use of the IPM method is an algorithm for determining the required submatrices of a g-inverse G of B. We can clearly apply any of existing explicit formulas for a g-inverse of the bordered matrix B. Such formulas are given, for example, by Pringle and Rayner (1971, Theorem 3.5), Rao (1972b, Theorem 2.5), and Hall and Meyer (1975, Theorem 3.1). However, the substitution of the submatrices obtained from any of these formulas into the estimators in Theorem 1.1 does not result in forms computationally simpler than those found in articles by Albert (1973), Mitra and Moore (1973), Werner (1985b), and others. The purpose of this chapter, therefore, is to have a further fresh look at the computational aspects of the IPM method and to reveal some new and simpler estimation formulas. Expressions for the Moore–Penrose inverse of B will also be derived.

2. NOTATION AND PRELIMINARY RESULTS

All matrices in this text are defined over the real number field. However, analogous results can be obtained assuming that the matrices are defined over the complex number field. The symbol I denotes the identity matrix and O the zero matrix of whatever size is appropriate to the context. For a given matrix A in the space $\mathscr{R}^{n,m}$ of all real $n \times m$ matrices, denote by A', $\mathscr{N}(A)$, $\mathscr{R}(A)$, rank(A), tr(A), $\mathscr{N}_c(A)$, and $\mathscr{R}_c(A)$, respectively, the transpose of A, the null space of A, the range space of A, the rank of A, the trace of A, the set of all direct complements of $\mathscr{N}(A)$, and the set of all direct complements of $\mathscr{R}(A)$. If $\mathscr{M}$ is a direct complement of the linear subspace $\mathscr{N}$ in the n-dimensional real space $\mathscr{R}^n$ (i.e., $\mathscr{R}^n = \mathscr{M} \oplus \mathscr{N}$ where $\oplus$ indicates a direct sum), then $P_{\mathscr{M},\mathscr{N}}$ will denote the well-defined (possibly *oblique*) projector onto $\mathscr{M}$ along $\mathscr{N}$, and if $\mathscr{N}$ is the orthogonal complement of $\mathscr{M}$ in $\mathscr{R}^n$ with respect to the usual inner product (i.e., $\mathscr{N} = \mathscr{M}^{\perp}$), $P_{\mathscr{M}}$ will also be used instead of $P_{\mathscr{M},\mathscr{M}^{\perp}}$ [see, e.g., Rao and Mitra (1971, pp. 106–113)].

Now let $A \in \mathscr{R}^{n,m}$, let $\mathscr{M} \in \mathscr{N}_c(A)$, let $\mathscr{S} \in \mathscr{R}_c(A)$, and consider the matrix equations

$$AZA = A \tag{G1}$$

$$ZAZ = Z \tag{G2}$$

$$ZA = P_{\mathscr{M},\mathscr{N}(A)} \tag{$G\mathscr{M}$}$$

$$AZ = P_{\mathscr{R}(A),\mathscr{S}} \tag{$G\mathscr{S}$}$$

Suppose that η is a nonempty subset of $\{1, 2, \mathscr{M}, \mathscr{S}\}$. Then let $A\eta$ denote the set of all those matrices Z that satisfy equations (Gi) for all $i \in \eta$. Any $Z \in A\eta$ is called an *η-inverse* of A, and is also denoted by A^{η}. $\{1\}$-inverses are usually called *g-inverses* and are also denoted by A^{-}.

For every matrix A, for every $\mathscr{M} \in \mathscr{N}_c(A)$, and for every $\mathscr{S} \in \mathscr{R}_c(A)$ we shall use the following well-known facts [cf. Werner (1977, 1985a, 1985b)]:

P2.1. The $\{2, \mathscr{M}, \mathscr{S}\}$-inverse of A exists uniquely.

P2.2. Any $\{\mathscr{M}\}$-inverse of A and likewise any $\{\mathscr{S}\}$-inverse of A is always a $\{1\}$-inverse of A. Conversely, for any $\{1\}$-inverse A^{-} of A, there uniquely exist an $\mathscr{M} \in \mathscr{N}_c(A)$ and an $\mathscr{S} \in \mathscr{R}_c(A)$ such that $A^{-} \in A\{\mathscr{M}, \mathscr{S}\}$.

P2.3. If $Z = A^{\{2,\mathscr{M},\mathscr{S}\}}$, then $\mathscr{S} = \mathscr{N}(Z)$ and $\mathscr{M} = \mathscr{R}(Z)$. If $Z \in A\{\mathscr{M}, \mathscr{S}\}$, then $Z\mathscr{S} \subseteq \mathscr{N}(A)$ and $Z\mathscr{R}(A) = \mathscr{M}$.

P2.4. The $\{2, \mathscr{R}(A'), \mathscr{N}(A')\}$-inverse of A coincides with its *Moore–Penrose inverse*, which is usually denoted A^+.

P2.5. If A is nonsingular, then its g-inverses all coincide with its usual (*regular*) inverse A^{-1}.

3. COMPUTATIONAL ASPECTS

Theorem 1.1 indicates that the problem of inference in the GGM model (1.1) can be completely solved once one has obtained a g-inverse for the bordered matrix B of (1.2). For this reason, a great deal of attention has been given to the problem of obtaining a particular form for an entire B^- [see, e.g., Pringle and Rayner (1971)]. However, as pointed out by Rao (1971), there is still a need for efficient algorithms for computing g-inverses of B. Hall and Meyer (1975) first observed that the submatrices G_i ($i = 1, \ldots, 4$) of any g-inverse G for B may be computed separately and independently. Their observation, of course, opened the door to more efficient algorithms because the sizes of the matrices involved in a computational scheme for a g-inverse of B may be greatly reduced. Following Baksalary and Kala (1980), we now have a fresh look at the computational aspects of the IPM method. It turns out that the geometry of g-inversion will readily lead to new characterizations of the sets Γ_i ($i = 1, \ldots, 4$).

In what follows, we write $(G_1, \ldots, G_4) \in \Gamma$ whenever G_1, G_2, G_3, and G_4 constitute an entire g-inverse G of B. Further let $Q := P_{\mathscr{N}(X')}$, $P := P_{\mathscr{N}(X') \cap \mathscr{N}(V)}$, and $\overline{V} := VQ$. We first remark that

$$\mathscr{R}(V, X) = \mathscr{R}(\overline{V}, X) \tag{3.1a}$$

$$\mathscr{R}(\overline{V}) \cap \mathscr{R}(X) = \{0\} \tag{3.1b}$$

[i.e., $\mathscr{R}(V, X) = \mathscr{R}(\overline{V}) \oplus \mathscr{R}(X)$] hold true; see, for example, Lemma 3.2 in Werner (1985b). Suppose now that the vector $z' = (v', w')$ is partitioned according to the matrix B given by (1.2). Evidently, $z \in \mathscr{N}(B)$ iff $v \in \mathscr{R}(Q)$, $Vv = -Xw$. In view of (3.1b) we thus obtain

$$\mathscr{N}(B) = \mathscr{R}(P) \times \mathscr{N}(X) \tag{3.2}$$

wherein $\times$ indicates a cartesian product. From (3.2) we further get $\mathscr{R}(B) = \mathscr{N}(B)^{\perp} = [\mathscr{R}(P) \times \mathscr{N}(X)]^{\perp} = \mathscr{R}(P)^{\perp} \times \mathscr{N}(X)^{\perp} = \mathscr{R}(X, V) \times \mathscr{R}(X')$, that is,

$$\mathscr{R}(B) = \mathscr{R}(X, V) \times \mathscr{R}(X') \tag{3.3}$$

Since $G \in B\{1\}$ iff $BGx = x$ for all $x \in \mathscr{R}(B)$ (note P2.2), the following characterization is an easy consequence of (3.3).

Theorem 3.1. Let B be as given in (1.2). Then $(G_1, \ldots, G_4) \in \Gamma$ (i.e., $G \in B\{1\}$) iff

$$(VG_1 + XG_3)(V, X) = (V, X) \tag{3.4a}$$

$$X'G_1(V, X) = 0 \tag{3.4b}$$

$$(VG_2 + XG_4)X' = 0 \tag{3.4c}$$

$$X'G_2X' = X' \tag{3.4d}$$

This theorem admits a corollary; its proof is clear from (3.4) and (3.1a).

Corollary 3.1. For B as given in (1.2), $(G_1, \ldots, G_4) \in \Gamma$ iff (3.4c) and (3.4d) together with

$$(VG_1 + XG_3)(\bar{V}, X) = (\bar{V}, X) \tag{3.4a$'$}$$

$$X'G_1(\bar{V}, X) = 0 \tag{3.4b$'$}$$

are satisfied.

We next show that VG_2X' is invariant to the choice of $G_2 \in \Gamma_2$. Premultiplying (3.4c) by Q yields

$$QVG_2X' = 0 \tag{3.5}$$

because $QX = 0$. As a consequence of (3.1a) we further have

$$V = \bar{V}T + XU \quad \text{for some } T \text{ and } U \tag{3.6}$$

Observe that in view of (3.1b) $\bar{V}T$ and XU are uniquely determined matrices. Substitution of (3.6) into VG_2X' results in $VG_2X' = T'QVG_2X' + U'X'G_2X' = U'X'$, where use is made of (3.5) and (3.4d). This completes the proof of

Theorem 3.2. For B as given in (1.2), the matrix VG_2X' is invariant among all choices for $G_2 \in \Gamma_2$. Precisely, $VG_2X' = F'$, where $F := P_{\mathscr{R}(X), \mathscr{R}(\bar{V}) \oplus \mathscr{R}(P)}V$.

These observations lead us to our first main result.

Theorem 3.3. Let B be as given in (1.2). Then $(G_1, \ldots, G_4) \in \Gamma$ iff $G_i \in \Gamma_i$ for $i = 1, \ldots, 4$, or equivalently, $\Gamma = \times_1^4 \Gamma_i$. Moreover,

$$G_1 \in \Gamma_1 \quad \text{iff} \left(\begin{array}{c} V \\ \hline X' \end{array}\right) G_1(\overline{V}, X) = \left(\begin{array}{c|c} \overline{V} & 0 \\ \hline 0 & 0 \end{array}\right) \tag{3.7a}$$

$$G_2 \in \Gamma_2 \quad \text{iff } XG_2'(\overline{V}, X) = (0, X) \tag{3.7b}$$

$$G_3 \in \Gamma_3 \quad \text{iff } XG_3(\overline{V}, X) = (0, X) \tag{3.7c}$$

$$G_4 \in \Gamma_4 \quad \text{iff } XG_4X' = -F' \tag{3.7d}$$

where F is as before.

Proof. First suppose that the matrix equations under (3.7a)–(3.7d) all hold true for $(G_1, \ldots, G_4)$. The equation under (3.7b) implies $F = XU = XG_2'(\overline{V}T + XU) = XG_2'V$, where T and U are chosen as in (3.6). Then it is not difficult to see that $(G_1, \ldots, G_4)$ satisfies the conditions (3.4a′), (3.4b′), (3.4c), and (3.4d) as well. In view of Corollary 3.1 we hence arrive at $(G_1, \ldots, G_4) \in \Gamma$. Conversely, let $(G_1, \ldots, G_4) \in \Gamma$, that is, let $(G_1, \ldots, G_4)$ be a solution of (3.4a′), (3.4b′), (3.4c), and (3.4d). Condition (3.4b′) implies $\mathscr{R}(G_1\overline{V}) \subseteq \mathscr{R}(Q)$. Hence $\mathscr{R}(VG_1\overline{V}) \subseteq \mathscr{R}(\overline{V})$, from which, in view of (3.1b), we see at once that the left block equation in (3.4a′) can be true only if $VG_1\overline{V} = \overline{V}$ and $XG_3\overline{V} = 0$. The conditions $VG_1X = 0$ and $XG_3X = X$ are obtained by similar arguments from the right block equations in (3.4b′) and (3.4a′). From (3.4c) we further obtain $XG_4X' = -F'$ and $XG_2'\overline{V} = 0$ by means of Theorem 3.2 and (3.5), respectively. The matrix equations under (3.7a)–(3.7d) are thus all satisfied for $(G_1, \ldots, G_4)$. These observations also establish the amazing fact that the blocks of the members of Γ are completely independent of each other, which in turn implies that $\Gamma_1 \times \cdots \times \Gamma_4 \subseteq \Gamma$. Since the converse inclusion is trivially true, the proof is complete. ∎

Note. We remark that alternative but more complicated proofs of most of the preceding results can be found in articles by Hall and Meyer (1975) and Baksalary and Kala (1980). In their study of the IPM method, Hall and Meyer present a characterization of the set Γ_4 that looks similar to ours. We shall see that they are indeed identical. The characterizations of the sets Γ_1, Γ_2, and Γ_3 are due to Baksalary and Kala; however, observe that (3.7b) corrects a misprint in the paper of those authors. The observation that the sets Γ_i $(i = 1, \ldots, 4)$ can be characterized independently one of each other dates back to Hall and Meyer.

Note. A comparison of (3.7c) with (3.8) in Rao (1978) [or Theorem 3.1 in Werner (1985b)] shows that in the framework of the GGM model (1.1) Γ_3, in fact, *coincides* with the complete class of all those matrices D leading to *traditional* BLUEs $\hat{\gamma} = LDy$ for all estimable linear functions $\gamma = L\beta$. In passing, we should remark that in the literature there exist several extensions of the *traditional* BLUE approach. We mention here only the Rao–BLUE approach [see Rao (1973)] and the more comprehensive W–BLUE approach [see Schönfeld and Werner (1986)].

By (3.7b) and (3.7c) we know that the members of Γ_2 and Γ_3 are particular g-inverses of X' and X, respectively. The next corollary gives a new necessary and sufficient condition for a matrix G_1 to be a member of Γ_1 showing, in particular, that these matrices are also special g-inverses of $\overline{V}$. It is readily obtained from (3.7a) and the fact that $X'G_1(\overline{V}, X) = 0$ iff $QG_1(\overline{V}, X) = G_1(\overline{V}, X)$.

Corollary 3.2. Let B be as given in (1.2). Then

$$G_1 \in \Gamma_1 \text{ iff } \begin{pmatrix} \overline{V} \\ \hline X' \end{pmatrix} G_1(\overline{V}, X) = \left(\begin{array}{c|c} \overline{V} & 0 \\ \hline 0 & 0 \end{array}\right) \tag{3.7a$'$}$$

We proceed with some characterizations of the Γ_i that make use of the geometry of g-inversion. The matrix $\overline{V}$ is defined as before.

Theorem 3.4. For B as given by (1.2), each of the following statements is true.

1. $G_1 \in \Gamma_1$ iff $G_1 \in \overline{V}\{\mathscr{M}, \mathscr{S}\}$ for some pair $(\mathscr{M}, \mathscr{S}) \in \mathscr{N}_c(\overline{V}) \times \mathscr{R}_c(\overline{V})$ such that $\mathscr{M} \oplus \mathscr{R}(P) = \mathscr{N}(X')$, $\mathscr{S} \supseteq \mathscr{R}(X)$, and $G_1\mathscr{R}(X) \subseteq \mathscr{R}(P)$.
2. $G_2 \in \Gamma_2$ iff $G_2 \in X'\{\mathscr{M}\}$ for some $\mathscr{M} \in \mathscr{N}_c(X')$ such that $V\mathscr{M} \subseteq \mathscr{R}(X)$.
3. $G_3 \in \Gamma_3$ iff $G_3 \in X\{\mathscr{S}\}$ for some $\mathscr{S} \in \mathscr{R}_c(X)$ such that $\mathscr{R}(\overline{V}) \subseteq \mathscr{S}$.

Proof. Since the proofs of the three parts are similar, we only prove (1). First, let $G_1 \in \Gamma_1$. Then by (3.7a$'$), $G_1 \in \overline{V}\{1\}$, $\mathscr{R}(G_1X) \subseteq \mathscr{N}(\overline{V}) \cap \mathscr{N}(X')$ and $\mathscr{R}(G_1\overline{V}) \subseteq \mathscr{N}(X')$. By P2.2 we further know that $G_1 \in \overline{V}\{\mathscr{M}, \mathscr{S}\}$ for some pair $(\mathscr{M}, \mathscr{S}) \in \mathscr{N}_c(\overline{V}) \times \mathscr{R}_c(\overline{V})$. Thus $\mathscr{R}(G_1\overline{V}) = \mathscr{M}$ in view of P2.3. Since $\mathscr{N}(\overline{V}) = \mathscr{R}(X) \oplus [\mathscr{N}(X') \cap \mathscr{N}(V)] = \mathscr{R}(X) \oplus \mathscr{R}(P)$, we then obtain $\mathscr{M} \oplus \mathscr{R}(P) = \mathscr{N}(X')$ by virtue of $\mathscr{M} = \mathscr{R}(G_1\overline{V}) \subseteq \mathscr{N}(X') \in \mathscr{R}_c(X)$ and $\mathscr{M} \in \mathscr{N}_c(\overline{V})$. Because $\mathscr{N}(\overline{V}) \cap \mathscr{N}(X') = \mathscr{R}(P)$, we also have $G_1\mathscr{R}(X) \subseteq \mathscr{R}(P)$. From $\mathscr{R}(G_1X) \subseteq \mathscr{N}(\overline{V})$ and $\mathscr{R}(X) \cap \mathscr{R}(\overline{V}) = \{0\}$ [note (3.1b)] we finally obtain $\mathscr{S} \supseteq \mathscr{R}(X)$ by virtue of P2.3. This completes the proof of the "only if" part. Conversely, suppose that G_1 is such a g-inverse of $\overline{V}$ as described in (1). Using P2.3, it is a simple matter to show that G_1 satisfies the matrix equation given in (3.7a′), so that G_1 is indeed a member of Γ_1. ■

Note. We are now in the position to prove that (3.7d) is equivalent to the characterization of set Γ_4 given by Hall and Meyer (1975), that is, to

$$G_4 \in \Gamma_4 \quad \text{iff } XG_4X' = -(V - VG_1V) \text{ for any } G_1 \in \Gamma_1 \qquad (3.7\text{d}')$$

For this purpose, let $G_1 \in \Gamma_1$. Then $V - VG_1V = V - \overline{V}G_1V$, by virtue of (3.1a) and (3.7a). Moreover, $V - VG_1V = [I - \overline{V}\overline{V}^{\{\mathscr{M},\mathscr{S}\}}]V$ for some $(\mathscr{M}, \mathscr{S})$ as described in Theorem 3.4(1). Hence $[I - \overline{V}\overline{V}^{\{\mathscr{M},\mathscr{S}\}}]V = P_{\mathscr{S},\mathscr{R}(\overline{V})}V$. Since $\mathscr{R}(X) \subseteq \mathscr{S}$, we now arrive at $V - VG_1V = F$. The matrix F, which does not depend on the particular $G_1 \in \Gamma_1$, must be symmetric because $G_1 \in \Gamma_1$ implies that $G_1' \in \Gamma_1$ (note $BGB = B \Rightarrow BG'B = B$). This observation completes the proof.

Note. In this context, note that $-G_3VG_2 \in \Gamma_4$ holds true whenever $G_2 \in \Gamma_2$ and $G_3 \in \Gamma_3$. This may be seen as follows. Since $G_2 \in \Gamma_2$, there exists an $\mathscr{M} \in \mathscr{N}_c(X')$ with $G_2 \in X'\{\mathscr{M}\}$ and $V\mathscr{M} \subseteq \mathscr{R}(X)$ [see Theorem 3.4(2)]. Also by Theorem 3.4(3), $G_3 \in X\{1\}$ showing that XG_3 is a projector onto $\mathscr{R}(X)$. Since $\mathscr{R}(G_2X') = \mathscr{M}$ (note P2.3), we then obtain $-XG_3VG_2X' = -VG_2X' = -F'$ by noting Theorem 3.2. The desired result thus follows in view of (3.7d).

Corollary 3.3. Let B, $\overline{V}$, P, and the Γ_i be defined as before. Further let $(\mathscr{M}, \mathscr{S}) \in \mathscr{N}_c(X') \times \mathscr{R}_c(X)$ be such that $V\mathscr{M} \subseteq \mathscr{R}(X)$ and $\mathscr{R}(\overline{V}) \subseteq \mathscr{S}$. Then

1. $G_{10} := \overline{V}^{\{2,\mathscr{R}(\overline{V}'),\mathscr{N}(\overline{V})\}} \in \Gamma_1$.
2. $G_{20}^{\mathscr{M}} := (X')^{\{2,\mathscr{M},\mathscr{N}(X)\}} \in \Gamma_2$.
3. $G_{30}^{\mathscr{S}} := X^{\{2,\mathscr{R}(X'),\mathscr{S}\}} \in \Gamma_3$.
4. $G_{40}^{\mathscr{M},\mathscr{S}} := -G_{30}^{\mathscr{S}}VG_{20}^{\mathscr{M}} \in \Gamma_4$.

Proof. (1) First observe that $\mathscr{R}(\overline{V}') \in \mathscr{N}_c(\overline{V})$. From $\mathscr{N}(\overline{V}) = \mathscr{R}(X) \oplus \mathscr{R}(P)$ we further obtain $\mathscr{N}(\overline{V}) \in \mathscr{R}_c(\overline{V})$ by noting $\mathscr{R}(X) \oplus \mathscr{R}(\overline{V}) = \mathscr{R}(P)^{\perp}$. Then by P2.1 it is clear that G_{10} is well and uniquely defined. Since by construction $\mathscr{R}(G_{10}) = \mathscr{R}(\overline{V}') \subseteq \mathscr{N}(X')$, $\mathscr{N}(G_{10}) = \mathscr{N}(\overline{V}) \supseteq \mathscr{R}(X)$ (note P2.3) and $\mathscr{R}(\overline{V}') \oplus \mathscr{R}(P) = \mathscr{N}(X')$, the desired result now follows by virtue of Theorem 3.4(1).

(2) and (3) These results are direct consequences of Theorem 3.4(2) and (3), respectively [observe that $\mathscr{N}(X) \in \mathscr{R}_c(X')$ or equivalently $\mathscr{R}(X') \in \mathscr{N}_c(X)$].

(4) Combining (2) and (3) of this corollary with the previous note gives (4). ■

We next derive parametric representations of the sets Γ_i $(i = 1,\dots,4)$.

Corollary 3.4. Under the assumptions of Corollary 3.3,

1. $G_1 \in \Gamma_1$ iff $G_1 = G_{10} + U - (I - P)U(I - P)$ for some U.
2. $G_2 \in \Gamma_2$ iff $G_2 = G_{20}^{\mathscr{M}} + U - (I - P)UP_{\mathscr{R}(X')}$ for some U.
3. $G_3 \in \Gamma_3$ iff $G_3 = G_{30}^{\mathscr{S}} + U - P_{\mathscr{R}(X')}U(I - P)$ for some U.
4. $G_4 \in \Gamma_4$ iff $G_4 = G_{40}^{\mathscr{M},\mathscr{S}} + U - P_{\mathscr{R}(X')}UP_{\mathscr{R}(X')}$ for some U.

Proof. To prove (1), we may first observe that by Corollary 3.3(1), $G_{10} \in \Gamma_1$. Use of, for example, Theorem 2.3.2 of Rao and Mitra (1971) then readily shows that the general solution to the matrix equation in (3.7a′) can be written as claimed. [In this context, note that $\mathscr{N}(\bar{V}) \cap \mathscr{N}(X') = \mathscr{N}(V) \cap \mathscr{N}(X')$ and $\mathscr{R}(X, \bar{V}) = \mathscr{R}(X, V)$. Thus in view of P2.4, $I - P = (\bar{V}, X)(\bar{V}, X)^+ = (\bar{V}', X)'^+(\bar{V}', X)'$.] The other statements are similarly obtained from Theorem 3.3 and Corollary 3.3. ■

Corollary 3.4 enables us now to answer the interesting question of when Γ_i $(i = 1,\dots,4)$ is a singleton.

Corollary 3.5. For B as given in (1.2),

1. Γ_1 is a singleton iff $\operatorname{rank}(X, V) = n$.
2. Γ_2 is a singleton iff $\operatorname{rank}(X, V) = n$, $\operatorname{rank}(X) = m$.
3. Γ_3 is a singleton iff $\operatorname{rank}(X, V) = n$, $\operatorname{rank}(X) = m$.
4. Γ_4 is a singleton iff $\operatorname{rank}(X) = m$.

Moreover,

5. $B\{1\}$ is a singleton iff $\operatorname{rank}(X, V) = n$, $\operatorname{rank}(X) = m$.

Proof. The results (1)–(4) are easily obtained from the corresponding representations in Corollary 3.4. For example, it is seen from Corollary 3.4(1) that $\Gamma_1 = \{G_{10}\}$ iff $I - P = I$, which in turn happens iff $\mathscr{R}(I - P) = \mathscr{R}^n$. Since $\mathscr{R}(I - P) = \mathscr{R}(X, V)$, the proof of (1) should be clear. As a consequence of (1)–(4) and the fact that $\Gamma = \Gamma_1 \times \cdots \times \Gamma_4$ (note Theorem 3.3), we finally obtain (5). ■

Note. It is evident that $B\{1\}$ is a singleton iff B is nonsingular, that is, $\operatorname{rank}(B) = n + m$ (cf. Section 2). Part (5) of Corollary 3.5 thus corrects Theorem 2.2 by Rao (1972b). We further note that our necessary and sufficient condition for Γ_1 to be a singleton appears to be simpler than the equivalent conditions in Theorem 3.4 by Hall and Meyer (1975).

The special cases in which 0 is a member of Γ_1, Γ_2, Γ_3, and Γ_4, respectively, may be of some interest. These situations are characterized in the next corollary.

Corollary 3.6. Let B be as given in (1.2). Then

1. $0 \in \Gamma_1$ iff $\mathscr{R}(V) \subseteq \mathscr{R}(X)$.
2. $0 \in \Gamma_2$ iff $X = 0$.
3. $0 \in \Gamma_3$ iff $X = 0$.
4. $0 \in \Gamma_4$ iff $\mathscr{R}(X) \cap \mathscr{R}(V) = \{0\}$.

Proof. In view of (3.7a′), $0 \in \Gamma_1$ iff $\overline{V} = 0$ iff $\mathscr{N}(X') \subseteq \mathscr{N}(V)$. Since the last inclusion is equivalent to $\mathscr{R}(V) \subseteq \mathscr{R}(X)$, the proof of (1) is complete. The results (2) and (3) are direct consequences of (3.7b) and (3.7c), respectively. From (3.7d) we finally know that $0 \in \Gamma_4$ iff $F = 0$. But from the definition of F [note also (3.1)], it is then clear that this happens iff $\mathscr{R}(V) = \mathscr{R}(\overline{V})$, that is, iff $\mathscr{N}(X') + \mathscr{N}(V) = \mathscr{R}^n$. Thus $0 \in \Gamma_4$ iff $\mathscr{R}(X) \cap \mathscr{R}(V) = [\mathscr{N}(X') + \mathscr{N}(V)]^\perp = [\mathscr{R}^n]^\perp = \{0\}$. ■

Note. In this context, some remarks are worth mentioning. First, alternative necessary and sufficient conditions for $0 \in \Gamma_4$ are found in Theorem 4.3 by Hall and Meyer (1975). Second, Corollary 3.6 implies that $0 \in B\{1\}$ iff $B = 0$. Third, if $\mathscr{R}(X) \cap \mathscr{R}(V) = \{0\}$, then $0 \in \Gamma_4$, so that by Theorem 1.1(1), for any estimable $\gamma = L\beta$, one has $\hat{\gamma} = \gamma$ with probability 1.

Since there is a sizable literature relating only to the computation of Moore–Penrose inverses, we are now interested in conditions under which $\overline{V}^+$, X'^+, and X^+ are contained in Γ_1, Γ_2, and Γ_3, respectively.

Corollary 3.7. Let B, $\overline{V}$, P, and the Γ_i be as before. Then the following statements are equivalent:

1. $\overline{V}^+ \in \Gamma_1$.
2. $X'^+ \in \Gamma_2$.
3. $X^+ \in \Gamma_3$.
4. $V\mathscr{R}(X) \subseteq \mathscr{R}(X)$.
5. $V\mathscr{N}(X') \subseteq \mathscr{N}(X')$.

Proof. (4) ⇔ (5) This follows from the following chain of implications: $V\mathscr{R}(X) \subseteq \mathscr{R}(X) \Rightarrow V\mathscr{R}(X)^\perp = \mathscr{N}(X') \subseteq [V\mathscr{R}(X)]^\perp \Rightarrow V\mathscr{N}(X') \subseteq V[V\mathscr{R}(X)]^\perp \subseteq \mathscr{R}(X)^\perp = \mathscr{N}(X') \Rightarrow \mathscr{N}(X')^\perp = \mathscr{R}(X) \subseteq [V\mathscr{N}(X')]^\perp \Rightarrow V\mathscr{R}(X) \subseteq V[V\mathscr{N}(X')]^\perp \subseteq \mathscr{N}(X')^\perp = \mathscr{R}(X)$.

(1) ⇔ (4) We first observe that by P2.4, $\overline{V}^+ = \overline{V}^{\{2, \mathscr{R}(\overline{V}'), \mathscr{N}(\overline{V}')\}}$. Now let $\overline{V}^+ \in \Gamma_1$. Use of Theorem 3.4(1) then shows that $\mathscr{R}(X) \subseteq \mathscr{N}(\overline{V}') =$

$\mathcal{N}(QV)$ must hold, which happens iff $V\mathcal{R}(X) \subseteq \mathcal{N}(Q) = \mathcal{R}(X)$. Conversely, let $V\mathcal{R}(X) \subseteq \mathcal{R}(X)$. Then $\mathcal{R}(X) \subseteq \mathcal{N}(\overline{V}')$ and hence in view of P2.3, $\overline{V}^+\mathcal{R}(X) = \{0\} \subseteq \mathcal{R}(P)$. As was realized in the proof of Corollary 3.3, $\mathcal{R}(\overline{V}') \oplus \mathcal{R}(P) = \mathcal{N}(X')$. $\overline{V}^+ \in \Gamma_1$ thus follows by virtue of Theorem 3.4(1).

(2) ⇔ (4) By P2.4, $X'^+ = X^{\{2, \mathcal{R}(X), \mathcal{N}(X)\}}$. From Theorem 3.4(2) it is then clear that $X'^+ \in \Gamma_2$ iff $V\mathcal{R}(X) \subseteq \mathcal{R}(X)$.

(3) ⇔ (4) Using $X^+ = X^{\{2, \mathcal{R}(X'), \mathcal{N}(X')\}}$, the desired result follows directly from Theorem 3.4(3) (note P2.3 and P2.4). ■

Corollaries 3.6 and 3.7 describe situations in which the IPM method leads to simple estimation formulas. For the discussion of further special cases of the GGM model (1.1) in which the estimation formulas become simple, we refer, for example, to Baksalary and Kala (1980). Since in the general case of model (1.1) there is, however, still a need for simple estimation formulas, we demonstrate in Section 4 how the concept of *weak bicomplementarity* can be used to fill this gap.

4. WEAK BICOMPLEMENTARITY

From Chipman (1964) we have

Definition 4.1. Let E, F be matrices of the same column number. Then F is said to be *complementary* to E, and vice versa, iff $\mathcal{R}(F') \in \mathcal{R}_c(E')$.

Werner (1984, 1985a, 1986) introduced the following modifications of this concept.

Definition 4.2. Let E, F be conformable matrices. Then F is said to be

1. *weakly complementary* to E, and vice versa, iff $\mathcal{R}(E') \cap \mathcal{R}(F') = \{0\}$.
2. (*weakly*) *bicomplementary* to E, and vice versa, iff F and F' are (weakly) complementary to E and E', respectively.

Note. Matrices E and F, for which $\mathcal{R}(E) \cap \mathcal{R}(F) = \{0\}$ and $\mathcal{R}(E') \cap \mathcal{R}(F') = \{0\}$, are sometimes called *disjoint* matrices.

For later use, we cite the following theorem.

Theorem 4.1 (Werner, 1986). Let E, F be a pair of weakly bicomplementary matrices. Then

1. $(E + F)\{1\} \subseteq E\{1\} \cap F\{1\}$.
2. $E(E + F)^-F = 0$, $F(E + F)^-E = 0$ for every g-inverse.

Moreover, if $\mathscr{M} \in \mathscr{N}_c(E+F)$ and $\mathscr{S} \in \mathscr{R}_c(E+F)$, then

3. $(E+F)\{\mathscr{M}, \mathscr{S}\} \subseteq E\{\mathscr{M} \cap \mathscr{N}(F), \mathscr{S} \oplus \mathscr{R}(F)\} \cap F\{\mathscr{M} \cap \mathscr{N}(E), \mathscr{S} \oplus \mathscr{R}(E)\}$.
4. $(E+F)^{\{2,\mathscr{M},\mathscr{S}\}} = E^{\{2,\mathscr{M} \cap \mathscr{N}(F), \mathscr{S} \oplus \mathscr{R}(F)\}} + F^{\{2,\mathscr{M} \cap \mathscr{N}(E), \mathscr{S} \oplus \mathscr{R}(E)\}}$.

In particular,

5. $(E+F)^+ = E^+ + F^+$ iff $EF' = 0$, $E'F = 0$.

In what follows, let B, $\overline{V}$, $P, \ldots$ be defined as in Section 3. The following result, which is a direct consequence of (3.1b) and the fact that $\mathscr{R}(\overline{V}') \subseteq \mathscr{N}(X') = \mathscr{R}(X)^\perp$, plays a key role in this section.

Lemma 4.1. The matrices $\overline{V}$ and $Q\overline{V}$ are both weakly bicomplementary to the matrix XX'.

Theorem 4.1 enables us now to represent members of the sets Γ_i $(i = 1, \ldots, 4)$ in a fashion that appears to be, in general, computationally simpler than the forms already discussed in the literature. In particular, they need fewer g-inversions.

Theorem 4.2. Let $T := \overline{V} + XX'$. Then for every g-inverse of T,

$$T^- \overline{V} T^- \in \Gamma_1 \tag{4.1a}$$

$$(X'T^-)' \in \Gamma_2 \tag{4.1b}$$

$$X'T^- \in \Gamma_3 \tag{4.1c}$$

$$-X'T^- V(X'T^-)' \in \Gamma_4 \tag{4.1d}$$

Proof. From Lemma 4.1 we know that $\overline{V}$ is weakly bicomplementary to XX'. Theorem 4.1(2) thus guarantees that

$$XX'T^- \overline{V} = 0 \tag{4.2a}$$

$$\overline{V} T^- XX' = 0 \tag{4.2b}$$

hold true for every g-inverse of T. As is well known, $EFF' = 0$ iff $EF = 0$. Equations (4.2a) and (4.2b) are therefore equivalent to

$$X'T^- \overline{V} = 0 \tag{4.2a$'$}$$

$$\overline{V} T^- X = 0 \tag{4.2b$'$}$$

By Theorem 4.1(1), $T^- \in \overline{V}\{1\}$, which in turn implies that $\tilde{T} := T^- \overline{V} T^- \in \overline{V}\{1,2\}$. Then $\tilde{T}X = 0$ as well as $X'\tilde{T} = 0$ by means of (4.2a′) and (4.2b′). (4.1a) now follows by virtue of (3.7a′). Since again by Theorem 4.1(1), $T^- \in (X'X)\{1\}$, we also know that $XX'T^-$ is a projector onto $\mathscr{R}(XX') = \mathscr{R}(X)$ (note P2.2 and the definition of η-inversion). Hence $XX'T^-X = X$, so that $X'T^- \in X\{1\}$. This together with (4.2b′) gives (4.1b) and (4.1c) by virtue of (3.7b) and (3.7c). (4.1d) is finally obtained from (4.1b) and (4.1c) by recalling the note before Corollary 3.3. ■

Theorem 4.3. Let T be as defined in Theorem 4.2. Then

$$B^+ = \left[\begin{array}{c|c} T^+\overline{V}T^+ & T^{+\prime}X \\ \hline X'T^+ & -X'T^+VT^{+\prime}X \end{array}\right] \tag{4.3}$$

or equivalently

$$B^+ = \left[\begin{array}{c|c} G_{10} & G_{20}^{\mathscr{M}} \\ \hline G_{30}^{\mathscr{S}} & G_{40}^{\mathscr{M},\mathscr{S}} \end{array}\right] \tag{4.3'}$$

where

$$\mathscr{M} := \mathscr{N}(P) \cap \mathscr{N}(\overline{V}') =: \mathscr{S}^{\perp} \tag{4.4}$$

and G_{10}, $G_{20}^{\mathscr{M}}$, $G_{30}^{\mathscr{S}}$, and $G_{40}^{\mathscr{M},\mathscr{S}}$ are defined as in Corollary 3.3.

Proof. Let G denote the matrix given on the right-hand side of the equality sign in (4.3). We prove $G = B^+$ by checking the defining relations of B^+. For example, we obtain

$$BG = \left[\begin{array}{c|c} VT^+\overline{V}T^+ + XX'T^+ & (I - XX'T^+)VT^{+\prime}X \\ \hline 0 & X'T^{+\prime}X \end{array}\right]$$

$$= \left[\begin{array}{c|c} \overline{V}T^+\overline{V}T^+ + XX'T^+ & [X'T^+V(I - XX'T^+)']' \\ \hline 0 & [X'T^+X]' \end{array}\right]$$

$$= \left[\begin{array}{c|c} TT^+ & 0 \\ \hline 0 & P_{\mathscr{R}(X')} \end{array}\right]$$

$$= \left[\begin{array}{c|c} P_{\mathscr{R}(X,V)} & 0 \\ \hline 0 & P_{\mathscr{R}(X')} \end{array}\right]$$

$$= P_{\mathscr{R}(B)},$$

by means of (4.2a′), Theorem 4.1(1)–(3), P2.4, $\mathscr{R}(T') = \mathscr{R}(T) = \mathscr{R}(V, X) = \mathscr{R}(\overline{V}, X) = \mathscr{N}(P)$, $\mathscr{N}(T) = \mathscr{N}(T') = \mathscr{R}(P)$, $\mathscr{R}(B) = \mathscr{R}(B')$, and (3.3). Use was also made of the well-known fact that, since $I - XX'T^+$ is a projector along $\mathscr{R}(XX') = \mathscr{R}(X)$, $(I - XX'T^+)'$ is a projector onto $\mathscr{R}(X)^\perp = \mathscr{N}(X')$, which in view of (4.2a′) implies that $X'T^+V(I - XX'T^+)' = 0$. We obtain by similar arguments, $GB = P_{\mathscr{R}(B')} = P_{\mathscr{R}(B)}$ and $GBG = G$. Hence $G = B^+$, by P2.4 and P2.2. That the blocks in (4.3) coincide with the corresponding blocks in (4.3′) can be seen by applying Theorem 4.1(3) and P2.4. The details, however, are left to the reader. ■

Note. The representation (4.3) reduces to

$$B^+ = \left[\begin{array}{c|c} \overline{V}^+ & X^{+\prime} \\ \hline X^+ & -X^+VX^{+\prime} \end{array}\right] \tag{4.5}$$

iff

$$V\mathscr{R}(X) \subseteq \mathscr{R}(X) \tag{4.6}$$

This can be verified as follows. Observe first that $\overline{V}^+ = \overline{V}^{\{2, \mathscr{R}(\overline{V}'), \mathscr{N}(\overline{V}')\}}$. Hence $G_{10} = \overline{V}^+$ iff $\mathscr{N}(\overline{V}) = \mathscr{N}(\overline{V}')$, that is, iff $V\mathscr{R}(X) \subseteq \mathscr{R}(X)$, so showing necessity. Conversely, let (4.6) be true. Then $V\mathscr{N}(X') \subseteq \mathscr{N}(X')$ by Corollary 3.7, so that $XX'\overline{V} = 0$ and $\overline{V}XX' = 0$. So in view of Theorem 4.1(5), $T^+ = \overline{V}^+ + (XX')^+$. Since $(XX')^+ = X^{+\prime}X^+$ [cf. Theorem 1.2 in Pringle and Rayner (1971)], it is now not difficult to see that (4.3) indeed reduces to (4.5) (note P2.1–P2.4).

Note. Theorem 4.2 indicates that for performing estimation in the GGM model (1.1) by means of the IPM method it suffices to know an arbitrary g-inverse of T (e.g., T^+); T^- behaves like a brick which contains all information in a compact manner. In this context it should also be remarked that Werner (1986) gives a procedure for computing any $\{1,2\}$-inverse of a given matrix $A (A \in \mathscr{R}^{s,t})$ which, on the whole, requires the computation of the northwest $t \times s$ submatrix in the regular inverse of an appropriate square (nonsingular) matrix being only of order $\max\{s, t\}$; a determinantal formula for any $\{1,2\}$-inverse is given in Werner (1984).

If a particular relation between V and X exists, then Theorem 4.2 may be simplified slightly.

Theorem 4.4. Let V be weakly complementary to X'. Then with $S := V + XX'$,

$$S^-VS^- \in \Gamma_1 \tag{4.7a}$$

$$S^-X \in \Gamma_2 \tag{4.7b}$$

$$X'S^- \in \Gamma_3 \tag{4.7c}$$

$$0 \in \Gamma_4 \tag{4.7d}$$

Proof. The assumption, of course, implies that V is weakly bicomplementary to XX'. (4.7a)–(4.7d) may be established similar as (4.1a)–(4.1d) by applying Theorem 4.1 and observing that $S^{-\prime} \in S\{1\}$. So the details are left to the reader. ■

Note. We remark that under the assumption of this corollary,

$$B^+ = \left[\begin{array}{c|c} S^+VS^+ & S^+X \\ \hline X'S^+ & 0 \end{array}\right] \tag{4.8}$$

This can be seen, again by checking the corresponding defining relations of B^+ [compare the proof of (4.5)]. (4.8), (4.7a)–(4.7d) and also (4.1a)–(4.1d) thus extend result (3.47) by Pringle and Rayner (1971) to more general situations. In this context, recall that $0 \in \Gamma_4$ is not only a necessary but also a sufficient condition for V to be weakly complementary to X' [see Corollary 3.6(4)].

Note. If the inclusion $V\mathcal{N}(X') \subseteq \mathcal{N}(X')$ holds true, then $\overline{V} = Q\overline{V}$, so that $\overline{V}$ is a *nnds* matrix. Use of Theorem 4.2 in the IPM method then only requires the computation of an arbitrary g-inverse of the *nnds* matrix $T = \overline{V} + XX'$, which from a computational point of view might be preferable to an approach as indicated by Corollary 3.7.

Note. The preceding note could lead one to believe that, even in the general case, Theorem 4.2 remains true when we replace T by the *nnds* matrix $U := QVQ + XX'$. That this, however, is erroneous may be seen as follows. Let G be an arbitrary but fixed g-inverse of U. By P2.2, $G \in U\{\mathcal{S}\}$ for some unique $\mathcal{S} \in \mathcal{R}_c(U)$. Applying Theorem 4.1(1)–(3) (note Lemma 4.1) then readily yields $X'G = X^{\{2, \mathcal{R}(X'), \mathcal{S} \oplus \mathcal{R}(\overline{V}')\}}$. In view of (3.7c) we therefore have $X'G \in \Gamma_3$ iff $\mathcal{R}(\overline{V}) \subseteq \mathcal{R}(\overline{V}') \oplus \mathcal{S}$ (note P2.3). Since $\mathcal{R}(U) = \mathcal{R}(\overline{V}', X) = \mathcal{R}(\overline{V}, X) = \mathcal{R}(V, X)$ and $\mathcal{S}$ is a direct complement of $\mathcal{R}(U)$, it turns out that this happens iff $\mathcal{R}(\overline{V}) \subseteq \mathcal{R}(\overline{V}')$, that is, iff $V\mathcal{N}(X') \subseteq \mathcal{N}(X')$. So $X'U^- \in \Gamma_3$ only if we are back in the situation described in the preceding note. That, contrary to this, $U^-Q \in \Gamma_1$ holds true in any case can be seen by means of Theorem 4.1(1) and the fact that $\mathcal{N}(Q\overline{V}) = \mathcal{N}(\overline{V})$.

Note. We conclude this paper with a remark showing the connection of some of our results (especially Corollary 3.7 and some of the foregoing notes) with the classical problem: when is each traditional BLUE representation of $X\beta$ under $(y, X\beta, \sigma^2 I)$ also a traditional BLUE representation of $X\beta$ under $(y, X\beta, \sigma^2 V)$? As a consequence of the note just before Corollary 3.2, this happens iff $\Gamma_3(I) \subseteq \Gamma_3(V)$, where, for example, $\Gamma_3(I)$ denotes the Γ_3 set under $(y, X\beta, \sigma^2 I)$. Letting $\overline{I} := Q$ yields $\mathcal{R}(\overline{I}) = \mathcal{N}(X') \in \mathcal{R}_c(X)$. Combining these observations we now directly reobtain from Theorem 3.4(3) as a solution to the above problem that again (4.6), or equivalently,

$\mathscr{R}(\bar{V}) \subseteq \mathscr{R}(\bar{I})$, is a necessary and sufficient condition. By similar arguments, it is also easy to reobtain the following well-known more general result [cf. Mitra and Moore (1973)]. The traditional BLUE of $X\beta$ under $(y, X\beta, \sigma^2 V)$, irrespective of the linear representation used in its expression, is also a traditional BLUE of $X\beta$ under $(y, X\beta, \sigma^2 W)$ iff

$$\mathscr{R}(\overline{W}) \subseteq \mathscr{R}(\bar{V}) \tag{4.9}$$

holds true, where $\overline{W} := WQ$.

REFERENCES

Albert, A. (1973). The Gauß–Markoff theorem for regression models with possibly singular covariances. *SIAM J. Appl. Math.* **24**, 182–187.

Baksalary, J. K. and Kala, R. (1980). On estimation problems in a general Gauss–Markov model. In *Data Analysis and Informatics* (E. Diday et al., Eds.), Amsterdam: North-Holland, pp. 163–167.

Chipman, J. S. (1964). On least squares with insufficient observations. *J. Amer. Statist. Assoc.* **59**, 1078–1111.

Hall, F. J. and Meyer Jr., C. D. (1975). Generalized inverses of the fundamental bordered matrix used in linear estimation. *Sankhyā A* **37**, 428–438.

Mitra, S. K. and Moore, B. J. (1973). Gauß–Markoff estimation with an incorrect dispersion matrix. *Sankhyā A* **35**, 139–152.

Pringle, R. M. and Rayner, A. A. (1971). *Generalized Inverse Matrices with Applications to Statistics*, London: Griffin.

Rao, C. R. (1971). Unified theory of linear estimation. *Sankhyā A* **33**, 371–394.

Rao, C. R. (1972a). Some recent results in linear estimation. *Sankhyā B* **34**, 369–378.

Rao, C. R. (1972b). A note on the IPM method in the unified theory of linear estimation. *Sankhyā A* **34**, 285–288.

Rao, C. R. (1973). Representations of the best linear unbiased estimators in the Gauß–Markoff model with a singular dispersion matrix. *J. Multiv. Anal.* **3**, 276–292.

Rao, C. R. (1978). Choice of best linear estimators in the Gauß–Markoff model with a singular dispersion matrix. *Commun. Statist. Theor. Meth. A* **7**, 1199–1208.

Rao, C. R. and Mitra, S. K. (1971). *Generalized Inverse of Matrices and Its Applications*, New York: Wiley.

Schönfeld, P. and Werner, H. J. (1986). A note on C. R. Rao's wider definition BLUE in the general Gauss–Markov model, *SFB 303 Research Report*, Bonn.

Werner, H. J. (1977). *g*-Inverse und monotone Matrizen. Dissertation, Bonn.

Werner, H. J. (1984). On extensions of Cramer's rule for solutions of restricted linear systems. *Linear Multilinear Alg.* **15**, 319–330.

Werner, H. J. (1985a). Weak complementarity and missing observations. *Meth. Oper. Res.* **50**, 421–439.

Werner, H. J. (1985b). More on BLU estimation in regression models with possibly singular covariances. *Linear Alg. Appl.* **67**, 207–214.

Werner, H. J. (1986). Generalized inversion and weak bi-complementarity. *Linear Multilinear Alg.* **19**, 357–372.

Partial Parameter Instability and Small Sample Prediction Bias

Anders H. Westlund
Department of Economic Statistics
Stockholm School of Economics
Stockholm, Sweden

SUMMARY

In general, prediction of an economic system presupposes that the structure of the system is constant over sample and prediction periods. Such an assumption, however, is often very unrealistic. In this chapter we give an analysis of a prediction strategy, especially adapted to situations where structural instabilities are concentrated to a subset of the system relations. Partial structural instability, represented by discrete parameter shifts, is studied. As it often is difficult to identify and characterize structural instabilities, it is worthwhile to consider prediction strategies where the reduced form is estimated indirectly, based on just the restrictions given by the stable structural relations. A small-σ technique is used to analyze the prediction bias of such a strategy for cases with varying relative degree of structural instability. The prediction strategy is expected to be robust in the sense that it reduces the negative consequences of partial parameter instability.

1. INTRODUCTION

The prediction of an economic or social system is generally based on an assumption that the structure of the system will remain constant over the historical sample period, as well as over the prediction period. In spite of the fact that such an assumption is made explicitly or implicitly in nearly all prediction processes, that assumption is often very restrictive, not to say unrealistic. Economic and social structures are actually often characterized by certain instabilities, something that now is relatively commonly accepted [see bibliographies, e.g., by Johnson (1977, 1980) and Hackl and Westlund (1985)]. However, it is usually ignored that the instability often has such dimensions that a stability approximation is inappropriate, and that predictions should be determined according to strategies that allow for structural instability.

In this chapter we give an analysis of such a strategy. It is especially adapted to situations where structural instabilities are restricted to, or at least concentrated to, a subset of the relations of the system. Such partial structural instability very often characterizes economic and social systems. The structure of an economic or social system is often represented by the structural parameters of an econometric model. Structural instabilities may then be characterized by various kinds of parameter-instability models. In this chapter, instability models in the form of discrete parameter shifts are studied. Such instabilities are often consequences of changes in policy variables within the system or of exogenous shocks.

As predictions are functions of estimated reduced forms, a prediction strategy should take into consideration the choice of reduced-form estimation technique. As it always is very difficult to identify and characterize structural instabilities, it might be worthwhile to consider a strategy that attempts to reduce negative effects due to an incorrect handling of the parameter instabilities. The reduced form should in principle be estimated by indirect techniques based on the structural form restrictions. However, it is obviously tempting to avoid an inclusion of structural restrictions from those structural relations that actually are assumed being structurally unstable. Such a strategy for robust prediction is presented in Section 2.

Prediction and estimation strategies should, if possible, be finite sample evaluated, rather than studied with respect to asymptotic properties. One way to indicate finite sample properties is to use numerical Monte Carlo studies. However, in order to increase the possibility for more general conclusions, theoretical analyses are preferable. These are, however, in general extremely complicated and mathematically intractable. Thus, we have here, as a compromise, applied an approximate theoretical small-sample technique in evaluating the prediction strategy, namely, the so-called small-σ technique. Section 3 gives an analysis of the prediction bias according to that evaluation technique for different cases with varying relative degrees of structural instability. The analysis indicates that the prediction strategy is robust in the sense that it reduces the bias effects from the parameter instabilities. Finally, Section 4 summarizes the analysis, and gives a few suggestions for further research within the area of partial structural instability.

2. PARTIAL PARAMETER INSTABILITY AND ROBUST PREDICTION

The purpose of the present analysis is to study certain aspects of the prediction of p variables y_t, related through an interdependent system to q exogenous variables x_t. We here assume that a correct specification of the system presupposes that some parameters, related to the exogenous variables, will shift from one level to another, for example, between $t = t^*$ and $t = t^* + 1$. Such a model may be written

$$Y\Gamma + (X;\, X^*)(\Delta';\, \Delta^{*\prime})' + \sigma\Psi = 0 \tag{2.1}$$

where Y and X are matrices including T observations on p and q variables, respectively;

$$X^* = \left(O'_{t^*/q};\, X'_{T-t^*/q}\right)'$$

denotes the $T - t^*$ last rows of X; Γ (assumed nonsingular), Δ, and Δ^* are parameter matrices; Ψ is a residual matrix ($E\psi_{ti} = 0$, all i, t; $E\psi_{ti}\psi_{t'i'} = \sigma_{ii'}\delta_{tt'}$; $\delta_{tt'} = 0$ if $t \neq t'$ and $\delta_{tt'} = 1$ for $t = t'$); and σ is a small positive scalar.

The following analysis concentrates on the evaluation of a strategy for robust prediction of one of the endogenous variables, without loss of generality, the first variable in $Y = (y, Y_1, Y_2)$. In this partition of Y, the matrices are of order $T/1$, T/p_1, and T/p_2, where $p = 1 + p_1 + p_2$. The first structural relation of (2.1) is then

$$y = Y_1\gamma + X_1\delta + X^*\delta^* + \sigma\psi \tag{2.2}$$

if

$$\Gamma_{(1)} = \left(-1;\, \gamma';\, O'_{p_2}\right)'$$

$$\Delta_{(1)} = \left(\delta';\, O'_{q_2}\right)'$$

$$\Delta^*_{(1)} = \delta^*$$

$$\Psi_{(1)} = \psi$$

($C_{(1)}$ denotes the first column of C), and finally

$$X_1 = X\left(I_{q_1};\ O'_{q_2/q_1}\right)'$$

where $q = q_1 + q_2$.

Predictions are determined by the reduced form:

$$y = (X;\ X^*)\{-(\Delta';\ \Delta^{*\prime})'\Gamma_0^{-1}\} + \sigma\Psi\left(-\Gamma_0^{-1}\right) \tag{2.3a}$$

$$Y_1 = (X;\ X^*)\{-(\Delta';\ \Delta^{*\prime})'\Gamma_1^{-1}\} + \sigma\Psi\left(-\Gamma_1^{-1}\right) \tag{2.3b}$$

$$Y_2 = (X;\ X^*)\{-(\Delta';\ \Delta^{*\prime})'\Gamma_2^{-1}\} + \sigma\Psi\left(-\Gamma_2^{-1}\right) \tag{2.3c}$$

where Γ^{-1} is partitioned conformably with Y above, that is, as

$$\Gamma^{-1} = \left(\Gamma_0^{-1};\ \Gamma_1^{-1};\ \Gamma_2^{-1}\right)$$

When parameter instabilities in (2.1) are concentrated to just some of the structural relations, the following reduced form estimation strategy is suggested, and expected to be robust:

(a) (2.3a) is estimated by a direct unrestricted reduced form estimator (e.g., Least Squares No Restrictions), if (2.2) involves parameter instabilities.

(b) (2.3a) is estimated by a combined partially restricted estimator (see below), if (2.2) does not involve parameter instabilities.

The combined partially restricted reduced form estimator is derived as follows. Substituting $X_1 = X(I_{q_1};\ O'_{q_2/q_1})'$ and (2.3b) in (2.2), assuming $\delta^* = 0$, gives

$$y = \Big\{(X;\ X^*)\{-(\Delta';\ \Delta^{*\prime})'\Gamma_1^{-1}\};$$

$$X\left(I_{q_1};\ O'_{q_2/q_1}\right)'\Big\}\ (\gamma';\ \delta')' + \sigma\left(\psi - \Psi\Gamma_1^{-1}\gamma\right) \tag{2.4}$$

If some of the structural relations (not the first relation, however), are assumed characterized by structural instability, but incorrectly not considered in model specification, we suggest a combined estimator of π in $y = X\pi + \nu$ according to

$$\hat{\pi}_k = \left\{\hat{\Pi}_1;\left(I_{q_1};\ O'_{q_2/q_1}\right)'\right\}\left(\hat{\gamma}'_k,\ \hat{\delta}'_k\right)' \tag{2.5}$$

where $\hat{\gamma}_k$, $\hat{\delta}_k$ denote k-class structural form estimators of γ and δ, and $\hat{\Pi}_1$ is an LSNR estimator of Π_1 in $Y_1 = X\Pi_1 + V$.

The use of (2.5) allows us to consider the structural form restrictions of the parameter-stable first relation, but we avoid the incorrect restrictions of the structurally unstable relations. Thus the parameter-instability problem only partially influences the reduced-form estimator $\hat{\pi}_k$, and as a consequence, the prediction of y.

We now consider cases with varying degree of parameter instability, and analyze the prediction of y with respect to bias effects. To simplify the presentation below, the following notations will be used:

$$w = (\gamma'; \delta')', \qquad \hat{w}_k = (\hat{\gamma}_k'; \hat{\delta}_k')'$$

$$\nabla w_k = \hat{w}_k - w$$

$$\theta = \left\{\Pi_1; \left(I_{q_1}; O'_{q_2/q_1}\right)'\right\}$$

$$\hat{\theta} = \left\{\hat{\Pi}_1; \left(I_{q_1}; O'_{q_2/q_1}\right)'\right\}$$

$$\nabla\theta = \hat{\theta} - \theta$$

3. A SMALL-σ ANALYSIS OF PREDICTION BIAS

The purpose of the present analysis is to study the prediction bias, subject to various sizes of partial parameter instability. One way to vary the parameter instability is to relate it to the size of random disturbances, for example, in the following way:

$$\Delta^* = \sigma^{\lambda}(\Delta_0^* + O(1)), \tag{3.1}$$

where Δ_0^* is a constant matrix not dependent on σ, and $\lambda \geq 0$ indicates the relative importance of parameter instability. Increasing λ decreases the importance of parameter shifts.

The analysis is based on the small-σ approximation theory [see Kadane (1971)]. It involves asymptotic series in powers of σ. As $\sigma \to 0$, the residuals are forced to 0 in probability. The sample size T is a parameter in the small-σ approach, and by letting T increase, the large-sample asymptotic results are obtained. The small-σ technique, thus, includes a sample-size correction to the large-sample theory.

If $\hat{y}^p = X^p\hat{\pi}_k$ denotes predictions of y (where X^p are predictions of X) it is easily seen that the prediction bias is determined by

$$\text{Bias}(\hat{y}^p) = X^p\{\theta E(\nabla w_k) + E(\nabla\theta)w + E(\nabla\theta\nabla w_k)\} \tag{3.2}$$

Let us first consider the case in which parameter instabilities may remain substantial, as σ approaches 0 [corresponding to $\lambda = 0$ in (3.1)]. Although our basic interest is related to cases in which $\delta^* = 0$, that is, (2.2) is parameter stable, we derive the prediction bias for this case as a corollary of the more general theorem, assuming $\delta^* \neq 0$.

Incorrect assumption of the parameter stability in (2.2) implies

$$y = Y_1\gamma + X_1\delta + \sigma\bar{\psi}$$

If estimating γ and δ by k-class estimators, we shall have the following estimation errors:

$$\nabla w_k = \{(Y_1, X_1)'(I - k\bar{P}_X)(Y_1, X_1)\}^{-1}(Y_1, X_1)'(I - k\bar{P}_X)(X^*\delta^* + \sigma\psi) \tag{3.3}$$

where

$$\bar{P}_X = I - X(X'X)^{-1}X'$$

which is rewritten

$$\nabla w_k = \left\{Z'(I - k\bar{P}_X)Z + \sigma\langle Z'(I - k\bar{P}_X)\Lambda\rangle + O_p(\sigma^2)\right\}^{-1}$$
$$\times\left\{Z'(I - k\bar{P}_X)X^*\delta^* + \sigma\left\{Z'(I - k\bar{P}_X)\psi + \Lambda'(I - k\bar{P}_X)X^*\delta^*\right\} + O_p(\sigma^2)\right\} \tag{3.4}$$

where

$$Z = \left\{(X; X^*)\left\{-(\Delta'; \Delta^{*\prime})'\Gamma_1^{-1}\right\}; X_1\right\}$$

$\langle C\rangle$ denotes $C + C'$, and $\Lambda = \Psi(-\Gamma_1^{-1}, O_{p/q_1})$ are introduced to simplify notations. Furthermore, $C = O_p(\sigma^2)$, if for each $\varepsilon > 0$, there exist constants σ_0 and c, such that

$$P\left\{\left|\frac{C}{\sigma^2}\right| > c\right\} < \varepsilon, \qquad \forall\sigma < \sigma_0$$

As is seen from (3.2), we now have to determine $E(\nabla w_k)$, $E(\nabla\theta)$, and

$E(\nabla\theta\nabla w_k)$. The series expansion of ∇w_k (in powers of σ) is then

$$\nabla w_k = \nabla w_k^{(0)} + \sigma\nabla w_k^{(1)} + O_p(\sigma^2) \tag{3.5}$$

where

$$\nabla w_k^{(0)} = W_k Z'(I - k\bar{P}_X) X^*\delta^*$$

$$\nabla w_k^{(1)} = W_k Z'(I - k\bar{P}_X)\psi + W_k \Lambda'(I - k\bar{P}_X) X^*\delta^*$$
$$- W_k\langle Z'(I - k\bar{P}_X)\Lambda\rangle W_k Z'(I - k\bar{P}_X) X^*\delta^*$$

and where

$$W_k = \{Z'(I - k\bar{P}_X)Z\}^{-1}$$

Thus, we have

$$E(\nabla w_k) = W_k Z'(I - k\bar{P}_X) X^*\delta^* + O(\sigma^2) \tag{3.6}$$

The reduced form estimation error will now be

$$\nabla\theta = \Omega + (X'X)^{-1}X'\Lambda \tag{3.7}$$

where

$$\Omega = \{(X'X)^{-1}X'X^*(-\Delta^*\Gamma_1^{-1}); 0\}$$

that is,

$$E(\nabla\theta) = \Omega \tag{3.8}$$

Furthermore, it is found that

$$E(\nabla\theta\nabla w_k) = \Omega\nabla w_k^{(0)} + O(\sigma^2) \tag{3.9}$$

Substituting (3.6), (3.8), and (3.9) into (3.2) proves Theorem 3.1.

Theorem 3.1. The approximate small-sample prediction bias to the order of $O(\sigma^2)$, when parameter instabilities are assumed substantial [i.e., $\lambda = 0$ in (3.1)], is

$$\text{Bias}(\hat{y}^p) = X^p\{(\theta + \Omega)\nabla w_k^{(0)} + \Omega w + O(\sigma^2)\} \tag{3.10}$$

where $\nabla w_k^{(0)}$ is defined in (3.5).

Corollary 3.1, corresponding to the case of a parameter stable first relation, follows directly.

Corollary 3.1. The approximate small-sample prediction bias to the order of $O(\sigma^2)$, when parameter instabilities are assumed essential, but the first relation is stable [i.e., $\lambda = 0$ in (3.1), and $\delta^* = 0$], is

$$\text{Bias}(\hat{y}^p) = X^p\{\Omega w + O(\sigma^2)\} \tag{3.11}$$

In the following we shall consider cases with gradually decreasing importance of parameter instabilities. When $\lambda > 0$ in (3.1), Z as well as Λ will be affected, that is, these quantities could be separated into terms of order $O_p(1)$ and $O_p(\sigma^\lambda)$. The crucial effect, however, is on Z.

Thus, if $\lambda > 0$,

$$Z = Z(0) + \sigma^\lambda Z(\lambda) \tag{3.12}$$

where $Z(0)$ is of order $O_p(1)$ and $\sigma^\lambda Z(\lambda)$ is of order $O_p(\sigma^\lambda)$. It should be noted that $Z(0)$ lies in a subspace of $\eta(X)$ (the column space of X), so that $\bar{P}_X Z(0)$ vanishes.

It may now be shown that

$$\begin{aligned}\nabla w_k = \Big\{ & Z^{(0)\prime}Z^{(0)} + \sigma^\lambda\langle Z^{(\lambda)\prime}Z^{(0)}\rangle \\ & + \sigma\langle \Lambda' Z^{(0)}\rangle + O_p\big(\sigma^{2(\lambda \wedge 1)}\big)\Big\}^{-1} \\ & \times \Big\{\sigma^\lambda Z^{(0)\prime}X^*\delta^*\sigma^{-\lambda} \\ & \quad + \sigma Z^{(0)\prime}\psi + \sigma^{2\lambda}Z^{(\lambda)\prime}(I - k\bar{P}_X)X^*\delta^*\sigma^{-\lambda} \\ & \quad + \sigma^{1+\lambda}\big\{Z^{(\lambda)\prime}(I - k\bar{P}_X)\psi + \Lambda'(I - k\bar{P}_X)X^*\delta^*\sigma^{-\lambda}\big\} \\ & \quad + \sigma^2\Lambda'(I - k\bar{P}_X)\psi + O_p\big(\sigma^{(1+2\lambda)\wedge(2+\lambda)}\big)\Big\}\end{aligned} \tag{3.13}$$

where $c_1 \wedge c_2$ denotes the minimum of c_1 and c_2.

As it is seen that

$$Z^{(0)} = \{X(-\Delta\Gamma_1^{-1});\ X_1\}$$

Z reduces to $Z(0)$ when no parameter instabilities do exist. To simplify notation, let us use

$$W = (Z^{(0)\prime}Z^{(0)})^{-1}$$

$$\overline{P}_k = \overline{P}_{Z^{(0)}} - k\overline{P}_X$$

$$\delta_1^* = \overline{P}_{Z^{(0)}}X^*\delta^*\sigma^{-\lambda}$$

$$\Lambda_x = -\Lambda W Z^{(0)\prime}X^*$$

We may now modify (3.5) as follows:

$$\nabla w_k = \sigma^{\lambda}\nabla w_k^{(\lambda)} + \sigma\nabla w_k^{(1)} + \sigma^{2\lambda}\nabla w_k^{(2\lambda)} + \sigma^{1+\lambda}\nabla w_k^{(1+\lambda)} + \sigma^2\nabla w_k^{(2)} + O_p(\sigma^{3(\lambda\wedge 1)}) \tag{3.14}$$

where

$$\nabla w_k^{(\lambda)} = WZ^{(0)\prime}X^*\delta^*\sigma^{-\lambda}$$

$$\nabla w_k^{(1)} = WZ^{(0)\prime}\psi$$

$$\nabla w_k^{(2\lambda)} = WZ^{(\lambda)\prime}\overline{P}_k\delta_1^* - WZ^{(0)\prime}Z^{(\lambda)}WZ(0)'X^*\delta^*\sigma^{-\lambda}$$

$$\nabla w_k^{(1+\lambda)} = WZ^{(\lambda)\prime}\overline{P}_k\psi + W\Lambda'\overline{P}_k\delta_1^* + WZ^{(0)\prime}\Lambda_x\delta^*\sigma^{-\lambda} - WZ^{(0)\prime}Z^{(\lambda)}WZ^{(0)\prime}\psi$$

and

$$\nabla w_k^{(2)} = W\Lambda'\overline{P}_k\psi - WZ^{(0)\prime}\Lambda WZ^{(0)\prime}\psi$$

Thus, if $0 < \lambda < 1$, that is, an incorrect handling of parameter instabilities may still be relatively important (in relation to σ), the following theorem is easily proved.

Theorem 3.2. The approximate small-sample prediction bias to the order of $O(\sigma^2)$, when parameter instabilities are assumed relatively important [i.e., $0 < \lambda < 1$ in (3.1)], is

$$\text{Bias}(\hat{y}^p) = X^p\Big\{\Omega w + \sigma^{\lambda}(\Omega + \theta)WZ^{(0)\prime}X^*\delta^*\sigma^{-\lambda} + \sigma^{2\lambda}(\Omega + \theta) \times\big(WZ^{(\lambda)\prime}\overline{P}_kX^*\delta^*\sigma^{-\lambda} - WZ^{(0)\prime}Z^{(\lambda)}WZ^{(0)\prime}X^*\delta^*\sigma^{-\lambda}\big) + O(\sigma^2)\Big\} \tag{3.15}$$

Corollary 3.2 follows immediately.

Corollary 3.2. The approximate small-sample prediction bias to the order of $O(\sigma^2)$, when parameter instabilities are assumed relatively important, but the first relation is stable [i.e., $0 < \lambda < 1$ in (3.1), and $\delta^* = 0$], is

$$\text{Bias}(\hat{y}^p) = X^p\{\Omega w + O(\sigma^2)\} \tag{3.16}$$

Proof (Theorem 3.2 and Corollary 3.2). First, we see that

$$\begin{aligned} E(\nabla w_k) &= \sigma^\lambda W Z^{(0)\prime} X^* \delta^* \sigma^{-\lambda} \\ &\quad + \sigma^{2\lambda}\left(W Z^{(\lambda)\prime} \overline{P}_k X^* \delta^* \sigma^{-\lambda} - W Z^{(0)\prime} Z^{(\lambda)} W Z^{(0)\prime} X^* \delta^* \sigma^{-\lambda}\right) \\ &\quad + O(\sigma^2) \end{aligned} \tag{3.17}$$

As

$$E(\nabla\theta) = \Omega$$

we also find that

$$\begin{aligned} E(\nabla\theta\nabla w_k) &= \sigma^\lambda \Omega W Z^{(0)\prime} X^* \delta^* \sigma^{-\lambda} \\ &\quad + \sigma^{2\lambda}\Omega\left(W Z^{(\lambda)\prime} \overline{P}_k X^* \delta^* \right. \\ &\quad \left. - W Z^{(0)\prime} Z^{(\lambda)} W Z^{(0)\prime} X^* \delta^* \sigma^{-\lambda}\right) + O(\sigma^2) \end{aligned} \tag{3.18}$$

That proves the Theorem 3.2, and setting $\delta^* = 0$ gives Corollary 3.2. ∎

A situation in which incorrect handling of parameter instabilities may be seen as "equally important" to the random errors of the model, corresponds to $\lambda = 1$ in (3.1). The following theorem is now given for that situation:

Theorem 3.3. The approximate small-sample prediction bias when parameter instabilities may be assumed equally important to random errors of the structural model [i.e., $\lambda = 1$ in (3.1)], is

$$\begin{aligned} \text{Bias}(\hat{y}^p) &= X^p\left\{\Omega w + \sigma(\Omega + \theta) W Z^{(0)\prime} X^* \delta^* \sigma^{-\lambda} + \sigma^2(\Omega + \theta)\right. \\ &\quad \times \left\{W Z^{(\lambda)\prime} \overline{P}_k X^* \delta^* \sigma^{-\lambda} - W Z^{(0)\prime} Z^{(\lambda)} W Z^{(0)\prime} X^* \delta^* \sigma^{-\lambda}\right. \\ &\quad \left.\left. + \{(1-k)\operatorname{tr}\overline{P}_X + \operatorname{tr}\overline{P}_1 - 1\} W E(\Lambda'\psi)/n\right\}\right\} + O(\sigma^3) \end{aligned} \tag{3.19}$$

When (2.2) is parameter stable, the bias will be

Corollary 3.3. The approximate small-sample prediction bias, when parameter instabilities are assumed equally important to random errors of the structural model, but the first relation is stable [i.e., $\lambda = 0$ in (3.1), and $\delta^* = 0$], is

$$\mathrm{Bias}(\hat{y}^p) = X^p\{\Omega w + \sigma^2\{(1-k)\operatorname{tr}\bar{P}_X + \operatorname{tr}\bar{P}_1 - 1\} \times(\Omega+\theta)WE(\Lambda'\psi)/n\} + O(\sigma^3) \quad (3.20)$$

Proof (Theorem 3.3 and Corollary 3.3). If $\lambda = 1$ in (3.1), we see that

$$E(\nabla w_k) = \sigma WZ^{(0)\prime}X^*\delta^*\sigma^{-\lambda} + \sigma^2\{WZ^{(1)\prime}\bar{P}_k X^*\delta^*\sigma^{-\lambda} - WZ^{(0)\prime}Z^{(1)}WZ^{(0)\prime}X^*\delta^*\sigma^{-\lambda} + \{(1-k)\operatorname{tr}\bar{P}_X + \operatorname{tr}\bar{P}_1 - 1\}WE(\Lambda'\psi)/n\} + O(\sigma^3) \quad (3.21)$$

and

$$E(\nabla\theta\nabla w_k) = \Omega E(\nabla w_k) + O(\sigma^2) \quad (3.22)$$

which proves the theorem.

Furthermore, setting $\delta^* = 0$ in (3.20) gives the corollary. ■

Finally, we shall consider cases in which effects of incorrect handling of parameter instabilities are assumed to be relatively less important. Such a case presupposes that $\lambda > 1$. The prediction bias results are summarized below.

Theorem 3.4. The approximate small-sample prediction bias to the order of $O(\sigma^{1+\lambda})$, when parameter instabilities are assumed to be relatively less important [i.e., $\lambda > 1$ in (3.1)], is

$$\mathrm{Bias}(\hat{y}^p) = X^p\{\Omega w + \sigma^\lambda(\Omega+\theta)WZ^{(0)\prime}X^*\delta^*\sigma^{-\lambda} + \sigma^2(\Omega+\theta) \times\{(1-k)\operatorname{tr}\bar{P}_X + \operatorname{tr}\bar{P}_1 - 1)\}WE(\Lambda'\psi)/n\} + O(\sigma^{1+\lambda}) \quad (3.23)$$

When (2.2) is parameter stable, the bias will be

Corollary 3.4. The approximate small-sample prediction bias to the order of $O(\sigma^{1+\lambda})$, when parameter instabilities are assumed to be relatively less important, and the first equation is parameter stable [i.e., $\lambda > 1$ in (3.1), and $\delta^* = 0$], is

$$\mathrm{Bias}(\hat{y}^p) = X^p\{\Omega w + \sigma^2(\Omega+\theta)\{(1-k)\operatorname{tr}\bar{P}_X + \operatorname{tr}\bar{P}_1 - 1\}WE(\Lambda'\psi)/n\} + O(\sigma^{1+\lambda}) \quad (3.24)$$

Proof (Theorem 3.4 and Corollary 3.4). If $\lambda > 1$ in (3.1), we see that

$$E(\nabla w_k) = \sigma^{\lambda} W Z^{(0)\prime} X^* \delta^* \sigma^{-\lambda}$$

$$+\sigma^2\{(1-k)\operatorname{tr}\bar{P}_X + \operatorname{tr}\bar{P}_1 - 1\} W E(\Lambda'\psi)/n + O(\sigma^{1+\lambda}) \quad (3.25)$$

and, furthermore, that

$$E(\nabla\theta\nabla w_k) = \Omega E(\nabla w_k) + O(\sigma^2) \quad (3.26)$$

which completes the proof of the theorem. The corollary is obtained by assuming $\delta^* = 0$ in (3.23). ■

A comparison of Corollaries 3.1–3.4 shows that $\text{Bias}(\hat{y}^p) = X^p\{\Omega w + O(\sigma^2)\}$, irrespective of λ in $\Delta^* = \sigma^{\lambda}(\Delta_0^* + O(1))$. Thus, the hypothesis of robustness of the prediction strategy described in Section 2 is at least partially supported.

4. CONCLUDING COMMENTS

Structural instabilities in interdependent (ID) systems are often restricted to just a subset of the relations, and to a subset of the parameters of each relation. Furthermore, such structural instabilities often manifest themselves as discrete parameter shifts. A prediction and reduced-form estimation strategy is suggested to be applied for ID systems characterized by such partial structural instabilities.

The strategy, assumed to be robust against problems owing to incorrect handling of the structural instabilities, is analyzed with respect to prediction biases for finite samples by the small-σ approach.

It is shown that, irrespective of degree of parameter instability, the approximate finite-sample prediction bias may be unchanged, if the reduced form is estimated indirectly, using just structural form restrictions connected with relations characterized by parameter stability.

The present analysis should be generalized to other types of parameter instability models. Furthermore, it is urgent to determine the corresponding approximate finite-sample prediction variances and to design simulation studies with the purpose of indicating degree of approximation in bias results given in this chapter.

REFERENCES

Hackl, P. and Westlund, A. H. (1985). Statistical analysis of "structural change"; an annotated bibliography. International Institute for Applied Systems Analysis, CP-85-31, Laxenburg.

Johnson, L. W. (1977). Stochastic parameter regression: an annotated bibliography. *Int. Statist. Rev.* **45**, 257–272.

Johnson, L. W. (1980). Stochastic parameter regression: an additional annotated bibliography. *Int. Statist. Rev.* **48**, 95–102.

Kadane, J. B. (1971). Comparison of *k*-class estimators when disturbances are small. *Econometrica* **39**, 723–737.

Asymptotically Optimal Sequential Point Estimation in Three Stages

Michael Woodroofe
Department of Statistics
The University of Michigan
Ann Arbor, Michigan

SUMMARY

A three-stage procedure for estimating the mean of an exponential family with weighted squared error loss and a constant cost for each observation is proposed. The procedure is shown to be asymptotically optimal to second order among all stopping times.

1. INTRODUCTION

Let Ω denote an open interval; and let F_w, $w \in \Omega$, denote a one-parameter exponential family of probability distributions on $R = (-\infty, \infty)$, say,

$$dF_w(x) = \exp[wx - \psi(w)]\, dL(x)$$

for $x \in R$ and $w \in \Omega$, where L is a nondegenerate σ-finite measure on R and $\exp[\psi(w)]$ is the normalizing constant. Recall that the mean and

Research supported by the National Science Foundation under DMS84-13452.

variance of F_w are

$$\theta = \psi'(w) = \int_{-\infty}^{\infty} x \, dF_w(x)$$

and

$$\psi''(w) = \int_{-\infty}^{\infty} (x - \theta)^2 \, dF_w(x)$$

for each $w \in \Omega$; and observe that θ is increasing in w, since $\psi''(w) > 0$ for all w.

It is assumed throughout that there are potential observations $X_1, X_2, \ldots$ that are independent and identically distributed with common distribution F_w under a probability measure P_w for some unknown $w \in \Omega$. In addition, it is assumed that the probability space supports a random variable U that is independent of $X_1, X_2, \ldots$ and uniformly distributed, under P_w, for all w.

Suppose that $X_1, X_2, \ldots$ may be observed sequentially for as long as desired; suppose further that observation must cease at some time n; and suppose finally that when observation is terminated the population mean θ is estimated by the sample mean $\bar{X}_n = (X_1 + \cdots + X_n)/n$ and the loss

$$L_a(n, w) = a^2 \gamma(w)^2 (\bar{X}_n - \theta)^2 + n \tag{1.1}$$

is incurred, where $a > 0$ and γ is a positive twice continuously differentiable function on Ω.

Thus, the formulation assumes that the population mean θ is to be estimated by its maximum-likelihood estimator $\bar{X}_n$, so that the only problem is to choose the sample size n. The loss (1.1) has been so normalized that each observation costs one unit. The parameter a and the function γ are determined by the cost of estimation error, relative to the cost of single observation. The main theorems require $a \to \infty$.

If $t = t(U, X_1, X_2, \ldots)$ is a stopping time, then the risk function and regret of t are defined to be

$$R_a(t; w) = E_w\{L_a(t, w)\}$$

and

$$r_a(t; w) = R_a(t, w) - 2ag(\theta)$$

where

$$g(\theta) = \gamma(w)\sqrt{\psi''(w)}$$

and $\theta = \psi'(w)$ for $w \in \Omega$ and $a > 0$; and the problem is to find a stopping time $t = t_a$ for which the regret is as small as possible. Here the term $2ag(\theta)$ is the minimum risk over all fixed sample sizes.

If J is a compact subinterval of Ω, let

$$M_a(t; J) = \sup_{w \in J} r_a(t; w)$$

and

$$M_a(J) = \inf_s M_a(s; J)$$

where the infimum extends over all stopping times s. Then a family $t = t_a$, $a > 0$, of stopping times is said to have *asymptotic local minimax regret* iff

$$\lim_{J \downarrow \omega} \limsup_{a \to \infty} [M_a(t; J) - M_a(J)] = 0$$

for all $w \in \Omega$. This notion of asymptotic optimality was studied recently by Woodroofe (1985), who showed that

$$\liminf_{a \to \infty} M_a(J) \geq \sup_{w \in J} G(w) \tag{1.2}$$

for every compact subinterval J of Ω, where

$$G = \frac{\psi'' g'(\theta)^2}{g(\theta)^2} + \gamma^2 \left(\frac{d^2}{dw^2} \right) \left(\frac{1}{g(\theta)^2} \right)$$

and the prime denotes differentiation. See also Section 6. So, $t = t_a$, $a > 0$, has asymptotic local minimax regret if

$$\lim r_a(t; w) = G(w)$$

uniformly on compacts as $a \to \infty$.

Woodroofe (1985) did not give any general results on the construction of stopping times that have asymptotic local minimax regret. Here it is shown that asymptotic local minimax regret may be obtained by procedures that take observations in three stages. The procedure studied takes an initial sample of size $m \geq 1$, a second sample to bring the sample size to

$$s = \max\{m, \lfloor (1 - \eta) a\hat{g}_m \rfloor + 1\} \tag{1.3a}$$

where $\hat{g}_m = \hat{g}_m(X_1, \ldots, X_m)$ denotes an estimate of g, $0 < \eta < 1$, and $[x]$

denotes the greatest integer that does not exceed x, and then a third sample to bring the total sample size to

$$t = \max\{s, \lfloor a\hat{g}_s \rfloor + D\} \tag{1.3b}$$

where

$$D = I\{U \le a\hat{g}_s - \lfloor a\hat{g}_s \rfloor\}$$

and $I\{\cdot\}$ denotes the indicator function of $\{\cdot\}$. Here $m = m(a)$ and $\eta = \eta(a)$ are design parameters that must approach ∞ and 0 at suitable rates. It develops that m and $t - s$ are both small compared to s. So, the three batches may reasonably be called *the pilot study*, *the main study*, and *fine tuning*.

The procedure is similar to one introduced by Hall (1981) for fixed-width interval estimation of a normal mean with an unknown variance. The analysis differs from Hall's, however, in that the sample mean and the variance estimators may be highly dependent in this case. Martinsek (1985) also considers multistage estimation.

2. PRELIMINARIES

The function g may be estimated as follows in (1.3): let Ω_n be compact subintervals of Ω for which $1/n \le g \le n$ on Ω_n; let g_n be continuous on R and $g_n = g$ on Ω_n; and let

$$\hat{g}_n = g_n(\overline{X}_n) \tag{2.1}$$

where $\overline{X}_n = S_n/n$ with $S_n = X_1 + \cdots + X_n$, for $n \ge 1$.

If J is a compact subinterval of Ω, write

$$E_J(Y) = \sup_{w \in J} E_w(Y)$$

for non-negative random variables Y, possibly depending on w; and write $P_J(B) = E_J(I_B)$ for events B, possibly depending on Ω.

Lemma 2.1. Let J be any compact subinterval of Ω, and let h be continuously differentiable on some compact neighborhood K of $\psi'(J)$. If h_n, $n \ge 1$, are continuous functions on R for which $h_n = h$ on K for all $n \ge n_0$, for some integer n_0, then there is a constant $c = c(J, h) > 0$ for which

$$P_J\{|h_k(\overline{X}_k) - h(\theta)| > y, \quad \exists k \ge n\} \le 2\exp(-cny^2) \tag{2.2}$$

for all sufficiently small $y > 0$ for all $n \ge n_0$.

Proof. Let $\varepsilon > 0$ denote the distance from $\psi'(J)$ to the complement of K; and let B denote an upper bound for $|h'|$ on K. If $w \in J$, $k \geq n_0$, and $|\overline{X}_k - \theta| < \varepsilon$, then $|h(\overline{X}_k) - h(\theta)| \leq B|\overline{X}_k - \theta|$. So, it suffices to prove the lemma when h_n are replaced by the identity function.

Then since $\overline{X}_n$, $n \geq 1$, is a reverse martingale with respect to P_w for each $w \in J$,

$$P_w\{\overline{X}_k - \theta > y, \exists k \geq n\} \leq e^{-ry}E_w\{\exp[r(\overline{X}_n - \theta)]\}$$

$$= \exp\left\{n\left[\psi\left(w + \frac{r}{n}\right) - \psi(w) - \frac{r}{n}(\theta + y)\right]\right\}$$

for all $r > 0$ for which $w + r/n \in \Omega$, all $y > 0$, and all $n \geq 1$. See, for example, Doob (1953, pp. 314, 342). The lemma now follows by minimizing with respect to $r > 0$, maximizing with respect to $w \in J$, and applying the resulting inequality to both X and $-X$. ■

Lemma 2.2. Let J be a compact subinterval of Ω. If h_n, $n \geq 1$, are as in Lemma 4.1 and $\sup |h_n| = O(n^q)$ for some $q > 0$, then

$$\lim E_J\left\{\sup_{k>n} |h_k(\overline{X}_k) - h_k(\theta)|^p\right\} = 0$$

as $n \to \infty$ for all $p > 0$.

Proof. Given $\varepsilon > 0$, there is a $\delta > 0$ for which $|h(x) - h(\theta)| < \varepsilon$ whenever $w \in J$ and $|x - \theta| < \delta$; and δ may be made as small as desired. So, letting $y = \delta$ in Lemma 4.1 shows that the expected supremum is at most

$$\varepsilon^p + \sum_{k=n}^{\infty} Ck^{pq}\exp(-ck\delta^2)$$

for some constants c and C. Lemma 2.2 follows easily. ■

In particular, the lemmas are applicable to the estimators $\hat{g}_n = g_n(\overline{X}_n)$, $n \geq 1$, in (1.3).

3. THE DISTRIBUTION OF *s*

In the remainder of the paper it is assumed that the design parameters $m = m(a)$ and $\eta = \eta(a)$ satisfy the conditions

$$m/\sqrt{a} \to 0, \qquad \eta \to 0, \qquad (\log^2 a)/m\eta^2 \to 0 \tag{3.1}$$

as $a \to \infty$. Observe that this requires $m\eta^2 \to \infty$.

Lemma 3.1. If J is any compact subinterval of Ω, then there is a constant $c = c_J$ for which

$$P_J\left\{\left|\frac{s}{a} - (1-\eta)g(\theta)\right| > y\right\} \le 2\exp(-cmy^2)$$

$$P_J\{s = m\} \le 2\exp(-cm)$$

and

$$P_J\{t = s\} \le 2\exp(-cm\eta^2)$$

for all sufficiently small $y > 0$ and sufficiently large a.

Proof. First observe that the first assertion clearly holds if $y < 3/a$. Let b denote a lower bound for g on $\psi'(J)$. If $w \in J$, $3/a \le y \le b$, $|\hat{g}_m - g(\theta)| \le y/2$, and a is sufficiently large, then $s = [a(1-\eta)\hat{g}_m] + 1 > m$ and $|s/a - (1-\eta)g(\theta)| \le y$. So, the first two assertions of the lemma follows directly from Lemma 2.1.

Next, if $w \in J$, $|\hat{g}_k - g(\theta)| \le b\eta/4$ for all $k \ge m$, and a is sufficiently large, then

$$\begin{aligned} s = \left[a(1-\eta)\hat{g}_m\right] + 1 &\le a(1 - 3\eta/4)g(\theta) \\ +1 < a(1-\eta/2)g(\theta) &< \cdots < [a\hat{g}_s], \end{aligned}$$

so that $t > s$. So, the second assertion follows from Lemma 2.1, too. ■

Lemma 3.2. If J is any compact subinterval of Ω, then

$$\lim E_J\left|\frac{s}{a} - g(\theta)\right|^p = 0$$

and

$$\lim E_J\left|\frac{a}{s} - \frac{1}{g(\theta)}\right|^p = 0$$

as $a \to \infty$ for all $p \ge 1$.

Proof. Since $a\hat{g}_m \le s \le a\hat{g}_m + 1$, Lemma 3.2 follows by applying Lemma 2.2 to g_n and $1/g_n$. ■

In the next lemma D_n denotes the σ algebra generated by $X_1, \ldots, X_n$, that is,

$$D_n = \sigma\{X_1, \ldots, X_n\} \quad \text{for } n = 1, 2, \ldots$$

Lemma 3.3. For all $w \in \Omega$,

$$E_w\left[(\bar{X}_s - \theta)^2 | D_m\right] = \frac{1}{s^2}\left[m^2(\bar{X}_m - \theta)^2 + (s - m)\psi''(w)\right]$$

$$E_w\left[(\bar{X}_s - \theta)^3 | D_m\right] = \frac{1}{s^3}\left[m^3(\bar{X}_m - \theta)^3 + 3m(s - m) \times\psi''(w)(\bar{X}_m - \theta) + (s - m)\psi'''(w)\right]$$

and

$$E_w\left[(\bar{X}_s - \theta)^4 | D_m\right] = \frac{1}{s^4}\left[m^4(\bar{X}_m - \theta)^4 + 6m^2(s - m)\psi''(w)(\bar{X}_m - \theta)^2 + 4m(s - m)\psi'''(w)(\bar{X}_m - \theta) + 3(s - m)^2\psi''(w)^2 + (s - m)\psi(w)''''\right]$$

Proof. The lemma follows easily from the independence of $X_1, X_2, \ldots$ and the fact that s is D_m measurable by writing

$$\bar{X}_s - \theta = \frac{1}{s}\left[m(\bar{X}_m - \theta) + (X_{m+1} - \theta + \cdots + X_s - \theta)\right]$$

expanding the powers, and computing the conditional expectations. ■

Lemma 3.4. If J is any compact subinterval of Ω, then

$$\lim E_w\left[s(\bar{X}_s - \theta)^2\right] = \psi''(w)$$

$$\lim E_w\left[a(\bar{X}_s - \theta)^2\right] = \frac{\psi''(w)}{g(\theta)}$$

$$\lim E_w\left[s^2(\bar{X}_s - \theta)^3\right] = \psi'''(w)$$

and

$$\lim E_w\left[s^2(\bar{X}_s - \theta)^4\right] = 3\psi''(w)^2$$

uniformly on J as $a \to \infty$.

Proof. By Lemma 3.3,

$$\left|E_w\{s(\bar{X}_s - \theta)^2 - \psi''(w)\}\right| \le E_w\left\{\left(\frac{m}{s}\right)\left[m(\bar{X}_m - \theta)^2 + \psi''(w)\right]\right\}$$

which tends to 0 as $a \to \infty$ uniformly with respect to $w \in J$, by Lemma 3.2 and Schwarz's inequality. This establishes the first assertion of the lemma, and the last two may be proved similarly.

The second assertion follows from the first, the fourth, Lemma 3.2, and Schwarz's inequality, since

$$a(\bar{X}_s - \theta)^2 - \psi''(w)/g(\theta) = \left[\frac{a}{s} - \frac{1}{g(\theta)}\right]s(\bar{X}_s - \theta)^2 + \left[\frac{1}{g(\theta)}\right]\left[s(\bar{X}_s - \theta)^2 - \psi''(w)\right]$$

4. THE DISTRIBUTION OF t

The distribution of t is studied in this section.

Lemma 4.1. Let J be a compact interval of Ω; let h be a twice continuously differentiable function on Ω; and let h_n, $n \ge 1$, be a sequence of continuous functions on R for which $h_n = h$ on some compact neighborhood K of $\psi'(J)$ for all sufficiently large n and $\sup|h_n| = O(n^q)$ for some $q > 0$. Define R_n by

$$R_n = h_n(\bar{X}_n) - h(\theta) - h'(\theta)(\bar{X}_n - \theta) - \tfrac{1}{2}h''(\theta)(\bar{X}_n - \theta)^2$$

for $n \ge 1$ and $w \in J$. Then

$$E_J\left\{\sup_{k>n} |R_k|^p\right\} = o(1/n^p)$$

and

$$E_J\{|R_s|^p\} = o(1/a^p)$$

as $n \to \infty$ and as $a \to \infty$ for all $p > 1$.

Proof. Given $\varepsilon > 0$, there is $\delta > 0$ for which $|h''(y) - h''(z)| < \varepsilon$ and $z \in K$ whenever $y \in \psi'(J)$ and $|y - z| < \delta$; let n be so large that $h_n(z) = h(z)$ whenver $|z - \psi'(J)| \le \delta$; and let $B = B(n, w)$ denote the event

$$B(n, w) = \{|\bar{X}_k - \theta| < \delta \text{ for all } k \ge n\}$$

Then B implies that $|R_k| \le \varepsilon(\bar{X}_k - \theta)^2$ for all $k > n$, by Taylor's theorem. So, by Doob's (1953, p. 317) maximal inequality (applied to the reverse submartingale $\bar{X}_k$, $k \ge 1$), and elementary considerations,

$$\int_B \sup_{k>n} |R_k|^p \, dP_w \le \varepsilon^p E_w\left\{ \sup_{k>n} |\bar{X}_k - \theta|^{2p} \right\}$$

$$\le C\varepsilon^p E_w\{|\bar{X}_n - \theta|^{2p}\} \le \frac{C'\varepsilon^p}{n^p} \tag{4.1}$$

for some constants C and C', which are independent of ε; and

$$\int_{B'} - \, dP_w \le \sqrt{P_w}(B')\sqrt{E_w}\left\{ \sup_{k>n} |R_k|^{2p} \right\}$$

which is $o(1/n^p)$ uniformly in $w \in J$, by Lemmas 2.1 and 2.2 and the maximal inequality. The first assertion of the lemma follows easily.

For the second assertion, let $B = B(m, w)$. If δ is sufficiently small, then there is a $b > 0$ for which $s \ge ab$ on B for all sufficiently large a. So

$$E_J(|R_s|^p) \le \sqrt{P_J}(B')\sqrt{E_J}\left\{ \sup_{k \ge m} |R_k|^{2p} \right\}$$

$$+ E_J\left\{ \sup_{k > ab} |R_k|^p \right\} = o\left(\frac{1}{a^p}\right)$$

as $a \to \infty$ for all $p > 1$. ■

Theorem 4.1. If J is any compact subinterval of Ω, then

$$t^* := \frac{t - ag(\theta)}{\sqrt{ag(\theta)}} \Rightarrow N\left[0, \Delta(w)^2\right]$$

and

$$\lim E_w(t^{*2}) = \Delta(w)^2$$

where

$$\Delta(w)^2 = \frac{g'(\theta)^2 \psi''(w)}{g(\theta)^2}$$

and $\Rightarrow$ denotes convergence in distribution, uniformly with respect to $w \in J$ as $a \to \infty$.

Proof. By (1.3)

$$t - ag(\theta) = a[\hat{g}_s - g(\theta)] + D^* \quad \text{on } \{t > s\}$$

where $|D^*| \le 1$; and

$$\hat{g}_s - g(\theta) = g'(\theta)(\bar{X}_s - \theta) + R_{s,0},$$

where $E_J(R_{s,0}^2) = o(1/a)$ as $a \to \infty$, by an easy application of Lemmas 3.2, 3.4, and 4.1.

Since $P_w\{t > s\} \to 1$ by Lemma 3.1 and $s/a \to g(\theta)$ in probability by Lemma 3.2, it follows directly from Anscombe's (1952) theorem that $\sqrt{s}(\bar{X}_s - \theta)$ is asymptotically normal with mean 0 and variance $\psi''(w)$ and, therefore, that t^* is asymptotically normal with mean 0 and variance $\Delta^2(w)$ for fixed $w \in J$. Uniformity with respect to ω follows from the uniformity in Lemmas 3.1 and 3.2, the uniformity of the convergence in distribution of $\sqrt{n}(\bar{X}_n - \theta)$, and an examination of the proof of Anscombe's theorem. Since this result is not needed for the asymptotic local minimaxity, the details are omitted.

For the second assertion (which is needed below), it suffices to show that $\sqrt{g(\theta)}t^* - g'(\theta)\sqrt{a}(\bar{X}_s - \theta) \to 0$ in mean square, uniformly with respect to $w \in J$, by Lemma 3.4; and this follows easily from Lemmas 3.1, 3.4, and 4.1, since

$$|\sqrt{g(\theta)}t^* - g'(\theta)\sqrt{a}(\bar{X}_s - \theta)|$$

$$\le \frac{1}{\sqrt{a}} + \sqrt{a}R_{s,0} + C\sqrt{a}\left[1 + (\bar{X}_s - \theta)^2\right] I\{t = s\}$$

for all sufficiently large a for some C. ■

The following corollary is needed in the proof of asymptotic local minimaxity.

Corollary 4.1. If J is any compact subinterval of Ω, then

$$\lim E_w \frac{(t - ag(\theta))^2}{t} = \Delta(w)^2$$

uniformly with respect to $w \in \Omega$.

Proof. Given $\varepsilon > 0$, let A be the event $A = \{t > (1 - \varepsilon)ag(\theta)\}$. Then $P_J(A') \le 2\exp(-cm\eta^2)$ for all sufficiently large a by Lemmas 2.1 and 2.2. (If A' occurs, then either $t = s$ or $\hat{g}_s - g(\theta) > \varepsilon/2$.) Since $t \le as + 1$ for large a, it follows easily that $E_J[(t - ag(\theta))^2 I_{A'}] \to 0$ and

$$E_w \frac{(t - ag(\theta))^2}{t} \le E_w \frac{t_a^{*2}}{1 - \varepsilon} + o(1)$$

$$\to \frac{\Delta^2(w)}{1 - \varepsilon}$$

uniformly in $w \in J$ as $a \to \infty$. Since $\varepsilon > 0$ was arbitrary, it follows that $\limsup E_w[(t - ag(\theta))^2/t] \le \Delta^2(w)$ uniformly with respect to $w \in J$; and a corresponding lower bound may be established similarly (and more simply). ∎

5. ASYMPTOTIC LOCAL MINIMAXITY

The main result is presented next.

Theorem 5.1. Suppose that condition (3.1) is satisfied, and define t by (1.3). If J is any compact subinterval of Ω, then

$$\lim r_a(t; w) = G(w)$$

uniformly with respect to $w \in J$ as $a \to \infty$. Consequently, t has asymptotic local minimax regret.

Proof. The proof consists of two parts: first the regret is written in a form convenient for analysis; then that form is analyzed.

For the algebra, write (as in Lemma 3.3),

$$E_w\left[(\overline{X}_t - \theta)^2 \mid U, D_s\right] = \left(\frac{s}{t}\right)^2 (\overline{X}_s - \theta)^2 + \frac{(t-s)\psi''(w)}{t^2}$$

$$= \frac{\psi''(w)}{t} + \left[(S_s^2 - s\theta)^2 - \frac{s\psi''(w)}{t^2}\right]$$

and

$$r_a(t; w) = E_w\left[a^2\gamma(w)^2(\overline{X}_t - \theta)^2 + t\right] - 2ag(\theta)$$

$$= E_w\left[\frac{a^2 g(\theta)^2}{t} - 2ag(\theta) + t\right]$$

$$+a^2\gamma(w)^2 E_w\left[\frac{(S_s^2 - s\theta)^2 - s\psi''(w)}{t^2}\right]$$

$$= E_w\left[\frac{(t - ag(\theta))^2}{t} + \gamma(w)^2(Q_1 + Q_2 + Q_3)\right] \tag{5.1}$$

where

$$Q_1 = \frac{(S_s - s\theta)^2 - s\psi''(w)}{g(\theta)^2}$$

$$Q_2 = \left[(S_s - s\theta)^2 - s\psi''(w)\right]\left[\frac{1}{\hat{g}_s^2} - \frac{1}{g(\theta)^2}\right]$$

$$Q_3 = \left[(S_s - s\theta)^2 - s\psi''(w)\right]\left[\frac{a^2}{t^2} - \frac{1}{\hat{g}_s^2}\right]$$

The first term in the last line of (5.1) was computed in Corollary 4.1; the remaining three are considered separately.

The first term, Q_1, is simplest. Since s is a stopping time with a finite expectation,

$$E_w(Q_1) = E_w\left[\frac{(S_s - s\theta)^2 - s\psi''(w)}{g(\theta)^2}\right] = 0$$

for all a and w by Wald's lemma.

The last term, Q_3, is also simple. First, if $t = s$, then

$$|Q_3| \le \left[s(\bar{X}_s - \theta)^2 + \psi''(w)\right]\left[\frac{a^2}{s} + s^3\right]$$

So

$$E_J\{|Q_3| I\{t = s\}\} \le Ca^3\sqrt{P_J}\{t = s\} \to 0,$$

as $a \to \infty$, by Lemmas 2.1, 3.2, 3.4, and Schwarz's inequality. On the set where $t > s$, write

$$\frac{1}{t^2} - \frac{1}{a^2\hat{g}_s^2} = -\frac{2}{(a\hat{g}_s)^3}(t - a\hat{g}_s) + \frac{3}{r^4}(t - a\hat{g}_s)^2$$

where r denotes an intermediate point between t and $a\hat{g}_s$. Then

$$E_w[t - a\hat{g}_s | D_s] = 0 \quad \text{on } t > s \tag{5.2}$$

by definition of D in (1.3); and $(3/r^4)(t - a\hat{g}_s)^2 \le 3/s^4$. So,

$$\left|\int_{t>s} Q_3 \, dP_w\right| \le 3\int_{t>s}\left[s(\bar{X}_s - \theta)^2 + \psi''(w)\right]\frac{a^2}{s^3}\, dP_w \to 0$$

uniformly in $w \in J$ as $a \to \infty$, by Lemmas 3.2 and 3.4.

The remaining term, Q_2, is more complicated. To simplify the writing, let

$$h = \frac{1}{g^2} \quad \text{and} \quad h_n = \frac{1}{g_n^2}$$

for $n \ge 1$. If a is so large that $g_n = g$ on some neighborhood of $\psi'(J)$ for all $n \ge m$, then

$$\frac{1}{\hat{g}_s^2} - \frac{1}{g(\theta)^2} = h_s(\bar{X}_s) - h(\theta)$$

$$= h'(\theta)(\bar{X}_s - \theta) + \tfrac{1}{2}h''(\theta)(\bar{X}_s - \theta)^2 + R_s$$

where R_s is as in Lemma 4.1 [and, therefore, $E_J(|R_s|^p) = o(1/a^p)$ as

$a \to \infty$ for all $p > 0$]. It follows that

$$\begin{aligned} Q_2 &= s^2 h'(\theta)(\overline{X}_s - \theta)^3 + \tfrac{1}{2}s^2 h''(\theta)(\overline{X}_s - \theta)^4 \\ &\quad - s\psi''(w)h'(\theta)(\overline{X}_s - \theta) - \tfrac{1}{2}s\psi''(w)h''(\theta)(\overline{X}_s - \theta)^2 \\ &\quad + a\frac{s}{a}\left[s(\overline{X}_s - \theta)^2 - \psi''(w)\right]R_s \\ &= Q_{21} + Q_{22} + Q_{23} + Q_{24} + Q_{25} \quad \text{(say)} \end{aligned}$$

These five terms are analyzed separately.

The middle and final terms, Q_{23} and Q_{25}, contribute negligibly, since

$$E_w[s(\overline{X}_s - \theta)] = E_w[S_s - s\theta] = 0$$

and

$$\begin{aligned} E_J|Q_{25}| &\le \sqrt{E_J\left\{\left[s(\overline{X}_s - \theta)^2 - \psi''(w)\right]^2\right\}} \\ &\quad \times \sqrt{E_J\left[\left(\frac{s}{a}\right)^4 (aR_s)^4\right]} \to 0 \end{aligned}$$

by Wald's lemma, and Lemmas 3.1, 3.2, and 3.4.

For the remaining three terms, a direct application of Lemma 3.4 shows that

$$\lim E_w(Q_{21}) = h'(\theta)\psi'''(w)$$

$$\lim E_w(Q_{22}) = \tfrac{1}{2}\left[3h''(\theta)\psi''(w)^2\right]$$

and

$$\lim E_w(Q_{24}) = \tfrac{1}{2}\left[-h''(\theta)\psi''(w)^2\right]$$

uniformly on J as $a \to \infty$.

Collecting terms and recalling that $h = 1/g^2$ now shows that

$$\begin{aligned} \lim r_a(t; w) &= \frac{\psi''(w)g'(\theta)^2}{g(\theta)^2} \\ &\quad + \gamma^2(w)\left[h'(\theta)\psi'''(w) + h''(\theta)\psi''(w)^2\right] \\ &= G(w) \end{aligned}$$

uniformly on J as $a \to \infty$, to complete the proof. ■

6. REMARKS ON THE RANDOMIZATION

The use of randomization in (1.3) is troubling on several counts.

First, there is a general presumption that randomization is undesirable in practice, though perhaps less so for design than analysis.

Next, Woodroofe's (1985) formulation did not explicitly include randomized stopping times; so, (1.2) is not a direct consequence of its main result. However, (1.2) does follow from Theorem 1 of Woodroofe (1985) in conjunction with Theorem 5.3 of Chow et al. (1971, p. 111).

The randomization was used in the proof only in (5.2); the remainder of the proof used only $0 \leq D \leq 1$.

If the distribution F_w, $w \in \Omega$, are sufficiently smooth, then the randomization may be eliminated and D may be replaced by $D = 1$. In this case $\sqrt{s}(\bar{X}_s - \theta)$ and the fractional part of $a\hat{g}_s$ are asymptotically independent as $a \to \infty$, and the latter has a uniform distribution. This fact may be used in place of (5.2). The details are omitted, but a related calculation will appear elsewhere.

REFERENCES

Anscombe, F. (1952). Large sample theory of sequential estimation. *Proc. Camb. Philos. Soc.* **48**, 600–607.

Chow, Y. S., Robbins, H. and Siegmund, D. (1971). *Great Expectations*, Reading, Massachusetts: Houghton-Mifflin.

Doob, J. (1953). *Stochastic Processes* New York: Wiley.

Hall, P. (1981). Asymptotic theory of triple sampling for sequential estimation of a mean. *Ann. Statist.* **9**, 1229–1238.

Martinsek, A. (1985). Multistage estimation: optimal and asymptotically optimal policies. *Sequential Analysis*, **5**, 1–18.

Woodroofe, M. (1985). Asymptotic local minimaxity in sequential point estimation. *Ann. Statist.* **13**, 676–688.

A Partial Sequential Test for the Parameters of Simple Regression

N. R. Arghami
Department of Mathematics, Statistics, and Computer Science
Faculty of Science
Mashhad University
Mashhad, Iran

L. Billard
Department of Statistics
University of Georgia,
Athens, Georgia

SUMMARY

Let

$$Y_i = a + bX_i + \varepsilon_i, \qquad i = 1, 2, \ldots \tag{1}$$

where X_i, $i = 1, 2, \ldots$, are fixed levels of a variable X, a and b are unknown parameters, and ε_i, $i = 1, 2, \ldots$, are independent and identically distributed (i.i.d.) normal random variables with mean 0 and unknown variance σ^2. In this chapter, we introduce sequential procedures for testing simple hypotheses about the slope b and the intercept a (1). In existing sequential procedures, the hypotheses concerning a and b are defined in terms of units of unknown standard deviation σ. We shall define our hypotheses independently of σ.

L. Billard acknowledges research support from the National Institutes of Health Grant No. 2 R01 GM 30325-02A1.

Sections 1 and 2 contain extensions of the test introduced in Arghami and Billard (1981) to the case of the slope and the intercept of the regression line, respectively.

1. PARTIAL SEQUENTIAL TEST FOR THE SLOPE OF THE REGRESSION LINE

Let Y_i, $i = 1, 2, \ldots$, be as in (1) and suppose we wish to test the hypotheses

$$H_0\colon b = b_0 \text{ against } H_1\colon b = b_1 \tag{1.1}$$

We know that the Y_i are independent and that

$$Y_i \approx N\left(a + bX_i, \sigma^2\right), \qquad i = 1, 2, \ldots$$

Here, b is the parameter of interest, while a and σ^2 are nuisance parameters. Using a transformation similar to that used in Arghami and Billard (1981) we can eliminate the nuisance parameters a and σ^2 and perform an SPRT (Sequential Probability Ratio Test) based on the transformed variables.

To do this, we take n_0 (≥ 3) pairs of initial observations $(X_1, Y_1), \ldots, (X_{n_0}, Y_{n_0})$ and compute the minimum variance unbiased estimator of σ^2,

$$S^2 = \frac{(1 - r_0^2)\Sigma_{i=1}^{n_0}(Y_i - \overline{Y}_0)^2}{n_0 - 2} \tag{1.2}$$

where r_0 is the sample correlation coefficient of X and Y based on the n_0 initial observations and $\overline{Y}_0 = \Sigma_{i=1}^{n_0} Y_i / n_0$. We take n_1 additional pairs of observations, where n_1 is the smallest integer (≥ 2) such that

$$\sum_{i=1}^{n^*} x_i^2 > \frac{S^2}{z_1} \tag{1.3}$$

where

$$\begin{aligned} n^* &= n_0 + n_1 \\ x_i &= \begin{cases} X_i - \overline{X}_0, & i = 1, \ldots, n_0 \\ X_i - \overline{X}_1, & i = n_0 + 1, \ldots, n^* \end{cases} \\ \overline{X}_0 &= \sum_{i=1}^{n_0} \frac{X_i}{n_0} \\ \overline{X}_1 &= \sum_{i=n_0+1}^{n^*} \frac{X_i}{n_1} \end{aligned} \tag{1.4}$$

and z_1 is a positive number independent of Y_i, $i = 1, 2, \ldots$, which may depend on X_i, $i = 1, 2, \ldots$. We then find real numbers $\alpha_1, \ldots, \alpha_{n^*}$, in a prespecified way, such that

1. α_i is proportional to $x_i / \sum_{i=1}^{n_0} x_i^2$, $i = 1, \ldots, n_0$.
2. $\sum_{i=1}^{n^*} \alpha_i = 0$.
3. $\sum_{i=1}^{n^*} \alpha_i X_i = 1$.
4. $\sum_{i=1}^{n^*} \alpha_i^2 = z_1 / S^2$.

Then, we let

$$U_1 = \sum_{i=1}^{n^*} \frac{\alpha_i Y_i}{\sqrt{z_1}} \tag{1.5}$$

Next, we take n_2 pairs of observations where n_2 is the smallest integer (> 2) such that

$$\sum_{i=n^*+1}^{n^*+n_2} x_i^2 > \frac{S^2}{z_2} \tag{1.6}$$

where

$$x_i = X_i - \overline{X}_2, \qquad i = n^* + 1, \ldots, n^* + n_2 \tag{1.7}$$

$$\overline{X}_2 = \sum_{i=n^*+1}^{n^*+n_2} \frac{X_i}{n_2}$$

and z_2 is a positive number independent of Y_i, $i = 1, 2, \ldots$, which may depend on X_i, $i = 1, 2, \ldots$. We then find real numbers $\beta_1, \ldots, \beta_{n_2}$ in a predetermined way such that

1. $\sum_{i=1}^{n_2} \beta_i = 0$.
2. $\sum_{i=n^*+1}^{n^*+n_2} \beta_{i-n^*} X_i = 1$.
3. $\sum_{i=1}^{n_2} \beta_i^2 = z_2 / S^2$.

Then, we let

$$U_2 = \sum_{i=n^*+1}^{n^*+n_2} \frac{\beta_{i-n^*} Y_i}{\sqrt{z_2}} \tag{1.8}$$

Similarly, we compute U_j, $j = 3, 4, \ldots$, using positive numbers z_j, $j = 3, 4, \ldots$, which are independent of the Y_i but may depend on the X_i.

It can then be shown that the joint density of $U_1, \ldots, U_m$ is

$$f(u_1, \ldots, u_m) = (n_0 - 2)^{(n_0-2)/2}\Gamma[(n_0 + m - 2)/2]\Pi^{-m/2}\left(\Gamma\frac{n_0 - 2}{2}\right)^{-1}$$

$$\times\left\{(n_0 - 2) + \sum_{j=1}^{m}(u_j - \delta_j)^2\right\}^{-(n_0+m-2)/2},$$

$$-\infty < u_j < \infty, \quad j = 1, \ldots, m \quad (1.9)$$

where $\delta_j = b/\sqrt{z_j}$.

Hence, to perform an SPRT with error probabilities α and β for the hypotheses in (1.1), we proceed as follows. We continue taking observations as long as

$$B^{2/(n_0+m-2)} < \lambda < A^{2/(n_0+m-2)} \quad (1.10)$$

and accept or reject H_0 according to whether the lower or the upper inequality is violated, respectively, where

$$\lambda = \frac{n_0 - 2 + \Sigma_{j=1}^{m}(U_j - \delta_{j0})^2}{n_0 - 2 + \Sigma_{j=1}^{m}(U_j - \delta_{j1})^2}$$

$$\delta_{ji} = \frac{b_i}{\sqrt{z_j}}, \qquad i = 0, 1, \quad j = 1, 2, \ldots$$

$$A = \frac{1 - \beta}{\alpha}, \qquad B = \frac{\beta}{1 - \alpha}$$

We note that the OC (Operating Characteristic) function of this test is completely free of the nuisance parameters a and σ^2. The ASN (Average Sample Number) function, although independent of a, depends on σ^2.

2. PARTIAL SEQUENTIAL TEST FOR THE INTERCEPT OF THE REGRESSION LINE

Assume the model represented by (1) and suppose we wish to test the hypotheses,

$$H_0\colon a = a_0 \text{ against } H_1\colon a = a_1 \quad (2.1)$$

where b and σ^2 are nuisance parameters.

After taking $n_0(\geq 3)$ initial pairs of observations, we compute S^2 as in (1.2). We then take n_1 additional pairs of observations, where n_1 is the smallest integer (≥ 2) such that

$$n_0 \frac{\sum_{i=1}^{n_0} x_i^2}{\sum_{i=1}^{n_0} X_i^2} + n_1 \frac{\sum_{i=n_0+1}^{n^*} x_i^2}{\sum_{i=n_0+1}^{n^*} X_i^2} > \frac{S^2}{z_1} \tag{2.2}$$

where x_i, $i = 1, \ldots, n^*$, are as in (1.4). By analogy with Section 1, we find real numbers $\alpha_1, \ldots, \alpha_{n^*}$ such that

$$\begin{aligned}
&1.\ \alpha_i \text{ is proportional to } (1/n_0 - \overline{X}_0 x_i / \textstyle\sum_{i=1}^{n_0} x_i^2), \quad i = 1, \ldots, n_0.\\
&2.\ \textstyle\sum_{i=1}^{n^*} \alpha_i = 1.\\
&3.\ \textstyle\sum_{i=1}^{n^*} \alpha_i X_i = 0.\\
&4.\ \textstyle\sum_{i=1}^{n^*} \alpha_i^2 = z_1/S^2.
\end{aligned} \tag{2.3}$$

We define U_1 as in (1.5) [but note that the actual values of n_1 and hence the α_i now satisfy (2.2) and (2.3)].

Next, we let n_2 be the smallest integer (> 2) such that

$$n_2 \frac{\sum_{i=n^*+1}^{n^*+n_2} x_i^2}{\sum_{i=n^*+1}^{n^*+n_2}} X_i^2 > \frac{S^2}{z_2} \tag{2.4}$$

where x_i, $i = n^* + 1, \ldots, n^* + n_2$, are as in (1.7), and let $\beta_1, \ldots, \beta_{n_2}$ be real numbers satisfying

1. $\sum_{i=1}^{n_2} \beta_i = 1$.
2. $\sum_{i=n^*+1}^{n^*+n_2} \beta_{i-n^*} X_i = 0$.
3. $\sum_{i=1}^{n_2} \beta_i^2 = z_2/S^2$.

Then, we let U_2 be defined as in (1.8). Similarly, we compute U_j, $j = 3, 4, \ldots$, using positive numbers z_j, $j = 3, 4, \ldots$.

We can then obtain the joint density of $U_1, \ldots, U_m$ and hence develop an SPRT procedure. It can be shown that this will be similar in format to that given in equations (1.9) and (1.10) but with δ_j and δ_{ji} replaced by

$$\delta_j = \frac{a}{\sqrt{z_j}}, \qquad j = 1, \ldots, m$$

and

$$\delta_{ji} = \frac{a_i}{\sqrt{z_j}}, \qquad i = 0, 1, \quad j = 1, 2, \ldots$$

3. CONCLUSION

The sequential tests for hypotheses concerning the slope and intercept parameters of the simple regression model have been described very briefly here. More complete details including derivation of the stated results will be provided elsewhere.

REFERENCE

Arghami, N. R. and Billard, L. (1981). A sequential test for a normal mean with unknown variance. *Bull. Int. Statist. Inst.* **43**, 113–115.

PART 6

Probability Theory

Remarks on the Foundations of Measures of Dependence

Richard C. Bradley
Department of Mathematics
Indiana University
Bloomington, Indiana

Włodzimierz Bryc
Department of Mathematics
University of California at San Diego
La Jolla, California

Svante Janson
Department of Mathematics
Uppsala University
Uppsala, Sweden

SUMMARY

Comparisons between measures of dependence are studied. Special emphasis is given to measures of dependence based on B-valued (and in particular H-valued) random variables and their connection to the absolute regularity condition for stochastic processes.

R. C. Bradley's research was partially supported by NSF Grant DMS 84-01021 and AFOSR Grant F49620 82 C 0009; work done partly at the Center for Stochastic Processes in Chapel Hill.

1. INTRODUCTION

In this note, for any number p, $1 \le p \le \infty$, the conjugate exponent will be denoted by p' $(1 \le p' \le \infty)$; that is, $1/p + 1/p' = 1$.

Let $(\Omega, \mathscr{M}, P)$ be a fixed probability space. By a *measure of dependence* we mean any function d mapping pairs of sub-σ-fields of $\mathscr{M}$ into $\overline{\mathbb{R}}_+ := \{0\} \cup \mathbb{R}_+ \cup \{\infty\}$ and satisfying the following two natural conditions:

Condition 1.1. For any two σ-fields $\mathscr{F}$ and $\mathscr{G} \subset \mathscr{M}$, $d(\mathscr{F}, \mathscr{G}) = \sup d(\mathscr{F}_0, \mathscr{G}_0)$ where the sup is taken over all pairs of finite σ-fields $\mathscr{F}_0 \subset \mathscr{F}$ and $\mathscr{G}_0 \subset \mathscr{G}$.

Condition 1.2. $d(\mathscr{F}, \mathscr{G}) = 0$ if and only if $\mathscr{F}$ and $\mathscr{G}$ are independent σ-fields.

Note that Condition 1.1 implies that d is *nondecreasing*, that is, if $\mathscr{F}_1 \subset \mathscr{F}$ and $\mathscr{G}_1 \subset \mathscr{G}$ then $d(\mathscr{F}_1, \mathscr{G}_1) \le d(\mathscr{F}, \mathscr{G})$.

In this chapter we are concerned with *dominations* of measures of dependence in the following sense: we say that a measure of dependence d_1 is *dominated* by another measure of dependence d_2 if there exists a function $\Phi\colon \overline{\mathbb{R}}_+ \to \overline{\mathbb{R}}_+$ such that $\Phi(0) = 0$, Φ is continuous at 0, and $d_1(\mathscr{F},\mathscr{G}) \le \Phi(d_2(\mathscr{F},\mathscr{G}))$ for all pairs of σ-fields $\mathscr{F}$ and $\mathscr{G} \subset \mathscr{M}$. By a trivial argument, d_2 dominates d_1 if and only if $d_1(\mathscr{F}_n, \mathscr{G}_n) \to 0$ as $n \to \infty$ for every sequence of pairs of σ-fields $(\mathscr{F}_n, \mathscr{G}_n)$, $n = 1, 2, \ldots$ such that $d_2(\mathscr{F}_n, \mathscr{G}_n) \to 0$. [To see why, consider the function $\Phi(t) := \sup\{d_1(\mathscr{F},\mathscr{G}) : d_2(\mathscr{F},\mathscr{G}) \le t\}$.] We say that two measures of dependence are *equivalent* if each one dominates the other.

Throughout this chapter, for measures of dependence d_1 and d_2, the equation $d_1 = d_2$ means that $d_1(\mathscr{F},\mathscr{G}) = d_2(\mathscr{F},\mathscr{G})$ for all pairs of σ-fields $\mathscr{F}$ and $\mathscr{G}$, and the equation $d_1 \le d_2$ means that $d_1(\mathscr{F},\mathscr{G}) \le d_2(\mathscr{F},\mathscr{G})$ for all $\mathscr{F}, \mathscr{G}$. The inequality $d_1 \le d_2$ is of course stronger than the statement that d_2 dominates d_1.

Dominations between measures of dependence play, directly or indirectly, an important role in central limit theory for stochastic processes satisfying strong mixing conditions. For example, an important role in this theory is played by *moment inequalities* involving covariances [see, e.g., Chapters 17 and 18 of Ibragimov and Linnik (1971)], and many of the most important of these moment inequalities have natural formulations in terms of dominations between measures of dependence. The measures of dependence involved here are mostly ones that are defined as norms of the bilinear form *covariance*. For more details on this for real or complex-valued random variables, see Section 1 of Bradley and Bryc (1985). All the above comments essentially carry over to the case where the random variables are H-valued (Hilbert space valued) or, more generally, B-valued (Banach space valued). Some of the measures of dependence involved in this case are norms of a bilinear form, analogous to covariance, involving B-valued (or H-valued) random variables. This chapter is devoted to a study of these particular measures of dependence. Their definition will be given in Section 2.

For now, let us quickly review the ones based on real-valued random variables. For $1 \leq p, q \leq \infty$, define the measure of dependence $R_{p,q}(\mathbb{R}) := R_{p,q}(\cdot, \cdot; \mathbb{R})$ as follows: for any two σ-fields $\mathscr{F}$ and $\mathscr{G}$,

$$R_{p,q}(\mathscr{F}, \mathscr{G}; \mathbb{R}) := \sup \frac{|Efg - EfEg|}{\|f\|_p \|g\|_q} \tag{1.1}$$

where this sup is taken over all f and g such that f is simple, real-valued, and $\mathscr{F}$-measurable and g is simple, real-valued, and $\mathscr{G}$-measurable. (In this chapter, $0/0$ is interpreted to be 0.) $R_{p,q}(\mathbb{R})$ is simply a norm of the bilinear form *covariance*. The restriction to simple functions is for convenience only; obviously it can be weakened. For $0 \leq r,\ s \leq 1$, the measure of dependence $\alpha_{r,s}$ is defined as follows: for any two σ-fields $\mathscr{F}$ and $\mathscr{G}$,

$$\alpha_{r,s}(\mathscr{F}, \mathscr{G}) := \sup \frac{|P(A \cap B) - P(A)P(B)|}{[P(A)]^r [P(B)]^s}, \qquad A \in \mathscr{F}, \quad B \in \mathscr{G} \tag{1.2}$$

This is simply an analogous *restricted* norm, the restriction being to indicator functions.

Some of these measures of dependence have been used frequently in limit theory; for example,

$$\alpha(\mathscr{F}, \mathscr{G}) := \alpha_{0,0}(\mathscr{F}, \mathscr{G}) \tag{1.3}$$

$$\phi(\mathscr{F}, \mathscr{G}) := \alpha_{1,0}(\mathscr{F}, \mathscr{G}) \tag{1.4}$$

$$\psi(\mathscr{F}, \mathscr{G}) := \alpha_{1,1}(\mathscr{F}, \mathscr{G}) \tag{1.5}$$

$$\rho(\mathscr{F}, \mathscr{G}) := R_{2,2}(\mathscr{F}, \mathscr{G}; \mathbb{R}) \tag{1.6}$$

These are the bases for, respectively, the strong mixing, ϕ-mixing, ψ-mixing, and ρ-mixing conditions for stochastic processes [see, e.g., Peligrad (1982)].

Another prominent mixing condition, absolute regularity, is based on a measure of dependence that is *not* of the form $\alpha_{r,s}$ or $R_{p,q}(\mathbb{R})$:

$$\beta(\mathscr{F},\mathscr{G}) := \sup \frac{1}{2} \sum_{i=1}^{I} \sum_{j=1}^{J} |P(A_i \cap B_j) - P(A_i)P(B_j)| \tag{1.7}$$

where this sup is taken over all pairs of partitions $\{A_1, \dots, A_I\}$ and $\{B_1, \dots, B_J\}$ of Ω such that $A_i \in \mathscr{F}$ for all i and $B_j \in \mathscr{G}$ for all j. To put this another way, $\beta(\mathscr{F},\mathscr{G}) = \frac{1}{2}\operatorname{var}(P_{\mathscr{F}\times\mathscr{G}} - P_{\mathscr{F}} \times P_{\mathscr{G}})$ where "var" denotes total variation, $P_{\mathscr{F}\times\mathscr{G}}$ is the restriction to $\mathscr{F}\times\mathscr{G}$ of the measure on $\Omega \times \Omega$ induced by P and the diagonal mapping $\omega \mapsto (\omega, \omega)$, and $P_{\mathscr{F}}$, $P_{\mathscr{G}}$ are, respectively, the restrictions to $\mathscr{F}$ and $\mathscr{G}$ of the measure P. In Section 3 we shall briefly discuss the absolute regularity condition and clarify a nice relationship between β and the measures of dependence based on B-valued random variables.

Here are some simple dominations between the five measures of dependence in equations (1.3)–(1.7):

$$\alpha \le \beta \le \phi \le \psi \quad \text{and} \quad \alpha \le \rho \tag{1.8}$$

The domination

$$\rho \le 2 \cdot \phi^{1/2} \tag{1.9}$$

is well known [see, e.g., Theorem 17.2.3 of Ibragimov and Linnik (1971)]. It has been improved independently by Denker and Keller (1983) and by Peligrad (1983).

Among the measures of dependence $\alpha_{r,s}$ $(0 \le r, s \le 1)$ and $R_{p,q}(\mathbb{R})$ $(1 \le p, q \le \infty)$ the following dominations are trivial

1. $\alpha_{r,s} \le R_{1/r,1/s}(\mathbb{R})$.
2. $\alpha_{r,s} \le \alpha_{r_1,s_1}$ if $r \le r_1$ and $s \le s_1$.
3. $R_{p,q}(\mathbb{R}) \le R_{p_1,q_1}(\mathbb{R})$ if $p \ge p_1$ and $q \ge q_1$.

The methods and results in interpolation theory [see, e.g., Bergh and Löfström (1976)] have been useful in obtaining other domination results and moment inequalities involving these (or closely related) measures of dependence [see Rosenblatt (1970, 1971), Lifshits (1984), Bradley and Bryc (1985), and Bradley, Bryc and Janson (1985)]. [In the latter three references, *complex*-valued functions are considered, but that affects the quantity in equation (1.1) by no more than a factor of 4.] In Remark 4.1 of Bradley and Bryc (1985) it was explained that among the measures of dependence $\alpha_{r,s}$

and $R_{p,q}(\mathbb{R})$ in the "lower triangle" ($r + s \le 1, 1/p + 1/q \le 1$) there are only four equivalence classes:

1. $\alpha_{r,s}$, $r + s < 1$, $R_{p,q}(\mathbb{R})$, $1/p + 1/q < 1$.
2. $\alpha_{r,1-r}$, $0 < r < 1$, $R_{p,p'}(\mathbb{R})$, $1 < p < \infty$.
3. $\alpha_{1,0}$, $R_{1,\infty}(\mathbb{R})$.
4. $\alpha_{0,1}$, $R_{\infty,1}(\mathbb{R})$.

In Section 1.2 of Bradley, Bryc and Janson (1985), it was shown that in the "upper triangle" ($r + s > 1, 1/p + 1/q > 1$), aside from the elementary equality $\alpha_{1,1} = R_{1,1}(\mathbb{R})$, there are no equivalencies, and the only dominations are ones that are either trivial or follow immediately from interpolation theory on Lorentz spaces.

The measures of dependence based on B-valued random variables (to be defined in Section 2) will be analogous to the measures $R_{p,q}(\mathbb{R})$; they will be denoted $R_{p,q}(B)$. The main tasks of this note are as follows:

1. To clarify the role that basic results and techniques in functional analysis for B-valued functions play in the study of these measures $R_{p,q}(B)$ (Sections 2–4).
2. To compare the measures $R_{p,q}(B)$ with the ones $R_{p,q}(\mathbb{R})$ [Theorems 2.2 and 3.2(2); Theorem 4.1 for the special case where B is a Hilbert space].
3. To compare the measure β in equation (1.7) to the ones $\alpha_{r,s}$ and $R_{p,q}(\mathbb{R})$ [Section 3, especially Theorem 3.2(1) with Example 3.1], and to show that, although β is not equivalent to any of the measures $\alpha_{r,s}$ and $R_{p,q}(\mathbb{R})$, it is equal to a constant multiple of a particular measure $R_{p,q}(B)$ (Theorem 3.1).

This chapter is part of Bradley, Bryc and Janson (1985). That reference gives a further discussion of measures of dependence and dominations between them.

We conclude Section 1 with a remark that will be needed later on.

REMARK 1.1. Every measure of dependence d studied in this chapter is given by a well-defined formula that is applicable on any probability space. With this in mind, the statement "d_2 fails to dominate d_1" will be used to mean that there exists a probability space on which d_2 fails to dominate d_1. (It is easily seen that in such cases d_2 will fail to dominate d_1 on the probability space $[0,1]$ with Lebesgue measure, or on any other atomless probability space.) To prove that d_2 fails to dominate d_1, it suffices to show that for different choices of pairs of σ-fields $\mathscr{F}$ and $\mathscr{G}$, on different choices of probability spaces, one can have $d_2(\mathscr{F}, \mathscr{G})$ arbitrarily small without $d_1(\mathscr{F}, \mathscr{G})$ being (arbitrarily) small (for the same $\mathscr{F}$ and $\mathscr{G}$). For then by taking the product of countably many such probability spaces, one can define a single probability space on which d_2 fails to dominate d_1 [i.e., on which there are σ-fields $\mathscr{F}_n$ and $\mathscr{G}_n$, $n = 1, 2, 3, \ldots$ such that $d_2(\mathscr{F}_n, \mathscr{G}_n) \to 0$ as $n \to \infty$ but $d_1(\mathscr{F}_n, \mathscr{G}_n)$ fails to converge to 0].

2. PRELIMINARIES

In what follows, for simplicity, we shall consider only real Banach spaces. But all results here are valid for complex Banach spaces as well. In fact, every complex Banach space is also a real Banach space (by restricting the scalars), and a simple argument will show that the results for the complex case follow from the real case.

We shall first give some standard background comments on Banach spaces, then define the measures of dependence $R_{p,q}(B)$, and finally state, in our language of measures of dependence, some elementary theorems from functional analysis for B-valued functions.

Suppose B is a Banach space (with just real scalar multiplication). The norm of an element $x \in B$ will be denoted by $\|x\|_B$. If $x \in B$ and g is a real bounded linear functional on B, then we shall use the notation $\langle x, g \rangle := g(x)$ (by analogy with the case of a Hilbert space, where each bounded linear functional can be expressed as the inner product with a fixed element). We shall let B^* denote the dual space of B, that is, the Banach space of all real bounded linear functionals on B, with norm

$$\|g\|_{B^*} := \sup \frac{|\langle x, g \rangle|}{\|x\|_B}, \qquad x \in B$$

It is well known that B is isometric to a subspace of B^{**} [$= (B^*)^*$, the "second dual" of B].

Now let us return to our probability space $(\Omega, \mathscr{M}, P)$. For each σ-field $\mathscr{A} \subset \mathscr{M}$ and each Banach space B, define $\mathscr{S}(\mathscr{A}, B)$ to be the set of all $\mathscr{A}$-measurable simple B-valued random variables, that is, the set of random variables of the form $\sum_{i=1}^{N} x_i I_{A(i)}$, where N is a positive integer, $x_1, \ldots, x_N$ are elements of B, and $\{A_1, \ldots, A_N\}$ is a partition of Ω with $A_i \in \mathscr{A}$ $\forall i$. Here I denotes the indicator function. For each $X \in \mathscr{S}(\mathscr{A}, B)$ and each p,

$1 \le p \le \infty$, define the norm

$$\|X\|_p := \| \|X\|_B \|_p = \begin{cases} [E(\|X\|_B^p)]^{1/p} & \text{if } 1 \le p < \infty \\ \text{ess sup} \|X\|_B & \text{if } p = \infty \end{cases}$$

That is, if $X = \sum_{i=1}^N x_i I_{A(i)}$ with $N, x_1, \ldots, x_N, A_1, \ldots, A_N$ as above, then $\|X\|_p = [\sum_{i=1}^N \|x_i\|_B^p \cdot P(A_i)]^{1/p}$ if $1 \le p < \infty$, and $\|X\|_\infty = \max\{\|x_i\|_B : P(A_i) > 0\}$.

For each p, q with $1 \le p, q \le \infty$ and each Banach space B, define the measure of dependence $R_{p,q}(B) := R_{p,q}(\cdot, \cdot; B)$ as follows: for any two σ-fields $\mathscr{F}$ and $\mathscr{G}$,

$$R_{p,q}(\mathscr{F}, \mathscr{G}; B) := \sup \frac{|E\langle X, Y\rangle - \langle EX, EY\rangle|}{\|X\|_p \cdot \|Y\|_q},$$

$$X \in \mathscr{S}(\mathscr{F}, B), \quad Y \in \mathscr{S}(\mathscr{G}, B^*)$$

In the special case $B = \mathbb{R}$, this quantity coincides with that in equation (1.1) by a trivial argument.

Now five elementary theorems will be stated, without proofs. Their proofs, or sketches of them, can be found, for example, in Section 2.1 of Bradley, Bryc and Janson (1985).

Proposition 2.1. Suppose $1 \le p, q \le \infty$ and B is a Banach space. Then for any two σ-fields $\mathscr{F}$ and $\mathscr{G}$ the following three statements hold:

1. For any Banach space B_1 that is isometric to B, $R_{p,q}(\mathscr{F}, \mathscr{G}; B_1) = R_{p,q}(\mathscr{F}, \mathscr{G}; B)$.
2. For any subspace B_1 of B (or Banach space B_1 isometric to a subspace of B), $R_{p,q}(\mathscr{F}, \mathscr{G}; B_1) \le R_{p,q}(\mathscr{F}, \mathscr{G}; B)$.
3.

$$R_{p,q}(\mathscr{F}, \mathscr{G}; B) = \left[\frac{\sup R_{p,q}(\mathscr{F}, \mathscr{G}; B_1)}{\text{finite-dimensional subspaces } B_1 \text{ of } B} \right]$$

Proposition 2.2. If $1 \le p, q \le \infty$ and B is a Banach space, then for any two σ-fields $\mathscr{F}$ and $\mathscr{G}$, $R_{p,q}(\mathscr{F}, \mathscr{G}; B^*) = R_{q,p}(\mathscr{G}, \mathscr{F}; B)$.

Corollary 2.1. If $1 \le p, q \le \infty$ and B is a Banach space, then $R_{p,q}(B^{**}) = R_{p,q}(B)$.

For $1 \le p \le \infty$, let l^p denote the Banach space of all sequences $x := (\xi_1, \xi_2, \xi_3, \dots)$ of real numbers with summable absolute pth powers (all bounded real sequences if $p = \infty$), with the usual norm

$$\|x\|_{l^p} := \begin{cases} \left[\sum_{i=1}^{\infty} |\xi_i|^p \right]^{1/p} & \text{if } 1 \le p < \infty \\ \sup_i |\xi_i| & \text{if } p = \infty \end{cases}$$

Theorem 2.1. If $1 \le p, q \le \infty$ and B is a nontrivial Banach space, then $R_{p,q}(\mathbb{R}) \le R_{p,q}(B) \le R_{p,q}(l^\infty) = R_{p,q}(l^1)$.

For given p and q with $1 \le p, q \le \infty$, Theorem 2.1 suggests the interpretation of $R_{p,q}(l^\infty)$ $[= R_{p,q}(l^1) = \sup_B R_{p,q}(B)]$ as $R_{p,q}$ ("Banach space").

For the next theorem, recall equation (1.4).

Theorem 2.2. For every nontrivial Banach space B, the following three statements hold:

1. $R_{1,q}(B) = R_{1,q}(\mathbb{R})$ $\forall q$, $1 \le q \le \infty$.
2. $R_{p,1}(B) = R_{p,1}(\mathbb{R})$ $\forall p$, $1 \le p \le \infty$.
3. In particular, $R_{1,\infty}(B) = R_{1,\infty}(\mathbb{R}) = 2\phi$.

Let us quickly comment on the last equality here: $R_{1,\infty}(\mathscr{F}, \mathscr{G}; \mathbb{R}) = 2\phi(\mathscr{F}, \mathscr{G})$ for any two σ-fields $\mathscr{F}$ and $\mathscr{G}$. The inequality $R_{1,\infty}(\mathscr{F}, \mathscr{G}; \mathbb{R}) \le 2\phi(\mathscr{F}, \mathscr{G})$ is simple and follows, for example, from Lemma 1.1.9 of Iosifescu and Theodorescu (1969, pp. 11–12), and the opposite inequality follows from a simple argument based on the identity

$$E[I_A(I_B - I_{B^c})] - EI_A E(I_B - I_{B^c})$$
$$= 2 \cdot [P(B|A) - P(B)] \cdot \|I_A\|_1 \cdot \|I_B - I_{B^c}\|_\infty$$

(say with $A \in \mathscr{F}$, $B \in \mathscr{G}$, $P(A) > 0$). Here I denotes the indicator function and c denotes the complement.

REMARK 2.1. Interpolation theory applies nicely to B-valued functions; see Chapter 5 of Bergh and Löfström (1976). By the Riesz (or Riesz–Thorin) interpolation theorem, there are only four equivalence classes for the measures of dependence $R_{p,q}(\mathbb{R})$ in the lower triangle $1/p + 1/q \le 1$, represented, respectively, by, say, $R_{\infty,\infty}(\mathbb{R})$, $R_{1,\infty}(\mathbb{R})$, $R_{\infty,1}(\mathbb{R})$, and

$R_{2,2}(\mathbb{R})$. [This comment is contained in Remark 4.1 of Bradley and Bryc (1985), and a key part of it came from Rosenblatt (1970, 1971).] Similarly, by corresponding interpolation theorems for B-valued functions, for any given Banach space B, there are at most four equivalence classes for the measures of dependence $R_{p,q}(B)$ in the lower triangle $1/p + 1/q \le 1$, represented, respectively, by, say, $R_{\infty,\infty}(B)$, $R_{1,\infty}(B)$, $R_{\infty,1}(B)$, and $R_{2,2}(B)$. By Theorem 2.2, $R_{1,\infty}(B)$ and $R_{\infty,1}(B)$ do not depend on the choice of (nontrivial) B; but as we shall see in Theorem 3.2(2), $R_{\infty,\infty}(B)$ and $R_{2,2}(B)$ can significantly vary with B.

3. ABSOLUTE REGULARITY

The measure of dependence β in equation (1.7) is the basis of the well-known *absolute regularity* condition for stochastic processes [see Volkonskii and Rozanov (1959)]. Now β is not equivalent to any of the other measures of dependence given in equations (1.3)–(1.6), or to any of the measures of dependence $\alpha_{r,s}$ $(0 \le r, s \le 1)$ or $R_{p,q}(\mathbb{R})$ $(1 \le p, q \le \infty)$ at all. (More on this shortly.) Also, it seems that many limit theorems for real-valued (or even H-valued) random variables under the *strong mixing* condition [which is based on the measure of dependence $\alpha = \alpha_{0,0}$ in equation (1.3)] cannot be extended to more general B-valued random variables unless the assumption of strong mixing is replaced by the (stronger) assumption of absolute regularity. [See, e.g., the discussion and results in Dehling and Philipp (1982).] To some extent, absolute regularity appears to be the most natural mixing condition for limit theory for B-valued random variables. Indeed, the measure of dependence β has a quite natural interpretation in terms of the measures of dependence $R_{p,q}(B)$; in Theorem 3.1, it will be shown with a totally elementary argument that $\beta = \frac{1}{2} \cdot R_{\infty,\infty}(l^\infty)$ $(= \frac{1}{2} \cdot R_{\infty,\infty}(l^1))$.

Now let us clarify the pattern of dominations between β and the measures of dependence $\alpha_{r,s}$, $R_{p,q}(\mathbb{R})$. Recall the inequalities $\alpha \le \beta \le \phi$ from equation (1.8). Also, β does not dominate ρ [see equation (1.6)]. [For a σ-field $\mathcal{F} = \{\Omega, A, A^c, \phi\}$ with $P(A) = \varepsilon$, say with $\varepsilon > 0$ small, one has $\beta(\mathcal{F}, \mathcal{F}) = 2\varepsilon(1 - \varepsilon)$ and $\rho(\mathcal{F}, \mathcal{F}) = 1$.] From these facts and, for example, Remark 4.1 of Bradley and Bryc (1985), one has that β dominates $\alpha_{r,s}$ only for $r + s < 1$, that β dominates $R_{p,q}(\mathbb{R})$ only for $1/p + 1/q < 1$, and that β is dominated by $\alpha_{r,s}$ if either $r = 1$ or $s = 1$, and hence by $R_{p,q}(\mathbb{R})$ if either $p = 1$ or $q = 1$. To complete this discussion of β, in Theorem 3.2 we shall show that β is not dominated by $\alpha_{r,s}$ if $r < 1$ and $s < 1$, or even by $R_{p,q}(\mathbb{R})$ if $p > 1$ and $q > 1$. As a corollary of these facts, β is not equivalent to any of the measures $\alpha_{r,s}$, $R_{p,q}(\mathbb{R})$.

Theorem 3.1. $R_{\infty,\infty}(l^\infty) = R_{\infty,\infty}(l^1) = 2\beta$.

Proof. By Theorem 2.1 it suffices to prove that $R_{\infty,\infty}(l^1) = 2\beta$. Since these measures of dependence satisfy condition 1.1, it suffices to show that if $\mathscr{F}$ and $\mathscr{G}$ are finite σ-fields, then $R_{\infty,\infty}(\mathscr{F}, \mathscr{G}; l^1) = 2 \cdot \beta(\mathscr{F}, \mathscr{G})$.

Suppose $\mathscr{F}$ is a finite σ-field, generated by a partition $\{A_1, \dots, A_I\}$ of Ω, and $\mathscr{G}$ is a finite σ-field, generated by a partition $\{B_1, \dots, B_J\}$ of Ω. Without loss of generality we assume that $P(A_i) > 0$ for all i and $P(B_j) > 0$ for all j.

Recall the elementary fact that $(l^1)^*$ is isometric to l^∞, the standard isometry being (informally described) as follows: if $x := (\xi_1, \xi_2, \dots) \in l^1$ and $y := (\eta_1, \eta_2, \dots) \in l^\infty$, then $y(x) = \langle x, y \rangle := \sum_{i=1}^\infty \xi_i \eta_i$.

If $X \in \mathscr{S}(\mathscr{F}, l^1)$ and $Y \in \mathscr{S}(\mathscr{G}, l^\infty)$, then one can represent these random variables by $X := \sum_{i=1}^I x_i I_{A(i)}$ and $Y := \sum_{j=1}^J y_j I_{B(j)}$ (with each $x_i \in l^1$ and each $y_j \in l^\infty$), and one has

$$\begin{aligned} |E\langle X, Y\rangle - \langle EX, EY\rangle| &= \left| \sum_{i=1}^I \sum_{j=1}^J \langle x_i, y_j \rangle [P(A_i \cap B_j) - P(A_i)P(B_j)] \right| \\ &\le \left(\sup_i \|x_i\|_{l^1} \right) \cdot \left(\sup_j \|y_j\|_{l^\infty} \right) \\ &\quad \cdot \sum_{i=1}^I \sum_{j=1}^J |P(A_i \cap B_j) - P(A_i)P(B_j)| \\ &= \|X\|_\infty \cdot \|Y\|_\infty \cdot 2\beta(\mathscr{F}, \mathscr{G}) \end{aligned} \tag{3.1}$$

Hence $R_{\infty,\infty}(\mathscr{F}, \mathscr{G}; l^1) \le 2\beta(\mathscr{F}, \mathscr{G})$. Conversely, in equation (3.1), equality is achieved if we take

$$x_i := (0, \dots, 0, 1, 0, 0, \dots) \qquad \text{(where the 1 is the } i\text{th coordinate)}$$

$$y_j := (y_{j1}, y_{j2}, y_{j3}, \dots)$$

where

$$y_{jk} := \begin{cases} \text{sign of } [P(A_k \cap B_j) - P(A_k)P(B_j)] & \text{if } 1 \le k \le I \\ 0 & \text{if } k > I \end{cases}$$

Hence $R_{\infty,\infty}(\mathscr{F}, \mathscr{G}; l^1) = 2\beta(\mathscr{F}, \mathscr{G})$. This completes the proof. ■

Corollary 3.1. For every Banach space B, $R_{\infty,\infty}(B) \le 2\beta$.

This just follows from Theorems 2.1 and 3.1.

Theorem 3.2. Suppose that $1 < p, q \le \infty$. Then (on some probability space) the following two statements hold:

1. $R_{p,q}(\mathbb{R})$ fails to dominate β.
2. For each p_1 and q_1 such that $1 \le p_1, q_1 \le \infty$, $R_{p,q}(\mathbb{R})$ fails to dominate $R_{p_1,q_1}(l^\infty)$.

Here (2) is a trivial corollary of (1) and Theorem 3.1 [since $R_{\infty,\infty}(l^\infty) \le R_{p_1,q_1}(l^\infty)$]. To prove (1) we shall show in Example 3.1 that for different choices of pairs of σ-fields $\mathscr{F}$ and $\mathscr{G}$, on different choices of probability spaces, one can (simultaneously) have $\beta(\mathscr{F},\mathscr{G}) = \frac{1}{2}$ and $R_{p,q}(\mathscr{F},\mathscr{G};\mathbb{R})$ arbitrarily small. Then by Remark 1.1, (1) will follow.

Example 3.1. Let p and q be arbitrary but fixed, such that $1 < p, q \le \infty$. We shall show that $R_{p,q}(\mathscr{F},\mathscr{G};\mathbb{R})$ can be arbitrarily small for σ-fields $\mathscr{F}$ and $\mathscr{G}$ satisfying $\beta(\mathscr{F},\mathscr{G}) = \frac{1}{2}$. Now for any $\mathscr{F}$ and $\mathscr{G}$, $R_{p,q}(\mathscr{F},\mathscr{G};\mathbb{R})$ is nonincreasing as p and/or q increases; without loss of generality we impose the following further restrictions on p and q:

$$1 < p \le 2 \tag{3.2}$$

$$1 < q < \infty \tag{3.3}$$

(Thus q' satisfies $1 < q' < \infty$.)

The following elementary fact will be used later on:

$$\forall a_1,\dots,a_m \in \mathbb{R}, \quad \left(\sum_{k=1}^{m} |a_k|^p\right)^{1/p} \ge \left(\sum_{k=1}^{m} |a_k|^2\right)^{1/2} \tag{3.4}$$

Now let n be an arbitrary but fixed positive integer. Construct the probability space $(\Omega, \mathscr{M}, P)$ as follows:

Define $\Omega := \{1,2,\dots,2n\} \times [0,1]$. Define $\mathscr{M}$ to be the family of Borel subsets of Ω. Define the probability measure P_1 on $\{1,2,\dots,2n\}$ (with the standard discrete σ-field) by $P_1(\{j\}) = (2n)^{-1}$, $j = 1,2,\dots,2n$. Define the probability measure P_2 on $[0,1]$ (with the Borel σ-field) to be Lebesgue measure. Finally, define the probability measure P on $(\Omega,\mathscr{M})$ to be absolutely continuous with respect to $P_1 \times P_2$ such that

$$\frac{dP}{d(P_1 \times P_2)}(j,x) = \begin{cases} 1 + h_j(x) & \text{if } 1 \le j \le n \\ 1 - h_{j-n}(x) & \text{if } n+1 \le j \le 2n \end{cases} \tag{3.5}$$

where $h_1, h_2, \ldots$ are the Rademacher functions on [0, 1] [see, e.g., Zygmund (1959)].

Note that the marginals of P on $\{1, 2, \ldots, 2n\}$ and [0, 1] are P_1 and P_2, respectively.

Let us define the "marginal" σ-fields

$$\mathscr{F} := \{A \times [0,1] : A \text{ is a subset of } \{1, 2, \ldots, 2n\}\}$$

$$\mathscr{G} := \{\{1, 2, \ldots, 2n\} \times B : B \text{ is a Borel subset of } [0,1]\}$$

By a well-known elementary argument,

$$R_{p,q}(\mathscr{F}, \mathscr{G}; \mathbb{R}) = \sup \frac{\|E(f|\mathscr{G}) - Ef\|_{q'}}{\|f\|_p}, \qquad f \in \mathscr{S}(\mathscr{F}, \mathbb{R}) \quad (3.6)$$

Let f be an arbitrary but fixed simple real-valued $\mathscr{F}$-measurable random variable. Our first task is to obtain an upper bound on $\|E(f|\mathscr{G}) - Ef\|_{q'}/\|f\|_p$.

This function $f = f(j, x)$ depends only on j; we can use the notation $f(j) := f(j, x)\ \forall j = 1, 2, \ldots, 2n$, $\forall x \in [0, 1]$.

Define the functions f_1 and f_2 on $\{1, 2, \ldots, 2n\}$ as follows:

$$f_1(j) := -f_1(j+n) := [f(j) - f(j+n)]/2 \qquad \forall j = 1, \ldots, n \quad (3.7)$$

$$f_2(j) := f_2(j+n) := [f(j) + f(j+n)]/2 \qquad \forall j = 1, \ldots, n \quad (3.8)$$

Then $\forall j = 1, \ldots, 2n$, $f(j) = f_1(j) + f_2(j)$.

For a.e. $x \in [0, 1]$ (and every $l \in \{1, 2, \ldots, 2n\}$), we have by equation (3.5) and some simple calculations, $E(f|\mathscr{G})(l, x) - Ef = \sum_{j=1}^n f_1(j) \cdot h_j(x)/n$. Therefore

$$\|E(f|\mathscr{G}) - Ef\|_{q'} = \frac{1}{n} \cdot \left[\int_0^1 \left| \sum_{j=1}^n f_1(j) h_j(x) \right|^{q'} dx\right]^{1/q'}$$

By equation (3.7) and the convexity of the function u^p (for non-negative u), we have that $\|f_1\|_p \leq \|f\|_p$. Also, $\|f_1\|_p = [(1/n) \cdot \sum_{j=1}^n |f_1(j)|^p]^{1/p}$ by a trivial calculation. Hence by equation (3.4) and Khinchin's inequality [see, e.g., Theorem (8.4) in Chapter 5 of Zygmund (1959)],

$$\frac{\|E(f|\mathscr{G}) - Ef\|_{q'}}{\|f\|_p} \le \frac{n^{(1/p)-1}\left[\int_0^1 |\Sigma_{j=1}^n f_1(j) h_j(x)|^{q'}\,dx\right]^{1/q'}}{\left[\Sigma_{j=1}^n |f_1(j)|^p\right]^{1/p}}$$

$$\le \frac{n^{(1/p)-1} C(q') \cdot \left[\Sigma_{j=1}^n (f_1(j))^2\right]^{1/2}}{\left[\Sigma_{j=1}^n |f_1(j)|^2\right]^{1/2}}$$

$$= n^{-1/p'} \cdot C(q')$$

where the constant $C(q')$ depends only on (our fixed value of) q'.

Since f was arbitrary, we have by equation (3.6),

$$R_{p,q}(\mathscr{F}, \mathscr{G}; \mathbb{R}) \le n^{-1/p'} \cdot C(q')$$

By taking n sufficiently large to begin with, we can make $R_{p,q}(\mathscr{F}, \mathscr{G}; \mathbb{R})$ arbitrarily small. Also, by equation (3.5) and an elementary argument, regardless of the value of n,

$$\beta(\mathscr{F}, \mathscr{G}) = \frac{1}{2} \cdot \int_\Omega \left| \frac{dP}{d(P_1 \times P_2)} - 1 \right| d(P_1 \times P_2)$$

$$= \frac{1}{2} \cdot \int_\Omega 1\, d(P_1 \times P_2) = \tfrac{1}{2}$$

This completes the discussion for Example 3.1, and thereby completes the proof of Theorem 3.2. ■

Remark 3.1. Gastwirth and Rubin (1975) defined for each t, $1 \le t \le \infty$, the following measure of dependence Δ_t:

$$\Delta_t(\mathscr{F}, \mathscr{G}) := \|\text{var}[P_{\mathscr{G}}(\cdot|\mathscr{F}) - P_{\mathscr{G}}(\cdot)]\|_t$$

where "var" denotes total variation and $P_{\mathscr{G}}$ is the restriction of the measure P to events in $\mathscr{G}$. For finite σ-fields $\mathscr{F}$ and $\mathscr{G}$, the definition of $\Delta_t(\mathscr{F}, \mathscr{G})$ is clear; and for general σ-fields $\mathscr{F}$ and $\mathscr{G}$, where a measure-theoretic ambiguity may have to be settled, we shall simply stipulate that condition 1.1 be satisfied by Δ_t. It is easy to see that $\Delta_1 = 2\beta$, $\Delta_\infty = 2\phi$, and Δ_t is equivalent to β if $1 \le t < \infty$. By a simple argument similar to the proof of Theorem 3.1, one has that for each t, $1 \le t \le \infty$,

$$\Delta_t = R_{t',\infty}(l^\infty) = R_{t',\infty}(l^1)$$

4. THE HILBERT-SPACE CASE

Let us now take a look at measures of dependence based on H-valued random variables. We shall let H be an arbitrary real Hilbert space, and its inner product will be denoted by $\langle \cdot , \cdot \rangle$.

For each p, $1 \leq p < \infty$, the quantity $Z_p := [(2\pi)^{-1/2}\int_{-\infty}^{\infty}|x|^p \exp(-x^2/2)\,dx]^{1/p}$ [the p norm of a standard $N(0,1)$ r.v.] will be useful.

Theorem 4.1. If H is a nontrivial Hilbert space, $1 \leq p, q \leq \infty$, and $\mathcal{F}, \mathcal{G} \subset \mathcal{M}$, then $R_{p,q}(\mathcal{F}, \mathcal{H}; H) \leq A \cdot R_{p,q}(\mathcal{F}, \mathcal{G}; \mathbb{R})$, where $A = A(p,q)$ is a function only of p and q. Further, if $1 \leq p, q < \infty$, then one can take $A(p,q) = Z_{p_1} \cdot Z_{q_1}$ for any choice of p_1 and q_1 satisfying $p \leq p_1 < \infty$, $q \leq q_1 < \infty$, and $p_1^{-1} + q_1^{-1} \leq 1$; and in particular, if $1 \leq p, q \leq 2$, then $R_{p,q}(\mathcal{F}, \mathcal{G}; H) = R_{p,q}(\mathcal{F}, \mathcal{G}; \mathbb{R})$.

Thus for a given (p,q), $1 \leq p$, $q \leq \infty$, the measure of dependence $R_{p,q}(H)$ is "linearly" equivalent to $R_{p,q}(\mathbb{R})$. For the case $p = q = \infty$, this was shown by Dehling and Philipp (1982, p. 692, Lemma 2.2), who used the Grothendieck inequality in their proof. For the other cases satisfying $p^{-1} + q^{-1} \leq 1$, a simple proof was given in Bradley and Bryc (1985, Theorem 4.2). The main purpose here is to remove that extra restriction, $p^{-1} + q^{-1} \leq 1$. Theorem 4.1 is not a particularly new result, but rather an adaptation to our context of results well known in functional analysis.

The very last case of Theorem 4.1 ($1 \leq p, q \leq 2$) follows from the rest of the theorem by taking $p_1 = q_1 = 2$.

The proof of Theorem 4.1 will be based on the following lemma:

Lemma 4.1. Suppose $1 \leq p \leq p_1 \leq \infty$, $1 \leq q \leq q_1 \leq \infty$, and $p_1^{-1} + q_1^{-1} \leq 1$. Suppose $X_1, \ldots, X_I$ and $Y_1, \ldots, Y_J$ are real-valued random variables. Then

$$E\left(\|(X_i)_{i=1}^I\|_{l^p} \cdot \|(Y_j)_{j=1}^J\|_{l^q}\right) \leq \|(\|X_i\|_{p_1})_{i=1}^I\|_{l^p} \cdot \|(\|Y_j\|_{q_1})_{j=1}^J\|_{l^q}$$

Proof of Lemma 4.1

$$E\left(\|(X_i)_{i=1}^I\|_{l^p} \cdot \|(Y_j)_{j=1}^J\|_{l^q}\right) \leq \|\,\|(X_i)_{i=1}^I\|_{l^p}\|_{p_1} \cdot \|\,\|(Y_j)_{j=1}^J\|_{l^q}\|_{q_1}$$

$$\leq \|(\|X_i\|_{p_1})_{i=1}^I\|_{l^p} \cdot \|(\|Y_j\|_{q_1})_{j=1}^J\|_{l^q}$$

Here the first inequality comes from Hölder's inequality (and the fact that our measure is a probability measure). To show the second inequality, it is enough to show that

$$\| \|(X_i)_{i=1}^{I}\|_{l^p}\|_{p_1} \le \|(\|X_i\|_{p_1})_{i=1}^{I}\|_{l^p}$$

If $p_1 = \infty$, then this is simple. (Also, if $p = 1$, then this is simply Minkowski's inequality.) If $p_1 < \infty$, then

$$\begin{aligned}
\| \|(X_i)_{i=1}^{I}\|_{l^p}\|_{p_1} &= \left[E\left(\sum_{i=1}^{I}|X_i|^p\right)^{p_1/p}\right]^{1/p_1} \\
&= \left[\left\|\sum_{i=1}^{I}|X_i|^p\right\|_{p_1/p}^{p_1/p}\right]^{1/p_1} = \left[\left\|\sum_{i=1}^{I}|X_i|^p\right\|_{p_1/p}\right]^{1/p} \\
&\le \left[\sum_{i=1}^{I}\||X_i|^p\|_{p_1/p}\right]^{1/p} = \left[\sum_{i=1}^{I}\left[E|X_i|^{p\cdot p_1/p}\right]^{p/p_1}\right]^{1/p} \\
&= \left[\sum_{i=1}^{I}\|X_i\|_{p_1}^{p}\right]^{1/p} = \|(\|X_i\|_{p_1})_{i=1}^{I}\|_{l^p}
\end{aligned}$$

This completes the proof of Lemma 4.1. ■

Proof of Theorem 4.1. For the cases in which $p = \infty$ or $q = \infty$, the reader is referred to the references mentioned after the statement of Theorem 4.1. Here we consider only the case in which $1 \le p < \infty$ and $1 \le q < \infty$. As in the proof of Theorem 4.2 of Bradley and Bryc (1985), we can (and do) restrict our attention to an arbitrary finite-dimensional real Hilbert space H; and we use the same gaussian probability measure γ on H as was used there (the "standard" one with mean 0 and identity covariance matrix, if we interpret H as $\mathbb{R}^d$ where d is the dimension of H). The use of a gaussian measure here is similar to its use in Rietz (1974).

Suppose $X := \sum_{i=1}^{I} x_i I_{A(i)}$ and $Y := \sum_{j=1}^{J} y_j I_{B(j)}$, where $\{A_1, \dots, A_I\}$ and $\{B_1, \dots, B_J\}$ are each a partition of Ω with $A_i \in \mathscr{F}$ $\forall i$ and $B_j \in \mathscr{G}$ $\forall j$, and $x_1, \dots, x_I$ and $y_1, \dots, y_J$ are elements of H. To prove Theorem 4.1 it suffices to prove that

$$|E\langle X, Y\rangle - \langle EX, EY\rangle| \le R_{p,q}(\mathscr{F}, \mathscr{G}; \mathbb{R}) \cdot Z_{p_1} \cdot Z_{q_1} \cdot \|X\|_p \|Y\|_q \quad (4.1)$$

holds for every choice of p_1 and q_1 meeting the specifications $p \le p_1 < \infty$, $q \le q_1 < \infty$, and $p_1^{-1} + q_1^{-1} \le 1$.

Before we prove equation (4.1), a couple of preliminary observations will be needed. For any $x, y \in H$, one has $\langle x, y\rangle = \int_H \langle x, u\rangle \cdot \langle y, u\rangle\, \gamma(du)$. [This is elementary and was used in the proof of Theorem 4.2 of Bradley and Bryc (1985).] For any $x \in H$, the random variable $V: H \to \mathbb{R}$ defined by $V(u) := \langle x, u\rangle$ is a $N(0, \|x\|_H^2)$ r.v. [on the probability space (H, γ)], and hence for any t, $1 \le t < \infty$, $\int_H |\langle x, u\rangle|^t \gamma(du) = \|x\|_H^t \cdot Z_t^t$. These observations, Fubini's theorem, and Lemma 4.1 [on the probability space (H, γ)] will now be used to prove equation (4.1), under the given specifications on p_1 and q_1, as follows:

$$|E\langle X, Y\rangle - \langle EX, EY\rangle|$$

$$= \left| \int_H E(\langle X, u\rangle \cdot \langle Y, u\rangle)\,\gamma(du) - \int_H \langle EX, u\rangle \cdot \langle EY, u\rangle\,\gamma(du) \right|$$

$$= \left| \int_H [E(\langle X, u\rangle \cdot \langle Y, u\rangle) - E\langle X, u\rangle \cdot E\langle Y, u\rangle]\,\gamma(du) \right|$$

$$\le \int_H R_{p,q}(\mathscr{F}, \mathscr{G}; \mathbb{R}) \cdot [E|\langle X, u\rangle|^p]^{1/p} \cdot [E|\langle Y, u\rangle|^q]^{1/q}\,\gamma(du)$$

$$= R_{p,q}(\mathscr{F}, \mathscr{G}; \mathbb{R}) \int_H \|(\langle x_i, u\rangle \cdot P^{1/p}(A_i))_{i=1}^I\|_{l^p} \cdot \|$$

$$\cdot (\langle y_j, u\rangle \cdot P^{1/q}(B_j))_{j=1}^J\|_{l^q}\,\gamma(du)$$

$$\le R_{p,q}(\mathscr{F}, \mathscr{G}; \mathbb{R}) \cdot \left\| \left(\left(\int_H |\langle x_i, u\rangle \cdot P^{1/p}(A_i)|^{p_1}\,\gamma(du) \right)^{1/p_1} \right)_{i=1}^I \right\|_{l^p}$$

$$\cdot \left\| \left(\left(\int_H |\langle y_j, u\rangle \cdot P^{1/q}(B_j)|^{q_1}\,\gamma(du) \right)^{1/q_1} \right)_{j=1}^J \right\|_{l^q}$$

$$= R_{p,q}(\mathscr{F}, \mathscr{G}; \mathbb{R}) \cdot \left\| \left(P^{1/p}(A_i) \cdot \left(\int_H |\langle x_i, u\rangle|^{p_1}\,\gamma(du) \right)^{1/p_1} \right)_{i=1}^I \right\|_{l^p}$$

$$\cdot \left\| \left(P^{1/q}(B_j) \cdot \left(\int_H |\langle y_j, u\rangle|^{q_1}\,\gamma(du) \right)^{1/q_1} \right)_{j=1}^J \right\|_{l^q}$$

$$= R_{p,q}(\mathscr{F}, \mathscr{G}; \mathbb{R}) \cdot \left[\sum_{i=1}^I P(A_i) \cdot \left(\|x_i\|_H^{p_1} \cdot Z_{p_1}^{p_1}\right)^{p/p_1} \right]^{1/p}$$

$$\cdot \left[\sum_{j=1}^J P(B_j) \cdot \left(\|y_j\|_H^{q_1} \cdot Z_{q_1}^{q_1}\right)^{q/q_1} \right]^{1/q}$$

$$= R_{p,q}(\mathscr{F}, \mathscr{G}; \mathbb{R}) \cdot Z_{p_1} \cdot Z_{q_1} \cdot \|X\|_p \cdot \|Y\|_q$$

Thus equation (4.1) holds. This completes the proof of Theorem 4.1. ■

ACKNOWLEDGMENT

The authors thank Allan Gut, whose hospitality and encouragement led to this collaboration.

REFERENCES

Bergh, J. and Löfström, J. (1976). *Interpolation Spaces*, Berlin: Springer-Verlag.

Bradley, R. C. and Bryc, W. (1985). Multilinear forms and measures of dependence between random variables. *J. Multivar. Anal.* **16**, 335–367.

Bradley, R. C., Bryc, W., and Janson, S. (1985). Remarks on the foundations of measures of dependence. Tech. Report No. 105 (June), Center for Stochastic Processes, University of North Carolina, Chapel Hill.

Dehling, H. and Philipp, W. (1982). Almost sure invariance principles for weakly dependent vector-valued random variables. *Ann. Probab.* **10**, 689–701.

Denker, M. and Keller, G. (1983). On *U*-statistics and v. Mises' statistics for weakly dependent processes. *Z. Wahrsch. verw. Gebiete* **64**, 505–522.

Gastwirth, J. L. and Rubin, H. (1975). The asymptotic distribution theory of the empiric c.d.f. for mixing stochastic processes. *Ann. Statist.* **3**, 809–824.

Ibragimov, I. A. and Linnik, Y. V. (1971). *Independent and Stationary Sequences of Random Variables*, Gröningen: Wolters-Noordhoff.

Iosifescu, M. and Theodorescu, R. (1969). *Random Processes and Learning*, Berlin: Springer-Verlag.

Lifshits, B. A. (1984). Invariance principle for weakly dependent variables. *Theory Probab. Appl.* **29**, 33–40.

Peligrad, M. (1982). Invariance principles for mixing sequences of random variables. *Ann. Probab.* **10**, 968–981.

Peligrad, M. (1983). A note on two measures of dependence and mixing sequences. *Adv. Appl. Probab.* **15**, 461–464.

Rietz, R. E. (1974). A proof of the Grothendieck inequality. *Israel J. Math.* **19**, 271–276.

Rosenblatt, M. (1970). Density estimators and Markov sequences. In *Nonparametric Techniques in Statistical Inference* (M. L. Puri, Ed.), London: Cambridge University Press, pp. 199–210.

Rosenblatt, M. (1971). *Markov Processes, Structure and Asymptotic Behavior*, Berlin: Springer-Verlag.

Volkonskii, V. A. and Rozanov, Y. A. (1959). Some limit theorems for random functions I. *Theory Probab. Appl.* **4**, 178–197.

Zygmund, A. (1959). *Trigonometric Series*, Volumes I and II. Cambridge: Cambridge University Press.

Semigroups and Poisson Approximation

Paul Deheuvels
Institute de Statisique de Paris
Université de Paris VI
Paris, France

D. Pfeifer
Institut für Statistik
RWTH
Aachen, Federal Republic of Germany

SUMMARY

In this Chapter we extend previous work of the authors on *Poisson approximation* for (general) independent Bernoulli summands with respect to the total variation distance, without imposing any conditions on the underlying parameters. This enables one to study also the case of unbounded means, without asymptotic uniform "smallness" of the individual summands, provided that the variance increases with the same rate as the mean. An important practical situation in which such an asymptotic behavior occurs is described by Ross's Markov chain model for the *simplex algorithm in linear programming*, which will be discussed as an example of possible application.

The main tool for the derivation of the results is the same Poisson *convolution semigroup* approach as used formerly, allowing again for a

---

D. Pfeifer's research was supported in part by AFOSR Contract No. F49620 85 C 0144 while visiting the Center for Stochastic Processes, U.N.C. at Chapel Hill, U.S.A.

treatment of optimal choice problems for the Poisson parameter, in different asymptotic settings.

1. INTRODUCTION

Let $X_1, \ldots, X_n$ be independent Bernoulli random variables with means $p_i = P(X_i = 1) = 1 - P(X_i = 0)$, $0 \le p_i \le 1$, $i = 1, 2, \ldots, n$, and $T(\mu)$ a Poisson random variable with mean $\mu > 0$. We are concerned with the *approximation of the partial sums* $S_n = \Sigma_{k=1}^n X_k$ by $T(\mu)$, where the goodness of fit is measured by the *total variation distance*

$$d(S_n, T(\mu)) = \sup_{M \subseteq \mathbf{Z}^+} |P(S_n \in M) - P(T(\mu) \in M)|$$

$$= \frac{1}{2} \sum_{k=0}^{\infty} |P(S_n = k) - P(T(\mu) = k)| \tag{1.1}$$

Estimations and asymptotic expansions for the distance d have been given by different authors, for instance, LeCam (1960), Kerstan (1964), Chen (1974, 1975), Serfling (1975, 1978), Barbour and Hall (1984), and most recently by Deheuvels and Pfeifer (1986). Whereas in earlier approaches, mainly the choices $\mu = \Sigma_{k=1}^n p_k$ (LeCam, Kerstan, Chen, and Barbour and Hall) or $\mu = -\Sigma_{k=1}^n \log(1 - p_k)$ (Serfling) have been considered, which possess certain optimality characteristics in that they asymptotically minimize the distance $d(S_n, T(\mu))$ if $\Sigma_{k=1}^n p_k \gg 1$ and $\Sigma_{k=1}^n p_k^2 \ll 1$ (Deheuvels and Pfeifer, 1986) or $\Sigma_{k=1}^n p_k \ll 1$, resp. (Serfling, 1978; Deheuvels and Pfeifer, 1986), it is more generally of importance to consider also the cases where

$$\mu = \sum_{k=1}^{n} p_k + \gamma \sum_{k=1}^{n} p_k^2 \equiv \mu(\gamma)$$

with $0 \le \gamma \le \frac{1}{2}$, where $\gamma = \frac{1}{2}$ (asymptotically)

corresponds to Serfling's approach. This is because for large n, if $\Sigma_{k=1}^n p_k$ is close to some positive value a and $\Sigma_{k=1}^n p_k^2 \ll 1$, then there always exists some $\gamma = \gamma(a) \in [0, \frac{1}{2}]$ such that the choice $\mu = \mu(\gamma(a))$ asymptotically minimizes $d(S_n, T(\mu))$ (Deheuvels and Pfeifer, 1986). For instance, in the range $0 < a < 2$, we have

$$\text{if } 0 < a \leq 1: \qquad \gamma(a) = \tfrac{1}{2},\ d(S_n, T(\mu(\gamma(a)))) \approx \tfrac{1}{2}e^{-a} \sum_{k=1}^{n} p_k^2$$

$$\text{if } 1 < a \leq \sqrt{2}: \qquad \gamma(a) = \tfrac{1}{2},\ d(S_n, T(\mu(\gamma(a)))) \approx \tfrac{1}{2}ae^{-a} \sum_{k=1}^{n} p_k^2$$

$$\text{if } \sqrt{2} < a \leq \sqrt[3]{6}: \qquad \gamma(a) = \frac{1}{2} - \frac{3}{2a}\frac{2-a}{3-a},$$

$$d(S_n, T(\mu(\gamma(a)))) \approx \frac{1}{2}\left[a + \left(1 - \frac{a^2}{2}\right)\frac{3}{a}\frac{2-a}{3-a}\right]e^{-a}\sum_{k=1}^{n} p_k^2$$

$$= \frac{1}{2}\left[\frac{a^2}{2} + \left(1 - \frac{a^3}{6}\right)\frac{3}{a}\frac{2-a}{3-a}\right]e^{-a}\sum_{k=1}^{n} p_k^2$$

$$\text{if } \sqrt[3]{6} < a < 2: \qquad \gamma(a) = 0,$$

$$d(S_n, T(\mu(\gamma(a)))) \approx \frac{1}{2}\left[\frac{a^2}{2} + \left(1 - \frac{a^3}{6}\right)\right]e^{-a}\sum_{k=1}^{n} p_k^2$$

This follows from Theorem 1.3 and Section 3 in Deheuvels and Pfeifer (1986) [note that for $0 < a \leq \sqrt{2}$ in (1.2), we could as well have used Serfling's choice for μ].

Whereas in all former investigations the condition $\Sigma_{k=1}^{n} p_k^2 \ll 1$ is (more or less) necessary to obtain asymptotically sharp results, we should also like to deal with more general cases in which $\Sigma_{k=1}^{n} p_k^2$ may be arbitrary. It turns out again that the semigroup technique as developed in Deheuvels and Pfeifer (1986) can be fruitfully applied, as will be worked out in Section 2.

2. THE SEMIGROUP TECHNIQUE

In order to facilitate the understanding for readers who are not familiar with the language of functional analysis and operator theory, we shall first state the main results, without appealing to the concept of semigroups.

Theorem 2.1. Suppose that $X_1, \ldots, X_n$ are independent Bernoulli random variables with means $p_i = P(X_i = 1) \in [0,1]$, $i = 1, 2, \ldots, n$ and $T(\mu)$ is a Poisson random variable with mean $\mu > 0$. For abbreviation, let

$$\lambda_j = \sum_{k=1}^{n} p_k^j, \quad j = 1,2,3 \quad \text{and} \quad \mu(\gamma) = \lambda_1 + \gamma\lambda_2 \quad \text{for } 0 \le \gamma \le \tfrac{1}{2}$$

Then

$$d(S_n, T(\mu(\gamma))) = \tfrac{1}{2}\lambda_2 K(\gamma, \lambda_1) + r_n \tag{2.1}$$

where

$$K(\gamma, \lambda) = \left\{ \frac{\lambda^{\delta-1}(\delta - (1-2\gamma)\lambda)}{\delta!} + \frac{\lambda^{\eta-1}((1-2\gamma)\lambda - \eta)}{\eta!} \right\} e^{-\lambda}$$

$$(\lambda > 0) \quad (2.2)$$

with

$$\delta = \left[\!\left[(1-\gamma)\lambda + \tfrac{1}{2} + \sqrt{\gamma^2\lambda^2 + (1-\gamma)\lambda + \tfrac{1}{4}} \right]\!\right]$$

$$\eta = \left[\!\left[(1-\gamma)\lambda + \tfrac{1}{2} - \sqrt{\gamma^2\lambda^2 + (1-\gamma)\lambda + \tfrac{1}{4}} \right]\!\right]$$

where $[\![\ ]\!]$ means integer part and

$$|r_n| \le e^{22\lambda_2}\left[\frac{18\lambda_3}{\sqrt{\lambda_1^3}} + 3168\left(\frac{\lambda_2}{\lambda_1}\right)^2 \right] + \gamma^2 \frac{\lambda_2^2}{\lambda_1} \tag{2.3}$$

Especially,

$$K(0, \lambda) = \left\{ \frac{\lambda^{\delta-1}(\delta - \lambda)}{\delta!} + \frac{\lambda^{\eta-1}(\lambda - \eta)}{\eta!} \right\} e^{-\lambda} \approx \frac{2}{\sqrt{2\pi e}} \frac{1}{\lambda} \quad \text{for } \lambda \to \infty$$

$$K(\tfrac{1}{2}, \lambda) = \frac{\lambda^{[\![\lambda]\!]}}{[\![\lambda]\!]!} e^{-\lambda} \approx \frac{1}{\sqrt{2\pi\lambda}} \quad \text{for } \lambda \to \infty \tag{2.4}$$

Theorem 2.2. Under the conditions of Theorem 2.1, we have

1. If $\lambda_2 \ll 1$, then also $|r_n/\lambda_2| \ll 1$, and

$$d(S_n, T(\mu(\gamma))) \approx \tfrac{1}{2}\lambda_2 K(\gamma, \lambda_1) \tag{2.5}$$

   which can be minimized (asymptotically) by choosing the corresponding value for γ that minimizes (2.2) for $\lambda = \lambda_1$. Here, the (asymptotically) optimal value for γ for $\lambda_1 \ll 1$ is $\gamma = \frac{1}{2}$, whereas the (asymptotically) optimal value for $\lambda_1 \gg 1$ is $\gamma = 0$.
2. If λ_2 remains bounded and $\lambda_1 \gg 1$, then $\gamma = 0$ is asymptotically optimal, and

$$d(S_n, T(\mu(\gamma))) = d(S_n, T(\lambda_1)) \approx \frac{1}{\sqrt{2\pi e}} \frac{\lambda_2}{\lambda_1} \tag{2.6}$$

Note that the condition under 2 implies that $\mathrm{Var}(S_n) = \lambda_1 - \lambda_2 \approx \lambda_1 = E(S_n)$ for $\lambda_1 \gg 1$, and that by Theorem 2.1, no Poisson convergence takes place unless $\lambda_2/\lambda_1 \to 0$, that is, for λ_2 bounded (and bounded away from 0), $\lambda_1 \gg 1$ is a necessary condition for Poisson convergence.

Theorems 2.1 and 2.2, which extend results of Deheuvels and Pfeifer (1985), will be proved by the following operator semigroup technique, paralleling (but improving) LeCam's (1960) approach.

Consider the Banach space l^1 of all absolutely summable sequences $f = (f(0), f(1), \ldots)$ with norm

$$f = \sum_{k=0}^{\infty} |f(k)| \tag{2.7}$$

The convolution $f * g$ for $f, g \in l^1$ is defined by

$$f * g(n) = \sum_{k=1}^{n} f(k)g(n-k), \qquad n \geq 0 \tag{2.8}$$

being again an element of l^1 with

$$\|f * g\| \leq \|f\| \, \|g\| \tag{2.9}$$

Henceforth we shall think of discrete probability measures ν over $\mathbb{Z}^+$ as elements $(\nu(\{0\}), \nu(\{1\}), \ldots) \in l^1$. Specifically, ε_k will denote the Dirac measure concentrated in $k \in \mathbb{Z}^+$. The *infinitesimal generator* of the Poisson convolution semigroup is defined by

$$Ag(n) = \begin{cases} g(n-1) - g(n), & n \geq 1 \\ -g(0), & n = 0 \end{cases} \qquad g \in l^1 \tag{2.10}$$

such that

$$e^{tA}g = \sum_{k=0}^{\infty} \frac{t^k}{k!} A^k g = e^{-t} \sum_{k=0}^{\infty} \frac{t^k}{k!} \varepsilon_k * g = \Pi(t) * g, \qquad t \geq 0, \quad g \in l^1 \tag{2.11}$$

where $\Pi(t)$ denotes the Poisson distribution with mean t. Also,

$$(I + p_k A)g = ((1 - p_k)\varepsilon_0 + p_k \varepsilon_1) * g = B(p_k) * g, \qquad g \in l^1 \tag{2.12}$$

where $B(p_k)$ denotes the binomial distribution with mean p_k over $\{0, 1\}$ and I denotes the identity operator. Because of (1.1), we can now rewrite

$$d(S_n, T(\mu)) = \frac{1}{2} \left\| \prod_{k=1}^{n} (I + p_k A) g_0 - e^{\mu A} g_0 \right\| \tag{2.13}$$

where $g_0 = (1, 0, 0, \ldots) \in l^1$. Since $I + p_k A$ is the first-order Taylor expansion of the semigroup e^{tA} for $t = p_k$ [see, e.g., Butzer and Berens (1967)], it seems natural that the right-hand side of (2.13) could be attacked by general operator semigroup theory, in combination with the well-known fact that for commuting operators $U_1, \ldots, U_n, V_1, \ldots, V_n$ we always have

$$\prod_{k=1}^{n} U_k - \prod_{k=1}^{n} V_k = \sum_{k=1}^{n} U_{k+1} \cdots U_n (U_k - V_k) V_1 \cdots V_{k-1} \tag{2.14}$$

as can be easily proved, for example, by induction.

Consider first the case $\mu = \lambda_1$. From what has been said above, we should have, in some sense,

$$e^{\lambda_1 A} - \prod_{k=1}^{n} (I + p_k A) \simeq \frac{1}{2} \sum_{k=1}^{n} p_k^2 e^{\lambda_1 A} A^2 \tag{2.15}$$

A more precise statement is given below.

Theorem 2.3. We have, for $\lambda_1^{(k)} = \sum_{i \neq k} p_i$,

$$\left\| e^{\lambda_1 A} - \prod_{k=1}^{n} (I + p_k A) - \frac{1}{2} \sum_{k=1}^{n} p_k^2 e^{\lambda_1^{(k)} A} A^2 \right\|$$

$$\leq e^{22\lambda_2} \left[\tfrac{3}{2}\lambda_3 \| e^{\lambda_1 A} A^3 \| + 99\lambda_2^2 \| e^{\lambda_1 A} A^4 \| \right] \qquad (2.16)$$

Proof. From (2.14) we obtain, as in the proof of Theorem 2.1 in Deheuvels and Pfeifer (1986),

$$\left\| e^{\lambda_1 A} - \prod_{k=1}^{n} (I + p_k A) - \frac{1}{2} \sum_{k=1}^{n} p_k^2 e^{\lambda_1^{(k)} A} A^2 \right\|$$

$$\leq \sum_{k=1}^{n} \left[\prod_{i=1}^{k-1} \| e^{-p_i A}(I + p_i A) \| \left\| e^{\lambda_1^{(k)} A} \left(e^{p_k A} - \left(I + p_k A + \frac{p_k^2}{2} A^2 \right) \right) \right\| \right.$$

$$\left. + \frac{p_k^2}{2} \left\| e^{\lambda_1^{(k)} A} A^2 \left\{ \prod_{i=1}^{k-1} e^{-p_i A}(I + p_i A) - I \right\} \right\| \right]$$

But

$$e^{-p_i A}(I + p_i A) = (I + p_i A)\left\{ I - p_i A + \int_0^{p_i} (p_i - u)\, e^{-uA} A^2\, du \right\}$$

$$= I - p_i^2 A^2 + (I + p_i A) \int_0^{p_i} (p_i - u)\, e^{-uA} A^2\, du \qquad (2.17)$$

hence

$$\| e^{-p_i A}(I + p_i A) \| \leq 1 + p_i^2 \|A\|^2 + \tfrac{1}{2} p_i^2\, e^{p_i \|A\|} \|A\|^2 \qquad (2.18)$$

where $\|A\| = 2$, such that, since $p_i \leq 1$, the left-hand side of (2.16) can be estimated by

$$e^{22\lambda_2} \sum_{k=1}^{n} \left[\frac{p_k^3}{6} \| e^{\lambda_1^{(k)} A} A^3 \| + 11 p_k^2 \sum_{i=1}^{n} \| e^{\lambda_1^{(k)} A} A^4 \| p_i^2 \right] \qquad (2.19)$$

But

$$\| e^{\lambda_1^{(k)} A} A^j \| \leq e^{-p_k A} \| \, \| e^{\lambda_1 A} A^j \| \leq 9 \| e^{\lambda_1 A} A^j \|, \qquad j \geq 1$$

Hence the result. ■

In a similar manner, the following result can be obtained.

Theorem 2.4. We have

$$\left\| e^{\mu(\gamma)A} - \prod_{k=1}^{n} (I + p_k A) \right\| = \tfrac{1}{2}\lambda_2 \|(2\gamma A + A^2)\, e^{\lambda_1 A}\| + R_n \quad (2.20)$$

with

$$|R_n| \le e^{22\lambda_2}\left[6\lambda_3\|e^{\lambda_1 A}A^3\| + 99\lambda_2^2\|e^{\lambda_1 A}A^4\|\right] + 2\gamma^2\frac{\lambda_2^2}{\lambda_1} \quad (2.21)$$

Proof. The proof is analogous to the proof of Theorem 2.1 in Deheuvels and Pfeifer (1986).

From relation (3.19) in Pfeifer (1985a) we see that

$$\|e^{\lambda_1 A}A^3\| \le \min\left(8, 6/\sqrt{\lambda_1^3}\right) \quad (2.22)$$

but also

$$\|e^{\lambda_1 A}A^4\| \le \|e^{(1/2)\lambda_1 A}A^2\|^2 \le 16\min\left(1, 4/\lambda_1^2\right) \quad (2.23)$$

Theorem 2.1 now follows from Theorem 2.4 and Theorem 1.2 in Deheuvels and Pfeifer (1986). Theorem 2.2 follows from Theorem 2.1 and the fact that for $\gamma > 0$, we have $\|(2\gamma A + A^2)\, e^{\lambda_1 A}\| = O(1/\sqrt{\lambda_1})$, but $\|A^2 e^{\lambda_1 A}\| = O(1/\lambda_1)$ for $\lambda_1 \gg 1$.

REMARK 2.1. An application of the above technique directly to the norm term

$$\left\| e^{\lambda_1 A} - \prod_{k=1}^{n} (I + p_k A) \right\| \text{ yields}$$

$$d(S_n, T(\lambda_1)) \le e^{22\lambda_2}\left\{ \tfrac{1}{2}\lambda_2 K(0, \lambda_1) + 27\frac{\lambda_3}{\sqrt{\lambda_1^3}} \right\} \quad (2.24)$$

which is considerably sharper than the corresponding bound in Barbour and Hall (1984) for moderate and large values of λ_1 when $\lambda_2 \ll 1$.

3. APPLICATION

Here we shall consider Ross's (1983) Markov chain model for the simplex algorithm in linear programming and the *average-case analysis* corresponding to it. Characteristically, the simplex algorithm moves through the n

(say) extreme points of the feasible region defined by the linear constraints in such a way that at each step of the algorithm, all points visited and those that cannot increase the value of the goal function are excluded from further consideration. If we assume that the extreme points are ranked such that 1 corresponds to the optimal and n to the worst point, then the ranks visited can be considered as a homogeneous Markov chain with uniform initial distribution, absorbing state 1 and uniform transition probabilities for the remaining higher ranks [see Ross (1983, Chapter 4)]. If we assume that the random variable X_k takes the value 1 if rank k was visited and 0 otherwise ($k = 1, \dots, n$), then it is possible to show that by the structure of the underlying Markov chain, $X_1, \dots, X_n$ are independent with $p_k = 1/k$ (note that 1 is absorbing, hence $p_1 = 1$, and the first selected extreme point is the worst with probability $1/n$). Then S_n denotes the number of steps required for termination. Here,

$$\lambda_1 \approx \log n + C, \qquad \lambda_2 \sim \frac{\pi^2}{6} \quad \text{for } n \to \infty \tag{3.1}$$

(where $C = 0.5772$ denotes Euler's constant). Hence by Theorem 2.2, S_n is asymptotically Poisson with mean $\lambda_1 \approx \log n$ and

$$d(S_n, T(\lambda_1)) = \frac{\pi^2}{6\sqrt{2\pi e}} \frac{1}{\log n} + O\left((\log n)^{-3/2}\right) \quad \text{for } n \to \infty \tag{3.2}$$

Note that by methods similar to those above, we obtain only

$$d(S_n, T(\log n)) = \frac{C}{\sqrt{2\pi \log n}} + O\left((\log n)^{-1}\right) \quad \text{for } n \to \infty \tag{3.3}$$

which means that S_n is indeed (asymptotically) closer to a Poisson random variable with mean λ_1 than to a Poisson random variable with mean $\log n$.

4. FURTHER COMMENTS

It seems reasonable that the generalizations presented in this chapter should also apply to distance measures other than total variation, for instance, the Kolmogorov distance or Wasserstein distances [see, e.g., Deheuvels and Pfeifer (1985)], since these can also be treated in the operator semigroup framework, and Theorems 2.3 and 2.4 do not heavily depend on the underlying semigroup or the underlying Banach space.

A generalization to Poisson point process approximation by the semigroup approach seems also to be tractable quite well (Pfeifer, 1985b), although in this area we are still far from sufficient results.

Note that the constants which appear in our theorems can be precised. For instance the constant 22 in (2.24) originates from (2.18) and the crude inequalitites

$$1 + p^2\|A\|^2 + \tfrac{1}{2}p^2 e^{p\|A\|}\|A\|^2 \leq 1 + (4 + 2e^2)p^2 \leq 1 + 22p^2 \leq e^{22p^2}.$$

If $p \to 0$, it follows that 22 can be replaced by any constant $c > 6$, for asymptotic validity.

REFERENCES

Barbour, A. D. and Hall, P. (1984). On the rate of Poisson convergence. *Math. Prob. Camb. Phil. Soc.* **95**, 473–480.

Butzer, H. L. and Berens, H. (1967). *Semi-Groups of Operators and Approximation*, New York: Springer.

Chen, L. H. Y. (1974). On the convergence of Poisson binomial to Poisson distributions. *Ann. Prob.* **2**, 178–180.

Chen, L. H. Y. (1975). Poisson approximation for dependent trials. *Ann. Prob.* **3**, 534–545.

Deheuvels, P. and Pfeifer, D. (1985). A semigroup approach to Poisson approximation. II Tech. Report, L.S.T.A., Université Paris VI.

Deheuvels, P. and Pfeifer, D. (1986). A semigroup approach to Poisson approximation. *Ann. Prob.* **14**, 663–676.

Kerstan, J. (1964). Verallgemeinerung eines Satzes von Prochorow und LeCam. *Z. Wahrscheinlichkeitsth. verw. Geb.* **2**, 173–179.

LeCam, L. (1960). An approximation theorem for the Poisson binomial distribution. *Pacific J. Math.* **10**, 1181–1197.

Pfeifer, D. (1985a). On the distance between mixed Poisson and Poisson distributions. Center for Stochastic Processes, University of North Carolina at Chapel Hill, Technical Report No. 115.

Pfeifer, D. (1985b). A semigroup approach to Poisson point process approximation. Personal communication.

Ross, S. M. (1983). *Stochastic Processes*. New York: Wiley.

Serfling, R. J. (1975). A general Poisson approximation theorem. *Ann. Prob.* **3**, 726–731.

Serfling, R. J. (1978). Some elementary results on Poisson approximation in a sequence of Bernoulli trials. *SIAM Rev.* **20**, 567–679.

Nonstable Limit Laws in Extreme Value Theory

M. E. Graça Martins and Dinis D. Pestana
Departamento de Estatística
Centro de Estatística e Aplicações
Faculdade de Ciências
Universidade de Lisboa
Lisbon, Portugal

SUMMARY

In this chapter we investigate the class of all possible limiting distributions of $\max_{1 \le i \le n}\{a_{i,n}^{-1} Y_i\}$, where $Y_1, Y_2, \ldots$ is an independent and identically distributed sequence of random variables, and where $a_{i,n}$ are normalizing constants that satisfy general uniformity conditions.

1. INTRODUCTION AND PRELIMINARIES

Let $\{X_i\}_{i \in N}$ be a sequence of independent random variables, $F_i(x) = P[X_i \le x]$. The limit distribution function of $M_n = \max_{1 \le i \le n} X_i$

$$G(x) = \lim_{n \to \infty} G_n(x) = \lim_{n \to \infty} \prod_{i=1}^{n} F_i(x) \tag{1.1}$$

is, under the uniformity assumption for the maximum [Galambos (1978, 3.10)] in Mejzler's M or max-self-decomposable class. In other words, either $G(x)$ or $G(R - \exp(-x))$, where in this last case the right-endpoint $R = \sup\{x\colon\ F(x) < 1\} < \infty$, is log-concave, $x > 0$ (Mejzler, 1956; Galambos, 1978; Graça Martins and Pestana, 1984, 1985).

In this chapter we deal with the case in which

$$X_i = a_{i,n}^{-1}(Y_i - b_{i,n}) \tag{1.2}$$

where $\{Y_i\}_{i \in N}$ is a sequence of independent and identically distributed random variables, $a_{i,n} > 0$ and $b_{i,n}$ are coefficients such that Mejzler's uniformity condition holds (Mejzler, 1956). This amounts to

$$\lim_{n\to\infty} \left\{ \max_{1 \le i \le n} (1 - F(a_{i,n}x + b_{i,n})) \right\} = 0, \tag{1.3a}$$

and

$$\lim_{n\to\infty} \sum_{i=1}^{[nt]} (1 - F(a_{i,n}x + b_{i,n})) = w(t,x) \text{ exists } \quad \forall t \in (0,1] \tag{1.3b}$$

We shall assume throughout the rest of this chapter that $a_{i,n} > 0$ and $b_{i,n}$ are admissible normalizing constants, in the sense that (1.3a) and (1.3b) are satisfied.

Definition 1.1. M_F is the class of all possible limit distributions of maxima of suitably normed independent identically distributed random variables with common distribution function $F(x)$:

$$M_F = \left\{ G(x) \colon G(x) = \lim_{n\to\infty} \prod_{i=1}^{n} F(a_{i,n}x + b_{i,n}) \right\} \subset M \tag{1.4}$$

for some choice of admissible constants $a_{i,n} > 0$, $b_{i,n}$.

For the sake of simplicity, we shall limit ourselves to the case $R = \infty$. This corresponds to a subclass closed under weak convergence of $\{G(x) = \exp[-K(x)]\colon K(x) \text{ convex}\}$, and implies by (1.3a) that

$$\lim_{n\to\infty} \left\{ \min_{1 \le i \le n} a_{i,n} \right\} = \infty \tag{1.5}$$

It may be that $G(x) = \lim_{n\to\infty} F^n(a_n x + b_n)$ exists and is nondegenerate for some choice $a_n > 0$, b_n. This special case $a_{i,n} = a_n$, $b_{i,n} = b_n$, $i = 1, \ldots, n$, is the classical convergence of types scheme investigated by Gnedenko (1943). We say then that $G(x)$ is an extreme-value (or max-stable) distribution, that F belongs to the domain of attraction of G, and denote this fact by $F \in \mathcal{D}(G)$.

Under the assumption $R = \infty$, if F belongs to the domain of attraction of an extreme-value distribution of the form $G_\gamma(x) = \exp[-(1 + \gamma x)^{-1/\gamma}]$, then we must have $\gamma \geq 0$. This corresponds to the Gumbel distribution ($\Lambda(x) = \exp[-e^{-x}]$) for $\gamma = 0$, and to the Fréchet distribution ($\Phi_\alpha(x) = \exp[-x^{-\alpha}]$) for $\gamma = 1/\alpha > 0$.

Note that $F \in \mathscr{D}(\Phi_\alpha)$ iff $\tilde{F}(x) = 1 - F(x) = x^{-\alpha}L(x)$, where $L(\cdot)$ is a slowly varying function.

The motivation for this chapter comes from the fact that in the more general scheme introduced above, it may happen that a nondegenerate limit distribution does exist although F is not in the domain of attraction of an extreme-value distribution. If this is the case, when does there exist sequences $a_{i,n} > 0$ and $b_{i,n}$ such that (1.1) holds? We intend in the sequel to give answers to this question. Related problems on this topic are to be found in Resnick (1972), Schreiber (1977) and Graça Martins and Pestana (1985).

2. THE CLASS M_F

In the sequel we shall assume that the hypotheses of Section 1 are satisfied.

Definition 2.1. Let

$$A_F(r, x) = x^r[1 - F(x)] = x^r\tilde{F}(x) \tag{2.1}$$

Assume that $R = \sup\{x\colon F(x) < 1\} = \infty$. The *lower index* α_1 of $F(x)$ is defined as

$$\alpha_1 = \sup\left\{ r \geq 0 : \limsup_{x\to\infty} \sup_{t>1} \frac{A_F(r, tx)}{A_F(r, x)} < \infty \right\} \tag{2.2}$$

and the *upper index* α_2 of $F(x)$ is defined as

$$\alpha_2 = \inf\left\{ r \geq 0 : \liminf_{x\to\infty} \inf_{t>1} \frac{A_F(r, tx)}{A_F(r, x)} > 0 \right\} \tag{2.3}$$

Note that if $F \in \mathscr{D}(\Phi_\alpha)$, $\alpha_1 = \alpha_2 = \alpha$. The converse, however, is not true. An example of this is the distribution function

$$F(x) = 1 - \frac{1}{x[4 + \sin(\log x) - \cos(\log x)]}, \qquad x \geq 1$$

Throughout this section, we assume that $\alpha_2 < \infty$.

We first consider the case where $\alpha_1 < \alpha_2$.

Consider a sequence $x_n \uparrow \infty$. For any $\alpha \in (\alpha_1, \alpha_2)$, from (2.1) we may select corresponding $t_n < u_n < v_n < w_n$, for any $n = 1, 2, \ldots,$ such that

$$u_n = \inf\left\{ u > t_n : \frac{A_F(\alpha, u)}{A_F(\alpha, t_n)} \geq x_n \right\} \tag{2.4a}$$

and

$$w_n = \inf\left\{ w > v_n : \frac{A_F(\alpha, v_n)}{A_F(\alpha, w)} \geq x_n \right\} \tag{2.4b}$$

Definition 2.2. Let

$$y_n = \begin{cases} u_n & \text{if } A_F(\alpha, u_n) \geq A_F(\alpha, v_n) \\ v_n & \text{otherwise} \end{cases} \tag{2.5}$$

Define the sequences

$$l_n = \sup\{ x \leq y_n : A_F(\alpha, x) \geq \max[A_F(\alpha, t_n), A_F(\alpha, w_n)] \} \tag{2.6a}$$

and

$$r_n = \inf\{ x \geq y_n : A_F(\alpha, x) \geq \max[A_F(\alpha, t_n), A_F(\alpha, w_n)] \} \tag{2.6b}$$

Lemma 2.1. The double sequence $\{l_n, r_n\}_{n \in N}$ defined by (2.6) is such that $A_F(\alpha, x) > \min[A_F(\alpha, l_n), A_F(\alpha, r_n)] = m_n$ for any $x \in (l_n, r_n)$, and $l_n \to \infty$ and $r_n/l_n \to \infty$ as $n \to \infty$.

Proof. From the above construction, it is obvious that $l_n \to \infty$ as $n \to \infty$ and that $A_F(\alpha, x) > m_n$, $l_n < x < r_n$.

The last statement follows from $r_n/l_n > y_n/l_n$ and

$$\left(\frac{y_n}{l_n} \right)^{\alpha} > \frac{A_F(\alpha, y_n)}{A_F(\alpha, l_n)} \geq x_n \uparrow \infty \quad \blacksquare$$

Definition 2.3. Let $\{P_n\}_{n \in N}$ be a sequence of probability measures with support $[l_n, r_n]$ defined by

$$dP_n(t) = K_n \frac{A_F(\alpha, t)}{t} I_{[l_n, r_n]} \tag{2.7}$$

Here K_n is a sequence such that

$$K_n^{-1} = \int_{[l_n, r_n]} \frac{A_F(\alpha, t)}{t}\, dt > m_n \log \frac{r_n}{l_n} \underset{n \to \infty}{\to} \infty \tag{2.8}$$

Lemma 2.2. Let $\alpha \in (\alpha_1, \alpha_2)$. For $x > 0$,

$$K_n \int_{[l_n, l_n x]} \frac{A_F(\alpha, t)}{t}\, dt = O\left(\frac{1}{\log(r_n/l_n)}\right), \qquad n \to \infty \tag{2.9a}$$

and

$$K_n \int_{[r_n, r_n x]} \frac{A_F(\alpha, t)}{t}\, dt = O\left(\frac{1}{\log(r_n/l_n)}\right), \qquad n \to \infty \tag{2.9b}$$

Proof. For $x \in [1, z_n]$, where the sequence $\{z_n\}_{n \in N}$ will be specified below, we have

$$\int_{[l_n, l_n x]} \frac{A_F(\alpha, t)}{t}\, dt \le \sup_{l_n \le l \le l_n x} A_F(\alpha, l) \int_{[l_n, l_n x]} \frac{dt}{t}$$

$$\le x^{\alpha} \log x < z_n^{\alpha} \log z_n$$

Hence, from (2.8),

$$K_n \int_{[l_n, l_n x]} \frac{A_F(\alpha, t)}{t}\, dt < \frac{z_n^{\alpha} \log z_n}{\log(r_n/l_n)}$$

The result follows choosing $\{z_n\}_{n \in N}$ such that $\lim_{n \to \infty} z_n = \infty$ and

$$\lim_{n \to \infty} \frac{z_n^{\alpha} \log z_n}{\log(r_n/l_n)} = 0$$

On the other hand, for $x \in [z_n^*, 1]$, where the sequence $z_n^* \downarrow 0$ and such that $l_n z_n^* > 1$ will be specified below, we have that

$$\int_{[l_n x, l_n]} \frac{A_F(\alpha, t)}{t}\, dt \le \sup_{l_n x \le l \le l_n} A_F(\alpha, l) \log \frac{1}{x}$$

$$\le x^{\alpha - \alpha_2} \log \frac{1}{x} < z_n^{*\alpha - \alpha_2} \log \frac{1}{z_n^*}$$

and hence

$$K_n \int_{[l_n x, l_n]} \frac{A_F(\alpha, t)}{t}\, dt < \frac{z_n^{*\alpha - \alpha_2} \log(1/z_n^*)}{\log(r_n/l_n)}$$

The result follows choosing $\{z_n^*\}_{n\in N}$ such that $\lim_{n\to\infty} z_n^* = 0$ and

$$\lim_{n\to\infty} \frac{z_n^{*\alpha-\alpha_2}\log(1/z_n^*)}{\log(r_n/l_n)} = 0$$

The proof for

$$\int_{[r_n,\, r_n x]} \frac{A_F(\alpha, t)}{t}\, dt$$

is similar. ■

Theorem 2.1. Let $\alpha \in (\alpha_1, \alpha_2)$. For $x > 0$,

$$\psi_n(x) = \int_{[l_n,\, r_n]} \frac{\tilde{F}(tx)}{\tilde{F}(x)}\, dP_n(t) \underset{n\to\infty}{\to} x^{-\alpha} \tag{2.10}$$

Proof.

$$\psi_n(x) = \int_{[l_n,\, r_n]} \frac{\tilde{F}(tx)}{\tilde{F}(x)}\, dP_n(t)$$

$$= K_n \int_{[l_n,\, r_n]} t^{\alpha-1}\tilde{F}(tx)\, dt$$

$$= x^{-\alpha}\left\{K_n \int_{[l_n x,\, l_n]} \frac{A_F(\alpha, t)}{t}\, dt + 1 + K_n \int_{[r_n,\, r_n x]} \frac{A_F(\alpha, t)}{t}\, dt\right\}$$

The result follows from Lemma 2.2. ■

All the above results are readily extended for the case $\alpha_1 = \alpha_2 = \alpha$. More precisely:

Theorem 2.2. If $\alpha_1 = \alpha_2 = \alpha$, there exists a double sequence $\{l_n, r_n\}_{n\in N}$ such that $l_n \to \infty$ and $l_n/r_n \to \infty$ as $n \to \infty$, and $A_F(\alpha, x) > m_n = \min[A_F(\alpha, l_n), A_F(\alpha, r_n)]$ for every $x \in (l_n, r_n)$. For the above sequence,

$$K_n \int_{[l_n,\, l_n x]} \frac{A_F(\alpha, t)}{t}\, dt = O\left(\frac{1}{\log(r_n/l_n)}\right), \qquad n \to \infty, \quad x > 0 \tag{2.11a}$$

$$K_n \int_{[r_n,\, r_n x]} \frac{A_F(\alpha, t)}{t}\, dt = O\left(\frac{1}{\log(r_n/l_n)}\right), \qquad n \to \infty, \quad x > 0 \tag{2.11b}$$

and hence

$$\int_{[l_n, r_n]} \frac{\tilde{F}(tx)}{\tilde{F}(x)}\, dP_n(t) \underset{n\to\infty}{\to} x^{-\alpha}, \qquad x > 0 \tag{2.12}$$

The proof is similar to the ones presented above, by considering

$$l = \liminf_{x\to\infty} \inf_{t>1} \frac{A_F(\alpha, tx)}{A_F(\alpha, x)} \tag{2.13a}$$

and

$$L = \limsup_{x\to\infty} \sup_{t>1} \frac{A_F(\alpha, tx)}{A_F(\alpha, x)} \tag{2.13b}$$

Theorem 2.3. For every $\alpha \in (\alpha_1, \alpha_2]$ (or $\alpha \in [\alpha_1, \alpha_2]$, if $\alpha_1 > 0$), there exists $\{a_{i,n}(\alpha)\}_{1\le i\le \nu_n, n\in N}$ such that

$$\prod_{i=1}^{\nu_n} F(a_{i,n}(\alpha)x) = \exp(-x^{-\alpha}) = \Phi_\alpha(x) \tag{2.14}$$

Proof. The fact that

$$\int_{[l_n, r_n]} \frac{A_F(\alpha, t)}{t}\, dt \le \int_{[l_n, r_n]} t^{\alpha-1}\, dt \tag{2.15}$$

implies, for any n and $\alpha \in (\alpha_1, \alpha_2)$ (resp. $\alpha = \alpha_1 = \alpha_2$), the existence of an increasing sequence $\{a_{i,n}(\alpha)\}$, $1 \le i \le \nu_n + 1$, such that

$$a_{1,n}(\alpha) = l_n \tag{2.16a}$$

$$\int_{[a_{i,n}(\alpha),\, a_{i+1,n}(\alpha)]} t^{\alpha-1}\, dt = \int_{[l_n, r_n]} \frac{A_F(\alpha, t)}{t}\, dt$$

$$= K_n^{-1}, \qquad i = 1, \ldots, \nu_n \tag{2.16b}$$

$$a_{\nu_n, n}(\alpha) \le r_n \le a_{\nu_n+1, n}(\alpha) \tag{2.16c}$$

Let

$$\underline{G_n(x)} = \exp\left[-\sum_{i=1}^{\nu_n} \tilde{F}(a_{i,n}(\alpha)x)\right]$$

$$= \exp\left[-K_n \sum_{i=1}^{\nu_n} \tilde{F}(a_{i,n}(\alpha)x) \int_{[a_{i,n}(\alpha),\, a_{i+1,n}(\alpha)]} x^{\alpha-1}\, dx\right] \tag{2.17a}$$

and

$$\overline{G_n(x)} = \exp\left[-\sum_{i=1}^{\nu_n - 1} \tilde{F}(a_{i,n}(\alpha)x)\right] \tag{2.17b}$$

From

$$\underline{G_n(x)} \le \exp\left[-K_n \int_{[l_n, r_n]} \frac{A_F(\alpha, t)}{t}\, dt\right]$$

$$= \exp\left[-\int_{[l_n, r_n]} \frac{\tilde{F}(tx)}{\tilde{F}(x)}\, dP_n(t)\right] \le \overline{G_n(x)}$$

together with $\overline{G_n(x)} - \underline{G_n(x)} \to 0$ as $n \to \infty$, we get, in view of theorem 2.2 if $\alpha \in (\alpha_1, \alpha_2)$ and in view of Theorem 2.3 if $\alpha = \alpha_1 = \alpha_2$, that

$$\lim_{n \to \infty} \underline{G_n(x)} = \exp[-x^{-\alpha}] = \Phi_\alpha(x)$$

The result follows from

$$\prod_{i=1}^{\nu_n} F(a_{i,n}(\alpha)x) = \prod_{i=1}^{\nu_n} [1 - \tilde{F}(a_{i,n}(\alpha)x)]$$

$$\approx \exp\left[-\sum_{i=1}^{\nu_n} \tilde{F}(a_{i,n}(\alpha)x)\right]$$

and, for $\alpha = \alpha_1 > 0$ and for $\alpha = \alpha_2$, from the closedness of M_F. ■

Corollary 2.1. Let $L(x)$ denote the Laplace transform of a probability measure with support $S \subset [\alpha_1, \alpha_2]$. Then

$$G(x) = k \exp[-L(\log x)] \tag{2.18}$$

where k is a normalizing factor, is in M_F.

Proof. If $\Phi_{\alpha_k} \in M_F$, $k = 1, \ldots, n$, then $\prod_{k=1}^{n} \Phi_{\alpha_k}^{\lambda_k} \in M_F$, where $\lambda_k \ge 0$, $k = 1, \ldots, n$, and $\sum_{k=1}^{n} \lambda_k = 1$. This follows from Graça Martins and Pestana (1985) restatement in terms of maxima of Zinger's (1965) results on Gnedenko's conjecture. The result follows from the closedness of M_F. ■

3. FINAL COMMENTS

In case $R < \infty$, if F belongs to the domain of attraction of an extreme-value distribution of the form $G_\gamma(x) = \exp[-(1 + \gamma x)^{-1/\gamma}]$, then we must have

$\gamma \leq 0$. This corresponds to the Gumbel distribution for $\gamma = 0$, and to the Weibull distribution ($\Psi_\alpha(x) = \exp[-(-x)^\alpha]$) for $\gamma = -(1/\alpha) < 0$.

For this case, $R < \infty$, similar results to those presented in Section 2 may be established by using

$$A_F^*(r, x) = x^r \tilde{F}(R - 1/x) \tag{3.1}$$

The assumption $\alpha_2 < \infty$ in Section 2 may be removed using standard arguments on regular variation [cf. the appendix in Smith (1982)]. Convergence toward the Gumbel distribution, or to the generalized extreme-value distribution $G_\gamma(\cdot)$, is, however, trickier to establish, because non-null centering constants are needed.

ACKNOWLEDGMENTS

The authors are grateful to the National Institute for Scientific Research (INIC), Lisbon, for financial support to present this paper at the III IMSIBAC, Bilbao, and to the anonymous referee for constructive criticism that helped considerably in improving its final form.

REFERENCES

Galambos, J. (1978). *The Asymptotic Theory of Extreme Order Statistics*, New York: Wiley.

Gnedenko, B. V. (1943). Sur la distribution limite du terme maximum d'une série aléatoire. *Ann. Math.* **44**, 423–453.

Graça Martins, E. and Pestana, D. D. (1984). L_r classes for maxima of independent random variables. *Abs. 16th Europ. Meet. Statist.*, 122–123.

Graça Martins, E. and Pestana, D. D. (1985). The extremal limit problem—extensions. *V Pannonian Symp. Math. Statist.*, Viségrad (to appear).

Mejzler, D. (1956). On the problem of the limit distribution for the maximal term of a variational series. *L'vov Pol. Inst. Naucn Zp.* **38**, 90–109.

Resnick, S. I. (1972). Products of distribution functions attracted to extreme value laws. *J. Appl. Prob.* **8**, 781–793.

Schreiber, M. (1977). Dérivé convexe d'une loi de probabilité et lois stables. Application à un problème d'isomorphisme. *Israel J. Math.* **28**, 287–312.

Smith, R. L. (1982). Uniform rates of convergence in extreme-value theory. *Adv. Appl. Probab.* **14**, 600–622.

Zinger, A. A. (1965). On a class of limit distributions for normed sums of independent random variables. *Theor. Probab. Appl.* **10**, 607–620.

Almost Sure Asymptotic Behavior of the Record and Record Time Sequences of a Stationary Time Series

George Haiman
Institut de Recherche des Transports
Arcueil, France

SUMMARY

It has been shown (Haiman, 1985a, b), in the case of some stationary m-dependent sequences and in the case of gaussian sequences with a covariance function having rapidly decreasing tails that the records and record times defined with respect to a threshold could be identified by means of a translation of the time index to the corresponding elements defined on an independent and identically distributed sequence. We show in this chapter that the record times corresponding to the threshold we have used can also be identified by means of an appropriate shift to a sequence of record times defined in the usual way. We also extend our results to kth records and kth record times.

1. INTRODUCTION

Let $\{X_n; n \geq 1\}$ be a *stationary sequence* of real-valued random variables with a common continuous distribution function.

Let $\tau > 0$ and $\alpha > 0$ be given constants such that $\tau/2^\alpha < 1$ and denote by $\{u_n;\ n \geq 1\}$ the sequence defined by

$$u_1 = -\infty \tag{1.1}$$

and for any $n \geq 2$

$$P\{X_n \geq u_n\} = \tau/n^\alpha$$

For any fixed integer $r_0 \geq 1$, define the *record time sequence* $\{T_n;\ n \geq 1\}$ with respect to $\{u_n;\ n \geq 1\}$ and the corresponding *record sequence* $\{\theta_n;\ n \geq 1\}$ by

$$T_1 = \min\{k \geq 1:\ X_k > u_{r_0}\}, \qquad \theta_1 = \min\{r \geq 1 : X_{T_1} \leq u_r\}$$

and for any $n \geq 1$,

$$\begin{aligned} T_{n+1} &= T_n + \min\{k \geq 1 : X_{T_n + k} > u_{\theta_n}\} \\ \theta_{n+1} &= \min\{r \geq 1 : X_{T_{n+1}} \leq u_r\} \end{aligned} \tag{1.2}$$

When $\{X_n;\ n \geq 1\}$ is independent and identically distributed (i.i.d.), $\{(T_n, \theta_n);\ n \geq 1\}$ is such that [see Deheuvels (1974)].

Theorem A

1. For any integers $s \geq 1$ and $r \geq r_0$,

$$P\{T_1 = s, \theta_1 = r\} = P\{X_1 \leq u_{r_0}\}^{s-1} P\{u_{r-1} < X_1 \leq u_r\} \tag{1.3}$$

2. For any integers $n \geq 1$, $1 \leq s_1 < \cdots < s_{n+1}$ and $r_0 < r_1 < \cdots < r_{n+1}$ we have

$$\begin{aligned} P\{T_{n+1} - T_n &= s_{n+1} - s_n, \theta_{n+1} - \theta_n = r_{n+1} - r_n | T_1 = s_1, \\ \theta_1 &= r_1, \ldots, T_n = s_n, \theta_n = r_n\} \\ &= P\{T_{n+1} - T_n = s_{n+1} - s_n, \theta_{n+1} - \theta_n = r_{n+1} - r_n | \theta_n = r_n\} \\ &= P\{X_1 \leq u_{r_n}\}^{s_{n+1} - s_n - 1} P\{u_{r_{n+1}} < X_1 \leq u_{r_{n+1}}\} \end{aligned} \tag{1.4}$$

The asymptotic behavior of $\{(T_n, \theta_n);\ n \geq 1\}$ in the two following cases of dependent sequences were studied in Haiman (1985a, b).

CASE 1. $\{X_n : n \geq 1\}$ is m-dependent, $m \geq 2$, and there exists a $\beta > 0$ such that for any $1 < i \leq m$,

$$\limsup_{z \uparrow \infty} \left\{ \sup_{\substack{z \leq v \leq w \leq \infty \\ z < u < \infty}} \frac{P\{X_1 \geq u : v \leq X_i \leq w\}}{P\{X_1 \geq u\}^\beta P\{v \leq X_1 \leq w\}} \right\} < \infty \tag{1.5}$$

CASE 2. $\{X_n : n \geq 1\}$ is a centered *gaussian sequence* with covariance function

$$\rho(n) = E(X_i X_{i+n}), \qquad \rho(0) = 1, \quad n = 0, 1, \ldots \tag{1.6}$$

Put

$$\delta = \max_{n \geq 1} |\rho(n)| \leq 1 \tag{1.7}$$

we assume that

$$\lim_{n \uparrow \infty} |\rho(n)| = 0 \tag{1.8}$$

This clearly implies that $\delta < 1$. In addition, we shall assume that one among the following assumptions holds:

Assumption 1. $\{X_n : n \geq 1\}$ has a *spectral density* $f(\lambda) \geq 0$, $-\pi \leq \lambda \leq \pi$, such that

$$\int_{-\pi}^{\pi} \frac{d\lambda}{f(\lambda)} < \infty \tag{1.9}$$

and there exists an $\varepsilon > 0$ such that

$$\limsup_{n \to \infty} |\rho(n)| n^{[6(1+\delta)/(1-\delta)] + \varepsilon} < \infty \tag{1.10}$$

Assumption 2. We have

$$\sum_{n=1}^{+\infty} |\rho(n)| < \tfrac{1}{2} \tag{1.11}$$

and there exists a $\gamma > 0$ such that

$$\limsup_{n \to \infty} |\rho(n)| n^{[4(1+\delta)/(1-\delta)] + \gamma} < \infty \tag{1.12}$$

We have proved the following theorem.

Theorem B. Let $\{X_n;\ n \geq 1\}$ satisfy the assumptions of Case 1 or Case 2. Then, for any $\tau > 1$, there exists an $\alpha > 0$, an integer $r_0 \geq 1$, and two strictly increasing integer-valued sequences of random variables $\{S_n;\ n \geq 1\}$ and $\{R_n;\ n \geq 1\}$ defined on the same probability space as $\{X_n;\ n \geq 1\}$ such that:

1. $\{(S_n, R_n);\ n \geq 1\}$ satisfies Theorem A by replacing T_k by S_k and θ_k by R_k, $k = 1, 2, \ldots$ in (1.3) and (1.4).
2. There exist a.s. n_0 and q such that for any $n \geq n_0$, we have

$$S_n = T_{n-q} \quad \text{and} \quad R_n = \theta_{n-q} \tag{1.13}$$

We now complete Theorem B in the two following directions.

2. COMPARISON OF $\{T_n;\ n \geq 1\}$ WITH THE CLASSICAL RECORD TIME SEQUENCE IN THE i.i.d. CASE

Let $\{\nu_n;\ n \geq 1\}$ be the record time sequence associated with $\{X_n;\ n \geq 1\}$, that is,

$$\nu_1 = 1, \qquad \nu_k = \min\{m > \nu_{k-1} : X_m > X_{\nu_{k-1}}\}, \qquad k = 2, 3, \ldots \tag{2.1}$$

and the record sequence

$$\{\rho_k = X_{\nu_k};\ k \geq 1\} \tag{2.2}$$

Define the integer-valued sequence of random variables $\{\hat{\rho}_k;\ k \geq 1\}$ by

$$\{\hat{\rho}_k = r\} \Leftrightarrow \{u_r \leq \rho_k < u_{r+1}\}, \qquad r = 1, 2, \ldots, \qquad k = 1, 2, \ldots \tag{2.3}$$

Theorem 2.1. Let $\{X_n;\ n \geq 1\}$ be i.i.d. If $F(x) = P\{X_1 \leq x\}$ is continuous, then there exist a.s. two integers m and p such that for any $n > m$,

$$T_n = \nu_{n-p} \quad \text{and} \quad \theta_n = \hat{\rho}_{n-p} \tag{2.4}$$

Proof. Without loss of generality we can assume that the random variables X_k, $k = 1, 2, \ldots$, are uniformly distributed on $(0, 1)$.

We shall prove that

$$P\{\hat{\rho}_k = \hat{\rho}_{k+1} \text{ i.o.}\} = 0 \tag{2.5}$$

from which (2.4) follows in a straightforward way.

For any fixed $n \geq 1$, we have

$$\begin{aligned}
P\{\hat{\rho}_n = \hat{\rho}_{n+1}\} &= \sum_{p=1}^{\infty} P\{\hat{\rho}_n = \hat{\rho}_{n+1} | \hat{\rho}_n = p\} P\{\hat{\rho}_n = p\} \\
&= \sum_{p=1}^{\infty} \left(\int_{u_p}^{u_{p+1}} \frac{u_{p+1} - x}{1 - x} \, dP\{\rho_n = x | \hat{\rho}_n = p\} \right) P\{\hat{\rho}_n = p\} \\
&= \sum_{p=1}^{\infty} \int_{u_p}^{u_{p+1}} \frac{u_{p+1} - x}{1 - x} \, dP_{\rho_n}(x) \qquad (2.6)
\end{aligned}$$

From (2.6) we see, since $u_p = 1 - \tau/p^{\alpha}$, $p = 1, 2, \ldots$, that for any $n \geq 1$,

$$\begin{aligned}
P\{\hat{\rho}_n = \hat{\rho}_{n+1}\} &= 1 - \sum_{p=1}^{\infty} \frac{\tau}{(p+1)^{\alpha}} \int_{u_p}^{u_{p+1}} \frac{1}{1-x} \, dP_{\rho_n}(x) \\
&\leq 1 - \sum_{p=1}^{\infty} \frac{p^{\alpha}}{(p+1)^{\alpha}} P\{\rho_n \in (u_p, u_{p+1})\} \\
&= 1 - \sum_{p=1}^{n^2} \frac{p^{\alpha}}{(p+1)^{\alpha}} P\{\hat{\rho}_n = p\} \\
&\quad - \sum_{p=n^2+1}^{\infty} \frac{p^{\alpha}}{(p+1)^{\alpha}} P\{\hat{\rho}_n = p\} \\
&\leq 1 - \frac{1}{2^{\alpha}} \int_0^{u_n 2} dP_{\rho_n}(x) - \left(1 - \frac{1}{n^2 + 1}\right)^{\alpha} \int_{u_n^2}^{1} dP_{\rho_n}(x) \qquad (2.7)
\end{aligned}$$

The probability density of ρ_n is [see Chandler (1952)]

$$\frac{d}{dx} P_{\rho_n}(x) = \frac{1}{(n-1)!} (-\mathrm{Log}(1-x))^{n-1}, \qquad n = 2, 3, \ldots \qquad (2.8)$$

and this, in turn, implies by routine computations that

$$\int_{u_n^2}^{1} dP_{\rho_n}(x) = 1 - O\left(\left(\frac{\alpha \, \mathrm{Log} \, n}{n!} \right)^{n} \right), \qquad n \uparrow \infty \qquad (2.9)$$

Let us observe that for n large enough,

$$\frac{(\alpha \operatorname{Log} n)^n}{n!} \leq n^{-n/4} \tag{2.10}$$

and that

$$\left(1 - \frac{1}{n^2+1}\right)^{\alpha} = 1 + O\left(\frac{1}{n^2}\right) \quad \text{as } n \uparrow \infty \tag{2.11}$$

Hence, the last inequality in (2.7) combined with (2.9)–(2.11) implies that

$$P\{\hat{\rho}_n = \hat{\rho}_{n+1}\} = O(n^{-n/4}) + O\left(\frac{1}{n^2}\right) \quad \text{as } n \uparrow \infty$$

Since then

$$\sum_{n \geq n_0} P\{\hat{\rho}_n = \hat{\rho}_{n+1}\} < \infty$$

we obtain (2.5) by Borel–Cantelli. ■

3. kth RECORD TIME AND RECORD SEQUENCES ASSOCIATED WITH $\{u_n; n \geq 1\}$

Define the integer-valued sequence of random variables $\{I_n; n \geq 1\}$ by

$$I_n = i \quad \text{if and only if} \quad u_{i-1} < X_n \leq u_i, \quad i = 2, 3, \ldots, \quad n = 1, 2, \ldots \tag{3.1}$$

Let

$$I_{1,n} < I_{2,n} < \cdots < I_{s(n),n}, \qquad s(n) \leq n \tag{3.2}$$

be the order statistics associated with $\{I_j; 1 \leq j \leq n\}$ with $s(n)$ distinct values.

Let $J \geq 2$ and $r_0 \geq 1$ be fixed integers.

Put

$$\overline{T}_1 = \inf\{n : s(n) \geq J,\ I_{s(n)-J+1,n} \geq r_0\} \tag{3.3}$$

It is easy to show that a.s.

$$\lim_{n \uparrow \infty} s(n) = +\infty$$

and hence that $\overline{T}_1 < \infty$ with probability 1.

Put

$$\overline{\Theta}_1 = \left(I_{s(\overline{T}_1)-J, \overline{T}_1}, \dots, I_{s(\overline{T}_1), \overline{T}_1} \right) \tag{3.4}$$

and for any $k \geq 2$, let

$$\overline{T}_k = \inf\left\{ n > \overline{T}_{k-1} : \exists 1 \leq j \leq J \text{ such that } I_{s(n)-j+1, n} > I_{s(n-1)-j, n-1} \right\}$$

and

$$\overline{\Theta}_k = \left(I_{s(\overline{T}_k)-J, \overline{T}_k}, \dots, I_{s(\overline{T}_k), \overline{T}_k} \right)$$

$$= \left(\Theta_k^{(J)}, \Theta_k^{(J-1)}, \dots, \Theta_k^{(1)} \right) \tag{3.5}$$

If $\{X_n; n \geq 1\}$ is i.i.d., $\{(\overline{T}_k, \overline{\Theta}_k); k \geq \zeta\}$ is such that [see Deheuvels (1974)]

Theorem C. For any integers $n \geq 1$, $t_1 < t_2 < \cdots < t_n$, $r_0 \leq r_k^J < r_k^{J-1} < \cdots < r_k^1$, $k = 1, \dots, n$, put

$$\bar{r} = \left(r_k^J, r_k^{J-1}, \dots, r_k^1 \right) \tag{3.6}$$

if the following conditions hold:

$$t_{n+1} > t_n, \qquad r_n^1 < r_{n+1}^1, \qquad r_n^j < r_{n+1}^j < r_n^{j-1}$$

if

$$j = 2, \dots, J \text{ and } r_n^j \leq r_n^{j-1} - 2,$$

then, we have

$$P\left\{ \overline{T}_{n+1} = t_{n+1}, \Theta_{n+1}^{(j)} = r_{n+1}^j | \overline{T}_1, \overline{\Theta}_1 = \bar{r}_1, \dots, \overline{T}_n = t_n, \overline{\Theta}_n = \bar{r}_n \right\}$$

$$= \left(P\{ X_1 \leq u_{r_n^J} \} + \sum_{1 \leq i \leq J-1} P\left\{ u_{r_n^i - 1} \leq X_1 \leq u_{r_n^i} \right\} \right)^{t_{n+1} - t_n - 1}$$

$$\times P\left\{ u_{r_{n+1}^j - 1} \leq X_1 \leq u_{r_{n+1}^j} \right\} \tag{3.7}$$

The proof of Theorem B [see Haiman (1985a, b)] may be easily adapted in order to obtain the following:

Theorem 3.1. Let $\{X_n; n \geq 1\}$ satisfy the hypotheses of Theorem B. Then for any $\tau > 1$, there exist an $\alpha > 0$, an integer $r_0 \geq 1$, and two random sequences $\{\bar{S}_n; n \geq 1\}$ and $\{\bar{R}_n; n \geq 1\}$, defined on the same probability space as $\{X_n; n \geq 1\}$, such that:

1. $(\bar{S}_1, \bar{R}_1)$ has the same probability law as $(\bar{T}_1, \bar{\Theta}_1)$ in the i.i.d. case and $\{(\bar{S}_k, \bar{R}_k); k \geq 2\}$ satisfies Theorem C by replacing $\bar{T}_k$ by $\bar{S}_k$ and $\bar{\Theta}_k$ by $\bar{R}_k$ in (3.6) and (3.7).
2. There exist a.s. n_0 and q such that for any $n \geq n_0$,

$$\bar{S}_n = \bar{T}_{n-q} \quad \text{and} \quad \bar{R}_n = \bar{\Theta}_{n-q} \tag{3.8}$$

These results have obvious application in *extreme-value theory* by enabling us to reduce limiting theorems for dependent sequences to the i.i.d. case, by the bias of strong invariance principles.

REFERENCES

Chandler, K. N. (1952). The distribution and frequency of record values. *J. R. Stat. Soc. B* **14**, 120–228.

Deheuvels, P. (1974). Majoration et minoration presque sure optimale des éléments de la statistique ordonnée d'un échantillon croissant de variables aléatoires indépendantes. *Rendi Conti Acad. Nation. Lincei, Ser. VIII* **56**, 707–719.

Haiman, G. (1985a). Almost sure asymptotic behaviour of the extremes of some stationary *m*-dependent sequences. Technical Report, LSTA, Univ. Paris VI (to appear in *Ann. Inst. Henri Poincaré* in 1987).

Haiman, G. (1985b). Almost sure asymptotic behaviour of the extremes of some stationary gaussian sequences. Technical Report, LSTA, Univ. Paris VI.

Nonstandard Domains of Attraction and Rates of Convergence

M. Ivette Gomes and Dinis D. Pestana
Departamento de Estatística
Centro de Estatística e Aplicações
Faculdade de Ciências
Universidade de Lisboa
Lisbon, Portugal

SUMMARY

Max-stable distribution functions and their domains of attraction play a central role in extreme-value theory. These may be further refined by considering standard (or normal) and nonstandard domains of attraction. Penultimate phenomena, extensively studied for the Gumbel domain of attraction, are the rule, not the exception, in case the parent distribution is in some nonstandard domain of attraction. Numerical evidence is provided to support this fact.

1. INTRODUCTION AND PRELIMINARIES

Let $\{X_i\}_{i\geq 1}$ be a sequence of independent and identically distributed random variables with distribution function (d.f.) $F(x)$. Throughout this chapter we shall assume that $F(x)$ belongs to the *domain of attraction* (d.a.) of some *max-stable* (or *extreme-value*) d.f. $G_\theta(x)$, and denote this by $F \in \mathscr{D}(G_\theta)$. In other words, there exist attraction coefficients $\{a_n\}_{n\geq 1}$

$(a_n > 0)$ and $\{b_n\}_{n\geq 1}$ such that

$$F^n(a_n x + b_n) = P\left\{\frac{M_n - b_n}{a_n} \leq x\right\} \underset{n\to\infty}{\to} G_\theta(x) \tag{1.1}$$

where $M_n = \max(X_1, \ldots, X_n)$ and

$$G_\theta(x) = \exp\left(-(1-\theta x)^{1/\theta}\right) \tag{1.2}$$

(with support $\{x\colon 1 - \theta x > 0\}$), $G_0(x) = \Lambda(x) = \exp(-\exp(-x))$, $x \in \mathbb{R}$ (the Gumbel distribution). Characterization of domains of attraction are well known (Galambos, 1978).

In the sequel, we shall be concerned with *penultimate* phenomena. Such a situation occurs when $F \in \mathcal{D}(G_\theta)$ and when there exists a sequence θ_n such that G_{θ_n} gives a much closer approximation of $F^n(a_n x + b_n)$ than G_θ does. What is meant by a "much closer approximation" will be made clear later on. It will be seen among other facts that the Kolmogorov distance between the distribution functions can be dramatically reduced for any usable sample size by the use of a penultimate approximation G_{θ_n} instead of G_θ. It appears from the available results that penultimate phenomena are the rule, rather than the exception, in practice.

For the sake of simplicity we shall restrict ourselves to the case $\theta < 0$ (Fréchet types), similar in essence to the case $\theta > 0$ (Weibull types). The case $\theta = 0$ has been dealt with in detail in Gomes (1978, 1981, 1984), and will serve as a guide for the general case. Numerical evidence will be presented in addition to theoretical developments.

2. NONSTANDARD DOMAINS OF ATTRACTION, RATES OF CONVERGENCE, AND PENULTIMATE BEHAVIOR

In the sequel, we shall assume for simplicity that F has continuous derivative $f = F'$. If this is the case, then, a sufficient condition due to von Mises for $F \in \mathcal{D}(G_\theta)$, $\theta < 0$, is that

$$\lim_{x\to\infty} \frac{xf(x)}{1 - F(x)} = -\frac{1}{\theta} \tag{2.1}$$

[see, e.g., Hahn (1970, p. 109)]. Equivalently to the von Mises condition, we may say that

$$\lim_{x\to\infty} xk(x) = -\frac{1}{\theta} \tag{2.2}$$

where

$$k(x) = -f(x)/(F(x)\log F(x)) \tag{2.3}$$

Note that these assumptions are not excessively restrictive because of tail equivalence [see Balkema and Hahn (1972)]. A more general characterization is that $F \in \mathscr{D}(G_\theta)$, $\theta < 0$, if and only if $\sup\{x\colon F(x) < 1\} = +\infty$, and

$$1 - F(x) = (1 - \theta x)^{1/\theta} L(x), \qquad x > 0 \tag{2.4}$$

where $L(\cdot)$ is a slowly varying function at ∞ [i.e., $L(\lambda x)/L(x) \to 1$, as $x \to \infty$, for any $\lambda > 0$], i.e. F has a generalized paretian upper tail.

Admissible normalizing constants $\{\alpha_n\}_{n\geq 1}$ ($\alpha_n > 0$) and $\{\beta_n\}_{n\geq 1}$ such that $F^n(\alpha_n x + \beta_n) \to G_\theta(x)$, $\theta \leq 0$, are then asymptotically equivalent, in the sense of Khinchine's convergence of types theorem, to

$$\begin{aligned} b_n &: F(b_n) = \exp\left(-\frac{1}{n}\right), \qquad n \geq 1 \\ a_n &= \frac{1}{k(b_n)}, \qquad n \geq 1 \end{aligned} \tag{2.5}$$

[i.e., $a_n/\alpha_n \to 1$, $(b_n - \beta_n)/\alpha_n \to 0$, as $n \to \infty$]. This is a well-known result for $\theta = 0$ and is easily proved for $\theta \leq 0$ (see the first assertion of Theorem 3.2). In the rest of this chapter, the particular choice of attraction coefficients given in (2.5) will be considered.

Moreover, it follows easily from (2.4) that we may choose $\alpha_n = n^{-\theta}\psi(n)$, where $\psi(\cdot)$ is a slowly varying function at ∞.

We now introduce the following:

Definition 2.1. Let $F \in \mathscr{D}(G_\theta)$, $\theta \leq 0$, $\sup\{x\colon F(x) < 1\} = +\infty$. F is said to be in the *standard domain of attraction* of G_θ if $L(x) = (1 - \theta x)^{-1/\theta}(1 - F(x)) \to C \in]0, \infty[$ as $x \to \infty$. In all other cases we say that F is in the *nonstandard domain of attraction* of G_θ (hence our title), including the situation in which $L(x) \to 0$ and $L(x) \to \infty$, as $x \to \infty$.

Note that when F is in the *standard domain of attraction* of G_θ, we may choose $\alpha_n = (Kn)^{-\theta}$, K constant, as a scale-normalizing attraction coefficient.

Though $F^n(a_n x + b_n)$ is, in general, close to G_θ even for moderate values of n, there are cases in which the improvement in the approximation obtained by taking large values of n is very small (see the tables in the sequel). Moreover, some unexpected phenomenon occurs: Fisher and Tippett (1928) observed that $\Phi^n(a_n x + b_n)$, Φ the standard normal d.f., is nearer to a "penultimate" Weibull distribution than to the "ultimate" $\Lambda(x)$; Haldane and Jayakar (1963) and Daniels (1982) have shown that nonlinear transformations of the original normal random variables have a positive effect on the rate of convergence of Φ^n to its limiting law. We now define precisely the concepts of *rate of convergence* and *penultimate behavior*, as used here:

Definition 2.2. Let $F \in \mathscr{D}(G_\theta)$, with attraction coefficients $\{a_n\}_{n\geq 1}$ $(a_n > 0)$ and $\{b_n\}_{n\geq 1}$. For every $x \in \mathbb{R}$, let $D_{n,\theta}(x; a_n, b_n) = F^n(a_n x + b_n) - G_\theta(x)$.

If there exist $h(x)$ and $\{u_n\}_{n\geq 1}$ such that $D_{n,\theta}(x; a_n, b_n) = h(x)u_n + o(u_n)$, as $n \to \infty$, we say that the *rate of convergence* of $D_{n,\theta}(x; a_n, b_n)$ to 0 [or of $F^n(a_n x + b_n)$ to $G_\theta(x)$] is of the order of u_n.

If there exists a sequence $\{\theta_n\}_{n\geq 1}$, $\theta_n \to \theta$, as $n \to \infty$, such that the rate of convergence of $D_{n,\theta_n}(x; a_n, b_n)$ to 0 is of smaller order than u_n, we then say that $G_{\theta_n}(x)$ provides a *penultimate approximation* for $F^n(a_n x + b_n)$.

Important results on rates of convergence in extreme value theory are those in Anderson (1971), Galambos (1978, 1984), Dziubdziela (1978), Hall (1979), Reiss (1981), Smith (1982), and Deheuvels and Pfeifer (1986). Elaborating on expansions of F^n given by Uzgören (1954) and Anderson (1971), Gomes (1978, 1981, 1984) studied in great detail the class $\mathscr{C}$ of d.f.'s attracted to $\Lambda(x)$, and given in

Definition 2.3. $\mathscr{C}$ is the class of d.f.'s of the form

$$\{1 - A(\alpha x + \beta)^p \exp(-(\alpha x + \beta)^q)(1 + \varepsilon(x))\} I_{[y_0, \infty)}$$

where y_0 is the greatest real solution of $A(\alpha x + \beta)^p \exp(-(\alpha x + \beta)^q)(1 + \varepsilon(x)) = 1$, A chosen in such a way that such a solution does exist, $\beta, p \in \mathbb{R}$, $q, \alpha > 0$, $\varepsilon(x) \to 0$, as $x \to \infty$.

We shall use the notation $F(x; \alpha, \beta; A; p, q; \varepsilon(x))$, which is self-explanatory, to denote a member of the class $\mathscr{C}$.

Observe that d.f.'s in $\mathscr{C}$ exhibit penultimate behavior except for the case in which $q = 1$, $p = 0$. More precisely, as proved by Gomes (1984), with a slightly different terminology, there is a penultimate Weibull $(\theta_n \downarrow 0)$ approximation when $q > 1$ or $q = 1$, $p > 0$ and there is a penultimate Fréchet approximation $(\theta_n \uparrow 0)$ when $q \in (0,1)$ or $q = 1$, $p < 0$ (see Theorem 3.1).

Observe also that the normal d.f. $\Phi(x) = F(x; 1/\sqrt{2}, 0; 1/\sqrt{2\pi}; -1, 2; \Sigma_{n\geq 1}(-1)^n 1 \cdot 3 \cdot 5 \cdot \cdots \cdot (2n-1)x^{-2n})$ belongs to $\mathscr{C}$, given in Definition 2.3, and is in the nonstandard domain of attraction of G_0. Moreover, a penultimate Weibull behavior is in this case clearcut. Indeed, the penultimate phenomenon seems to occur only in nonstandard domains of attraction. The effect of Haldane and Jayakar (1963) and of Daniels (1982) transformations is to obtain random variables with d.f. "closer" to the standard domain of attraction of G_0.

The penultimate problem has also been studied by Cohen (1982).

Gomes and Pestana (1984) announced some penultimate behavior for the case $\theta \neq 0$, stressing then the importance of nonstandard domains of attraction.

3. THE CLASS $\mathscr{C}_\theta$

Consider

$$\mathscr{C}_0 = \{ F_0(x) = F(x; 1, 0; A; p, 1; \varepsilon(x)),\ F \in \mathscr{C} \} \tag{3.1}$$

and for $\theta < 0$, define

$$\mathscr{C}_\theta = \left\{ F_\theta(x) = F_0\left(\frac{-\log(1 - \theta x)}{\theta} \right), 1 - \theta x > 0,\ F_0 \in \mathscr{C}_0 \right\} \tag{3.2}$$

The d.f. $F_\theta(x)$ is obviously in the d.a. of G_θ. Let us, in the sequel, denote $k(x)$, defined in (2.1), by $k_\theta(x)$, $\theta \leq 0$, to enhance the parameter θ.

The following theorem contains results of Anderson (1971) and Gomes (1984), which are here presented in slightly different versions.

Theorem 3.1. Let $F_0 \in \mathscr{C}_0$, defined in (3.1), and let the attraction coefficients $\{a_{n,0}\}_{n \geq 1}$ ($a_{n,0} > 0$) and $\{b_{n,0}\}_{n \geq 1}$ of F_0 to the limiting law G_0 be defined by

$$F_0(b_{n,0}) = \exp\left(-\frac{1}{n} \right), \qquad a_{n,0} = \frac{1}{k_0(b_{n,0})} \tag{3.3}$$

Then, for $p \neq 0$, and $n \geq N_0$

$$F_0^n(a_{n,0}x + b_{n,0}) = \begin{cases} G_0(x) + \frac{1}{2}x^2 g_0(x)\theta_n + O(\theta_n^2) & \text{(3.4a)} \\ G_{\theta_n}(x) + O(\theta_n^2) & \text{(3.4b)} \end{cases}$$

uniformly on compact sets of $\mathbb{R}$, where $g_0(x) = G_0'(x)$, and

$$\theta_n = \frac{k_0'(b_{n,0})}{k_0^2(b_{n,0})} \tag{3.5}$$

that is, we do have penultimate behavior $G_{\theta_n}(x)$ for $F_0^n(a_{n,0}x + b_{n,0})$.

If we choose equivalent attraction coefficients $\{\alpha_{n,0} = 1\}_{n \geq 1}$, then for $p \in \mathbb{R}$ and $n \geq N_1$, we have uniformly on compact sets of $\mathbb{R}$

$$F_0^n(x + b_{n,0}) = \begin{cases} G_0(x) + R(x)g_0(x)\theta_n^* + O(\theta_n^{*2}) & \text{(3.6a)} \\ G_{\theta_n^*}^*(x) + \dfrac{R(x) - x^2}{2} g_0(x)\theta_n^* + O(\theta_n^{*2}) & \text{(3.6b)} \end{cases}$$

where

$$R(x) = \int_0^x \exp(Bt)\, dt, \qquad B = \lim_{x \to \infty} \frac{k_0'(x)}{k_0(x) - 1} \tag{3.7}$$

and

$$\theta_n^* = k_0(b_{n,0}) - 1 \tag{3.8}$$

that is, no penultimate behavior is detected. Moreover, $\theta_n = O(\theta_n^*)$, as $n \to \infty$, if $p = 0$, and $\theta_n = O(\theta_n^{*2})$, as $n \to \infty$, if $p \neq 0$.

Proof. The proof of this result is based on the fact that $F_0^n(a_{n,0}x + b_{n,0}) = G_0(x + d_n(x))$, where $d_n(x) = -\log(-\log(F_0(a_{n,0}x + b_{n,0}))) - \log(n) - x$.

For the choice of attraction coefficients in (3.3), $d_n(0) = d_n'(0) = 0$, and $d_n^{(j+1)}(0) = k_0^{(j)}(b_{n,0})/k_0^{j+1}(b_{n,0}) = O(\theta_n^j)$, $j \geq 1$, the last equality holding iff $p \neq 0$, and where $\theta_n = k_0'(b_{n,0})/k_0^2(b_{n,0})$ goes to 0, as $n \to \infty$. Consequently, for $p \neq 0$, $d_n(x) = x^2\theta_n/2 + O(\theta_n^2)$ uniformly on compact sets of $\mathbb{R}$. The relation $F_0^n(a_{n,0}x + b_{n,0}) = G_0(x + x^2\theta_n/2 + O(\theta_n^2))$ gives (3.4a), and this same relation written as $F_0^n(a_{n,0}x + b_{n,0}) = G_0(-\log(1 - \theta_n x)/\theta_n + O(\theta_n^2))$ provides (3.4b).

If we choose $\alpha_{n,0} = 1$, $n \geq 1$, then $d_n(0) = 0$, $d_n'(0) = \theta_n^* = k_0(b_{n,0}) - 1 \to 0$, as $n \to \infty$, and $d_n^{(j)}(0) = k_0^j(b_{n,0})$, $j \geq 2$. If $p \neq 0$, $k_0'(x) = O((k_0(x) - 1)^2)$, as $x \to \infty$ [i.e., $\theta_n = O(\theta_n^{*2})$], and consequently $B = \lim_{x \to \infty} k_0'(x)/(k_0(x) - 1) = 0$. Indeed, we have uniformly on compact sets of $\mathbb{R}$, $d_n(x) = x\theta_n^* + O(\theta_n^{*2})$. If $p = 0$, $k_0'(x) = O(k_0(x) - 1)$, as $x \to \infty$ [i.e., $\theta_n = O(\theta_n^*)$], and it is then necessary to prove the existence of $R(x)$, strictly increasing, such that $d_n(x) = R(x)(k_0(b_{n,0}) - 1) + O((k_0(b_{n,0}) - 1)^2)$, uniformly on compact sets of $\mathbb{R}$. The development of $d_n(x) = d_n(R^{-1}(y))$ around $y = 0$ gives $R(x) = \int_0^x e^{Bt}\, dt$, and consequently $F_0^n(x + b_{n,0}) = G_0(x + R(x)\theta_n^* + O(\theta_n^{*2}))$ gives (3.6a), while $F_0^n(x + b_{n,0}) = G_0(-\log(1 - \theta_n^* x)/\theta_n^* + (R(x) - x^2/2)\theta_n^* + O(\theta_n^{*2}))$ provides (3.6b). ■

More generally, we have

Theorem 3.2. If $F_\theta \in \mathscr{C}_\theta$, $\theta \le 0$, the attraction coefficients $\{a_{n,\theta}\}_{n\ge1}$ ($a_{n,\theta} > 0$) and $\{b_{n,\theta}\}_{n\ge1}$ of F_θ to G_θ may be chosen as

$$F_\theta(b_{n,\theta}) = \exp(-1/n), \qquad a_{n,\theta} = \frac{1}{k_\theta(b_{n,\theta})} \tag{3.9}$$

We then have, for large n and uniformly on compact sets of $\mathbb{R}$,

$$F_\theta^n(a_{n,\theta}x + b_{n,\theta}) = G_\theta(x) + H_{\theta,B}(x)g_\theta(x)(\theta_n - \theta) + o\big((\theta_n - \theta)^2\big) \tag{3.10}$$

where $B = \lim_{x\to\infty} k_0'(x)/(k_0(x) - 1)$, $k_0(x) = \exp(-\theta x)k_\theta((1 - \exp(-\theta x))/\theta)$, $\theta < 0$,

$$H_{\theta,B}(x) = \begin{cases} -\dfrac{x - (1-\theta x)\big[(1-\theta x)^{-B/\theta} - 1\big]\big/B}{B - \theta} & \text{if } B \ne 0 \\[2ex] \dfrac{x + (1-\theta x)\log(1-\theta x)/\theta}{\theta} \underset{\theta\to 0}{\to} \dfrac{x^2}{2} & \text{if } B = 0 \end{cases} \tag{3.11}$$

and

$$\theta_n = \frac{k_\theta'(b_{n,\theta})}{k_\theta^2(b_{n,\theta})} \tag{3.12}$$

Also uniformly over compact sets of $\mathbb{R}$, for $p \ne 0$ and large n, we have

$$F_\theta^n(a_{n,\theta}x + b_{n,\theta}) = G_{\theta_n}(x) + o\big((\theta_n - \theta)^2\big) \tag{3.13}$$

If $p = 0$, no penultimate behavior is detected.

Proof. The case $\theta = 0$ is covered by Theorem 3.1. Assume now that $\theta < 0$. The relation $F_0^n(a_{n,0}s + b_{n,0}) = F_\theta^n((1 - \exp(-\theta(a_{n,0}s + b_{n,0})))/\theta)$, for $s = -\log(1 - \theta x)/\theta$ provides that, with $a_{n,\theta} = a_{n,0}\exp(-\theta b_{n,0})$ and $b_{n,\theta} = (1 - \exp(-\theta b_{n,0}))/\theta$, that is, such that (3.9) holds, $F_\theta^n(a_{n,\theta}x + b_{n,\theta} + o(a_{n,\theta})) \to G_\theta(x)$ as $x \to \infty$. Hence the possible choice of attraction coefficients given in (3.9). Using the fact that with $F_0(x) = F_\theta((1 - \exp(-\theta x))/\theta)$, $\lim_{x\to\infty} k_0'(x)/(k_0(x) - 1) = B$ ($B = 0$ for all $p \ne 0$), we have from (3.6a) $F_0^n(s + b_{n,0}) = G_0(s) + R(s)g_0(s)(k_0(b_{n,0}) - 1) +$

$O((k_0(b_{n,0}) - 1)^2)$. Putting $s = -\log(1 - \theta x)/\theta$, and attending to the fact that $k_0(b_{n,0}) - 1 = (\theta_n - \theta)/(B - \theta) + O((\theta_n - \theta)^2)$, $\theta_n = k'_\theta(b_{n,\theta})/k^2_\theta(b_{n,\theta})$, we get $F^n_\theta(\alpha_{n,\theta}x + b_{n,\theta}) = G_\theta(x) + R(-\log(1 - \theta x)/\theta)(1 - \theta x)g_\theta(x)(\theta_n - \theta)/(B - \theta) + O((\theta_n - \theta)^2)$, where $\alpha_{n,\theta} = k_0(b_{n,0})/k_\theta(b_{n,\theta})$ and $F_\theta(b_{n,\theta}) = \exp(-1/n)$.

If we think now of the choice of attraction coefficients given in (3.9), and putting $F^n_\theta(a_{n,\theta}x + b_{n,\theta}) = G_\theta(x + d_n(x; a_{n,\theta}, b_{n,\theta}))$, $d_n(x; a_{n,\theta}, b_{n,\theta}) = \{1 - \exp(-\theta(-\log(-\log(F(a_{n,\theta}x + b_{n,\theta}))) - \log(n)))\}/\theta - x$, we have $d_n(x; a_{n,\theta}, b_{n,\theta}) = d_n(x_n; \alpha_{n,\theta}, b_{n,\theta}) + (A_{n,\theta} - 1)x$, where $A_{n,\theta} = a_{n,\theta}/\alpha_{n,\theta} = 1/k_0(b_{n,0})$. From this follows that $d_n(x; a_{n,\theta}, b_{n,\theta}) = d_n(x; \alpha_{n,\theta}, b_{n,\theta}) - x(\theta_n - \theta)/(B - \theta) + O((\theta_n - \theta)^2)$, and consequently (3.10) follows from the fact that $F^n_\theta(a_{n,\theta}x + b_{n,\theta}) = G_\theta(x + H_{\theta,B}(x)(\theta_n - \theta) + O((\theta_n - \theta)^2))$, $H_{\theta,B}(x)$ given by (3.11).

The development of $(1 - (1 - \theta_n x)^{\theta/\theta_n})/\theta$ around θ gives $\{1 - (1 - \theta_n x)^{\theta/\theta_n}\}/\theta = x + (\theta_n - \theta)\{x + (1 - \theta x)\log(1 - \theta x)/\theta\} + O((\theta_n - \theta)^2)$, and consequently $F^n_\theta(a_{n,\theta}x + b_{n,\theta}) = G_\theta((1 - (1 - \theta_n x)^{\theta/\theta_n})/\theta + O((\theta_n - \theta)^2))$ iff $p \neq 0$, from which (3.13) follows, once $G_{\theta_n}(x) = G_\theta((1 - (1 - \theta_n x)^{\theta/\theta_n})/\theta)$. For $p = 0$, the rate of convergence of $F^n_\theta(a_{n,\theta}x + b_{n,\theta}) - G_{\theta_n}(x)$ is still of the order of $(\theta_n - \theta)$. ∎

Note that the results in Theorems 3.1 and 3.2 hold, uniformly on compact sets of $\mathbb{R}$, for a large class of distributions attracted to G_θ, provided that we impose adequate restrictions on the function $k_\theta(x)$ [or equivalently to $k_0(x) = \exp(-\theta x)k_\theta((1 - \exp(-\theta x))/\theta)$]. Moreover, the results of Smith (1982) guarantee the validity of (3.10) and (3.13) uniformly on $\mathbb{R}$.

Observe also that while for $\theta = 0$ we have either $\theta_n > 0$ or $\theta_n < 0$ for large n, according as $p > 0$ or $p < 0$, the situation is not so clearcut for $\theta_n - \theta$, $\theta \neq 0$.

For the simple case,

$$F_\theta(x) = \left[1 - (1 - \theta x)^{1/\theta}\left(-\frac{\log(1 - \theta x)}{\theta}\right)^p\right]\chi_{[y_0,\infty)} \qquad (3.14)$$

we present Tables 1 and 2, exhibiting the penultimate behavior we have been discussing.

In Table 1, for $p = 2$ [where a penultimate $G_{\theta_n}(x)$, $\theta_n < \theta$ occurs] we present for each entry (θ, n), $\theta = -0.5, -1.0$ and $n = 10^i$, $i = 2(1)6$, the values of θ_n, KS_θ, KS_{θ_n}, and $P_n = [(\mathrm{KS}_\theta - \mathrm{KS}_{\theta_n})/\max(\mathrm{KS}_\theta, \mathrm{KS}_{\theta_n})] \times 100\%$ in the first, second, third, and fourth line, respectively, where KS_θ denotes the Kolmogorov–Smirnov distance between $F^n_\theta(a_{n,\theta}x + b_{n,\theta})$ and the ultimate $G_\theta(x)$, and KS_{θ_n} denotes the Kolmogorov–Smirnov distance between $F^n_\theta(a_{n,\theta}x + b_{n,\theta})$ and the penultimate $G_{\theta_n}(x)$.

Table 1. Penultimate Behavior of $F_\theta^n(a_{n,\theta}x + b_{n,\theta})$ for $p = 2$

θ	n: 10^2	10^3	10^4	10^5	10^6
-0.5	-0.5928	-0.5738	-0.5619	-0.5534	-0.5470
	0.01464	0.01191	0.01011	0.00881	0.00781
	0.00133	0.00081	0.00057	0.00043	0.00033
	90.89%	93.19%	94.36%	95.14%	95.73%
-1	-1.2089	-1.1631	-1.1341	-1.1143	-1.0997
	0.03081	0.02362	0.01915	0.01600	0.01401
	0.00236	0.00149	0.00102	0.00075	0.00057
	92.33%	93.69%	94.66%	95.33%	95.92%

Table 2. Penultimate Behavior of $F_\theta^n(a_{n,\theta}x + b_{n,\theta})$ for $p = -1$

θ	n: 10^2	10^3	10^4	10^5	10^6
-0.5	-0.4439	-0.4501	-0.4575	-0.4635	-0.4683
	0.00966	0.00837	0.00712	0.00613	0.00536
	0.00036	0.00052	0.00043	0.00033	0.00026
	96.29%	93.78%	94.01%	94.58%	95.13%
-1	-0.8478	-0.8802	-0.9032	-0.9192	-0.9309
	0.02068	0.01627	0.01317	0.01099	0.00939
	0.00195	0.00132	0.00091	0.00065	0.00049
	90.58%	91.89%	93.10%	94.04%	94.78%

Table 2 is devised like Table 1, but for $p = -1$; we have then a penultimate G_{θ_n}, $\theta_n > \theta$.

These tables show a very slow evolution of θ_n, together with a considerable reduction of the Kolmogorov distance in case of a penultimate approximation.

4. FINAL COMMENTS

Although the d.a. of the Gumbel distribution exhibits more pathologies than the others—because it amalgamates all types of negative exponential behavior of the upper tail, while in Fréchet or Weibull cases the upper tail must have an almost-generalized paretian behavior with the same index—penultimate phenomena do exist in all cases. If, on one hand, it is a nice mathematical problem, on the other hand from the practical standpoint it seems rather irrelevant. In fact, for $F \in \mathcal{D}(G_\theta)$, there are infinitely many d.f.'s closer to $F^n(a_n x + b_n)$ than the ultimate G_θ, except for the case in which $F \equiv G_\theta$.

Fisher's interest in the matter has no doubt cast upon it much more weight than it deserves; on the other hand, the terminology "type I," "type II," and "type III distribution of extremes" meaning "Gumbel type," "Fréchet *family* of types," and "Weibull *family* of types" contributed to hiding that penultimate behavior is a general phenomenon in extreme-value theory.

ACKNOWLEDGMENTS

The authors are grateful to the National Institute for Scientific Research (INIC), Lisbon, for financial support to attend the III IMSIBAC, and to the anonymous referee for very detailed comments that resulted in substantial improvement of the presentation.

REFERENCES

Anderson, C. W. (1971). *Contributions to the Asymptotic Theory of Extreme Values*. Ph.D. Thesis, Imperial College, London.

Balkema, A. A. and Hahn, L. de (1972). On R. von Mises' condition for the domain of attraction of $\exp(-\exp(-x))$. *Ann. Math. Statist.* **43**, 1352–1354.

Cohen, J. (1982). Convergence rates for the ultimate and the penultimate approximation in extreme value theory. *Adv. Appl. Probab.* **14**, 833–854.

Daniels, H. E. (1982). A transformation for normal extremes. *J. Appl. Probab.* **19A**, 201–206.

Deheuvels, P. and Pfeifer, D. (1986). A semi-group approach to Poisson approximation. *Ann. Probab.* **14**, 663–676.

Dziubdziela, W. (1978). On convergence rates in the limit laws for extreme order statistics. *Trans. 7th Prague Conf. Math. Statist. Probab.* **B**, 119–127.

Fisher, R. A. and Tippett, L. H. C. (1928). Limiting forms of the frequency of the largest or smallest member of a sample. *Proc. Cambridge Phil. Soc.* **24**, 180–190.

Galambos, J. (1978). *The Asymptotic Theory of Extreme Order Statistics*. New York: Wiley.

Galambos, J. (1984). Rates of convergence in extreme value theory. In *Statistical Extremes and Applications* (J. Tiago de Oliveira, Ed.) Dordrecht: Reidel, pp. 347–352.

Gomes, M. I. (1978). *Some Probabilistic and Statistical Problems in Extreme Value Theory*. Ph.D. Thesis, Univ. Sheffield.

Gomes, M. I. (1981). Closeness of penultimate approximations in extreme value theory. *Abs. 14th Europ. Meet. Statist.*, 149–150, Wrocław.

Gomes, M. I. (1984). Penultimate limiting forms in extreme value theory. *Ann. Inst. Statist. Math.* **36A**, 71–85.

Gomes, M. I. and Pestana, D. D. (1984). Domains of attraction and penultimate behaviour. *Abs. 16th Europ. Meet. Statist.*, 202–203, Marburg.

Hahn, L. de (1970). *On Regular Variation and Its Application to the Weak Convergence of Sample Extremes*. Amsterdam: Math. Centrum.

Haldane, J. B. and Jayakar, S. D. (1963). The distribution of extremal and nearly extremal values in samples from a normal population. *Biometrika* **50**, 89–94.

Hall, P. (1979). On the rate of convergence of normal extremes. *J. Appl. Probab.* **16**, 491–500.

Reiss, R.-D. (1981). Uniform approximation to distributions of extreme order statistics. *Adv. Appl. Probab.* **13**, 533–547.

Smith, R. L. (1982). Uniform rates of convergence in extreme value theory. *Adv. Appl. Probab.* **14**, 600–622.

Uzgören, N. T. (1954). The asymptotic development of the distribution of the extreme values of a sample. *Studies in Mathematics and Mechanics Presented to R. von Mises*. New York: Academic Press.

Asymptotic Expansions in Nonlinear Renewal Theory

Robert Keener
Department of Statistics
The University of Michigan
Ann Arbor, Michigan

SUMMARY

Let $\{S_n\}$ be a random walk, $\tau = \inf\{n \geq 1 : S_n > ah(n/a)\}$, and $R = S_\tau - ah(\tau/a)$. The main result of this chapter is a third-order asymptotic expansion for $P(\tau = m, R > r)$ as $a \to \infty$.

1. INTRODUCTION

Let $X, X_1, X_2, \ldots$ be independent and identically distributed (i.i.d.), and let $S_n = \sum_{i=1}^n X_i$. Define $\tau = \tau_a = \inf\{n \geq 1 : S_n > ah(n/a)\}$, $R_n = S_n - ah(n/a)$, and $R = R_\tau$. Nonlinear renewal theory is the study of approximations for joint and marginal distributions of τ and R as $a \to \infty$. Many of the main results hold when S_n differs from a random walk by a slowly changing process [see Lai and Siegmund (1977, 1979)] in which case h can be taken identically equal to 1. The monograph by Woodroofe (1982) surveys theoretical results and applications in this area. Stopping rules of this form arise in the analysis of repeated significance tests as described by Robbins (1952, 1970) and Armitage (1975), and tests using Schwarz's (1962) boundaries. See, for example, Woodroofe (1976a, b, 1978, 1979), Siegmund (1977, 1978, 1980), Siegmund and Gregory (1980), and Lalley (1984). These

Research supported by NSF grant MCS-8102080

rules also arise in sequential point and interval estimation as discussed in papers by Woodroofe (1977), Robbins (1959), Chow and Robbins (1965), Starr (1966a, b), Starr and Woodroofe (1972), Bickel and Yahev (1967, 1968), and Ghosh et al. (1976).

Expansions in nonlinear renewal theory have been studied by several authors. Takahashi and Woodroofe [Takahashi and Woodroofe (1981), Woodroofe and Takahashi (1982)] give second-order expansions when X is normally distributed and $h(x) = \sqrt{x}$. Hagwood (1982) has extended their results when X is not normally distributed, and Klein (1986) has extended their results to other boundaries. Woodroofe and Keener (1986) give second-order expansions for the marginal distributions of τ and $(S_\tau - EX)/\sqrt{\tau}$. Woodroofe (1986) gives third-order expansions that hold in a weak (function-analytic) sense. Finally, Takahashi (1986) gives a third-order expansion in the normal case with $h(x) = \sqrt{x}$. Also, expansions for moments are given by Zhang (1984).

The main result of this chapter is a third-order expansion for the joint distribution of τ and R as $a \to \infty$. Specifically, Theorem 1.1 gives an approximation to $P(\tau = m, R > r)$ with error $o(a^{-3/2})$ (the actual result is slightly stronger than this in the tails of τ). The assumptions used in this theorem are

A1. $EX^4 < \infty$, $EX = 0$, and for some n_0, $\int |Ee^{itx}|^{n_0}\, dt < \infty$.

A2. $h(1) = 0$, $h'(1) = -1$, h'' is continuous in some neighborhood of 1, and for some $0 < \mu_0 < 1$ and $\delta_0 > 0$, $h(x) - h(y) \geq \mu_0(y - x)$ for $0 \leq x \leq y \leq 1 + \delta_0$.

The conditions $h(1) = 0$, $h'(1) = -1$, and $EX = 0$ are arbitrary standardizations and entail no loss of generality. By Anscombe's (1953) theorem, they imply that $(\tau - a)/(\sigma\sqrt{a})$ converges weakly to a standard normal variate where $\sigma^2 = EX^2$. Absolute integrability of some power of the characteristic function of X is a smoothness condition that implies that X has no atoms and that S_n has a bounded continuous Lebesgue density for all $n \geq n_0$.

To state the main results, let $(T(\mu), H(\mu))$ be the first ascending ladder point for the random walk $\{S_n + n\mu\}_{n \geq 1}$. Let $\nu = \nu(\mu) = EH(\mu)$, $W(k, x) = W(k, x, \mu) = P\{T(\mu) = k,\ H(\mu) > x\}$, and define

$$Q_k(r, \mu) = \sum_j \int_0^\infty dx\, W(j, r + x, \mu) \left(\frac{j - x/\mu}{\sqrt{\sigma^2 \nu / \mu^3}} \right)^k$$

$$= \left(\frac{\mu^3}{\sigma^2 \nu} \right)^{k/2} \frac{\mu}{k+1} E\left[T^{k+1} - \left(T - \frac{H - r}{\mu} \right)^{k+1} ;\ H > r \right]$$

Also let

$$M_{jk} = M_{jk}(\mu) = \frac{\mu^{2k} E H^j E(\mu T - H)^k}{\nu^{j+k} \sigma^{2k} j! k!}$$

and define $C_i = C_i(\mu)$ for $i = 1, \ldots, 5$ as follows: $C_1 = -M_{11}$, $C_2 = -\frac{1}{2}M_{11} + M_{03}$, $C_3 = M_{20} - M_{12} + M_{11}^2$, $C_4 = \frac{1}{4}M_{20} - \frac{1}{2}M_{12} + M_{04} + M_{11}^2 - 2M_{11}M_{03}$, and $C_5 = \frac{1}{8}M_{11}^2 - \frac{1}{2}M_{11}M_{03} + \frac{1}{2}M_{03}^2$. Using A1 and A2, $J = (j_0, j_1)$ will be a neighborhood of one chosen so that h'' is continuous on J, $j_1 \leq 1 + \delta_0$, $-h'(J) \subset [\mu_0, \mu_1]$ where $\mu_1 < \infty$, and $\inf_{x \in J}\{h(x) - xh'(x)\} > 0$. $\mathbf{H}_k$ will denote the kth Hermite polynomial. In the statement of the theorem, the dependence of Q_k, C_k, T, and H on μ is suppressed.

Theorem 1.1. Suppose A1 and A2 hold. Let $\mu = \mu_a = -h'(m/a)$, $a^* = ah(m/a) - mh'(m/a)$,

$$\hat{m} = \frac{m - a}{\sigma\sqrt{a}}$$

and

$$\tilde{m}^* = \frac{m - a^*/\mu}{\sqrt{\sigma^2 a^*/\mu^3}}$$

Then

$$P(\tau = m, R > r) = \frac{\exp\{-(\tilde{m}^*)^2/2\}}{\nu\sqrt{2\pi a^* \sigma^2/\mu^3}}$$

$$\times\Bigg\{Q_0(r) + \sqrt{\frac{\nu}{a^*}}\,[(C_1\mathbf{H}_1(\tilde{m}^*)$$

$$+ C_2\mathbf{H}_3(\tilde{m}^*))\mathbf{Q}_0(r) + \mathbf{H}_1(\tilde{m}^*)Q_1(r)]$$

$$+ \frac{\nu}{a}\Bigg[(C_3\mathbf{H}_2(\tilde{m}^*) + C_4\mathbf{H}_4(\tilde{m}^*) + C_5\mathbf{H}_6(\tilde{m}^*))Q_0(r)$$

$$+ (C_1\mathbf{H}_2(\tilde{m}^*) + C_2\mathbf{H}_4(\tilde{m}^*))Q_1(r)$$

$$+ \frac{1}{2}\mathbf{H}_2(\tilde{m}^*)Q_2(r) + \frac{1}{\nu}E\big[(H - r)^2; H > r\big]$$

$$- \frac{\mu}{2\nu}h''(1)E[T^2; H > r]\Bigg]\Bigg\}$$

$$+ o\left(\frac{a^{-3/2}}{1 + |\hat{m}|^3}\right)$$

uniformly for $r \geq 0$ and $m/a \in J$ as $a \to \infty$.

To motivate later results let us sketch the ideas used to obtain this expansion. Define

$$\hat{h}(x) = h\left(\frac{m}{a}\right) + \left(x - \frac{m}{a}\right)h'\left(\frac{m}{a}\right)$$

and let

$$t = \inf\left\{n : S_n > a\hat{h}\left(\frac{n}{a}\right)\right\}$$

If $\tau = m$, then $t \leq m$ and vice versa. Using this we have the identity

$$\begin{aligned} P(\tau = m, R_m > r) = P(t = m, R_m > r) & \\ + P(\tau = m, t < m, R_m > r) & \\ - P(t = m, \tau < m, R_m > r). & \qquad (1.1) \end{aligned}$$

The last two terms are approximated by conditioning, followed by approximating $P(t \geq j, \tau \geq j | S_j)$ using the methods used to prove Theorem 5.2 of Woodroofe (1982).

To approximate $P(t = m, R_m > r)$, let

$$t(a, \mu) = \inf\{n : S_n + n\mu > a\}$$

Note that $t = t(ah(m/a) - mh'(m/a), -h'(m/a))$. Let $\{(T_n(\mu), H_n(\mu))\}_{n \geq 0}$ be the ascending ladder points for the random walk $\{S_n + n\mu\}_{n \geq 0}$. Then

$$\begin{aligned} P\{t(a, \mu) = m, S_m + m\mu > a + r\} & \\ = \sum_{k=0}^{\infty} P\{T_{k+1}(\mu) = m, H_k(\mu) \leq a, H_{k+1}(\mu) > a + r\} & \\ = \sum_{k=0}^{\infty} E[W(m - T_k(\mu), a + r - H_k(\mu)); H_k(\mu) < a] & \end{aligned}$$

This expression is a convolution of W times an indicator function with the renewal measure for the two-dimensional random walk $\{(T_n(\mu), H_n(\mu))\}_{n \geq 0}$. Expansions for renewal measures are studied in Keener (1986), and these results are used in Section 3 to obtain an expansion for $P\{t(a, \mu) = m, S_m + m\mu > r\}$. The expansion derived holds uniformly for μ in a compact subset of $(0, \infty)$. Theorem 1.1 is proved in Section 4. Section 2 gives a few preliminary results. Various technical details are deferred to an appendix.

2. PRELIMINARIES

Lemma 2.1. If $\mu < 0$, then as $a \to \infty$,

$$P\{t(a, \mu) < \infty\} = o(1/a^3)$$

Proof. Let $U(x) = E\#\{n \geq 0 : S_n + n\mu \in (x - 1, x)\}$. By Theorem 2 of Stone (1965), $U(x) = o(1/x^3)$ as $x \to \infty$ and $U(x) \sim -1/\mu$ as $x \to -\infty$. On $(-\infty, 0]$, U is continuous and positive, so there exists $\delta > 0$ such that $U(x) \geq \delta$ for all $x \leq 0$. For $x > 0$,

$$\begin{aligned} U(x) &= E\left[U\left(S_{t(x,\mu)} + t(x, \mu)\mu - x; t(x, \mu) < \infty\right]\right. \\ &\geq \delta P\{t(x, \mu) < \infty\} \end{aligned}$$

and the result follows. ∎

Lemma 2.2. If $\mu > 0$, $ET^4(\mu) < \infty$ and $EH^4(\mu) < \infty$.

Proof. This follows from Theorem 2.1 of Gut (1974). ∎

Lemma 2.3. If $0 < \mu_0 < \mu_1 < \infty$, the families $\{T^4(\mu) : \mu_0 \leq \mu \leq \mu_1\}$ and $\{H^4(\mu) : \mu_0 \leq \mu \leq \mu_1\}$ are uniformly integrable.

Proof. If $\mu > \mu_0$, then $T(\mu) \leq T(\mu_0)$ and $H(\mu) \leq H(\mu_0) + (\mu - \mu_0)T(\mu_0)$ and the result follows from Lemma 2.2. ∎

Let

$$\chi_\mu(p, q) = Ee^{ipH(\mu) + iqT(\mu)}$$

Lemma 2.4. If $0 < \mu_0 < \mu_1 < \infty$, then

$$\inf_{\mu_0 < \mu < \mu_1} EH(\mu) > 0$$

$$\inf_{\mu_0 < \mu < \mu_1} E[T(\mu) - H(\mu)/\mu]^2 > 0$$

and for any $\varepsilon > 0$,

$$\inf_{\mu_0 < \mu < \mu_1} \inf_{|p| \geq \varepsilon,\, \varepsilon \leq |q| \leq \pi} |1 - \chi_\mu(p, q)| > 0$$

The proof of this lemma is given in the appendix.

Define

$$\|f\|_r = \left\{ \int |f(x)|^r \, dx \right\}^{1/r}$$

and let $\mathscr{L}^r$ be the set of all complex valued functions f with $\|f\|_r < \infty$. For $f \in \mathscr{L}^r, 1 < r < \infty$, define the Hilbert transform

$$\mathscr{H}f(x) = \lim_{\varepsilon \downarrow 0} \frac{1}{\pi} \int_{|y| \geq \varepsilon} \frac{f(x-y)}{y} \, dy$$

This limit exist in $\|\cdot\|_r$ norm and $\mathscr{H}$ is a bounded operator from $\mathscr{L}^r \to \mathscr{L}^r, 1 < r < \infty$ [see Chapter 2 of Stein (1970)]. Let

$$\Lambda_\mu(p) = Ee^{ip(X+\mu)}$$

and

$$\beta_\mu(p, s) = Es^{T(\mu)} e^{ipH(\mu)}.$$

Lemma 2.5. If $\mu > 0$ and $\Lambda_\mu \in \mathscr{L}^r$ for some $1 < r < \infty$, then for $s \in [0, 1]$,

$$\log\{1 - \beta_\mu(\cdot, s)\} = \tfrac{1}{2}\{\log(1 - s\Lambda_\mu) + i\mathscr{H} \log(1 - s\Lambda_\mu)\} \quad (2.1)$$

If $0 < \mu_0 \leq \mu_1 < \infty$, $1 < r < \infty$, and $\Lambda_0 \in \mathscr{L}^r$, then

$$\sup_{\mu_0 \leq \mu \leq \mu_1} \|\beta_\mu(\cdot, 1)\|_r < \infty \quad (2.2)$$

The proof of this lemma will also be relegated to the appendix.

For reference, we next state a version of the central limit theorem with a good error rate in the tails of the distribution. Let g_n denote the density of S_n and $\hat{g}_n(s) = \exp\{-s^2/(2n\sigma^2)\}/\sqrt{2\pi n\sigma^2}$ its normal approximation. Let $\hat{s}_n = s/\sqrt{n\sigma^2}$.

Lemma 2.6. Uniformly for $s \in \mathbf{R}$, as $n \to \infty$,

$$g_n(s) = \hat{g}_n(s) + o\left(\frac{1}{\sqrt{n}\,(1 + \hat{s}_n^4)} \right)$$

and

$$P(S_n < s) = \Phi(\hat{s}_n) + o\left(\frac{1}{1 + |\hat{s}_n|^3} \right)$$

If $\{s_n\}_{n\geq 1}$ satisfies

$$\exp\left\{\frac{s_n^2}{2n\sigma^2}\right\} = O(n)$$

as $n \to \infty$, then

$$g_n(s_n) \sim \hat{g}_n(s_n)$$

as $n \to \infty$. Finally, if $\liminf|\hat{s}_n|/n^{\varepsilon} > 0$ for some $\varepsilon > 0$, then

$$g_n(s_n) = o\left(\frac{n^{-3/2}}{s_n^4}\right)$$

as $n \to \infty$.

These results follow from asymptotic expansions for the densities g_n, such as Theorem 19.2 of Bhattacharya and Rao (1976).

Lemma 2.7. Let $\chi(p,q)$ be the characteristic function for a random vector (H,T) where T is integer valued.

Then

$$\int_{-\pi}^{\pi} dq \int dp |\chi(p,q)|^2 \leq \frac{1}{2\pi} \int |\chi(p,0)|\, dp$$

Proof. Assuming without loss of generality that the right-hand side is finite, H has density

$$f(x) = \frac{1}{2\pi}\int e^{-ipx}\chi(p,0)\, dp$$

and $\sup f \leq \int|\chi(p,0)|\, dp/(2\pi)$. Let $f(n,x)$ denote the joint mass function/density of T and H. Define

$$g(q,x) = \sum_n e^{iqn} f(n,x)$$

Then

$$\chi(p,q) = \int e^{ipx} g(q,x)\, dx$$

so Plancheral's identity [Feller, 1971, equation (7.8), p. 638] gives

$$\frac{1}{2\pi}\int |\chi(p,q)|^2\, dp = \int |g(q,x)|^2\, dx$$

Using

$$|g(q, x)| \leq \sum_n |f(n, x)| = f(x)$$

the lemma follows from the bound for $\sup f$ since $\int f = 1$. ■

3. EXPANSIONS IN LINEAR RENEWAL THEORY

By Lemmas 2.5 and 2.7, for $n \geq 2n_0$, (H_n, T_n) has a density $f_n(x, t)$ with respect to Lebesgue cross counting measure. Define the renewal density

$$U(x, t) = \sum_{n=2n_0+4}^{\infty} f_n(x, t)$$

Let

$$\hat{z} = \frac{\mu t - x}{\sqrt{x\sigma^2/\mu}}$$

and define

$$\tilde{U}(x, t) = \frac{e^{-\hat{z}^2/2}}{\nu\sqrt{2\pi x\sigma^2/\mu^3}}\left\{1 + \sqrt{\frac{\nu}{x}}\,[C_1\mathbf{H}_1(\hat{z}) + C_2\mathbf{H}_3(\hat{z})] + \frac{\nu}{x}[C_3\mathbf{H}_2(\hat{z}) + C_4\mathbf{H}_4(\hat{z}) + C_5\mathbf{H}_6(\hat{z})]\right\}$$

The following result is a slight modification of Theorem 1.4 of Keener (1986) (exploiting the positivity mentioned in the remark following the theorem). It shows that U is well approximated by $\tilde{U}$.

Theorem 3.1. As $x \to \infty$,

$$U(x, t) = \tilde{U}(x, t) + o\left\{\frac{x^{-3/2}}{1 + |\hat{z}|^3}\right\}$$

uniformly in t.

Inspecting the proofs in Keener (1986), the conclusion in this theorem will hold uniformly for $\mu \in \mathscr{C}$ provided the following conditions hold.

B1. $\inf_{\mu \in \mathscr{C}} \nu > 0$ and $\inf_{\mu \in \mathscr{C}} \nu\sigma^2/\mu^3 > 0$.

B2. $H^2(\mu)$ and $[T(\mu) - H(\mu)/\mu]^4$ are uniformly integrable as μ varies over $\mathscr{C}$.

B3. $\sup_{\mu \in \mathscr{C}} \int_{-\pi}^{\pi} dq \int dp |\chi(p, q)|^{2n_0} < \infty$, where $\chi(p, q) = E\exp\{ipH + iqT\}$.

B4. For any $\varepsilon > 0$, $\inf_{\mu \in \mathscr{C}} \inf\{|1 - \chi(p, q)| : \varepsilon < |q| < \pi, |p| > \varepsilon\} > 0$.

Under assumptions A1 and A2, B1–B4 hold with $\mathscr{C}$ any compact subset of $(0, \infty)$ by Lemmas 2.3, 2.4, 2.5, and 2.7. In the sequel $\mathscr{C}$ will denote a compact subset of $(0, \infty)$. All of the asymptotic relations in this section will hold uniformly for $\mu \in \mathscr{C}$.

Recall from Section 1 that

$$P\{t(a, \mu) = m, S_m + m\mu > r + a\}$$

$$= \sum_{k=0}^{\infty} P\{T_{k+1}(\mu) = m, H_k(\mu) \le a, H_{k+1}(\mu) > a + r\}$$

The sum from 0 to $2n_0 + 3$ is bounded by $(2n_0 + 4)P\{H_{2n_0+4}(\mu) > a\}$, which is $o(1/a^4)$ by Lemma 2.3. The sum from $2n_0 + 4$ to ∞ equals

$$\int dx \sum_j W(j, r + x)U(a - x, m - j)$$

Let $m = m_a$ and

$$\tilde{m} = \frac{m - a/\mu}{\sqrt{\sigma^2 a/\mu^3}}$$

Lemma 3.1. If A1, A2, and $\tilde{m} = O(a^{1/6})$, then

$$P\{t(a, \mu) = m, S_m + m\mu > r + a\}$$

$$= \int_0^{\sqrt{a}} dx \sum_{j=1}^{\sqrt{a}} W(j, r + x)\tilde{U}(a - x, m - j) + o\left\{\frac{a^{-3/2}}{1 + |\tilde{m}|^3}\right\}$$

as $a \to \infty$ uniformly for $r \ge 0$ and $\mu \in \mathscr{C}$.

Proof. Since $\int_0^\infty dx \sum_{j \ge 1} W(j, r + x) \le EH(\mu)$ and U and $\tilde{U}$ differ in the region of summation/integration by $o\{a^{-3/2}/(1 + |\tilde{m}|^3)\}$, we need only to show that the conclusion holds with U replacing $\tilde{U}$. By the uniform integrability of $H^4(\mu)$, as $y \to \infty$,

$$\int_y^\infty dx \sum_j W(j, r + x) = \int_y^\infty P\{H(\mu) > r + x\}\, dx = o\left(\frac{1}{y^3}\right) \tag{3.1}$$

Since U is bounded uniformly in μ [see Keener (1986, Lemma 2.2)]

$$\int_{a/2}^\infty \sum_j W(j, r + x) U(a - x, m - j) = o\left(\frac{1}{a^3}\right)$$

as $a \to \infty$. Using the uniform integrability of $T^4(\mu)$ and $H^4(\mu)$, as $y \to \infty$,

$$\begin{aligned}\int_0^\infty dx \sum_{j \geq y} W(j, x) &= \sum_{j \geq y} E[H(\mu); T(\mu) = j] \\ &\leq \frac{1}{y^3} \sum_{j \geq y} E[H(\mu) T^3(\mu); T(\mu) = j] \\ &= o\left(\frac{1}{y^3}\right) \end{aligned} \tag{3.2}$$

By Theorem 3.1, $U(x, t) = O(1/\sqrt{x})$ as $x \to \infty$ uniformly in t. This fact, (3.1), and (3.2) give

$$\left\{\int_0^{a/2} dx \sum_{j \geq \sqrt{a}} + \int_{\sqrt{a}}^{a/2} dx \sum_{j \geq 1}\right\} W(j, r + x) U(a - x, m - j) = o\left(\frac{1}{a^2}\right)$$

and the lemma holds. ■

Viewing $\tilde{U}(x, t)$ as a function of $\hat{z}$ and x/a and Taylor expanding about $\hat{z} = \tilde{m}$, $x/a = 1$ gives the next lemma. The identity

$$\frac{d}{d\hat{z}}\left\{\mathbf{H}_k(\hat{z})\, e^{-\hat{z}^2/2}\right\} = -\mathbf{H}_{k+1}(\hat{z})\, e^{-\hat{z}^2/2}$$

is useful computing the derivatives.

Lemma 3.2. As $a \to \infty$,

$$
\begin{aligned}
&\tilde{U}(a - x, m - j)\\
&= \frac{e^{-m^2/2}}{\nu\sqrt{2\pi a\sigma^2/\mu^3}}\Bigg\{1 + \sqrt{\frac{\nu}{a}}\left[C_1\mathbf{H}_1(\tilde{m}) + C_2\mathbf{H}_3(\tilde{m}) + \mathbf{H}_1(\tilde{m})\frac{j - x/\mu}{\sqrt{\sigma^2\nu/\mu^3}}\right]\\
&\qquad + \frac{\nu}{a}\Bigg[C_3\mathbf{H}_2(\tilde{m}) + C_4\mathbf{H}_4(\tilde{m}) + C_5\mathbf{H}_6(\tilde{m}) + \frac{x}{2\nu}\\
&\qquad + (C_1\mathbf{H}_2(\tilde{m}) + C_2\mathbf{H}_4(\tilde{m}))\frac{j - x/\mu}{\sqrt{\sigma^2\nu/\mu^3}} + \tfrac{1}{2}\mathbf{H}_2(\tilde{m})\frac{(j - x/\mu)^2}{\sigma^2\nu/\mu^3}\Bigg]\Bigg\}\\
&\qquad + O\left\{\frac{1 + x^3 + j^3}{a^2}\right\}
\end{aligned}
$$

uniformly for $(j, x) \in [0, \sqrt{a}]^2$ and $\mu \in \mathscr{C}$.

Theorem 3.2. Assuming A1 and A2,

$$
\begin{aligned}
&P\{t(a, \mu) = m, S_m + m\mu > r + a\}\\
&= \frac{e^{-\tilde{m}^2/2}}{\nu\sqrt{2\pi a\sigma^2/\mu^3}}\Bigg\{Q_0(r) + \sqrt{\frac{\nu}{a}}\,[(C_1\mathbf{H}_1(\tilde{m})\\
&\qquad + C_2\mathbf{H}_3(\tilde{m}))Q_0(r) + \mathbf{H}_1(\tilde{m})Q_1(r)]\\
&\qquad + \frac{\nu}{a}\Bigg[(C_3\mathbf{H}_2(\tilde{m}) + C_4\mathbf{H}_4(\tilde{m}) + C_5\mathbf{H}_6(\tilde{m}))Q_0(r)\\
&\qquad + (C_1\mathbf{H}_2(\tilde{m}) + C_2\mathbf{H}_4(\tilde{m}))Q_1(r)\\
&\qquad + \tfrac{1}{2}\mathbf{H}_2(\tilde{m})Q_2(r) + \frac{1}{\nu}E\left[(H(\mu) - r)^2; H(\mu) > r\right]\Bigg]\Bigg\}\\
&\qquad + o\{a^{-3/2}/(1 + |\tilde{m}|^3)\}
\end{aligned}
$$

as $a \to \infty$ uniformly for $m/a > \varepsilon$, $r \geq 0$, and $\mu \in \mathscr{C}$.

Proof. Let $m = m_a$. If $\tilde{m} = O(a^{1/6})$, the conclusion follows immediately from Lemmas 3.1 and 3.2. Suppose $\liminf|\tilde{m}|/a^{1/6} > 0$ and that $\tilde{m} > 0$. Let $\overline{F}(x) = P(X > x)$. Conditioning on S_{m-1} and using the last assertion of Lemma 2.6 (with s_n^4 decreased to $|\hat{s}_n|^3$),

$$\begin{aligned}
P\{t(a,\mu) = m\} &\le P\{S_{m-1}(m-1)\mu \le a,\, S_m + m\mu > a\} \\
&= \int_0^\infty \overline{F}(s-\mu)\, g_{m-1}(a+\mu-m\mu-s)\, ds \\
&= o\{|m\mu - \mu - a|^3\} \\
&= o\left(a^{-3/2}|\tilde{m}|^{-3}\right) \qquad (3.3)
\end{aligned}$$

The case $\tilde{m} < 0$ is similar. ■

REMARK. If $m = o(a)$, then uniform integrability of S_m^4/m^2 and (3.3) gives $P\{t(a,\mu) = m) = o(m^2/a^4)$, which is not quite enough to make the expansion in this theorem hold uniformly for all m.

4. PROOF OF THEOREM 1.1

Theorem 1.1 will be proved through a series of lemmas. The goal in the theorem is an expansion with error $o\{a^{-3/2}/(1 + |\hat{m}|^3)\}$, and the first lemma shows there is no loss of generality taking $\hat{m} = O(a^{1/6})$.

Lemma 4.1. If $m = m_a$ is a sequence such that $m/a \in J$ and $\hat{m}^{-1} = O(a^{-1/6})$, then as $a \to \infty$,

$$P(\tau = m) = o\left\{\frac{a^{-3/2}}{1 + |\hat{m}|^3}\right\}$$

and

$$P(t = m) = o\left\{\frac{a^{-3/2}}{1 + |\hat{m}|^3}\right\}$$

Proof. Using bounds like those in (3.3), this result follows from Lemma 2.6. ■

Recall from Section 1 that $t = t(a^*, \mu^*)$ where $\mu^* = -h'(m/a)$ and $a^* = ah(m/a) - mh'(m/a)$. Theorem 3.2 provides an expansion for

$P\{t = m, R_m > r\}$ that holds uniformly for $m/a \in J$ because when $m/a \in J$, $\mu^* \in [\mu_0, \mu_1]$ and $a^* > a \inf_{x \in J}\{h(x) - xh'(x)\}$. To use equation (1.1), approximations are needed for $P(\tau = m, t < m, R_m > r)$ and $P(t = m, \tau < m, R_m > r)$. As a first step in this direction, the next two lemmas show that $P(t = m, \tau < m - k)$ and $P(\tau = m, t < m - k)$ are small when k is large.

Lemma 4.2. If $\hat{m} = O(a^{1/6})$ as $a \to \infty$, then

$$P(\tau < m - a^{2/3}, t = m) = o\left(\frac{1}{a^2}\right)$$

and

$$P(t < m - a^{2/3}, \tau = m) = o\left(\frac{1}{a^2}\right)$$

as $a \to \infty$.

Proof. If $n < m - a^{2/3}$ and $m/a \in J$, then by A2 and the way J was chosen,

$$h\left(\frac{n}{a}\right) \geq h\left(\frac{m - a^{2/3}}{a}\right) + \frac{\mu_0}{a}(m - a^{2/3} - n) \tag{4.1}$$

and

$$\hat{h}\left(\frac{n}{a}\right) \geq \hat{h}\left(\frac{m - a^{2/3}}{a}\right) + \frac{\mu_0}{a}(m - a^{2/3} - n) \tag{4.2}$$

Using (4.1),

$$\begin{aligned}
&P(\tau < m - a^{2/3}, t = m)\\
&\quad \leq P\left\{S_{m-1} < a\hat{h}\left(\frac{m-1}{a}\right), S_n > ah\left(\frac{n}{a}\right), \exists n < m - a^{2/3}\right\}\\
&\quad \leq P\left\{S_{m-1} < a\hat{h}\left(\frac{m-1}{a}\right), S_n > ah\left(\frac{m - a^{2/3}}{a}\right)\right.\\
&\qquad\qquad \left. + \mu_0(m - a^{2/3} - n), \exists n < m - a^{2/3}\right\}\\
&\quad \leq P\left\{S_j + j\mu_0 < a\left[\hat{h}\left(\frac{m-1}{a}\right) - h\left(\frac{m - a^{2/3}}{a}\right)\right]\right.\\
&\qquad\qquad \left. + \mu_0(a^{2/3} - 1), \exists j > 0\right\}
\end{aligned}$$

Since

$$a\left[\hat{h}\left(\frac{m-1}{a}\right)-h\left(\frac{m-a^{2/3}}{a}\right)\right]+\mu_0(a^{2/3}-1)\sim -a^{2/3}(1-\mu_0)$$

this probability is $o(1/a^2)$ by Lemma 2.1. The other half of the lemma follows from (4.2) using the same reasoning. ∎

Lemma 4.3. If $\hat{m}=O(a^{1/6})$, then there exists a positive constant K such that for a sufficiently large,

$$P(t=m,m-a^{2/3}<\tau<m-k)\le\frac{KE\left[\left(t\left(\tfrac{1}{2},1\right)+1\right)^3;t\left(\tfrac{1}{2},1\right)>k\right]}{a^{3/2}(1+|\hat{m}|^3)}$$

and

$$P(\tau=m,m-a^{2/3}<t<m-k)\le\frac{KE\left[\left(t\left(\tfrac{1}{2},1\right)+1\right)^3;t\left(\tfrac{1}{2},1\right)>k\right]}{a^{3/2}(1+|\hat{m}|^3)}$$

for all $k\ge 1$.

Proof. Continuity of h'' at 1 implies that for some $K_1>0$, for all sufficiently large a, $|\hat{h}(j/a)-h(j/a)|\le K_1(m-j)^2/a^2$. Using Lemma 2.6, for some $K_2>0$, if a is sufficiently large

$$P\left\{a\hat{h}\left(\frac{j}{a}\right)\le S_j\le ah\left(\frac{j}{a}\right)\text{ or }ah\left(\frac{j}{a}\right)\le S_j\le a\hat{h}\left(\frac{j}{a}\right)\right\}\le\frac{K_2(m-j)^2}{a^{3/2}(1+|\hat{m}|^3)}$$

whenever $j\in(m-a^{2/3},m)$. Conditioning on $X_1,\dots,X_j$ and using the fact that $\hat{h}'(j/a)$ and $h'(j/a)$ approach -1 uniformly for $j\in(m-a^{3/2},m)$ as $a\to\infty$, for a sufficiently large

$$P(t=m,\tau=j\text{ or }\tau=m,t=j)$$

$$\le\frac{K_2(m-j)^2}{a^{3/2}(1+|\hat{m}|^3)}P\left(S_n-S_j<1-\frac{n-j}{2}\text{ for }j<n<m\right)$$

$$=\frac{K_2(m-j)^2}{a^{3/2}(1+|\hat{m}|^3)}P\left\{t\left(\tfrac{1}{2},1\right)\ge m-j\right\}$$

Summing this bound over $j < m - k$ proves the lemma since

$$\sum_{n>k} n^2 P\{t(\tfrac{1}{2},1) \geq n\} \leq \frac{E\left[\left(t(\tfrac{1}{2},1)+1\right)^3; t(\tfrac{1}{2},1) > k\right]}{3} \qquad \blacksquare$$

Lemma 4.4. For fixed j,

$$P(\tau = m, t = m - j, R_m > r) - P(t = m, \tau = m - j, R_m > r)$$

$$= \frac{\mu h''(1) j^2 P\{T(1) = j, H(1) > r\} e^{-\hat{m}^2/2}}{2a\nu\sqrt{2\pi\sigma^2 a}} + o\left\{\frac{a^{-3/2}}{1 + |\hat{m}|^3}\right\}$$

uniformly for $r \geq 0$ as $a \to \infty$.

Combining the last three lemmas,

$$P(t = m, \tau < m, R_m > r) - P(\tau = m, t < m, R_m > r)$$

$$= \frac{\mu h''(1) E\left[T(1)^2; H(1) > r\right] e^{-\hat{m}^2/2}}{2a\nu\sqrt{2\pi a\sigma^2}} + o\left\{\frac{a^{-3/2}}{1 + |\hat{m}|^3}\right\}$$

Theorem 1.1 then follows using (1.1) and the expansion in Theorem 3.4.

Proof of Lemma 4.4. By Lemma 4.1 we can assume that $\hat{m} = O(a^{1/6})$. Then

$$h\left(\frac{m-j}{a}\right) - \hat{h}\left(\frac{m-j}{a}\right) \sim \frac{j^2}{a^2} h''(1)$$

as $a \to \infty$ continuity of h'' near 1. By Lemma 2.6,

$$P\left\{a\hat{h}\left(\frac{m-j}{a}\right) < S_{m-j} \leq ah\left(\frac{m-j}{a}\right)\right\}$$

$$-P\left\{ah\left(\frac{m-j}{a}\right) < S_{m-j} \leq a\hat{h}\left(\frac{m-j}{a}\right)\right\}$$

$$= \frac{j^2 h''(1) e^{-\hat{m}^2/2}}{a\sqrt{2\pi a\sigma^2}} + o\left\{\frac{a^{-3/2}}{1 + |\hat{m}|^3}\right\}$$

Both probabilities can be absorbed into the $o(\cdot)$ when $e^{-\hat{m}^2/2} < 1/a$, and we can assume without loss of generality that $e^{\hat{m}^2/2} = O(a)$. Conditioning on S_j, Lemma 4.4 follows easily from the next lemma about conditional probabilities. ■

Let

$$\tilde{h}(x) = \min\{\hat{h}(x), h(x)\}$$

Lemma 4.5. If $e^{\hat{m}^2/2} = O(a)$ and $\{\varepsilon_a\}$ is a sequence of constants approaching 0, then

$$P\left\{S_n \le a\tilde{h}\left(\frac{n}{a}\right) \text{ for } n < m - j | R_{m-j} = \varepsilon_a\right\} \to \frac{\mu}{\nu}$$

$$P\left\{S_m > r, S_n \le ah\left(\frac{n}{a}\right) \text{ for } m - j < n < m | R_{m-j} = \varepsilon_a\right\}$$

$$\to P\{T(1) = j, H(1) > r\}$$

and

$$P\left\{S_m > r, S_n \le a\hat{h}\left(\frac{n}{a}\right) \text{ for } m - j < n < m | R_{m-j} = \varepsilon_a\right\}$$

$$\to P\{T(1) = j, H(1) > r\}$$

as $a \to \infty$.

Proof. The last two assertions are clear because $ah(n/a) - S_{m-j} = ah(n/a) - ah((m-j)/a) + R_{m-j}$, which approaches $m - j - n$. The proof of the first assertion is similar to the proof of Theorem 5.1 of Woodroofe (1982). Let g_n denote the density of S_n, and let

$$\hat{g}_n(s) = \exp\left\{-\frac{s^2}{2n\sigma^2}\right\} \Big/ \sqrt{2\pi n\sigma^2}$$

Define

$$q_{n,k}(x_1, \ldots, x_k | x) = g_{n-k}\left(nx - \sum_{i=1}^{k} x_i\right) \Big/ g_n(nx)$$

$q_{n,k}$ is the density of $\mathscr{L}(X_1, \ldots, X_k | \overline{X}_n = x)$ with respect to $\mathscr{L}(X_1, \ldots, X_k)$. Let

$$x_a = \frac{1}{m-j}\left[ah\left(\frac{m-j}{a}\right) + \varepsilon_a\right]$$

Using Lemma 2.6, for fixed k,

$$q_{m-j,k}(x_1,\ldots,x_k|x_a) \to 1$$

as $a \to \infty$ uniformly for $x_1,\ldots,x_k$ in any compact set, and these densities converge strongly by Scheffé's (1947) theorem. It then follows that

$$\mathscr{L}\left(\left\{S_{m-j} - S_{m-j-i} + a\tilde{h}\left(\frac{m-j-i}{a}\right) - ah\left(\frac{m-j}{a}\right) - \varepsilon_a\right\}_{i=1}^{k} \Big/ R_{m-j} = \varepsilon_a\right) \to \mathscr{L}(\{S_i + i\}_{i=1}^{k})$$

weakly as $a \to \infty$. Consequently

$$\lim_{k\to\infty}\lim_{a\to\infty} P\left\{S_n \le a\tilde{h}\left(\frac{n}{a}\right) \text{ for } m-j-k \le n < m-j | R_{m-j} = \varepsilon_a\right\}$$

$$= P(S_j + j \ge 0 \text{ for } j \ge 1) = \frac{\mu}{\nu}$$

Using this, the lemma is an immediate consequence for the next lemma. ■

Lemma 4.6. With the conditions of Lemma 4.5,

$$\lim_{k\to\infty}\liminf_{a\to\infty} P\left(S_n \le a\tilde{h}\left(\frac{n}{a}\right) \text{ for } n < m-j-k | R_{m-j} = \varepsilon_a\right) = 1$$

To prove this, we need the bound in the next lemma.

Lemma 4.7. Let $p = p_a \sim \sqrt{\log(a)}$ as $a \to \infty$. There exists a constant K such that

$$\limsup_{a\to\infty} \sup_{(x_1,\ldots,x_p)\in R^p} \left\{q_{m-j,p}(x_1,\ldots,x_p|x_a) - \frac{K}{p^2}\left(\sum_{i=1}^{p} x_i\right)^4\right\} < \infty$$

Proof. By Lemma 2.6,

$$\sup_{(x_1,\ldots,x_p)\in R^p} q_{m-j,p}(x_1,\ldots,x_p|x_a) = O(a)$$

so for any $\varepsilon > 0$, the inner supremum in the lemma can be taken over

$|\Sigma_{i=1}^{p} x_i| \le \varepsilon\sqrt{a \log a}$ and we can approximate $q_{m-j,p}$ by $\hat{g}_{m-j-p}/\hat{g}_{m-j}$. By symmetry we can assume $x_a \ge 0$ and the worst case will be when x_a is as large as possible. Consequently we can take $x_a \sim \sqrt{2\sigma^2 \log(a)/a}$. With these assumptions we must show

$$\limsup_{a\to\infty} \sup_{0<\Sigma_{i=1}^{p} x_i<\varepsilon\sqrt{a \log a}} \left[\exp\left\{\frac{(m-j)x_a - \Sigma_{i=1}^{p} x_i}{(m-j-p)\sigma^2}\right\} - K\frac{(\Sigma_{i=1}^{p} x_i)^4}{p^2}\right] < \infty$$

Choosing ε and K suitably, this expression is less than

$$\sup_{a>0,0<z<\sqrt{a \log a}} \left[\exp\left\{z\sqrt{\frac{\log a}{a}}\right\} - \frac{z^4}{a \log^2(a)}\right]$$

$$= \sup_{a>0,0<z<1} (a^z - az^4)$$

$$= \sup_{0<z<1} \{z^{3z/(z-1)} - z^{(4z-1)/(z-1)}\}$$

which is finite because the function is continuous. ■

Proof of Lemma 4.6. Reversing the order of the X_i and using A2, for a sufficiently large

$$P\left(S_n \le a\tilde{h}\left(\frac{n}{a}\right) \text{ for } n < m-j-k | R_{m-j} = \varepsilon_a\right)$$

$$\ge P(S_n + n\mu_0 \ge 1 \text{ for } k < n < m-j | \bar{X}_{m-j} = x_a)$$

One minus this probability is less than or equal to

$$P(S_n + n\mu_0 < 1, \exists k < n \le p | \bar{X}_{m-j} = x_a)$$

$$+P(S_n + n\mu_0 < 1, \exists p < n < m-j | \bar{X}_{m-j} = x_a)$$

By Lemma 4.7, the first of these probabilities is less than

$$E\left[K + KS_p^4/p^2; S_n + n\mu_0 < 1, \exists k < n \le p\right]$$

As $k \to \infty$, this expression approaches 0 uniformly in a because the family $\{S_p^4/p^2\}_{p\ge1}$ is uniformly integrable and $P(S_n + n\mu_0 < 1, \exists k < n) \to 0$ by the strong law of large numbers. The other probability is less than a multiple of $aP(S_n + n\mu_0 < 1, \exists n > p)$ since $q_{m-j,p} = O(a)$. Conditioning on S_p and using Lemma 2.1, this term is bounded by a multiple of

$$aE \min\left\{1, (S_p + p\mu_0 - 1)_+^{-3}\right\} \leq a^{-1/8} + aP\left(S_p + p\mu_0 - 1 < a^{3/8}\right)$$

and this goes to 0 by Lemma 2.6. ■

5. APPENDIX

Proof of Lemma 2.4. For the first assertion of the lemma, by Wald's identity

$$\inf_{\mu_0 \leq \mu \leq \mu_1} EH(\mu) = \inf_{\mu_0 \leq \mu \leq \mu_1} \mu ET(\mu) \geq \mu_0$$

Next, let $Y \sim \mathscr{L}(X | X > -\mu_0)$ and $p_0 = P(X > -\mu_0) > 0$. Conditioning on $\{X_1 > -\mu_0\}$,

$$\inf_{\mu_0 \leq \mu \leq \mu_1} E[T(\mu) - H(\mu)/\mu]^2 \geq p_0 \inf_{\mu_0 \leq \mu \leq \mu_1} E\left(1 - \frac{Y + \mu}{\mu}\right)^2$$

$$= \frac{p_0 EY^2}{\mu_0^2}$$

Finally, let $\chi_\mu^{(1)}$ and $\chi_\mu^{(2)}$ be the joint characteristic functions for $(H(\mu), T(\mu))$ conditional on $\{X_1 > -\mu_0\}$ and $\{X_1 \leq -\mu_0\}$, respectively. If $\mu \geq \mu_0$, $T(\mu) = 1$ on $\{X_1 > -\mu_0\}$ and

$$\chi_\mu^{(1)}(p, q) = \varepsilon^{iq} E e^{ipY}$$

By smoothing $\chi_\mu = p_0 \chi_\mu^{(1)} + (1 - p_0)\chi_\mu^{(2)}$. Let $\delta = p_0(1 - \cos \varepsilon)/(2E|X|)$. Since $E|X| \geq p_0 E|Y|$ and

$$\left|\frac{d}{dp} E e^{ipY}\right| = |EiYe^{ipY}| \leq E|Y|$$

for $|p| < \delta$,

$$|\Im\{Ee^{ipY}\}| \leq \delta E|X|/p_0$$

$\Re$ and $\Im$ denote real and imaginary part. So for $|p| < \delta$ and $\varepsilon \leq |q| \leq \pi$,

$$
\begin{aligned}
|1 - \chi_\mu(p, q)| &\geq \Re\{1 - \chi_\mu(p, q)\} \\
&\geq p_0 \Re\{1 - e^{iq} E e^{ipY}\} \\
&= p_0 \left[1 - \cos(q) \Re\{E e^{ipY}\} + \sin(q) \Im\{E e^{ipY}\}\right] \\
&\geq p_0 \left(1 - \cos \varepsilon - \frac{\delta E|X|}{p_0}\right) \\
&= \frac{p_0(1 - \cos \varepsilon)}{2}
\end{aligned}
$$

If $|p| \geq \min\{\delta, \varepsilon\}$,

$$
\begin{aligned}
|\chi_\mu(p, q)| &\leq 1 - p_0 + p_0 |E e^{ipY}| \\
&\leq 1 - p_0 + p_0 \sup_{|r| \geq \min\{\delta, \varepsilon\}} |E e^{irY}|
\end{aligned}
$$

which is less than 1 uniformly for $\mu_0 \leq \mu \leq \mu_1$ because Y is a continuous variable. The lemma follows because $1 \leq |\chi_\mu| + |1 - \chi_\mu|$ by the triangle inequality. ■

Proof of Lemma 2.5. Let $\hat{f} \in \mathscr{L}^2$ denote the (inverse) Fourier transform of $f \in \mathscr{L}^2$, given when $f \in \mathscr{L}^1$ by

$$
\hat{f}(x) = \frac{1}{2\pi} \int e^{-ixy} f(y)\, dy
$$

and by continuity otherwise. By Theorem 3, Chapter 3 of Stein (1970), the Fourier transform of $\mathscr{H}f$ when $f \in \mathscr{L}^2$ is $-i\, \mathrm{sgn}(x) \hat{f}(x)$.

Let $S_n + n\mu \sim F_\mu^{*n}$. Assume $\Lambda_\mu \in \mathscr{L}^1$. Then F_μ^{*n} has a density Λ_μ^n and

$$
\begin{aligned}
\int_0^\infty e^{ipx} F_\mu^{*k}(dx) &= \frac{1}{2} \int e^{ipx} [1 + \mathrm{sgn}(x)] F_\mu^{*k}(dx) \\
&= \tfrac{1}{2}\{\Lambda_\mu^k(p) + i\mathscr{H}\Lambda_\mu^k(p)\}
\end{aligned}
$$

By Spitzer's identity [Theorem 2.5 of Woodroofe (1982)] for $s \in [0,1)$,

$$-\log\{1-\beta(\cdot,s)\} = \frac{1}{2}\sum_{k=1}^{\infty}\frac{1}{k}s^k\{\Lambda_\mu^k + i\mathscr{H}\Lambda_\mu^k\}$$

Since $0 \le s < 1$ and $\Lambda_\mu \in \mathscr{L}^r$,

$$\sum_{k=1}^{n}\frac{1}{k}s^k\Lambda_\mu^k \to -\log(1 - s\Lambda_\mu^k)$$

as $n \to \infty$ in $\|\cdot\|_r$ norm, and since $\mathscr{H}$ is linear and bounded,

$$\sum_{k=1}^{n}\frac{1}{k}s^k i\mathscr{H}\Lambda_\mu^k = i\mathscr{H}\left\{\sum_{k=1}^{n}\frac{1}{k}s^k\Lambda_\mu^k\right\}$$

$$\to -i\mathscr{H}\log(1 - s\Lambda_\mu^k)$$

as $n \to \infty$ in $\|\cdot\|_r$ norm. This proves (2.1) when $0 \le s < 1$ and $\Lambda_\mu \in \mathscr{L}^1$. If $\Lambda_\mu \notin \mathscr{L}^1$, let $\Lambda_\mu^{(\alpha)}(p) = \Lambda_\mu(p)\,e^{-\alpha p^2}$. By weak convergence,

$$\log\{1-\beta_\mu(\cdot,s)\} = \tfrac{1}{2}\lim_{\alpha\downarrow 0}\left\{\log(1 - s\Lambda_\mu^{(\alpha)}) + i\mathscr{H}\log(1 - s\Lambda_\mu^{(\alpha)})\right\}$$

and (2.1) follows for $0 \le s < 1$ because $\log(1 - s\Lambda_\mu^{(\alpha)}) \to \log(1 - s\Lambda_\mu)$ in $\|\cdot\|_r$ norm as $\alpha \downarrow 0$, by dominated convergence. To verify (2.1) when $s = 1$, first note that the right-hand side approaches $\log\{1 - \beta_\mu(\cdot,1)\}$ pointwise as $s \uparrow 1$ by dominated convergence. Since $\mathscr{H}$ is bounded, it is sufficient to show that $\log(1 - s\Lambda_\mu) \to \log(1 - \Lambda_\mu)$ as $s \uparrow 1$ in $\|\cdot\|_r$ norm, that is,

$$\lim_{s\uparrow 1}\int|\log\{1 - s\Lambda_\mu(p)\} - \log\{1 - \Lambda_\mu(p)\}|^r\,dp = 0$$

This follows by dominated convergence. Away from the origin, the integrand can be dominated by a multiple of $|\Lambda_\mu|^r$ since $|\log(1-x)| \le |x|/(1-|x|)$ for $|x| < 1$. In a sufficiently small neighborhood of the origin, the integrand can be dominated by a multiple of $|\log p|^r$ since by Taylor expansion, $\log\{1 - \Lambda_\mu(p)\} \sim \log(-i\mu p)$ as $p \to 0$ and $\log\{1 - s\Lambda_\mu(p)\} \sim \log(1 - s - i\mu sp)$ as $p \to 0$ uniformly for s near 1.

To prove (2.2), first note that

$$\sup_{\mu_0 \le \mu \le \mu_1} \|\log(1 - \Lambda_\mu)\|_r < \infty$$

Let

$$\delta = \sup_{\mu_0 \le \mu \le \mu_1} \sup_{|p| > \varepsilon} |\beta_\mu(p, 1)|$$

From Lemma 2.4, $\delta < 1$. Define

$$K = \inf_{|x| \le \delta} \frac{|\log(1 - x)|}{|x|}$$

Then $K > 0$ and for $|p| > \varepsilon$,

$$|\log\{1 - \beta_\mu(p, 1)\}| \ge K|\beta_\mu(p, 1)|$$

for $\mu_0 \le \mu \le \mu_1$. Since $|\beta_\mu(p, 1)| \le 1$ for all p, we have

$$\int |\beta_\mu(p, 1)|^r \, dp \le 2\varepsilon + K^{-r} \int |\log\{1 - \beta_\mu(p, 1)\}|^r \, dp$$

for $\mu_0 \le \mu \le \mu_1$, proving the lemma. ■

REFERENCES

Anscombe, F. J. (1953). Sequential estimation. *J. R. Statist. Soc. B* **15**, 1–21.

Armitage, P. (1975). *Sequential Medical Trials*, New York: Halsted.

Bhattacharya, R. N. and Rao, R. R. (1976). *Normal Approximation and Asymptotic Expansions*, New York: Wiley.

Bickel, P. and Yahav, J. (1967). Asymptotically pointwise optimal procedures in sequential analysis. *Proc. Fifth Berkeley Sympos. Math. Statist. Prob. 1*, University of California, 401–413.

Bickel, P. and Yahav, J. (1968). Asymptotically optimal Bayes and minimax procedures in sequential estimation. *Ann. Math. Statist.* **39**, 442–456.

Chow, Y. S. and Robbins, H. (1965). Asymptotic theory of fixed width confidence intervals for the mean. *Ann. Math. Statist.* **36**, 457–462.

Feller, W. (1971). *An Introduction to Probability Theory and Its Applications*, Vol. 2, New York: Wiley.

Ghosh, M., Sinha, B. and Mukhopadhyay, N. (1976). Multivariate sequential estimation. *J. Multivar. Anal.* **6**, 281–294.

Gut, Allan (1974). On the moments and limit distributions of some first passage times. *Ann. Probab.* **2**, 277–308.

Hagwood, R. C. (1982). Asymptotic expansions in non-linear renewal theory. University of Virginia.

Keener, R. W. (1986). Asymptotic expansions for renewal measures in the plane. Univ. of Mich. Technical report.

Klein, H.-D. (1986). Tangent approximation and nonlinear renewal theory. Univ. of Frankfurt.

Lai, T. L. and Siegmund, D. (1977). A non-linear renewal theory with applications to sequential analysis, I. *Ann. Statist.* **5**, 946–954.

Lai, T. L. and Siegmund, D. (1979). A non-linear renewal theory with applications to sequential analysis, II. *Ann. Statist.* **7**, 60–76.

Lalley, S. P. (1984). Limit theorems for first-passage times in linear and non-linear renewal theory. *Adv. App. Prob.* **16**, 766–803.

Robbins, H. (1952). Some aspects of the sequential design of experiments. *Bull. Amer. Math. Soc.* **58**, 527–535.

Robbins, H. (1959). Sequential estimation of the mean of a normal population. *Probability and Statistics* (the Harald Cramer volume), Stockholm: Almqvist and Wiksell.

Robbins, H. (1970). Statistical methods related to the law of the iterated logarithm. *Ann. Math. Statist.*, **41**, 1397–1409.

Scheffé, H. (1947). A useful convergence theorem for probability distributions. *Ann. Math. Statist.* **18**, 434–438.

Schwarz, G. (1962). Asymptotic shapes of Bayes sequential testing regions. *Ann. Math. Statist.* **33**, 224–236.

Siegmund, D. (1977). Repeated significance tests for a normal mean. *Biometrika* **64**, 177–189.

Siegmund, D. (1978). Estimation following sequential testing. *Biometrika* **65**, 341–349.

Siegmund, D. (1980). Sequential χ^2 and F tests and related confidence intervals. *Biometrika* **67**, 389–402.

Siegmund, D. and Gregory, P. (1980). A sequential clinical trial for testing $p_1 = p_2$. *Ann. Statist.* **8**, 1219–1228.

Starr, N. (1966a). The performance of a sequential procedure for the fixed width interval estimation of the mean. *Ann. Math. Statist.* **37**, 36–50.

Starr, N. (1966b). On the asymptotic efficiency of a sequential procedure for estimating the mean. *Ann. Math. Statist.* **37**, 1173–1185.

Starr, N. and Woodroofe, M. (1972). Further remarks on sequential point estimation: the exponential case, *Ann. Math. Statist.* **43**, 1147–1154.

Stein, E. (1970). *Singular Integrals and Differentiability Properties of Functions*, Princeton, New Jersey: Princeton University Press.

Stone, C. J. (1965). Characteristic functions and renewal theory. *Trans. Amer. Math. Soc.* **120**, 326–342.

Takahashi, H. (1986). Asymptotic expansions in Anscombe's theorem in a repeated significance tests and estimation after sequential test. *Ann. Stat.* (to appear).

Takahashi, H. and Woodroofe, M. (1981). Asymptotic expansions in non-linear renewal theory. *Comm. Statist.* *A*13, 2113–2136.

Woodroofe, M. (1976a). A renewal theorem for curved boundaries and moments of first passage times. *Ann. Prob.* **4**, 67–80.

Woodroofe, M. (1976b). Frequentist properties of Bayesian sequential tests. *Biometrika* **63**, 101–110.

Woodroofe, M. (1977). Second order approximations for sequential point and interval estimation. *Ann. Stat.* **5**, 984–995.

Woodroofe, M. (1978). Large deviations of the likelihood ratio statistic with applications to sequential testing. *Ann. Statist.* **6**, 72–84.

Woodroofe, M. (1979). Repeated likelihood ratio tests. *Biometrika* **66**, 453–463.

Woodroofe, M. (1982). *Non Linear Renewal Theory in Sequential Analysis*, Pennsylvania: S.I.A.M.

Woodroofe, M. (1986). Very weak expansions for sequential confidence levels. *Ann. Statist.* (to appear).

Woodroofe, M. and Keener, R. (1986). Asymptotic expansions in boundary crossing problems. *Ann. Prob.* (to appear).

Woodroofe, M. and Takahashi, H. (1982). Asymptotic expansions for the error probabilities of some repeated significance tests. *Ann. Statist.* **10**, 895–908.

Zhang, C. H. (1984). Random walk and renewal theory. Ph.D. Thesis, Columbia University, New York.

The Local Time of a Random Walk in a Random Environment

Pal Révész*
Mathematical Institute
Hungarian Academy of Sciences
Budapest, Hungary

SUMMARY

We obtain strong limiting bounds for the local time of a random walk in a random environment.

1. INTRODUCTION

Let $\ldots, X_{-2}, X_{-1}, X_0, X_1, X_2, \ldots$ be a sequence of independent and identically distributed (i.i.d.) random variables (r.v.'s) with

$$\mathbb{P}(X_0 < x) = \begin{cases} 0 & \text{if } x \leq 0 \\ F(x) & \text{if } 0 \leq x \leq 1 \\ 1 & \text{if } x \geq 1 \end{cases}$$

The sequence $\chi = \{X_i\}_{i=-\infty}^{+\infty}$ is called a *random environment*. For any fixed sample sequence of this environment, define a *random walk* $S_0, S_1, \ldots$ by

* Present address: Institut für Statistik und Wahrscheinlichkeits theorie, Technische Universität, Vienna, Austria.

$S_0 = 0$ and

$$\begin{aligned} \mathbb{P}_{\mathbf{X}}\{S_{n+1} = i+1 \mid S_n = i\} &= X_i \\ \mathbb{P}_{\mathbf{X}}\{S_{n+1} = i-1 \mid S_n = i\} &= 1 - X_i \end{aligned} \qquad (n = 0, 1, 2, \ldots;\ i = 0, \pm 1, \pm 2, \ldots)$$

Introduce the following notations:

$$M(a, b) = \max_{a \le k \le b} |S_k|, \qquad M^+(a, b) = \max_{a \le k \le b} S_k,$$

$$M^-(a, b) = -\min_{a \le k \le b} S_k$$

$$M(n) = M(0, n), \qquad M^+(n) = M^+(0, n), \qquad M^-(n) = M^-(0, n)$$

$$\xi(x, n) = \#\{k : 0 \le k \le n, S_k = x\}$$

$$\xi(n) = \xi(0, n)$$

$$\rho_0 = 0, \qquad \rho_1 = \min\{k : k > 0, S_k = 0\}$$

$$\rho_2 = \min\{k : k > \rho_1, S_k = 0\}, \ldots$$

$$\rho_{j+1} = \min\{k : k > \rho_j, S_k = 0\}$$

The process $\xi(\cdot, \cdot)$ will be called the local time of the random walk $\{S_k\}$.

In Deheuvels and Révész (1986) the following results were proved.

Theorem A. Assume that

1. $\mathbb{E} \log \dfrac{X_0}{1 - X_0} = 0,$
2. $0 < \mathbb{E} \log^2 \dfrac{X_0}{1 - X_0} = \sigma^2 < \infty.$

Then for any $\varepsilon > 0$ and $p \ge 1$, we have

$$M(\rho_n) \le \left(\log n \log_2 n \cdots \log_{p-1} n (\log_p n)^{1+\varepsilon}\right)^2 \quad \text{a.s.} \tag{1.1}$$

except finitely many n,

$$M(\rho_n) \geq (\log n \log_2 n \cdots \log_{p-1} n \log_p n)^2 \quad \text{i.o.} \quad \text{a.s.} \tag{1.2}$$

$$M(\rho_n) \leq \frac{1+\varepsilon}{2\sigma^2} \frac{(\log n)^2}{\log_3 n} \quad \text{i.o.} \quad \text{a.s.} \tag{1.3}$$

$$M(\rho_n) \geq \frac{1-\varepsilon}{2\sigma^2} \frac{(\log n)^2}{\log_3 n} \quad \text{a.s.} \tag{1.4}$$

except finitely many n, where $\log_j n$ stands for the jth iterated logarithm of n.

REMARK 1.1. The meaning of "a.s." in these and the following statement is the following: for almost all environment χ, the stated inequality holds with probability 1.

A simple consequence of Theorem A is

Theorem B. Under the conditions of Theorem A for any $\varepsilon > 0$, we have

$$M(n) \leq (\log \xi(n) \log_2 \xi(n) \cdots \log_{p-1} \xi(n) \log_p \xi(n))^2 \quad \text{a.s.} \tag{1.5}$$

except finitely many n,

$$M(n) \geq (\log \xi(n) \log_2 \xi(n) \cdots \log_{p-1} \xi(n) \log_p \xi(n))^2 \quad \text{i.o.} \quad \text{a.s.} \tag{1.6}$$

$$M(n) \leq \frac{1+\varepsilon}{2\sigma^2} \frac{(\log \xi(n))^2}{\log_3 \xi(n)} \quad \text{i.o.} \quad \text{a.s.} \tag{1.7}$$

$$M(n) \geq \frac{1-\varepsilon}{2\sigma^2} \frac{(\log \xi(n))^2}{\log_3 \xi(n)} \quad \text{a.s.} \tag{1.8}$$

except finitely many n.

Theorem C. Under the conditions of Theorem A for any $\varepsilon > 0$, we have

$$\xi(n) \geq \exp\left[\frac{\log n}{(\log\log n)^{1+\varepsilon}}\right] \quad \text{a.s.} \tag{1.9}$$

except finitely many n and

$$\lim_{n\to\infty} n^{-1}\xi(n) = 0 \quad \text{a.s.} \tag{1.10}$$

Theorems B and C together imply

Theorem D. Under the conditions of Theorem A for any $\varepsilon > 0$, we have

$$\frac{(\log n)^2}{(\log_2 n)^{2+\varepsilon}} \leq M(n) \leq \left(\log n \log_2 n \cdots \log_{p-1} n (\log_p n)^{1+\varepsilon}\right)^2 \quad \text{a.s.} \tag{1.11}$$

except finitely many n and

$$M(n) \leq \frac{(1+\varepsilon)}{2\sigma^2}\frac{(\log n)^2}{\log_3 n} \quad \text{i.o.} \quad \text{a.s.} \tag{1.12}$$

Remark 1.2. (1.12) was not formulated in Deheuvels and Révész (1985); however, it follows from (1.3) and the trivial inequality $M(n) \leq M(\rho_n)$.

In this Chapter we intend to study the further properties of $\xi(n)$. We prove that our results (1.9) and (1.10) are not far from the best possible ones. In fact we have

Theorem 1.1. Assume that

1. $\mathbb{E}\log[X_0/(1-X_0)] = 0$.
2. $0 < \sigma^2 = \mathbb{E}\log[X_0/(1-X_0)]^2 < \infty$.
3. For some $0 < a < \frac{1}{2}$, $\mathbb{P}\{a < X_0 < 1-a\} = 1$.

Then

$$\xi(n) \leq \exp\left(\frac{\log n}{(\log_2 n)^{1-\varepsilon}}\right) \quad \text{i.o.} \quad \text{a.s.} \tag{1.13}$$

and there exists a $C = C(a) > 0$ such that

$$\xi(n) \geq \exp\left(\left(1 - \frac{C}{\log_3 n}\right)\log n\right) \quad \text{i.o.} \quad \text{a.s.} \tag{1.14}$$

The proof of Theorem 1.1 is based on a slightly stronger form of Theorem A that will be presented in Section 2. Section 3 is devoted to the proof of Theorem 1.1.

2. SOME FURTHER PROPERTIES OF $M(\rho_n)$

Introduce the following notations:

$$U_j = X_j^{-1}(1 - X_j), \qquad U_{-j} = (1 - X_{-j})^{-1}X_{-j} \qquad (j = 1, 2, \ldots)$$

$$U_j^{(k)} = (1 - X_{k-j})^{-1}X_{k-j}, \qquad U_j^{(-k)} = X_{-(k-j)}^{-1}(1 - X_{-(k-j)})$$

$$(k = 2, 3, \ldots; j = 1, 2, \ldots, k - 1)$$

$$V_0 = V_0^{(k)} = 1 \qquad (k = \pm 2, \pm 3, \ldots)$$

$$V_l = U_1U_2 \cdots U_l, \qquad V_{-l} = U_{-1}U_{-2} \cdots U_{-l} \qquad (l = 1, 2, \ldots)$$

$$V_l^{(k)} = U_1^{(k)}U_2^{(k)} \cdots U_l^{(k)}, \qquad V_{-l}^{(-k)} = U_{-1}^{(-k)}U_{-2}^{(-k)} \cdots U_{-l}^{(-k)}$$

$$(k = 2, 3, \ldots; l = 1, 2, \ldots, k - 1)$$

$$W(k) = V_0 + V_1 + \cdots + V_{k-1}, \qquad W(-k) = V_0 + V_{-1} + \cdots + V_{-(k-1)}$$

$$(k = 2, 3, \ldots)$$

$$W^*(k) = V_0^{(k)} + V_1^{(k)} + \cdots + V_{(k-1)}^{(k)},$$

$$W^*(-k) = V_0^{(-k)} + V_{-1}^{(-k)} + \cdots + V_{-(k-1)}^{(-k)} \qquad (k = 2, 3, \ldots)$$

Let $I(t)$ be the inverse of $W(k)$, that is,

$$I(t) = k \quad \text{if } W(k) \le t < W(k + 1) \qquad (k = 2, 3, \ldots; t \ge W(2))$$

and

$$I(-t) = k \quad \text{if } W(-k) \le t < W(-k - 1)$$

$$(k = 2, 3, \ldots; t \ge W(-2))$$

In Deheuvels and Révész (1986) the following lemmas were proved

Lemma A. For any $m \geq 1$ and $k \geq 2$, we have

$$\mathbb{P}_{\mathcal{X}}\{\min\{l: l \geq m, S_l = 0\} < \min\{l: l \geq m, S_l = k\} | S_m = 1\} = 1 - (W(k))^{-1} \tag{2.1}$$

and similarly

$$\mathbb{P}\{\min\{l: l \geq m, S_l = k\} < \min\{l: l \geq m, S_l = 0\} | S_m = k - 1\} = 1 - (W^*(k))^{-1} \tag{2.2}$$

Lemma B. Under the conditions of Theorem A for any $\varepsilon > 0$, we have

$$W(n) \leq \exp\left((1+\varepsilon)\sigma(2n \log_2 n)^{1/2}\right) \quad \text{a.s.} \tag{2.3}$$

except finitely many n,

$$W(n) \geq \exp\left((1-\varepsilon)\sigma(2n \log_2 n)^{1/2}\right) \quad \text{i.o.} \quad \text{a.s.} \tag{2.4}$$

$$W(n) \leq \exp\left(n^{1/2}(\log n \log_2 n \cdots \log_{p-1} n \log_p n)^{-1}\right) \quad \text{i.o.} \quad \text{a.s.} \tag{2.5}$$

$$W(n) \geq \exp\left(n^{1/2}\left(\log n \log_2 n \cdots \log_{p-1} n (\log_p n)^{1+\varepsilon}\right)^{-1}\right) \quad \text{a.s.} \tag{2.6}$$

except finitely many n.

Lemma C. For any $k = 1, 2, \ldots,$ we have

$$\begin{aligned} \mathbb{P}_{\mathcal{X}}\{\xi(k, \rho_1) = 0\} &= 1 - X_0(W(k))^{-1} \\ \mathbb{P}_{\mathcal{X}}\{\xi(k, \rho_1) = l\} &= \frac{X_0(1 - X_k)}{W(k)W^*(k)}\left(1 - \frac{1 - X_k}{W^*(k)}\right)^{l-1} \end{aligned} \tag{2.7}$$

and consequently

$$\mathbb{P}_{\mathcal{X}}\{\xi(k, \rho_1) \geq L\} = \frac{X_0}{W(k)}\left(1 - \frac{1 - X_k}{W^*(k)}\right)^{L-1} \qquad (L = 1, 2, \ldots)$$

Lemma D. Under the conditions of Theorem A we have

$$\limsup_{n\to\infty} \frac{\log W^*(n)}{(2n\log_2 n)^{1/2}} = \sigma \quad \text{a.s.} \tag{2.8}$$

$$W^*(n) \le n \quad \text{i.o.} \quad \text{a.s.} \tag{2.9}$$

$$\liminf_{n\to\infty} \max_{2\le k\le n} \frac{\log W^*(k)}{(2n\log_2 n)^{1/2}} = 0 \quad \text{a.s.} \tag{2.10}$$

If in addition we assume that for some $a > 0$,

$$\mathbb{E}\left|\log\frac{X_0}{1-X_0}\right|^{2+a} < \infty$$

then

$$\max_{0\le k\le n} \frac{\log W^*(k)}{n^{1/2}} \le \left(\frac{\pi^2}{8}\frac{1+\varepsilon}{\log_2 n}\right)^{1/2} \quad \text{i.o.} \quad \text{a.s.} \tag{2.11}$$

and

$$\max_{0\le k\le n} \frac{\log W^*(k)}{n^{1/2}} \ge \left(\frac{\pi^2}{8}\frac{1-\varepsilon}{\log_2 n}\right)^{1/2} \quad \text{a.s.} \tag{2.12}$$

except finitely many n.

Theorem 2.1. Under the conditions of Theorem A, for any $K > 0$, there exists a $C = C(K) > 0$ such that

$$\mathbb{P}\left\{\left|\mathbb{P}_x\{M^+(\rho_n) < I(yn)\} - \exp\left(-\frac{X_0}{y}\right)\right| > \frac{C}{y}\frac{1}{\log yn}\right\} \le Cn^{-K}$$

$$(y > 0)$$

and similarly

$$\mathbb{P}\left\{\left|\mathbb{P}_x\{M^-(\rho_n) < I(-yn)\} - \exp\left(-\frac{X_0}{y}\right)\right| > \frac{C}{y}\frac{1}{\log yn}\right\} \le Cn^{-K}$$

$$(y > 0)$$

Consequently for any $\varepsilon > 0$, we have

$$I\left(n(\log n)^{-(1+\varepsilon)}\right) \le M^+(\rho_n) \le I\left(n(\log n)^{1+\varepsilon}\right) \quad \text{a.s.} \tag{2.13}$$

$$I\left(-n(\log n)^{-(1+\varepsilon)}\right) \le M^-(\rho_n) \le I\left(-n(\log n)^{1+\varepsilon}\right) \quad \text{a.s.} \tag{2.14}$$

$$\begin{aligned} &\max\left\{ I\left(n(\log n)^{-(1+\varepsilon)}\right), I\left(-n(\log n)^{-(1+\varepsilon)}\right)\right\} \\ &\qquad \le M(\rho_n) \\ &\qquad \le \max\left\{ I\left(n(\log n)^{1+\varepsilon}\right), I\left(-n(\log n)^{1+\varepsilon}\right)\right\} \quad \text{a.s.} \end{aligned} \tag{2.15}$$

except finitely many n.

Lemma 2.1. Let

$$\nu_n = \#\left\{ i : 0 \le i \le n-1, S_{\rho_i+1} = 1\right\}$$

that is, ν_n is the number of positive excursions among the first n excursions. Then for any $K > 0$, there exists a $C = C(K) > 0$ such that

$$\mathbb{P}_{\mathcal{X}}\left\{ |\nu_n - nX_0| \ge C\left(nX_0(1-X_0)\log n\right)^{1/2}\right\} \le n^{-K} \qquad (n = 1, 2, \dots)$$

Consequently

$$|\nu_n - nX_0| \le C\left(nX_0(1-X_0)\log n\right)^{1/2} \tag{2.16}$$

with probability 1 for all but finitely many n.

Proof of Theorem 2.1. By Lemma A we have

$$\mathbb{P}_{\mathcal{X}}\{ M^+(\rho_n) \le k | \nu_n\} = \left(1 - (W(k))^{-1}\right)^{\nu_n} \qquad (k = 2, 3, \dots; n = 1, 2, \dots)$$

Since by the definition of $I(t)$ and Lemma B

$$yn\left(1 - \frac{1}{2\log yn}\right) \le W(I(yn)) \le yn \le W(I(yn) + 1)$$

we get

$$\mathbb{P}_X\{M^+(\rho_n) \le I(yn)|\nu_n\}$$

$$= \left(1 - (W(I(yn)))^{-1}\right)^{\nu_n} \le \left(1 - \frac{1}{yn}\right)^{\nu_n}$$

$$\le \exp\left(-\frac{1}{yn}\cdot\left(nX_0 - C(nX_0(1 - X_0)\log n)^{1/2}\right)\right)$$

$$\le \exp\left(-\frac{X_0}{y}\right) + \frac{C}{y}\left(\frac{\log n}{n}\right)^{1/2}$$

provided that (2.16) holds and n is big enough.

Similarly

$$\mathbb{P}_X\{M^+(\rho_n) \le I(yn)|\nu_n\} = \left(1 - (W(I(yn)))^{-1}\right)^{\nu_n}$$

$$\ge \left(1 - \frac{1}{yn}\left(1 + \frac{1}{2\log yn}\right)\right)^{\nu_n}$$

$$\ge \exp\left(-\frac{1}{yn}\left(1 + \frac{1}{2\log yn}\right)\right.$$

$$\left.\times\left(nX_0 + C(n\log n)^{1/2}\right)\right)$$

$$= \exp\left(-\frac{X_0}{y} - \frac{X_0}{y}\frac{1}{2\log yn}\right.$$

$$\left.-\frac{C}{y}\left(\frac{\log}{n}\right)^{1/2}\left(1 + \frac{1}{2\log yn}\right)\right)$$

$$\ge \exp\left(-\frac{X_0}{y}\right) - \frac{X_0}{y}\frac{1}{2\log yn}$$

which together with Lemma 2.1 proves Theorem 2.1.

3. THE PROOF OF THEOREM 1.1

At first we prove (1.13). This proof is based on the following simple

Lemma 3.1. For almost all environment χ and for all $\varepsilon > 0$, there exist a sequence $n_1 = n_1(\chi, \varepsilon) < n_2 = n_2(\chi, \varepsilon) < \cdots$ and a $C = C(\chi, \varepsilon) > 0$ such that

$$\log U_1 + \log U_2 + \cdots + \log U_{n_k} \leq -C(2n_k \log_2 n_k)^{1/2}$$

and

$$\sup_{j \leq n_k} \left(\log U_1 + \log U_2 + \cdots + \log U_j\right) \leq n_k^{1/2}(\log n_k)^{-(1-\varepsilon)}$$

The proof of this lemma is straightforward and omitted. A trivial consequence of this lemma is

Lemma 3.2. For almost all environments χ and $\varepsilon > 0$, there exist a sequence $n_1 = n_1(\chi, \varepsilon) < n_2 = n_2(\chi, \varepsilon) < \cdots$ and a $C = C(\chi, \varepsilon) > 0$ such that

$$W(n_k) \leq \exp\left(\frac{n_k^{1/2}}{(\log n_k)^{1-\varepsilon}}\right), \qquad W^*(n_k) \geq \exp\left(C(n_k \log_2 n_k)^{1/2}\right)$$

Our next lemma says that either $\xi(k, \rho_1)$ is 0 or it is close to $W^*(k)$ if k is big enough. In fact we have

Lemma 3.3. There exists a positive constant C such that

$$\mathbb{P}_\chi\{\xi(k, \rho_1) \geq CW^*(k) \log k | \xi(k, \rho_1) > 0\} \leq Ck^{-2} \tag{3.1}$$

and

$$\mathbb{P}_\chi\left\{\xi(k, \rho_1) \leq k^{-2}W^*(k) | \xi(k, \rho_1) > 0\right\} \leq Ck^{-2} \tag{3.2}$$

Proof. Let μ_k be the number of negative excursions away from k between 0 and ρ_1. By (2.2)

$$\mathbb{P}_\chi\{\mu_k = l | \xi(k, \rho_1) > 0\} = (W^*(k))^{-1}\left(1 - (W^*(k))^{-1}\right)^l \tag{3.3}$$

Consequently

$$\mathbb{P}_\chi\{\mu_k > L | \xi(k, \rho_1) > 0\} = \left(1 - (W^*(k))^{-1}\right)^L \tag{3.4}$$

Hence

$$\begin{aligned}\mathbb{P}_{\mathrm{x}}\{\xi(k,\rho_1) \le k^{-2}W^*(k) | \xi(k,\rho_1) > 0\} \\ \le \mathbb{P}_{\mathrm{x}}\{\mu_k \le k^{-2}W^*(k) | \xi(k,\rho_1) > 0\} \\ \le 1 - \left(1 - (W^*(k))^{-1}\right)^{k^{-2}W^*(k)} \\ \le Ck^{-2}\end{aligned}$$

and we have (3.2). In order to prove (3.1) observe that for any $0 < \varepsilon < a$, there exists a $C_2 = C_2(\varepsilon) > 0$ such that

$$\mathbb{P}_{\mathrm{x}}\{\mu_k < (1 - X_k - \varepsilon)\xi(k,\rho_1) | \xi(k,\rho_1)\} \le \exp(-C_2\xi(k,\rho_1))$$

Hence by (3.2) we have

$$\begin{aligned}\mathbb{P}_{\mathrm{x}}\{\mu_k < (1 - X_k - \varepsilon)\xi(k,\rho_1) | \xi(k,\rho_1) > 0\} \\ \le \exp\left(-C_2k^{-2}W^*(k)\right) + Ck^{-2}\end{aligned} \tag{3.5}$$

and

$$\begin{aligned}&\mathbb{P}_{\mathrm{x}}\{\xi(k,\rho_1) \ge C_1W^*(k)\log k | \xi(k,\rho_1) > 0\} \\ &\quad\le \mathbb{P}_{\mathrm{x}}\{\mu_k \ge C_1(1 - X_k - \varepsilon)W^*(k)\log k | \xi(k,\rho_1) > 0\} \\ &\qquad+ \exp\left(-C_2k^{-2}W^*(k)\right) + Ck^{-2} \\ &\quad= \left(1 - (W^*(k))^{-1}\right)^A + \exp\left(-C_2k^{-2}W^*(k)\right) + Ck^{-2} \le C_3k^{-2}\end{aligned}$$

if $C_1(1 - X_k - \varepsilon) > 2$ where $A = C_1(1 - X_k - \varepsilon)W^*(k)\log k$. Hence we have Lemma 3.3. ■

Now we turn to the proof of (1.13). Define the random sequence $\{N_k\}$ by

$$I\left(N_k(\log N_k)^{-(1+\varepsilon)}\right) + 1 = n_k$$

where $\{n_k\}$ is the random sequence of Lemma 3.2. Then by Theorem 2.1

$$\begin{aligned}M^+(\rho_{N_k}) &\ge I\left(N_k(\log N_k)^{-(1+\varepsilon/2)}\right) \\ &\ge I\left(N_k(\log N_k)^{-(1+\varepsilon)}\right) + 1 = n_k\end{aligned} \tag{3.6}$$

that is, $\xi(n_k, \rho_{N_k}) > 0$, or in other words, there exists a $0 < j < N_k$ such that $\xi(n_k, (\rho_j, \rho_{j+1})) = \xi(n_k, \rho_{j+1}) - \xi(n_k, \rho_j) > 0$. By (3.2) of Lemma 3.3 (and the Borel–Cantelli lemma)

$$\xi\big(n_k, (\rho_j, \rho_{j+1})\big) \geq n_k^{-2} W^*(n_k) \geq n_k^{-2} \exp\Big(C\big(n_k(\log_2 n_k)\big)^{1/2}\Big)$$

Consequently

$$\rho_{N_k} \geq \rho_{j+1} - \rho_j \geq \xi\big(n_k, (\rho_j, \rho_{j+1})\big) \geq n_k^{-2} \exp\Big(C(n_k \log_2 n_k)^{1/2}\Big) \tag{3.7}$$

and

$$\frac{N_k}{(\log N_k)^{1+\varepsilon}} \leq W\Big(\Big(I\big(N_k(\log N_k)^{-(1+\varepsilon)}\big) + 1\Big)$$

$$= W(n_k) \leq \exp\left(\frac{n_k^{1/2}}{(\log n_k)^{1-\varepsilon}}\right)$$

that is,

$$n_k \geq (\log N_k)^2 (\log_2 N_k)^{2-\varepsilon} \tag{3.8}$$

(3.7) and (3.8) combined imply

$$\rho_{N_k} \geq \exp\Big((\log N_k)(\log_2 N_k)^{1-\varepsilon}\Big)$$

which in turn implies (1.13).

In order to prove (1.14) we give a slightly stronger form of Lemmas B and D [cf. (2.4) and (2.11)].

Lemma 3.4. For almost all environments χ, there exists a sequence $n_1 = n_1(\chi) < n_2 = n_2(\chi) < \cdots$ and there are universal constants $C_1 > 0$ and $C_2 > 0$ such that

$$\max_{0 \leq j \leq n_k} W^*(j) \leq \exp\left(C_1 \frac{n^{1/2}}{(\log_2 n_k)^{1/2}}\right)$$

and

$$W(n_k) \geq \exp\Big(C_2(n_k \log_2 n_k)^{1/2}\Big)$$

The proof is simple and omitted.

Now we present an analog of Lemma 3.3.

Lemma 3.5. There exist universal constants $C > 0$ and $D > 0$ (depending only on a of Theorem 1.1) such that

$$\mathbb{P}_X\{\xi(k,\rho_N) \geq CNW^*(k)\} \leq D2^{-N} \qquad (k = 1, 2, \dots)$$

Proof. Clearly for any $\lambda > 0$, we have

$$\begin{aligned}\mathbb{P}_X\{\xi(k,\rho_N) \geq CNW^*(k)\} &= \mathbb{P}_X\{\exp(\lambda\xi(k,\rho_N)) \geq \exp(\lambda CNW^*(k))\}\\ &\leq \exp(-\lambda CNW^*(k))\mathbb{E}_X\{\exp(\lambda\xi(k,\rho_N))\}\\ &= \left[\exp(-\lambda CW^*(k))\mathbb{E}_X\{\exp(\lambda\xi(k,\rho_1))\}\right]^N\end{aligned}$$

By (2.7)

$$\begin{aligned}\mathbb{E}_X(\exp(\lambda\xi(k,\rho_1))) &= 1 - X_0(W(k))^{-1}\\ &\quad + \sum_{l=1}^{\infty} e^{\lambda l}\frac{X_0(1 - X_k)}{W(k)W^*(k)}\left(1 - \frac{1 - X_k}{W^*(k)}\right)^{l-1}\\ &= 1 - X_0(W(k))^{-1}\\ &\quad + \frac{e^{\lambda}X_0(1 - X_k)}{W(k)W^*(k)}\sum_{l=0}^{\infty}\left(e^{\lambda}\left(1 - \frac{1 - X_k}{W^*(k)}\right)\right)^l\\ &= 1 - X_0(W(k))^{-1}\\ &\quad + \frac{e^{\lambda}X_0(1 - X_k)}{W(k)W^*(k)}\frac{1}{1 - e^{\lambda}[-(1 - X_k)/W^*(k)]}\end{aligned}$$

Choose $\lambda = (1 - X_k)/2W^*(k)$. Then

$$\mathbb{E}_X(\exp \lambda\xi(k,\rho_1)) = \frac{2\lambda e^{\lambda}X_0}{W(k)}\frac{1}{1 - e^{\lambda}(1 - 2\lambda)} + 1 - X_0(W(k))^{-1}$$

Observe that

$$\frac{2\lambda e^{\lambda}}{1 - e^{\lambda}(1 - 2\lambda)} \leq 4 \qquad (\lambda > 0)$$

Hence we get

$$\mathbb{P}_{\mathcal{X}}\{\xi(k,\rho_N) \geq CNW^*(k)\}$$

$$\leq \left[\exp\left(-\frac{C(1-X_k)}{2}\right)\left(1+3X_0(W(k))^{-1}\right)\right]^N$$

which, choosing C big enough, proves Lemma 3.5. ■

Define the random sequence $\{N_k\}$ by

$$I\left(N_k(\log N_k)^{1+\varepsilon}\right) = n_k \tag{3.9}$$

where $\{n_k\}$ is the random sequence of Lemma 3.4. Observe that

$$\exp\left(C_2(n_k \log_2 n_k)^{1/2}\right) \leq W(n_k) = W\left(I\left(N_k(\log N_k)^{1+\varepsilon}\right)\right) \leq N_k(\log N_k)^{1+\varepsilon}$$

Consequently

$$n_k \leq C\frac{(\log N_k)^2}{\log_3 N_k} \tag{3.10}$$

By Theorem 2.1 and (3.9) we have

$$M^+(\rho_{N_k}) \leq I\left(N_k(\log N_k)^{1+\varepsilon}\right) = n_k$$

and consequently

$$\xi(j,\rho_{N_k}) = 0 \quad \text{if } j \geq n_k \tag{3.11}$$

By Lemma 3.5

$$\mathbb{P}_{\mathcal{X}}\left\{\max_{1\leq j\leq n_k} \xi(j,\rho_{N_k}) \geq CN_k n_k \max_{1\leq j\leq n_k} W^*(j)\right\}$$

$$\leq n_k \max_{1\leq j\leq n_k} \mathbb{P}_{\mathcal{X}}\left\{\xi(j,\rho_{N_k}) \geq CN_k W^*(j)\right\} \leq Dn_k 2^{-N_k}$$

$$\leq C\frac{(\log N_k)^2}{\log_3 N_k}2^{-N_k}$$

Hence by the Borel–Cantelli lemma and (3.11) we get

$$\sum_{j=1}^{\infty} \xi(j, \rho_{N_k}) = \sum_{j=1}^{n_k} \xi(j, \rho_{N_k}) \le C N_k n_k^2 \max_{1 \le j \le n_k} W^*(j)$$

$$\le C N_k n_k^2 \exp\left(C \frac{n_k^{1/2}}{(\log_2 n_k)^{1/2}} \right)$$

$$\le C \frac{(\log N_k)^4}{(\log_3 N_k)^2} N_k \exp\left(C \frac{\log N_k}{\log_3 N_k} \right)$$

$$\le \exp\left(\left(1 + \frac{C}{\log_3 N_k} \right) \log N_k \right)$$

Since similar inequality can be obtained for the sum $\Sigma_{j=1}^{\infty} \xi(-j, \rho_{N_k})$ and $\rho_{N_k} = \Sigma_{j=-\infty}^{+\infty} \xi(j, \rho_{N_k})$ we have

$$\rho_{N_k} \le \exp\left(\left(1 + \frac{C}{\log_3 N_k} \right) \log N_k \right)$$

which implies (1.14) and we have Theorem 1.1.

Looking through the above proof carefully one can realize that a somewhat stronger theorem has been proved than stated. In fact we have proved

Theorem 3.1. Assume the conditions of Theorem 1.1. Then for almost all environments χ and for all $\varepsilon > 0$ and C big enough, there exist two sequences

$$n_1 = n_1(\chi, \varepsilon) < n_2 = n_2(\chi, \varepsilon) < \cdots$$

and

$$m_1 = m_1(\chi, C) < m_2 = m_2(\chi, C) < \cdots$$

such that

$$\xi(n_k) \le \exp\left(\frac{\log n_k}{(\log \log n_k)^{1-\varepsilon}} \right) \quad \text{a.s.}$$

and

$$\xi(m_k) \le \exp\left(\left(1 - \frac{C}{\log\log\log m_k}\right)\log m_k\right) \quad \text{a.s.}$$

for all but finitely many k.

REFERENCE

Deheuvels, P. and Révész, P. (1986). Simple random walk on the line in random environment. *Probab. Th. Rel. Fields* **72**, 215–230.

Stochastic Convergence of Weighted Sums

Vijay K. Rohatgi
Department of Mathematics and Statistics
Bowling Green State University
Bowling Green, Ohio

SUMMARY

Let X_k, $k = 1, 2, \ldots$, be a sequence of independent random variables with finite first absolute moments, $a_{n,k}$ for $n, k = 1, 2, \ldots$ be real numbers, and set $S_n = \sum_{k=1}^{\infty} a_{n,k}(X_k - \mathscr{E}X_k)$. A great deal of effort has gone into the investigation of the convergence of the sequences $\{S_n\}$ and $P\{|S_n| > \varepsilon\}$. Over the past fifteen years this work has been extended to the case in which $\{X_k\}$ take values in a normed linear space. Other extensions include random weight sequences $a_{n,k}$ and random number of terms in the sum S_n. In this chapter we survey the current state of the field and indicate direction for further work on convergence of $\{S_n\}$ when the weight sequence consists of linear operators.

1. INTRODUCTION

Let $\{X_k\}$ be a sequence of random variables on some probability space $(\Omega, \mathscr{S}, P)$ and let $\{a_{nk}\}$ be an array of real numbers. Define

$$A_n = n^{-1} \sum_{k=1}^{n} X_k$$

and

$$S_n = \sum_{k=1}^{\infty} a_{nk} X_k \quad \text{or} \quad \sum_{k=1}^{\infty} a_{nk}(X_k - \mathscr{E}X_k)$$

depending on whether $\mathscr{E}X_k$ is infinite or finite, respectively. When dealing with S_n we shall have made enough assumptions to ensure that the sum makes sense.

The stochastic convergence of $\{A_n\}$ has been the subject of a great deal of investigation. In this chapter we shall concern ourselves with the convergence of the weighted sum sequence $\{S_n\}$. The convergence of $\{S_n\}$ is of interest in some practical problems in statistics. For example, when the X_k are independent and identically distributed, Koopmans (1961) considered the linear process

$$Y_n = \sum_{k=-\infty}^{\infty} a_k X_{n-k}, \qquad n = 1, 2, \ldots$$

The successive averages of the Ys are of the form S_n and arise, for example, in estimating the mean of the $\{Y_n\}$ process. This process has also been used by Lai (1972) for detecting changes in the location of the distribution of a sequence of independent observations, such as in quality-control problems. Other applications include obtaining the consistency of the kernel density estimates (Taylor, 1982). We refer to Taylor (1978, Chapter 8) for further possible applications.

Over the past fifteen years the work on stochastic convergence of $\{S_n\}$ for the real-valued case has been extended to random elements $\{X_k\}$ in linear spaces.

In this chapter we shall briefly survey the current state of the field and indicate some possible extensions. Section 2 deals with the real-valued case and Section 3 with the case when the X_k take values in a separable Banach space E. In Section 4 we allow the weights to be bounded linear operators on E.

The bibliography at the end is not meant to be complete. Indeed, for work up to 1976–1977 we refer to Taylor (1978) for an exhaustive bibliography. We include some references since then in the Bibliography.

2. STOCHASTIC CONVERGENCE OF WEIGHTED SUMS OF REAL-VALUED RANDOM VARIABLES

Let $\{x_k\}$ be a sequence of real numbers and $\{a_{nk}\}$ an infinite matrix of real numbers. In the classical summability theory an important question of interest is to relate the convergence properties of the sequence $\{s_n\}$, $s_n = \sum_{k=1}^{\infty} a_{nk} x_k$, to those of the sequence $\{x_k\}$. More precisely, one seeks to characterize matrices $\{a_{nk}\}$ for which $x_k \to x$ implies $s_n \to x$. Such summability matrices are known as *regular* matrices, and it is known that a matrix $\{a_{nk}\}$ is regular if and only if it is *Toeplitz*; a matrix a_{nk} is said to be a Toeplitz matrix if

1. $a_{nk} \to 0$ as $n \to \infty$ for every k.
2. $\lim_{n\to\infty} \sum_k a_{nk} = 1$.
3. $\sum_k |a_{nk}| \le C$ for all n, where $C > 0$ is a constant.

In the stochastic summability theory one replaces $\{x_k\}$ by a sequence of independent and identically distributed real-valued random variables $\{X_k\}$ with $\mathscr{E}|X_k| < \infty$ and $\mathscr{E}X_k = \mu$. Suppose $\{a_{nk}\}$ is a Toeplitz matrix. Since

$$\mathscr{E}\sum_k |a_{nk}X_k| \le \mathscr{E}|X_1| \sum_k |a_{nk}| < \infty$$

it follows that $\sum_k a_{nk} X_k$ converges absolutely with probability 1. It is therefore of interest to study the convergence properties of the sequence $S_n = \sum_k a_{nk} X_k$ as $n \to \infty$. In the special case in which $a_{nk} = w_k/W_n$ if $1 \le k \le n$, $w_k > 0$ and $a_{nk} = 0$ if $k > n$ where $W_n = \sum_{k=1}^{n} w_k$, this was done by Jamison et al. (1965). In the general case, Pruitt (1966) obtained the following results.

Theorem 2.1. $S_n \to^P \mu$ if and only if $\max_k |a_{nk}| \to 0$ as $n \to \infty$.

Theorem 2.2. If $\max_k |a_{nk}| = O(n^{-\alpha})$, $\alpha > 0$, then $\mathscr{E}|X_k|^{1+1/\alpha} < \infty$ implies that $S_n \to \mu$ a.s. as $n \to \infty$.

The proofs of results of this type involve a double truncation first used by Erdös (1949) and improved and extended by Katz (1963), Pruitt (1966), Heyde and Rohatgi (1967), Franck and Hanson (1966), and Hanson and Wright (1969). For a comprehensive discussion of the problem of complete and almost sure convergence of $\{S_n\}$, we refer to Stout (1974), where an extension of Theorem 2.2 is also given.

Rohatgi (1971) extends Theorems 2.1 and 2.2 to the case of nonidentically distributed independent random variables. Let $\{X_k\}$ be a sequence of real random variables uniformly dominated by a real random variable X, that is,

$$P\{|X_k| \ge x\} \le P\{|X| \ge x\}$$

for every $x > 0$ and $k \ge 1$. The following results were obtained by Rohatgi

(1971) by using the double truncation method of Erdös (1949) as used by Pruitt (1966).

Theorem 2.3. Suppose $\mathscr{E}|X|^r < \infty$ for some $0 < r \leq 1$, $a_{nk} \to 0$ as $n \to \infty$ for every $k \geq 1$, and $\Sigma_k |a_{nk}|^r \leq C$ for every $n \geq 1$ and some positive constant C. If $\max_k |a_{nk}| \to 0$ as $n \to \infty$, then $S_n \to^P 0$. (If $r = 1$, each X_k is centered at $\mathscr{E}X_k$.)

Theorem 2.4. Let $\{X_k\}$ and X be as above and $\{a_{nk}\}$ satisfy $a_{nk} \to 0$ as $n \to \infty$ for every $k \geq 1$, $\Sigma_k |a_{nk}| \leq C$ for every $n \geq 1$. Assume $\mathscr{E}X_k = 0$ for every $k \geq 1$. If $\max_k |a_{nk}| = O(n^{-\alpha})$ and $\mathscr{E}|X|^{1+1/\alpha} < \infty$ for some $\alpha > 0$, then $S_n \to 0$ a.s.

3. RANDOM ELEMENTS IN BANACH SPACES

Under certain conditions, a stochastic process can be identified as a random element in a certain Banach space, and hence a limit theorem for a sequence of random elements in a Banach space may be viewed as a limit theorem for a sequence of stochastic processes. This fact has inspired the study of stochastic convergence properties of random elements over the past fifteen years. The two monographs Padgett and Taylor (1973) and Taylor (1978) give an exhaustive account of the work done in this field up to 1978. We follow the terminology used in Taylor (1978).

Let E be a real separable Banach space, and let $\{X_k\}$ be a sequence of random elements taking values in E. Let $\{a_{nk}\}$ be a matrix of real numbers. In the stochastic summability theory for random elements $\{X_k\}$, one relates the convergence properties of the weighted sum sequence $S_n = \Sigma_{k=1}^{\infty} a_{nk} X_k$ to those of the sequence $\{X_k\}$. Thus we seek conditions under which S_n converges stochastically (that is, in pth mean, or in probability, or almost surely, or completely).

The following results are due to Padgett and Taylor (1974) and concern the partial sum sequence $S_n = \Sigma_{k=1}^{n} a_{nk} X_k$.

Theorem 3.1. Let $\{X_k\}$ be a sequence of independent and identically distributed random elements in E. Let $\{a_{nk}\}$ be a double sequence of real numbers such that

1. $\lim_{n \to \infty} a_{nk} = 0$ for every k.
2. $\Sigma_{k=1}^{\infty} |a_{nk}| \leq C$ for all $n \geq 1$ and some constant $C > 0$.

Then $\max_k |a_{nk}| \to 0$ as $n \to \infty$ if and only if $S_n = \Sigma_{k=1}^{n} a_{nk} X_k \to^P 0$.

Theorem 3.2. Let $\{X_k\}$ be as in Theorem 3.1 and $\{a_{nk}\}$ satisfy (1) and (2) of Theorem 3.1. If $\max_k|a_{nk}| = O(n^{-\alpha})$ for some $\alpha > 0$ and $\mathscr{E}\|X_1\|^{1+1/\alpha} < \infty$, then $S_n \to 0$ a.s.

Padgett and Taylor (1974) also considered the problem of extending Theorems 2.3 and 2.4 of Rohatgi (1971) to separable Banach spaces for the $r = 1$ case. They noted that these results are not generalizable verbatim. They imposed several conditions on the Banach space E and on the random elements $\{X_k\}$ to obtain the desired conclusion. Bozorgnia and Bhaskara Rao (1979) succeeded in proving the following extensions of Theorems 2.3 and 2.4. We note that while Padgett and Taylor (1974) considered only triangular arrays $\{a_{nk}\}$ and partial sums $\Sigma_{k=1}^{n} a_{nk} X_k$, Bozorgnia and Bhaskara Rao (1979) considered full sums $\Sigma_{k=1}^{\infty} a_{nk} X_k$. Moreover, Bozorgonia and Bhaskara Rao (1979) give analogs of these results by relaxing the condition of mutual independence to pairwise independence.

Theorem 3.3. Let $\{X_k\}$ be a sequence of independent random elements taking values in a separable Banach space E and uniformly dominated in probability by a random element X in the sense that for all $x > 0$ and $k \geq 1$

$$P\{\|X_k\| \geq x\} \leq P\{\|X\| \geq x\}$$

Suppose $\mathscr{E}\|X\|^r < \infty$ for some $0 < r < 1$. Let $\{a_{nk}\}$ satisfy conditions of Theorem 2.3. If $\lim_{n\to\infty}\max_k|a_{nk}| = 0$, then $\Sigma_{k=1}^{\infty} a_{nk} X_k \to^{P} 0$.

Theorem 3.4. Let $\{X_k\}$ be a sequence of independent random elements taking values in E and uniformly dominated in probability by a random element X. Let $\{a_{nk}\}$ satisfy

1. $\lim_{n\to\infty}\Sigma_{k=1}^{\infty}|a_{nk}| = 0$.
2. $\max_k|a_{nk}| = O(n^{-\alpha})$ for some $\alpha > 0$.

If $\mathscr{E}\|X\|^{1+1/\alpha} < \infty$, then $\Sigma_{k=1}^{\infty} a_{nk} X_k \to 0$ a.s.

A natural approach to establish convergence in probability for the sequence $\Sigma_{k=1}^{\infty} a_{nk} X_k$ is to check the convergence in probability of the sequence of real random variables $f(\Sigma_{k=1}^{\infty} a_{nk} X_k)$ for every continuous linear functional f from E to $\mathbb{R}$ and then expect $\Sigma a_{nk} X_k$ to converge in probability. Although this approach does not always work, it frequently does. To obtain convergence almost surely we note that $\|a_{nk} X_k\| = |a_{nk}|\,\|X_k\|$ and if $|a_{nk}|$ and $\|X_k\|$ satisfy the corresponding hypotheses for the real-valued case then the double truncation technique of Erdös (1949) can frequently be applied.

For other extensions and recent work we refer to references in the Bibliography.

4. LINEAR OPERATOR WEIGHTS

In this section we allow the weight sequence to be a sequence of bounded linear operators. Let E be a real separable Banach space, and let $\{X_n\}$ be a sequence of independent and identically distributed random elements in E such that $\mathscr{E}\|X_1\| < \infty$. Let $\{A_{nk}\}$ be a double array of bounded linear operators on E. It is natural to require that $\{A_{nk}\}$ satisfy the following conditions:

1. $\lim_{n\to\infty} A_{nk} = 0$ for all k.
2. $\sum_{k=1}^{\infty}\|A_{nk}\| \le C$, $C > 0$.

Without loss of generality assume $\mathscr{E}X_1 = 0$. Define $S_n = \sum_k A_{nk}X_k$, where the sum may be partial or full. Now for each n

$$\mathscr{E}\|S_n\| \le \sum_k \|A_{nk}\|\mathscr{E}\|X_k\| = \mathscr{E}\|X_1\| \sum_{k=1}^{\infty} \|A_{nk}\| < \infty$$

and it follows that the series $S_n = \sum_{k=1}^{\infty} A_{nk}X_k$ converges with probability 1. It is therefore of interest to relate the convergence properties of S_n to that of the sequence $\{X_k\}$.

For example, in order to study the convergence in probability of $\sum_{k=1}^{n} A_{nk}X_k$, it would be natural to see if the standard approach of Taylor and Padgett (1975) will yield the desired conclusion. Suppose that E has a Schauder basis $\{b_k\}$ and let $\{f_k\}$ be the coordinate functionals for the basis $\{b_k\}$. Let $U_t(x) = \sum_{k=1}^{t} f_k(x)b_k$ be the partial sum operators and set $Q_t(x) = x - U_t(x)$, $x \in E$. While it is no problem to deal with $\sum_{k=1}^{n} U_t(A_{nk}X_k)$, since A_{nk} are no longer real numbers the problem is how to deal with $\sum_{k=1}^{n} Q_t(A_{nk}X_k)$.

Work on the stochastic convergence of $S_n = \sum_k A_{nk}X_k$ is in progress and will be reported elsewhere.

REFERENCES

Bozorgnia, A. and Bhaskara Rao, M. (1979). Limit theorems for weighted sums of random elements in separable Banach spaces. *J. Multivar. Anal.* **9**, 428–433.

Erdös, P. (1949). On a theorem of Hsu and Robbins. *Ann. Math. Statist.* **20**, 286–291.

Franck, W. E. and Hanson, D. L. (1966). Some results giving rates of convergence in the law of large numbers for weighted sums of independent random variables. *Trans. Amer. Math. Soc.* **124**, 347–359.

Hanson, D. L. and Wright, F. T. (1969). Some more results on rates of convergence in the law of large numbers for weighted sums of independent random variables. *Trans. Amer. Math. Soc.* **141**, 443–464.

Heyde, C. C. and Rohatgi, V. K. (1967). A pair of complementary theorems on convergence rates in the law of large numbers. *Proc. Camb. Phil. Soc.* **63**, 73–82.

Jamison, B., Orey, S. and Pruitt, W. (1965). Convergence of weighted averages of independent random variables. *Z. Wahr. Verw. Geb.* **4**, 40–44.

Katz, M. L. (1963). Probability in the tail of a distribution. *Ann. Math. Statist.* **34**, 312–318.

Koopmans, L. H. (1961). An exponential bound on the strong law of large numbers for linear stochastic processes with absolutely convergent coefficients. *Ann. Math. Statist.* **32**, 583–586.

Lai, T. L. (1972). Control charts based on weighted sums. *Ann. Statist.* **2**, 134–147.

Padgett, W. J. and Taylor, R. L. (1973). *Laws of Large Numbers for Normed Linear Spaces and Certain Fréchet Spaces*. Lecture Notes in Mathematics, Vol. 360, New York: Springer-Verlag.

Padgett, W. J. and Taylor, R. L. (1974). Convergence of weighted sums of random elements in Banach space and Fréchet spaces. *Bull. Inst. Math. Acad. Sinica* **2**, 385–400.

Pruitt, W. (1966). Summability of independent random variables. *J. Math. Mech.* **15**, 769–776.

Rohatgi, V. K. (1971). Convergence of weighted sums of independent random variables. *Proc. Camb. Phil. Soc.* **69**, 305–307.

Stout, W. F. (1974). *Almost Sure Convergence*, New York: Academic Press.

Taylor, R. L. and Padgett, W. J. (1975). Stochastic convergence of weighted sums in normed linear spaces. *J. Multivariate Anal.* **5**, 434–450.

Taylor, R. L. (1978). *Stochastic Convergence of Weighted Sums of Random Elements in Linear Spaces*. Lecture Notes in Math. Vol. 672, New York: Springer-Verlag.

Taylor, R. L. (1982). Convergence of weighted sums of arrays of random elements in Type p spaces with application to density estimation. *Sankhyā A* **44**, 341–351.

BIBLIOGRAPHY

Ahmad, I. A. (1979). On the WLLN for sums of random elements in Banach spaces with random weights. Memphis State University Technical Report 79-10 (preprint).

Daffer, P. Z. and Taylor, R. L. (1982). Tightness and strong laws of large numbers in Banach spaces. *Bull. Inst. Math. Acad. Sinica* **10**, 251–263.

Padgett, W. J. and Taylor, R. L. (1979). Convergence of weighted sum of independent random variables and an extension of Banach-space valued random variables. *Int. J. Math. Math. Sci.* **2**, 309–323.

Potocky, R. (1984). Convergence of weighted sums of random variables in vector lattices. *Math. Slovaca* **34**, 273–276.

Sarapa, N. (1979). A note on the law of large numbers. *Glasnik Matematicki* **14**, 175–180.

Taylor, R. L. (1979). Complete convergence for weighted sums of arrays of random elements. University of South Carolina Technical Report (preprint).

Taylor, R. L. and Wei, D. (1979). Laws of large numbers for tight random elements in normed linear spaces. *Ann. Probab.* **7**, 150–155.

Taylor, R. L. and Calhoun, C. A. (1983). On the almost sure convergence of randomly weighted sums of random elements. *Ann. Probab.* **11**, 795–797.

Taylor, R. L., Raina, C. C. and Daffer, D. Z. (1984). Stochastic convergence of randomly weighted sums of random elements. *Stoch. Anal. Appl.* **2**, 299–321.

Wang, Xiang Chen and Bhaskara Rao, M. (1984). Convergence in the p-th mean and some weak laws of large numbers for weighted sums of random elements in separable normed linear spaces. *J. Multivar. Anal.* **15**, 124–134.

A Population-Size-Dependent Branching Process

Pedro Saavedra Santana
Departamento de Estadistica E Invastigacion Operativa
Facultad de Matematicas
Universidad de la Laguna
Canary Islands, Spain

SUMMARY

Either the extinction or the explosion of the population happens in the Galton–Watson process with probability 1. The same occurs if the descendency distribution depends on the number of individuals present in the aforementioned population. If we also set up the hypothesis that, for large population sizes, the expected number of descendants per individual decreases to a quantity smaller than 1, there will be surely no explosion.

1. INTRODUCTION

The environment is an important factor in the development of every population. This environment is determined by several causes, among them the evolution of the population itself. If its size achieves certain levels, it is reasonable to suppose that the environment becomes degraded and that this has a repercussion in the further increasing. If this deterioration does not trespass a certain level, we can expect the environment to recover itself, thus rendering a relaunching of the populational increase possible. Thus a certain cyclic movement in the population size would occur. The question now is to see what happens in the mediate or remote future. In view of these ideas, we study a model in which the reproduction law depends on the population size. This model is a generalization of the Galton–Watson process. In the Galton–Watson process there is with certainty extinction or explosion. In view of the cyclical character of our model we expect the existence of a set of trajectories with positive probability, which would fluctuate within a state interval. Nevertheless, because of the transitive character of all the states except the zero, we prove this to be not. By imposing conditions relative to the degradation that occurs in the environment because of the excess of the increase, we show that the explosion does not happen, and there is with certainty extinction.

2. THE MODEL

We define the population-size-dependent branching process in the following way: We start with Z_0 individuals. Each of them lives during a time unit and then each of them divides into $X(Z_0)$ new individuals. The distribution of $X(Z_0)$ depends on the number of present individuals. Thus we have

$$Z_{n+1} = \sum_{j=1}^{Z_n} X_j(Z_n)$$

where $X_j(k)$ are independent and identically distributed (i.i.d.) random variables independent of Z_n and with k-dependent distribution. We suppose that Z_0 is degenerate and thus a homogeneous Markov process with state space $E = \mathbb{N} \cup \{0\}$ is defined:

$$\pi_k(Z_n) = P(X(Z_n) = k|Z_n)$$

This distribution effectively depends on the population size Z_n, but is independent of the step n.

Let

$$P_{ij} = P(Z_{n+1} = j|Z_n = i) = \begin{cases} \pi_{ij}^*(i), & i \neq 0 \quad j \in E \\ \delta_{0j}, & i = 0, \quad j \in N \end{cases}$$

$$\pi_{ij}^*(i) = \sum_{x_1 + \cdots + x_i = j} P(X_1(i) = x_1, X_2(i) = x_2, \ldots, X_i(i) = x_i)$$

3. CLASSIFICATION OF STATES

In the state space $E = \mathbb{N} \cup \{0\}$, 0 is absorbent for obvious reasons. All the other states are transitive.

In fact, let $k \in \mathbb{N}$. If k communicates with 0, k is nonessential for being 0 absorbent and therefore transitive.

If k does not communicate with 0, then, if we start with k individuals, none of them will evidently have zero descendants and therefore the population surely increases, that is to say that $\forall n,\ k \leq Z_n \leq Z_{n+1}$. Then $\{Z_{n+1} = k\} \subset \{Z_n = k\}$ and the probability of having started from k, returning to $k = f_{kk} = P(\bigcup_{n=1}^{\infty}\{Z_n = k\}|Z_0 = k) = P(Z_1 = k|Z_0 = k) = p_{kk} < 1 \Rightarrow k$ is transitive. It is clearly discardable in the case $p_{kk} = 1$.

4. EXTINCTION

The probability of extinction will depend on the state we start from. Let us write $q_i = f_{i0}$, the probability of having started from i individuals, eventual transition to zero. Then

$$q_i = P(Z_n \to 0|Z_0 = i) = P\left(\bigcup_{k=1}^{\infty} \bigcap_{n=k}^{\infty} \{Z_n = 0\}|Z_0 = i\right)$$

$$= \lim_{k\to\infty} P\left(\bigcap_{n=k}^{\infty} \{Z_n = 0\}|Z_0 = i\right) = \lim_{k\to\infty} P(Z_k = 0|Z_0 = i)$$

In our model, $\pi_0(k)^k$ is the probability that, being k individuals in the population, this becomes extinct in the following step. If the population increase degrades the environment, it is reasonable to suppose that $\pi_0(k)$ is increasing with k. In the particular case that $\pi_0(k) = \xi^{1/k}$ (ξ constant), it is trivial to prove that the extinction occurs with certainty.

$$P(Z_{n+1} = 0) = P(Z_n = 0) + \sum_{k=1}^{\infty} P(Z_{n+1} = 0|Z_n = k) \cdot P(Z_n = k)$$

$$= P(Z_n = 0) + \sum_{k=1}^{\infty} \pi_0(k)^k P(Z_n = k)$$

$$= P(Z_n = 0) + \xi \sum_{k=1}^{\infty} P(Z_n = k)$$

$$= P(Z_n = 0) + \xi(1 - P(Z_n = 0)) \Rightarrow q$$

$$= q + \xi(1 - q) \Rightarrow q = 1$$

In the extinction study, a fundamental equation system is $g = \mathbf{P}g$ when $\mathbf{P}$ is the transition matrix of probabilities of the Markov chain.

Proposition 4.1. The extinction probabilities $q_i = f_{i0}$ are obtained as the minimal solution of the system $g = \mathbf{P}g$ in $[0,1]$.

Proof. Obviously $q_i = f_{i0} = p_{i0} + \sum_{j=1}^{\infty} p_{ij} q_j$ $(q_0 = 1)$. Then $q = \mathbf{P}q$. Let us show that $q = (q_i)_{i=0}^{\infty}$ is the minimal solution in this system in $[0,1]$. Let g be any non-negative solution of the system. Obviously $\forall i,\ g_i \geq p_{i0}^{(1)}$, and therefore $g_i = p_{i0} + \sum_{j=1}^{\infty} p_{ij} g_j \geq p_{i0} + \sum_{j=1}^{\infty} p_{ij} p_{j0} = p_{i0}^{(2)}$; by iteration we get $g_i \geq p_{i0}^{(n)} \Rightarrow g_i \geq \lim_{n\to\infty} p_{i0}^{(n)}$ [the limit exists since $(p_{i0}^{(n)})$ is nondecreasing] and furthermore

$$\lim_{n\to\infty} p_{i0}^{(n)} = \lim_{n\to\infty} P(Z_n = 0|Z_0 = i) = P\left(\bigcup_{n=1}^{\infty} \{Z_n = 0\}|Z_0 = i\right) = f_{i0} \Rightarrow$$

$$\forall i,\ g_i \geq f_{i0} = q_i \quad \blacksquare$$

In Section 3 we saw that except for the state zero, all the states are transitive. Let us find the probability that our system will remain forever between transitive states.

Let $\sigma_i^{(n)} = \sum_{j=1}^{\infty} p_{ij}^{(n)}$ be the probability that starting from state i, the system remains until the nth step between transitive states. Then

$$\sum_{j=1}^{\infty} p_{ij}\sigma_j^{(n)} = \sum_{j=1}^{\infty} p_{ij} \sum_{k=1}^{\infty} p_{jk}^{(n)} = \sum_{k=1}^{\infty}\sum_{j=1}^{\infty} p_{ij}p_{jk}^{(n)} = \sum_{k=1}^{\infty} p_{ik}^{(n+1)} = \sigma_i^{(n+1)}$$

Now, $\lim_{n\to\infty} \sigma_i^{(n)} = \sigma_i$ be the probability of remaining forever between the transitive states.

Proposition 4.2. The probabilities of remaining forever between the transitive states is given by the largest solution of the system

$$y_i = \sum_{j=1}^{\infty} p_{ij} y_j, \qquad i = 1, 2, \ldots, \infty \quad \text{on } [0,1]$$

Proof. Obviously $\{\sigma_i\}_i$ is a system solution. Let us show that it is the largest one in $[0,1]$.

Let $\{y\}$ be such that $y_i \in [0,1]$; $\forall i,\ y_i = \sum_{j=1}^{\infty} p_{ij} y_j \leq \sum_{j=1}^{\infty} p_{ij} = \sigma_i^{(n)}$

$$y_i = \sum_{j=1}^{\infty} p_{ij} y_j \leq \sum_{j=1}^{\infty} p_{ij}\sigma_j^{(1)} = \sigma_i^{(2)} \Rightarrow \forall i, \forall n,\ y_i \leq \sigma_i^{(n)} \Rightarrow y_i \leq \sigma_i$$

Theorem 4.1. The process $\{Z_n\}$ does not visit the transitive states set infinitely often (i.o.).

Proof. Let a be arbitrary but fixed. Let us find $P(1 \leq Z_n \leq a,$ i.o. $Z_0 = 1)$. According to Borel's 0–1 law

$$\sum_{n=1}^{\infty} P(1 \leq Z_n \leq a | Z_0 = 1)$$

$$= \sum_{n=1}^{\infty} \sum_{k=1}^{[a]} P(Z_n = k | Z_0 = 1) = \sum_{k=1}^{[a]} \sum_{n=1}^{\infty} P(Z_n = k | Z_0 = 1)$$

$$= \sum_{k=1}^{[a]} \sum_{n=1}^{\infty} p_{1k}^{(n)} < \infty$$

$$\Rightarrow \forall a,\ P(1 \leq Z_n \leq a, \text{i.o.} | Z_0 = 1) = 0$$

Thus the set of trajectories visiting $[1, a]$ i.o. has probability 0.

Thus we arrive at one of the main conclusions of our work. A population whose reproduction law depends exclusively on its size cannot establish itself between two fixed values. The question of finding conditions under which this is possible remains open.

Let us finally study what happens in the case that the excessive increase of the population degrades the environment.

If in our model the reproduction law depends exclusively on the population size, the degradation will have to translate into $\lambda(n) \to \lambda < 1 n \uparrow \infty$ $[\lambda(Z_n) = E(X(Z_n))$, the expected numbers of descendants per individual when the population has size $Z_n]$. This condition is natural, for if the expected number of descendants per individual is larger or equal to 1, then the population increases, this standing in contradiction with the fact that the environment has deteriorated by saturation of individuals.

Let $H = \{\omega\colon\ Z_n(\omega) \to \infty\}$. Let us show that if $\lambda(n) \to \lambda < 1$ for $n \uparrow \infty$, H has mass 0. ∎

Theorem 4.2. $\lambda(n) \to \lambda < 1$ for $n \uparrow \infty \Rightarrow Z_n \to 0$ (a.s.) (the population becomes extinct).

Proof. $\{Z_n / \prod_{k=0}^{n-1} \lambda(Z_k)\}_n$ is a positive martingale relative to $\mathscr{F}_n = \sigma\{Z_0, Z_1, \ldots, Z_n\}$. In fact,

$$E\left(Z_{n+1} \Big/ \prod_{k=0}^{n} \lambda(Z_k) | Z_n\right) = Z_n \Big/ \prod_{k=0}^{n-1} \lambda(Z_k) \Rightarrow \exists W_\infty \text{ finite (a.s.)}$$

such that

$$Z_n \Big/ \prod_{k=0}^{n-1} \lambda(Z_k) \overset{\text{(a.s.)}}{\to} W_\infty$$

On H,

$$\lambda(Z_n) \to \lambda < 1 \Rightarrow \prod_{k=0}^{n-1} \lambda(Z_k) \to 0 \; n \uparrow \infty, \; Z_n / \prod_{k=0}^{n-1} \lambda(Z_k) \to \infty$$

Now, $W_\infty < \infty$ (a.s.) $\Rightarrow H$ has probability $0 \Rightarrow Z_n \to 0$ (a.s.).

ACKNOWLEDGMENT

I express my gratitude to Professor Miguel Sánchez García for his contribution to this work.

REFERENCES

Feller, W. (1971). *An Introduction to Probability Theory and Its Applications*, Vol. 1, New York: Wiley.

Jagers, P. (1975). *Branching Processes with Biological Applications*, New York: Wiley-Interscience.

Klebaner, F. C. (1984). On population-size-dependent branching processes. *Adv. Appl. Probab.* **1**, 30–55.

Author Index

Subject Index

Applied Probability and Statistics (Continued)

JUDGE, HILL, GRIFFITHS, LÜTKEPOHL and LEE • Introduction to the Theory and Practice of Econometrics
JUDGE, GRIFFITHS, HILL, LÜTKEPOHL and LEE • The Theory and Practice of Econometrics, *Second Edition*
KALBFLEISCH and PRENTICE • The Statistical Analysis of Failure Time Data
KISH • Statistical Design for Research
KISH • Survey Sampling
KUH, NEESE, and HOLLINGER • Structural Sensitivity in Econometric Models
KEENEY and RAIFFA • Decisions with Multiple Objectives
LAWLESS • Statistical Models and Methods for Lifetime Data
LEAMER • Specification Searches: Ad Hoc Inference with Nonexperimental Data
LEBART, MORINEAU, and WARWICK • Multivariate Descriptive Statistical Analysis: Correspondence Analysis and Related Techniques for Large Matrices
LINHART and ZUCCHINI • Model Selection
LITTLE and RUBIN • Statistical Analysis with Missing Data
McNEIL • Interactive Data Analysis
MAINDONALD • Statistical Computation
MANN, SCHAFER and SINGPURWALLA • Methods for Statistical Analysis of Reliability and Life Data
MARTZ and WALLER • Bayesian Reliability Analysis
MIKÉ and STANLEY • Statistics in Medical Research: Methods and Issues with Applications in Cancer Research
MILLER • Beyond ANOVA, Basics of Applied Statistics
MILLER • Survival Analysis
MILLER, EFRON, BROWN, and MOSES • Biostatistics Casebook
MONTGOMERY and PECK • Introduction to Linear Regression Analysis
NELSON • Applied Life Data Analysis
OSBORNE • Finite Algorithms in Optimization and Data Analysis
OTNES and ENOCHSON • Applied Time Series Analysis: Volume I, Basic Techniques
OTNES and ENOCHSON • Digital Time Series Analysis
PANKRATZ • Forecasting with Univariate Box-Jenkins Models: Concepts and Cases
PIELOU • Interpretation of Ecological Data: A Primer on Classification and Ordination
PLATEK, RAO, SARNDAL and SINGH • Small Area Statistics: An International Symposium
POLLOCK • The Algebra of Econometrics
PRENTER • Splines and Variational Methods
RAO and MITRA • Generalized Inverse of Matrices and Its Applications
RÉNYI • A Diary on Information Theory
RIPLEY • Spatial Statistics
RIPLEY • Stochastic Simulation
RUBIN • Multiple Imputation for Nonresponse in Surveys
RUBINSTEIN • Monte Carlo Optimization, Simulation, and Sensitivity of Queueing Networks
SCHUSS • Theory and Applications of Stochastic Differential Equations
SEAL • Survival Probabilities: The Goal of Risk Theory
SEARLE • Linear Models
SEARLE • Matrix Algebra Useful for Statistics
SPRINGER • The Algebra of Random Variables
STEUER • Multiple Criteria Optimization
STOYAN • Comparison Methods for Queues and Other Stochastic Models
TIJMS • Stochastic Modeling and Analysis: A Computational Approach
TITTERINGTON, SMITH, and MAKOV • Statistical Analysis of Finite Mixture Distributions
UPTON • The Analysis of Cross-Tabulated Data
UPTON and FINGLETON • Spatial Data Analysis by Example, Volume I: Point Pattern and Quantitative Data

(*continued from front*)